Advances in Intelligent Systems and Computing

Volume 1390

The series "Advances in Intelligent Systems and Computing" contains publications on theory, applications, and design methods of Intelligent Systems and Intelligent Computing. Virtually all disciplines such as engineering, natural sciences, computer and information science, ICT, economics, business, e-commerce, environment, healthcare, life science are covered. The list of topics spans all the areas of modern intelligent systems and computing such as: computational intelligence, soft computing including neural networks, fuzzy systems, evolutionary computing and the fusion of these paradigms, social intelligence, ambient intelligence, computational neuroscience, artificial life, virtual worlds and society, cognitive science and systems, Perception and Vision, DNA and immune based systems, self-organizing and adaptive systems, e-Learning and teaching, human-centered and human-centric computing, recommender systems, intelligent control, robotics and mechatronics including human-machine teaming, knowledge-based paradigms, learning paradigms, machine ethics, intelligent data analysis, knowledge management, intelligent agents, intelligent decision making and support, intelligent network security, trust management, interactive entertainment, Web intelligence and multimedia.

The publications within "Advances in Intelligent Systems and Computing" are primarily proceedings of important conferences, symposia and congresses. They cover significant recent developments in the field, both of a foundational and applicable character. An important characteristic feature of the series is the short publication time and world-wide distribution. This permits a rapid and broad dissemination of research results.

Indexed by DBLP, EI Compendex, INSPEC, WTI Frankfurt eG, zbMATH, Japanese Science and Technology Agency (JST).

All books published in the series are submitted for consideration in Web of Science.

More information about this series at http://www.springer.com/series/11156

Roman Szewczyk · Cezary Zieliński ·
Małgorzata Kaliczyńska
Editors

Automation 2021: Recent Achievements in Automation, Robotics and Measurement Techniques

Springer

Editors
Roman Szewczyk
ŁUKASIEWICZ Research Network
Industrial Research Institute for Automation and Measurements PIAP
Warsaw, Poland

Cezary Zieliński
ŁUKASIEWICZ Research Network
Industrial Research Institute for Automation and Measurements PIAP
Warsaw, Poland

Małgorzata Kaliczyńska
ŁUKASIEWICZ Research Network
Industrial Research Institute for Automation and Measurements PIAP
Warsaw, Poland

ISSN 2194-5357 ISSN 2194-5365 (electronic)
Advances in Intelligent Systems and Computing
ISBN 978-3-030-74892-0 ISBN 978-3-030-74893-7 (eBook)
https://doi.org/10.1007/978-3-030-74893-7

This Springer imprint is published by the registered company Springer Nature Switzerland AG
The registered company address is: Gewerbestrasse 11, 6330 Cham, Switzerland

Foreword

Recently, our civilization has been facing one the most severe challenges in modern history. Although now the worldwide vaccination process is progressing slowly, threats are still lurking. The COVID-19 pandemic devastated the global economy and significantly disrupted numerous areas of economic activity. These dramatic changes create the risk of not only poverty for large part of the world population, but also social unrest. Those problems have to be solved by politicians. However, we would like to draw attention to another problem. We should expect that this is not the last pandemic with which humanity will have to cope. This pandemic shows that we have been ill-prepared to handle it, mainly because social distancing in the pre-vaccine period was hard to implement. Production and services had to be delivered by teams of human workers susceptible to infection. Hence, reorganization of work is necessary, but this requires further automatization of production, delivery chains and services. Thus not only to mitigate the long-term economic and social risks disclosed by the COVID-19 pandemic, but also to prepare ourselves for future outbreaks of yet unknown viruses the scientific and industrial community have to intensify the efforts focused on the Industry 4.0 transformation. Only radical increase of efficiency and versatility of industrial production, with further limitation of human involvement, paralleled by the decrease of environmental burden, will enable us to cope with such challenges.

We hope that the presented volume provides input to the solution of at least some of the problems brought about by this challenge. It contains 38 papers authored by both scientists and practitioners focused on interdisciplinary approach to the development of cyber-physical systems. This approach relies on the development of measuring techniques, robotic and mechatronic systems, industrial automation, numerical modelling and simulation as well as application of artificial intelligence techniques required by the transformation leading to Industry 4.0. We strongly believe that the solutions and guidelines presented in this volume will be useful to both researchers and engineers tackling problems associated with the reconstruction of the world economy after this pandemic and preparing it for possible future outbreaks.

February 2021

Roman Szewczyk
Cezary Zieliński
Małgorzata Kaliczyń

Contents

Editors

Prof. Roman Szewczyk received both his Ph.D. and D.Sc. in the field of mechatronics. He specializes in modelling of properties of magnetic materials as well as in sensors and sensor interfacing, in particular magnetic sensors for security applications. He leads the development of a sensing unit for a mobile robot developed for the Polish Police Central Forensic Laboratory and methods of non-destructive testing based on magnetoelastic effect. He has been involved in over 10 European Union-funded research projects within the FP6 and FP7 as well as projects financed by the European Defence Organization. Moreover, he has led two regional and national scale technological foresight projects and was active in the organization and implementation of technological transfer between companies and research institutes. He is Secretary for Scientific Affairs in the Industrial Research Institute for Automation and Measurements PIAP. He is also Associate Professor at the Faculty of Mechatronics, Warsaw University of Technology, and Vice-Chairman of the Academy of Young Researchers of the Polish Academy of Sciences.

Prof. Cezary Zieliński received his M.Sc./eng., Ph.D. and habilitation in control and robotics from the Faculty of Electronics and Information Technology, Warsaw University of Technology (WUT), Warsaw, Poland. He is Full Professor of WUT, Head of the Robotics Research Group, Director of the Institute of Control and Computation Engineering and Professor of Industrial Research Institute for Automation and Measurement PIAP. His research concentrates on robot control and programming methods. His research interests focus on robotics, in general, and in particular include robot programming methods, formal approach to the specification of architectures of multi-effector and multi-receptor systems, robot kinematics, robot position-force control, visual servo control and design of digital circuits. He is the author/co-author of over 200 conferences and journal papers as well as books concerned with the above-mentioned research subjects.

Dr. Małgorzata Kaliczyńska received her M.Sc./eng. in cybernetics from the Faculty of Electronics, Wrocław University of Technology, and her Ph.D. in the field of fluid mechanics from the Faculty of Mechanical and Power Engineering in the same university. Now she is Assistant Professor in the Industrial Research Institute for Automation and Measurement PIAP and Editor of the scientific and technological magazine "Measurements Automation Robotics". Her areas of research interest include distributed control systems, Internet of Things, Industry 4.0, information retrieval and webometrics.

Control and Automation

New Sufficient Conditions of Global Stability of Nonlinear Positive Electrical Circuits

Tadeusz Kaczorek(✉)

Białystok University of Technology, Wiejska 45D, 15-351 Białystok, Poland
kaczorek@ee.pw.edu.pl

Abstract. The global stability of electrical circuits composed of positive linear parts and nonlinear static element with given characteristic and positive gain feedbacks is investigated. New sufficient conditions for the global stability of this class of nonlinear positive electrical circuits are established. The effectiveness of these new stability conditions are demonstrated on simples examples of positive nonlinear electrical circuits.

Keywords: Global stability · Sufficient conditions · Positive · Nonlinear · Electrical · Circuit

1 Introduction

A dynamical system is called positive if its trajectory starting from any nonnegative initial state remains forever in the positive orthant for all nonnegative inputs. An overview of state of the art in positive theory is given in the monographs [1, 4, 9, 16]. Variety of models having positive behavior can be found in engineering, especially in electrical circuits [16], economics, social sciences, biology and medicine, etc. [4, 9].

The positive electrical circuits have been analyzed in [6, 9–13] and positive descriptor systems and electrical circuits in [2, 21]. The controllability and observability of standard and positive electrical circuits has been addressed in [16]. A new class of normal positive linear electrical circuits has been introduced in [9]. Positive fractional linear electrical circuits have been investigated in [3, 5–7, 10–12, 15]. The superstabilization of positive linear electrical circuits by state-feedbacks has been analyzed in [14].

In this paper the global stability of nonlinear positive linear electrical circuits with positive state-feedbacks will be investigated.

The paper is organized as follows. In Sect. 2 the basic definitions and properties of positive electrical circuits are recalled. The main result of the paper the new global stability conditions of the nonlinear positive electrical circuits with state feedbacks are established in Sect. 3. Concluding remarks are given and some open problems are formulated in Sect. 4.

The following notation will be used: $\Re$ - the set of real numbers, $\Re^{n\times m}$ - the set of $n \times m$ real matrices, $\Re_{+}^{n\times m}$ - the set of $n \times m$ real matrices with nonnegative entries and $\Re_{+}^{n} = \Re_{+}^{n\times 1}$, M_n - the set of $n \times n$ Metzler matrices (real matrices with nonnegative off-diagonal entries), I_n - the $n \times n$ identity matrix.

R. Szewczyk et al. (Eds.): AUTOMATION 2021, AISC 1390, pp. 3–9, 2021.
https://doi.org/10.1007/978-3-030-74893-7_1

2 Positive Electrical Circuits

Consider the linear continuous-time electrical circuit described by the state equations

$$\dot{x}(t) = Ax(t) + Bu(t), \tag{1a}$$

$$y(t) = Cx(t) + Du(t), \tag{1b}$$

where $x(t) \in \Re^n$, $u(t) \in \Re^m$, $y(t) \in \Re^p$ are the state, input and output vectors and $A \in \Re^{n\times n}, B \in \Re^{n\times m}, C \in \Re^{p\times n}, D \in \Re^{p\times m}$.

It is well-known [16] that any linear electrical circuit composed of resistors, coils, capacitors and voltage (current) sources can be described by the state Eqs. (1a, 1b). Usually as the state variables $x_1(t), \ldots, x_n(t)$ (the components of the state vector $x(t)$) the currents in the coils and voltages on the capacitors are chosen.

Definition 2.1 [16]. The electrical circuit (1a, 1b) is called (internally) positive if $x(t) \in \Re^n_+$ and $y = y(t) \in \Re^p_+$, $t \in [0, +\infty]$ for any $x_0 = x(0) \in \Re^n_+$ and every $u(t) \in \Re^m_+$, $t \in [0, +\infty]$.

Theorem 2.1 [16]. The electrical circuit (1a, 1b) is positive if and only if

$$A \in M_n,\ B \in \Re^{n\times m}_+,\ C \in \Re^{p\times n}_+,\ D \in \Re^{p\times m}_+. \tag{2}$$

Theorem 2.2 [16]. The linear electrical circuit composed of resistors, coils and voltage sources is positive for any values of the resistances, inductances and source voltages if the number of coils is less or equal to the number of its linearly independent meshes and the direction of the mesh currents are consistent with the directions of the mesh source voltages.

Theorem 2.3 [16]. The linear electrical circuit composed of resistors, capacitors and voltage sources is not positive for all values of its resistances, capacitances and source voltages if each its branch contains resistor, capacitor and voltage source.

Theorem 2.4 [16]. The *R, L, C, e* electrical circuits are not positive for any values of its resistances, inductances, capacitances and source voltages if at least one its branch contains coil and capacitor.

Definition 2.2 [16]. The positive electrical circuit is called asymptotically stable if

$$\lim_{t\to\infty} x(t) = 0 \text{ for any } x_0 \in \Re^n_+. \tag{3}$$

Theorem 2.5 [16, 20]. The positive electrical circuit (1a, 1b) is asymptotically stable if and only if one of the equivalent conditions is satisfied:

1) All coefficient of the characteristic polynomial

$$\det[I_n s - A] = s^n + a_{n-1}s^{n-1} + \ldots + a_1 s + a_0 \tag{4}$$

are positive, i.e. $a_k > 0$ for $k = 0, 1, \ldots, n-1$.

2) All principal minors $\overline{M}_i$, $i = 1, \ldots, n$ of the matrix $-A$ are positive, i.e.

$$\overline{M}_1 = |-a_{11}| > 0, \overline{M}_2 = \begin{vmatrix} -a_{11} & -a_{12} \\ -a_{21} & -a_{22} \end{vmatrix} > 0, \ldots, \overline{M}_n = \det[-A] > 0. \quad (5)$$

3) There exists strictly positive vector $\lambda^T = [\lambda_1 \cdots \lambda_n]^T$, $\lambda_k > 0$, $k = 1, \ldots, n$ such that

$$A\lambda < 0 \text{ or } A^T\lambda < 0. \quad (6)$$

3 Global Stability of Standard Nonlinear Feedback Systems

Consider the nonlinear feedback electrical circuit shown in Fig. 1 which consists of the positive linear part, the nonlinear element with characteristic $u = f(e)$ and the feedback with positive gain h. The linear part is described by the equations

$$\begin{aligned} \dot{x} &= Ax + Bu \\ y &= Cx \end{aligned} \quad (7)$$

where $x = x(t) \in \Re_+^n$, $u = u(t) \in \Re_+$, $y = y(t) \in \Re_+$ is the state vector, input and output and $A \in M_n$, $B \in \Re_+^{n\times 1}$, $C \in \Re_+^{1\times n}$.

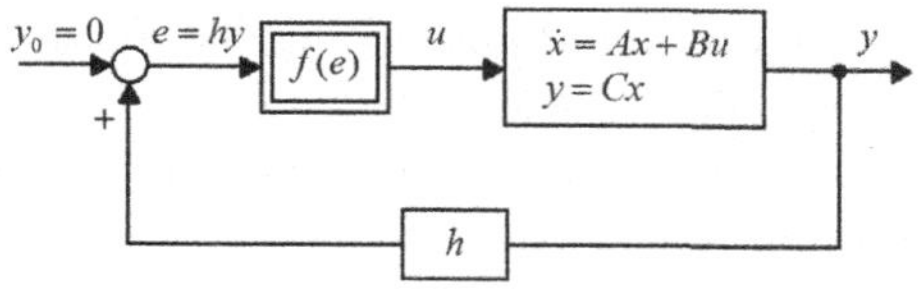

Fig. 1. The nonlinear feedback system

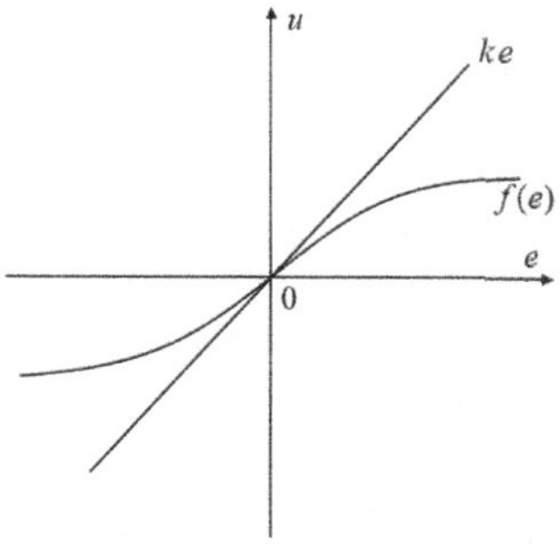

Fig. 2. Characteristic of the nonlinear element

The characteristic of the nonlinear element is shown in Fig. 2 and it satisfies the condition

$$0 \le \frac{f(e)}{e} \le k < \infty. \quad (8)$$

It is assumed that the positive linear part is asymptotically stable (the matrix $A \in M_n$ is Hurwitz).

Definition 3.1. The nonlinear positive electrical circuit is called globally stable if it is asymptotically stable for all nonnegative initial conditions $x(0) \in \Re_{+}$.

The following theorem gives sufficient conditions for the global stability of the positive nonlinear electrical circuit.

Theorem 3.1. The nonlinear electrical circuit consisting of the positive linear part, the nonlinear element satisfying the condition (8) and the feedback with positive gain h is globally stable if

$$A + khBC \in M_n. \tag{9}$$

Proof. The proof will be accomplished by the use of the Lyapunov method [18, 19]. As the Lyapunov function $V(x)$ we choose

$$V(x) = \lambda^T x \geq 0 \text{ for } x \in \Re_{+}^{n}, \tag{10}$$

where λ is strictly positive vector, i.e. $\lambda_k > 0$, $k = 1, \ldots, n$.

Using (10) and (7) we obtain

$$\dot{V}(x) = \lambda^T \dot{x} = \lambda^T (Ax + Bu) = \lambda^T (Ax + Bf(e)) \leq \lambda^T (A + khBC)x \tag{11}$$

since $u = f(e) \leq ke = khCx$.

From (11) it follows that $\dot{V}(t) < 0$ if the condition (9) is satisfied and the nonlinear positive electrical circuit is globally stable. □

To find the maximal value of k satisfying the condition (8) for the nonlinear positive electrical circuit the following procedure can be used.

Procedure 3.1

Step 1. Using the matrices A, B, C of the positive electrical circuit and h compute the matrix

$$A_1 = A + khBC. \tag{12}$$

Step 2. Compute the characteristic polynomial of the matrix (12) with coefficient depending on k

$$\det[I_n s - A_1]. \tag{13}$$

Step 3. Using the condition 1) of Theorem 2.5 find the maximal k for which all coefficients of (13) are positive.

Example 3.1. Consider the nonlinear positive electrical circuit shown in Fig. 3 with given resistances $R_1 = 3$, $R_2 = 2$, inductances $L_1 = L_2 = 1$, the characteristic $u = f(i)$ satisfying the condition

$$0 \leq \frac{f(i)}{i} \leq k \tag{14}$$

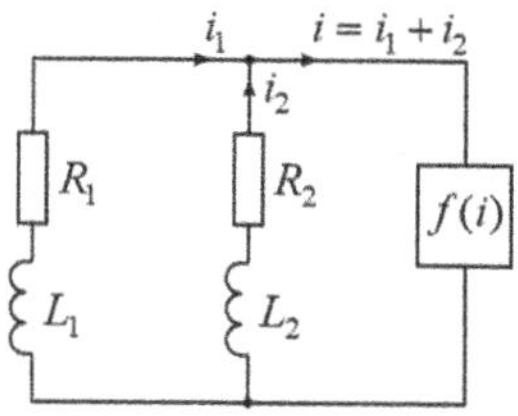

Fig. 3. Nonlinear electrical circuit

and with the feedback gain $h = 0.5$.

Find the maximal value of k for which the nonlinear electrical circuit is globally stable.

Using Procedure 3.1 we obtain the following.

Step 1. The positive linear part is described by the equations

$$\frac{d}{dt}\begin{bmatrix} i_1 \\ i_2 \end{bmatrix} = A\begin{bmatrix} i_1 \\ i_2 \end{bmatrix} + Bu, \quad y = C\begin{bmatrix} i_1 \\ i_2 \end{bmatrix}, \tag{15a}$$

where

$$A = \begin{bmatrix} -\frac{R_1}{L_1} & 0 \\ 0 & -\frac{R_2}{L_2} \end{bmatrix} = \begin{bmatrix} -3 & 0 \\ 0 & -2 \end{bmatrix}, \quad B = \begin{bmatrix} \frac{1}{L_1} \\ \frac{1}{L_2} \end{bmatrix} = \begin{bmatrix} 1 \\ 1 \end{bmatrix}, \quad C = [\,1\ 1\,]. \tag{15b}$$

The matrix (12) has the form

$$A_1 = A + khBC = \begin{bmatrix} -3 & 0 \\ 0 & -2 \end{bmatrix} + k\begin{bmatrix} 0.5 & 0.5 \\ 0.5 & 0.5 \end{bmatrix}. \tag{16}$$

Step 2. The characteristic polynomial of the matrix (16) has the form

$$\det[I_2 s - A_1] = \begin{vmatrix} s + 3 - 0.5k & -0.5k \\ -0.5k & s + 2 - 0.5k \end{vmatrix} = s^2 + (5 - k)s + 6 - 2.5k. \tag{17}$$

Step 3. Therefore, the maximal value of k for which all the coefficient of (17) are positive is $k < 2.4$.

Example 3.2. Consider the nonlinear positive electrical circuit shown in Fig. 4 with given resistances $R_1 = R_2 = 2$, capacitances $C_1 = 0.2$, $C_2 = 0.1$, the characteristic $i = f(u)$ satisfying the condition

$$0 \leq \frac{f(i)}{i} \leq k \tag{18}$$

and the feedback gain $h = 0.2$.

Check if the positive nonlinear electrical circuit is globally stable for $k = 0.5$.

The positive linear part of the nonlinear circuit is described by the equations

$$\frac{d}{dt}\begin{bmatrix} u_1 \\ u_2 \end{bmatrix} = A\begin{bmatrix} u_1 \\ u_2 \end{bmatrix} + Bu, \quad y = C\begin{bmatrix} u_1 \\ u_2 \end{bmatrix}, \tag{19a}$$

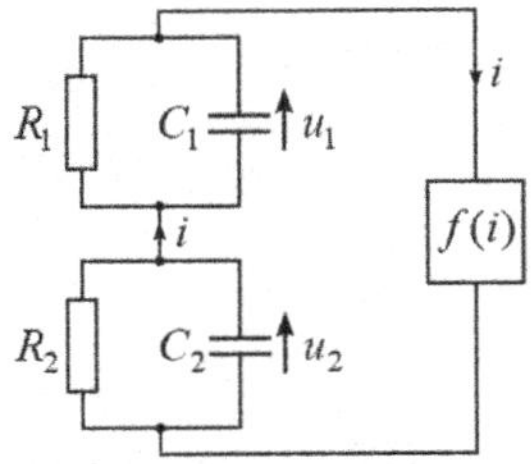

Fig. 4. Nonlinear electrical circuit

where

$$A = \begin{bmatrix} -\frac{1}{R_1C_1} & 0 \\ 0 & -\frac{1}{R_2C_2} \end{bmatrix} = \begin{bmatrix} -2.5 & 0 \\ 0 & -5 \end{bmatrix}, \quad B = \begin{bmatrix} \frac{1}{C_1} \\ \frac{1}{C_2} \end{bmatrix} = \begin{bmatrix} 5 \\ 10 \end{bmatrix}, \quad C = \begin{bmatrix} 1 & 1 \end{bmatrix}. \tag{19b}$$

In this case the matrix (12) has the form

$$A_1 = A + khBC = \begin{bmatrix} -2.5 & 0 \\ 0 & -5 \end{bmatrix} + 0.1 \begin{bmatrix} 5 \\ 10 \end{bmatrix} \begin{bmatrix} 1 & 1 \end{bmatrix} = \begin{bmatrix} -2 & 0.5 \\ 1 & -4 \end{bmatrix}. \tag{20}$$

and its characteristic polynomial

$$\det[I_2 s - A_1] = \begin{vmatrix} s+2 & -0.5 \\ -1 & s+4 \end{vmatrix} = s^2 + 6s + 7.5 \tag{21}$$

has positive coefficients.

Therefore, by condition 1 of Theorem 2.5 and Theorem 3.1 the nonlinear positive electrical circuit is globally stable.

4 Concluding Remarks

The global stability of electrical circuits composed of positive linear parts and nonlinear static element with given characteristic and positive feedbacks has been investigated. New sufficient conditions for the global stability of this class of nonlinear positive electrical circuits have been established. The effectiveness of these new stability conditions are demonstrated on simples examples of positive nonlinear electrical circuits. The considerations can be extended to fractional positive nonlinear electrical circuits and to discrete-time nonlinear positive systems. An open problem is an extension of these considerations by the use of the Kudrewicz approach [17].

Acknowledgment. This work was supported by National Science Centre in Poland under work No. 2017/27/B/ST7/02443.

References

1. Berman, A., Plemmons, R.J.: Nonnegative Matrices in the Mathematical Sciences. SIAM (1994)
2. Borawski, K.: Modification of the stability and positivity of standard and descriptor linear electrical circuits by state feedbacks. Electr. Rev. **93**(11), 176–180 (2017)
3. Busłowicz, M., Kaczorek, T.: Simple conditions for practical stability of positive fractional discrete-time linear systems. Int. J. Appl. Math. Comput. Sci. **19**(2), 263–169 (2009)
4. Farina, L., Rinaldi, S.: Positive Linear Systems; Theory and Applications. Wiley, New York (2000)
5. Kaczorek, T.: Absolute stability of a class of fractional positive nonlinear systems. Int. J. Appl. Math. Comput. Sci. **29**(1), 93–98 (2019)
6. Kaczorek, T.: Analysis of positivity and stability of discrete-time and continuous-time nonlinear systems. Comput. Prob. Electr. Eng. **5**(1), 11–16 (2015)
7. Kaczorek, T.: Analysis of positivity and stability of fractional discrete-time nonlinear systems. Bull. Pol. Acad. Sci. Techn. **64**(3), 491–494 (2016)
8. Kaczorek, T.: Global stability of nonlinear feedback systems with positive linear parts. . Int. J. Nonlinear Sci. Numer. Simul. **20**(5), 575–579 (2019)
9. Kaczorek, T.: Positive 1D and 2D Systems. Springer, London (2002)
10. Kaczorek, T.: Positive linear systems with different fractional orders. Bull. Pol. Acad. Sci. Techn. **58**(3), 453–458 (2010)
11. Kaczorek, T.: Positive linear systems consisting of n subsystems with different fractional orders. IEEE Trans. Circuits Syst. **58**(7), 1203–1210 (2011)
12. Kaczorek, T.: Positive fractional continuous-time linear systems with singular pencils. Bull. Pol. Acad. Sci. Techn. **60**(1), 9–12 (2012)
13. Kaczorek, T.: Selected Problems of Fractional Systems Theory. Springer, Berlin (2011)
14. Kaczorek, T.: Superstabilization of positive linear electrical circuit by state-feedbacks. Bull. Pol. Acad. Sci. Techn. **65**(5), 703–708 (2017)
15. Kaczorek, T., Borawski, K.: Stability of positive nonlinear systems. In: 22nd International Conference on Methods and Models in Automation and Robotics, Międzyzdroje, Poland (2017)
16. Kaczorek, T., Rogowski, K.: Fractional Linear Systems and Electrical Circuits. Springer, Cham (2015)
17. Kudrewicz, J.: Ustoicivost nieliniejnych sistem z obratnoj swjazju. Avtomatika i Telemechanika, vol. 25, no. 8 (1964)
18. Lyapunov, A.M.: Obscaja zadaca ob ustoicivosti dvizenija. Gostechizdat, Moskwa (1963)
19. Leipholz, H.: Stability Theory. Academic Press, New York (1970)
20. Mitkowski, W.: Dynamical properties of Metzler systems. Bull. Pol. Acad. Sci. Techn. **56**(4), 309–312 (2008)
21. Sajewski, L.: Stabilization of positive descriptor fractional discrete-time linear systems with two different fractional orders by decentralized controller. Bull. Pol. Acad. Sci. Techn. **65**(5), 709–714 (2017)

The Use of Multilayer ConvNets for the Purposes of Motor Imagery Classification

Szczepan Paszkiel[1](✉) and Paweł Dobrakowski[2]

[1] Faculty of Electrical Engineering, Automatic Control and Informatics, Opole University of Technology, Prószkowska 76, 45-271 Opole, Poland
s.paszkiel@po.edu.pl

[2] Humanitas University, Kilinskiego 43, 41-200 Sosnowiec, Poland

Abstract. An electroencephalographic signal is characterized by high complexity and character, which changes dynamically over time. At the same time, it should be noted that EEG signals in the field of Motor Imagery are increasingly more often used by scientists, among others, to help people with disabilities. Decoding these signals is important from the perspective of modern solutions based on the brain-computer interface technology. It is possible to isolate both spatial, as well as temporal features from an EEG signal using a ConvNet. This research paper suggests combining numerous ConvNet – Convolutional Neural Network models in terms of the EEG signal. The suggested multilayer ConvNets indicate high performance coefficients on a test data set. The method advocated in this article achieves a degree of accuracy on a data set at a precision level of 74.5%.

Keywords: EEG · Motor imagery · ConvNets · Brain · CNN

1 Introduction

One of the methods used to record human brain activity is electroencephalography [1]. It is distinguished by implementation easiness and is relatively inexpensive, which also translates to its frequent application in constructing brain-computer interfaces [2, 3]. An EEG signal is recorded in the form of time series, with many channels and low SNR value [4]. In devices based on the brain-computer technology, an EEG signal is recorded using numerous electrodes, which are placed on the head of the tested persons, as per the standardized 10–20 System developed by the International Federation for Clinical Neurophysiology [5]. During the feature extraction phase, the EEG signal obtained for motor imagery reveals specific traits both in terms of the spectrum, as well as space [6–8]. There are a few signal processing methods for extracting features from EEG signals currently in use, among others, wavelet transformation, Fourier transform and autoregressive model. The aforementioned methods have been reviewed, among others, by Pfurtscheller G, et al. in "Human brain-computer interface (BCI)" [9] and by Sagee GS et al. in "EEG feature extraction and classification in multiclass multiuser motor imagery brain computer interface using Bayesian Network and ANN" [10].

R. Szewczyk et al. (Eds.): AUTOMATION 2021, AISC 1390, pp. 10–19, 2021.
https://doi.org/10.1007/978-3-030-74893-7_2

The source literature increasingly more often addresses the issue of motor imagery in terms of its identification based on the EEG signal [11]. The EEG signal, generated in the scope of motor imagery, is identified in the sensorimotor cortex area, as a response to certain images. The automation of the motor imagery classification process is currently under research. It should be emphasized that conventional machine learning methods turned out effective in terms of classifying EEG signal data based on motor imagery, however, only to a certain extent. This is why the authors of this article proposed a combination of numerous ConvNet models for the purposes of motor imagery EEG signal identification [12]. Deep learning is finding an increasingly larger field of application in the field of classifying speech, image, etc. Motor imagery is the most frequently applied paradigm in BCI, which can be successfully used to support and improve sensorimotor functions of patients with disabilities. This has been proven by, among others, the research by Z.T. Al-Qaysi, et al. in the publication "Review of the EEG-based wheelchair control system: Consistent taxonomy, open challenges and recommendations" [13] and by R. Sitaram et al. in "Time classification of multichannel near-infrared spectroscopy signals for motor images for brain-computer interface development" [14]. Motor imagery is also used in entertainment and in smart solutions for healthy people, as evidenced by: A. Vourvopoulos, et al. in "EEG correlates of video game experience and user profile in motor imagery-based brain computer interaction" and by Z. Wang et al. in "Towards a hybrid BCI gaming paradigm based on motor imagery and SSVEP" [15]. It is worth noting that various ConvNet models can implement the extraction of various spatial and temporal features, which is why ConvNet architecture impacts their performance and accuracy [16]. So far, the outcome of combining ConvNet architectures and functions has not been tested in practice, using specific EEG signals. As indicated from the studies conducted for specific EEG signals, convolutional features depend directly on the ConvNet depth and filter size. At this time, it must be pointed out that contemporary machine learning techniques enable extracting information from EEG signals, which is implied by the fact that they play a key role in several important fields of research based on the electroencephalographic signal [17, 22, 23]. There are currently EEG-based BCI systems and machine learning techniques for clinical applications. This research can be successfully applied to the development of brain-computer technology in the field of automation and robotics. Especially in the field of development of the AR correlation with BCI [25], designing solutions for visualization of the brain's work [26] and in many different control processes [27].

2 Multilayer ConvNets

CNN is a kind of a deep learning structure, which has been designed for digital classification of images. It should be noted that CNN enables automatic extraction of an image function, by tuning the parameters in convolutional layers and by combining layers. Another advantage of CNN is also the fact that it is able to learn based on large data sets, in contrast to conventional machine learning algorithms, where a training set consists of hundreds of samples.

In many publications, researchers in the field of Motor Imagery classification used the support vector machine – SVM, as a classifier in practice. H. Cecotti et al. addressed this issue in the paper entitled: "Convolutional neural networks for P300 detection with application to brain-computer interfaces" [18]. The ICA – Independent Component Analysis and PCA – Principal Component Analysis methods were used for reducing dimensions and removing noise from an EEG signal [19]. This subject has already been addressed by one of the authors of this paper, in the publication: S. Paszkiel et al., "Methods of acquisition, archiving and biomedical data analysis of brain functioning" [20]. Currently, the ConvNet solution is popular for analysing the spatial features and classification of the EEG signal. Autoencoders and ConvNet were also used in studying emotion recognition using EEG signals. Despite numerous attempts in the 90s, scientists were unable to develop a method that would effectively train multilayer networks. It was not until 2006, when Hinton showed that deep networks can be trained in a different way. The suggested method assumed training networks layer by layer, followed by its supervised extra learning. Hinton also demonstrated increased quality of multilayer network mapping, compared to traditional, shallow MLP networks. At the moment, many researchers have distinguished convolutional features and numerous layers, and are combining them in order to increase solution effectiveness, proving that different features are obtained at each of the levels.

The methods for multilayer combination of ConvNet features suggested in this article are composed of a multilayer perceptron (MLP). Training stages involve two steps. ConvNets are initially trained individually on a set of Motor Imagery data in the first one, and the initially trained methods are trained on the target set of Motor Imagery data in the second one. After completing the training, ConvNet functions are transferred as input data to the MLP multilayer perceptron. MLP methods are then trained using the ConvNet functions, followed by activation with the Softmax function in terms of the MLP method and then tuning. The standard Softmax function (1) is defined by the formula:

$$\sigma(z)_i = \frac{e^{z_i}}{\sum_{j=1}^{K} e^{z_j}}$$

$$i = 1, \ldots, K; \quad z = (z_1, \ldots, z_k) \in R^K \tag{1}$$

Let us apply the standard exponential function to each element of the input vector z_i and normalize these values by dividing by the sum of all these exponentials; this normalization ensures that the sum of the components of the output vector $\sigma(z)$ is 1.

ConvNet models are composed of convolutions and layer block joining, and based on the well-known AlexNet architecture [21]. Many ConvNet models of various filter and layer sizes were developed during the tests. The number of convolutions and block joining gradually increased. Initial convolutions layers learn object edges and shapes, followed by learning more complex object shapes. Higher layers simultaneously use the convolution and non-linearity operations in order to obtain complex high-level functions, which are a combination of simple low-level functions. ConvNet is able to automatically learn hierarchical functions – layer by layer. ConvNet models with four convolution layers are shown in Fig. 1. The ConvNet model structure is shown in Table 1.

Currently, as indicated by a source literature review, many researchers are using ConvNet and deep learning models to classify EEG signals with shallow architecture

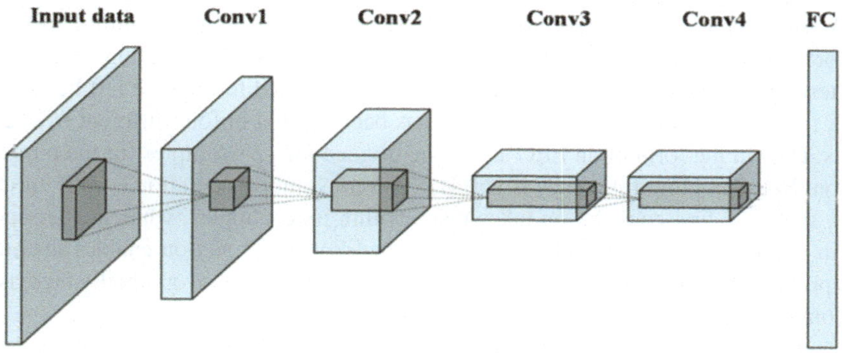

Fig. 1. ConvNet4 model with convolution layers

Table 1. ConvNet model structure used for the function, each of the model has 4 classes in Softmax assigned

ConvNet1	Conv (30 × 1.50 filters), Conv (1 × 22.50 filters), Pool (3 × 1), Dense (1024)
ConvNet2	Conv (25 × 1.50 filters), Conv (1 × 22.50 filters), Pool (3 × 1), Conv (10 × 1, 1 filters), Pool (3 × 1), Dense (1024)
ConvNet3	Conv (20 × 1.50 filters), Conv (1 × 22.50 filters), Pool (3 × 1), Conv (10 × 1.1 filters), Pool (3 × 1), Conv (10 × 1.1 filters), Pool (3 × 1), Dense (1024)
ConvNet4	Conv (10 × 1.50 filters), Conv (1 × 22.50 filters), Pool (3 × 1), Conv (10 × 1.1 filters), Pool (3 × 1), Conv (10 × 1.1 filters), Pool (3 × 1), Conv (10 × 1.1 filters), Pool (3 × 1), Dense (1024)

[12]. The first convolution operation is executed on each transmission channel or in some time intervals. The second convolution operation is simultaneously executed for all channels - one sample at a time. The outcome is a convolution in all input channels for many samples. Motor Imagery data is a two-dimensional array with channels as rows and time samples as columns. The models were tested using various training techniques and non-linearities, in order to train ConvNet in terms of general function. Data from the Institute for Knowledge Discovery, Laboratory of Brain-Computer Interfaces (Robert Leeb, Clemens Brunner, Gernot Müller-Putz, Alois Schlögl, Gert Pfurtscheller) were used for the purposes of the experiment – training and tests. Detailed data characteristics: 22 EEG channels, 0.5–100 Hz; notch filtered, 3 EOG channels, 250 Hz sampling rate, 4 classes, 9 subjects. Within the conducted conversions, the signals are processed to a minimum degree, only to remove interference.

3 Multilayer Network Training

The technique applied by scientists for training systems using EEG data is, among others, the division of EEG data into two sets, a training one, and a test one. Thus, the system is tested within a session, on data it did not previously have, but relating to the same topic. The data set used within the study is: BCI Competition IV Dataset IIa for 9 healthy people over two sessions, with each session consisting of a total of 280 tests. During each of the tests, the subjects were asked to imagine left hand, right hand, foot and tongue motions. A data set has been filtered in four different frequency bands. Data in the form of an EEG signal are divided into 2-s samples for the purposes of ConvNet. Sampling frequency is 250 Hz. The convolution operation is split into two parts, since the first convolution takes place within time samples, whereas the second one in all channels as a spatial filter. The applied Softmax function enables obtaining high probability values for each of the classes. MLP was trained by gradually increasing the convolution accumulating blocks at the time of increased efficiency. The best results were achieved using a four-layer ConvNet architecture – ConvNet4. It should be noted that none of the worldwide research reviewed by the authors in terms of deeper ConvNet models had achieved a good results in the scope of the EEG signal. Thus, it should be confirmed that a reasonable efficiency was exhibited by convolution summation blocks in terms of ConvNet1, ConvNet2, ConvNet3 and ConvNet4, with the filters of: 30, 25, 20 and 10 for ConvNet4, respectively. As part of the conducted study, input data is converted by the Softmax function into conditional probability for each class label for a given result. The network is trained in the scope of minimizing the sum of losses and assigning high probability to a correct label result. The completed experiment used a technique, which divides trainings into short intervals along with a validation procedure. The functions isolated by ConvNet layers act as input data for the classification layer.

The solution suggested for the purposes of the research include a method of linking MLP for the needs of combining ConvNet models. The main idea in this regard was the fact that various ConvNet architectures can be appropriate for extracting different features from an EEG signal, hence their combination may wise in terms of constructing general functions for EEG decoding. It is worth noting that given EEG signals are the records of time series, originating from numerous non-stationary channels, which translates into difficulties in isolating appropriate functions. Standard machine learning methods achieve correct results but an attempt to improve performance in this regard, through the application of deep learning, has not significant outcome. Nonetheless, when using the method of combined function, the specific knowledge extracted by different ConvNet models can be applied jointly. Combined ConvNet1, ConvNet2, ConvNet3 and ConvNet4 models are able to isolate the better functions and achieve higher accuracy than other ConvNet architectures. The function have been completed using MLP architecture. A multilayer ConvNet method was trained based on a data set described earlier in this article. MLP is constructed based on two hidden layers, each of which has 50 nodes. The suggested solution consists of a ConvNet and MLP combination. MLP methods are trained and output data sent to the Softmax algorithm within the experiment. The architecture of this solution is shown in Fig. 2.

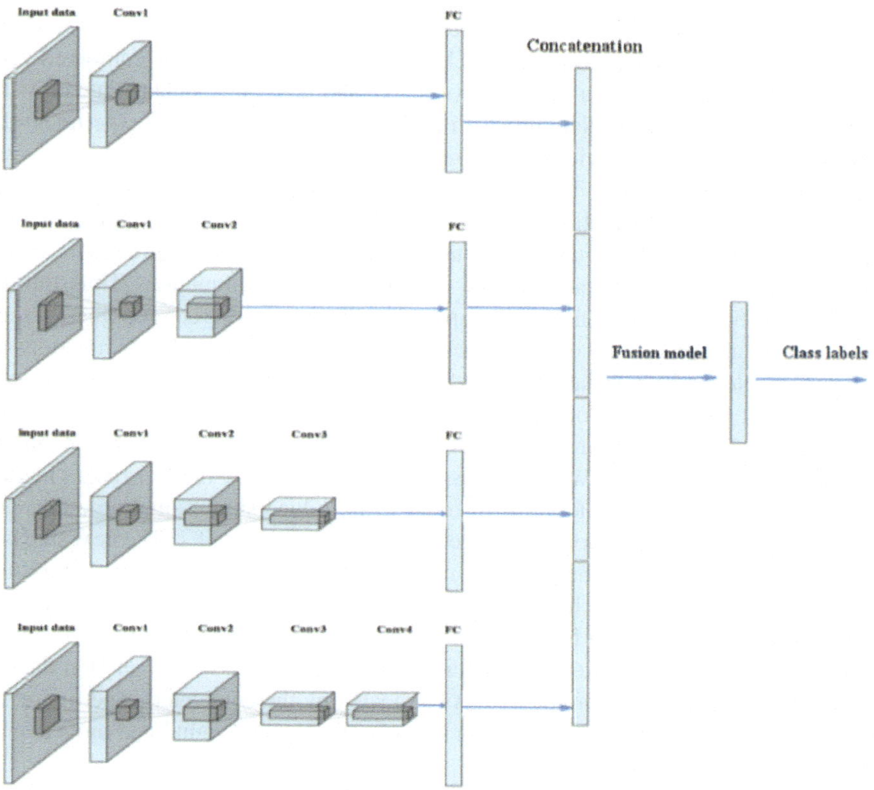

Fig. 2. Multilayer ConvNet

4 Research Experiment

The research experiment was conducted using a computer equipped with an Intel i7 processor and 32 GB of RAM. Initial processing of EEG data in the scope of Motor Imagery was executed with the use of the Python environment. In the case of ConvNet1, the source was an alpha band in the range of 7 to 14 Hz, ConvNet2 utilized the beta band in the range of 14–30 Hz, ConvNet3 in the range of: 7–30 Hz, whereas ConvNet4 from 0 to 40 Hz. The aforementioned results prove that ConvNet models with various depths enables obtaining correct accuracy, based on various frequency bands. Hence the conclusion that input data with specific frequencies for ConvNet models are important. It can be also observed that deeper models provide better results for a broad frequency spectrum, in the 0 to 40 Hz range. It is also important to correctly choose the filter, since it should also be noted that shallow models provide a better result the higher the filter size, as opposed to deep learning models. Using the data set presented in this article, the proposed method achieves an accuracy level of 74.5%.

However, it is worth mentioning that a ConvNet disadvantage is the fact that it is difficult to interpret and understand them in terms of information, what the network is learning and how outstanding results can be achieved. In order to find the functions used by ConvNet layers, one needs to determine whether the knowledge specific for the domain and class features is used by ConvNet receptive fields. The combined value function can be calculated for all receptive fields, and then one can calculate how that function value impact ConvNet output. In the course of the conducted experiment, the mean values for the frequencies of discussed bands were calculated so that they could be used as values of features. The developed correlation maps are a measure of the total amplitude spectrum. Correlation maps are compared with the layer output in order to see, which of the amplitude functions is used within the ConvNet. Mutual correlations have indicated that ConvNet can learn new information on these functions. Thus, through artificial modification of the data, their amplitude and features, one can see whether there was a change on the ConvNet output, hence confirming whether amplitude functions are used by ConvNet. If the correlations differ from the correlations of an untrained ConvNet, it may suggest that trained ConvNet applied these function to a greater degree than the untrained one. Classification results for ConvNet with a varying number of convolution pools are shown in Table 2. Whereas classification results for ConvNet fusion method combinations are shown in Table 3. In order to introduce signal amplitude changes, input training trials are converted to the frequency domain using FFT. The next step is to attempt random addition of the Gaussian noise, in order to implement signal amplitude changes. The outcome of the changes is the reconversion of the signal from the frequency domain into the time domain form.

Tables 4 give the confusion matrix for different output classes for an own dataset, for ConvNet with MLP fusion method. As indicated by the conducted experiment, the suggested method exhibits high decoding accuracy for left and right hand movement.

Table 2. Classification results for ConvNet with a varying number of convolution pools

ConvNet with a different number of Conv	ConvNet1	ConvNet2	ConvNet3	ConvNet4
Frequency bands	7–14 Hz	14–30 Hz	7–30 Hz	0–40 Hz
Conv Filter Size	30 × 1	22 × 1	15 × 1	10 × 1
Accuracy – IKD Data*	72.7%	70.4%	71.4%	71.8%

*IKD – Institute for Knowledge Discovery.

Table 3. Classification results for ConvNet fusion method combinations

Fusion stages	Accuracy
ConvNet1+ConvNet2	72.9%
ConvNet1+ConvNet3	71.1%
ConvNet1+ConvNet4	75.5%
ConvNet1+ConvNet2+ConvNet4	69.5%
ConvNet1+ConvNet3+ConvNet4	73.3%
ConvNet with MLP	**74.5%**

Table 4. Confusion matrix for ConvNet with MLP

	Targets for left hand	Targets for right hand	Targets for feet	Targets for tongue
	Value for the ConvNet with MLP method	Value for the ConvNet with MLP method	Value for the ConvNet with MLP method	Value for the ConvNet with MLP method
Prediction for left hand	**80.70**	11.11	4.2	3.1
Prediction for right hand	11.22	**81.06**	4.0	3.4
Prediction for feet	9.4	11.23	66.2	12.2
Prediction for tongue	8.5	12.1	7.2	71.08

5 Conclusions

As a result of the conducted research it should be concluded that conventional machine learning ensured a limited accuracy of Motor Imagery EEG signal classification improvement. It should also be stated that deep learning techniques show better results for the same task [24]. The completed research also enable a conclusion that the suggested multilayer ConvNet-based models are efficient in terms of various filter depths and sizes, and extracting various types of functions representing EEG data. The conducted experiment also demonstrated that the application of initially trained ConvNet models can help improve the accuracy and time of deep learning model training time. The solution suggested in this article shows a general EEG decoding efficiency for all output classes. The results obtained using the proposed methods prove that ConvNet characterized by various architectures, depths and filter sizes, highly impact the accuracy and can isolate other representation functions, which can increase classification precision when combined.

References

1. Mashat, M.E.M., Lin, C.T., Zhang, D.: Effects of task complexity on motor imagery based brain-computer interface. IEEE Trans. Neural Syst. Rehabil. Eng. **27**(10), 2178–2185 (2019)
2. Aldea, R., Fira, M.: Classifications of motor imagery tasks in brain computer interface using linear discriminant analysis. Int. J. Adv. Res. Artif. Intell. **3**(7), 5–9 (2014)
3. Vidaurre, C., Schlögl, A., Cabez, R., Scherer, R., Futtscheller, G.: Adaptive on-line classification for EEG-based brain computer interfaces with AAR parameters and band power estimates. Biomed. Tech. (Berlin) **50**(11), 350–354 (2005)
4. Kevric, J., Subasi, A.: Comparison of signal decomposition methods in classification of EEG signals for motor-imagery BCI system. Biomed. Signal Process. Control **31**, 398–406 (2017)
5. Wang, Z., Yu, Y., Xu, M., et al.: Towards a hybrid BCI gaming paradigm based on motor imagery and SSVEP. Int. J. Hum. – Comput. Interact. **35**(3), 97–205 (2019)

6. Athanasiou, A., Klados, M.A., Styliadis, C., Foroglou, N., Polyzoidis, K., Bamidis, P.D.: Study of the role of alpha and beta rhythms in functional motor networks. Neuroscience **378**, 54–70 (2018). https://doi.org/10.1016/j.neuroscience.2016.05.044
7. Courellis, H., Mullen, T., Poizner, H., Cauwenberghs, G., Iversen, J.R.: EEG-based quantification of cortical current density and dynamic causal connectivity generalized across subjects performing BCI-monitored cognitive tasks. Front. Neurosci. **11**, 180 (2017). https://doi.org/10.3389/fnins.2017.00180
8. Li, T., Xue, T., Wang, B., Zhang, J.: Decoding voluntary movement o f single hand based on analysis of brain connectivity by using EEG signals. Front. Hum. Neurosci. **12**, 381 (2018). https://doi.org/10.3389/fnhum.2018.00381
9. Pfurtscheller, G., Neuper, C., Birbaumer, N.: Human brain-computer interface (BCI). In: Riehle, A., Vaadia, E. (eds.) A Distributed System for Distributed Functions, pp. 367–401. Motor Cortex in Voluntary Movements (2005)
10. Sagee, G.S., Hema, S.: EEG feature extraction and classification in multiclass multiuser motor imagery brain computer interface using Bayesian Network and ANN. In: 2017 International Conference on Intelligent Computing, Instrumentation and Control Technologies (ICICICT). IEEE (2017)
11. Hanakawa, T., Immisch, I., Toma, K., Dimyan, M.A., Van Gelderen, P., Hallett, M.: Functional properties of brain areas associated with motor execution and imagery. J. Neurophysiol. **89**, 989–1002 (2003). https://doi.org/10.1152/jn.00132.2002
12. Schirrmeister, R.T., Springenberg, J.T., Fiederer, L.D.J., Glasstetter, M., Eggensperger, K., Tangermann, M., Ball, T.: Deep learning with convolutional and neural networks for EEG decoding and visualization. Hum. Brain Mapp. **38**, 5391–5420 (2017). https://doi.org/10.1002/hbm.23730
13. Al-Qaysi, Z.T., Zaidan, B.B., Zaidan, A.A., et al.: Review of the EEG-based wheelchair control system: consistent taxonomy, open challenges and recommendations. Comput. Methods Biomed Programs **164**, 221–237 (2018)
14. Sitaram, R., Zhang, H., Guan, C., et al.: Time classification of multichannel near-infrared spectroscopy signals for motor images for brain-computer interface development. Neuroimage **34**(4), 1416–1427 (2007)
15. Vourvopoulos, A., Badia, S.B.I., Liarokapis, F.: EEG correlates of video game experience and user profile in motor-imagery-based brain – computer interaction. Vis. Comput. **33**(4), 533–546 (2017)
16. Guler, N., Ubeyli, E., Guler, I.: Recurrent neural networks employing Lyapunov exponents for EEG signals classification. Expert Syst. Appl. **29**(3), 506–514 (2005). https://doi.org/10.1016/j.eswa.2005.04.011. ISSN 09574174
17. Amin, H.U., Malik, A.S., Ahmad, R.F., et al.: Feature extraction and classification for EEG signals using wavelet transform and machine learning techniques. Astralas. Phys. Eng. Sci. Med. **38**(1), 139–149 (2015)
18. Cecotti, H., Graser, A.: Convolutional neural networks for P300 detection with application to brain-computer interfaces. IEEE Trans. Pattern Anal. Mach. Intell. **33**(3), 433–445 (2011). https://doi.org/10.1109/TPAMI.2010.125. ISSN 01628828
19. Subasi, A., Gursoy, M.I.: EEG signal classification using PCA, ICA, LDA and support vector machines. Expert Syst. Appl. **37**(12), 8659–8666 (2010)
20. Paszkiel, S., Szpulak, P.: Methods of acquisition, archiving and biomedical data analysis of brain functioning. In: Hunek, W., Paszkiel, S. (eds.) Biomedical Engineering and Neuroscience, Proceedings of the 3rd International Scientific Conference on Brain-Computer Interfaces, BCI 2018, Opole, Poland, 13–14 March. Advances in Intelligent Systems and Computing, AISC, vol. 720, pp. 158–171. Springer (2018). https://doi.org/10.1007/978-3-319-75025-5_15

21. Lu, S., Lu, Z., Zhang, Y.-D.: Pathological brain detection based on AlexNet and transfer learning. J. Comput. Sci. (2018). https://doi.org/10.1016/j.jocs.2018.11.008
22. Paszkiel, S.: The use of facial expressions identified from the level of the EEG signal for controlling a mobile vehicle based on a State Machine. In: Automation 2020: Towards Industry of the Future. Advances in Intelligent Systems and Computing, pp. 227–238. Springer (2020). https://doi.org/10.1007/978-3-030-40971-5_21
23. Paszkiel, S.: Using BCI in IoT implementation. In: Studies in Computational Intelligence, vol. 852, pp. 111–128. Springer, Switzerland (2020). https://doi.org/10.1007/978-3-030-30581-9_13
24. Paszkiel, S.: Using neural networks for classification of the changes in the EEG signal based on facial expressions. In: Studies in Computational Intelligence, vol. 852, pp. 41–69. Springer, Switzerland (2020). https://doi.org/10.1007/978-3-030-30581-9_7
25. Paszkiel, S.: Augmented reality of technological environment in correlation with brain computer interfaces for control processes. In: International Conference on Automation. Advances in Intelligent Systems and Computing, vol. 267, pp. 197–203. Springer, Switzerland (2014). https://doi.org/10.1007/978-3-319-05353-0_20
26. Paszkiel, S., Hunek, W., Shylenko, A.: Project and simulation of a portable device for measuring bioelectrical signals from the brain for states consciousness verification with visualization on LEDs. In: International Conference Challenges in Automation, Robotics and Measurement Techniques (Automation). Advances in Intelligent Systems and Computing, vol. 440, pp. 25–39. Springer, Switzerland (2016). https://doi.org/10.1007/978-3-319-29357-8_3
27. Paszkiel, S.: Analysis and classification of EEG signals for brain-computer interfaces introduction. In: Studies in Computational Intelligence, vol. 852, Springer, Switzerland (2020). https://doi.org/10.1007/978-3-030-30581-9_1

Ordered–Fuzzy-Numbers-Driven Approach to Out-Plant Milk-Run Dynamic Routing and Scheduling

Grzegorz Bocewicz[1(✉)], Peter Nielsen[2], and Zbigniew Banaszak[1]

[1] Faculty of Electronics and Computer Science, Koszalin University of Technology, Koszalin, Poland
{grzegorz.bocewicz,zbigniew.banaszak}@tu.koszalin.pl
[2] Department of Materials and Production, Aalborg University, Aalborg, Denmark
peter@mp.aau.dk

Abstract. We consider a dynamic vehicle routing problem in which a fleet of service vehicles pay preventive maintenance visits to a set of spatially distributed customers over a given time horizon. Each vehicle follows a pre-planned separate route which links points defined by customer location and service periods. Customer orders and the feasible time windows for the execution of those orders are dynamically revealed over time. The objective is to maximize the number of new urgent requests, which are inserted dynamically throughout the assumed time horizon, but not at the expense of the already accepted orders. The question that must be considered is whether the newly reported service request and the correction of the fulfillment date of the already submitted request can be accepted or not. Problems of this type arise, for example, in systems in which preventive or corrective maintenance requests are scheduled to follow periodically performed inspections or repairs. The problem under consideration is formulated as a constraint satisfaction problem using the ordered fuzzy number formalism, which allows to handle the fuzzy nature of the variables involved through an algebraic approach. The computational results show that the proposed solution outperforms the commonly used computer simulation methods.

Keywords: Dynamic vehicle routing problem · Ordered fuzzy numbers formalism · Out-plant milk-run systems · Declarative modelling

1 Introduction

Milk-run routing and scheduling problems are usually recognized and formulated as special cases of the Vehicle Routing Problem (VRP), introduced over six decades ago as a generalization of the Traveling Salesman Problem (TSP) [6, 21, 31]. Consequently, milk-run driven problems concerning the distribution of components or commodities can be classified similarly to other, widely studied, extensions of the VRP aimed at finding the optimal set of routes for a fleet of vehicles delivering goods or services to various locations [6, 29].

R. Szewczyk et al. (Eds.): AUTOMATION 2021, AISC 1390, pp. 20–36, 2021.
https://doi.org/10.1007/978-3-030-74893-7_3

Just as some authors distinguish between inbound logistics, regarding the transport, storage and delivery of goods coming into a business, and outbound logistics, related to the same for goods going out of a business [23], other authors draw a distinction between in-plant milk-run, i.e. operations related to raw materials, work in process and distribution of finished goods, and out-plant milk-run, which involves supporting the transport of commodities and products between manufacturers and customers as well as service visits [9, 12]. In both cases, the same two kinds of intertwined decisions regarding vehicle routing policies are considered, i.e. determination of the routes along which customers are visited and determination of a schedule which will guarantee congestion-free movement of the vehicles.

Milk-run problems are usually concerned with planning routes that are cyclically repeated according to a fixed schedule in a fixed sequence and with fixed arrival times. The goal of this type of problems is to determine whom to serve, how much to deliver and which regularly repeated routes to travel using what fleet of vehicles. Moreover, since they belong to the class of NP-hard VRPs, an approximate solution is sought with the help of heuristic methods [16, 17, 27]. An exhaustive review of VRP taxonomy-inspired problems formulated for milk-run systems can be found in [6, 21].

In real-life settings, these kind of problems become more complex due to disturbances (changes in the execution of already planned requests/orders and the arrival of new ones, traffic jams, accidents, etc.) and the fuzzy nature of the parameters which determine the timeliness of the services/deliveries performed. In this paper we consider an out-plant dynamic milk-run routing problem (DMRP), which consists in designing vehicle routes (determined by customer visit sequences) in an online fashion as orders are revealed incrementally over time. The dynamic nature of the problem requires communicating to a vehicle which customer to serve next as soon as it has completed its previous visit. All related decisions are made without prior knowledge of upcoming orders. DMRPs commonly arise in connection with maintenance operations, where the ability to redirect a moving vehicle to a new request nearby affords additional savings [18, 20]. However, to re-route a vehicle in this way, a company needs to be able to track its position on an ongoing basis and have communication solutions for the quick assignment of a new destination.

The uncertainty of DMRP data connected with traffic disturbances as well as changes in service completion dates requires the use of a model based on the formalism of fuzzy sets. Because this uncertainty implies the uncertainty of the final result, a model that implements the formalism of fuzzy numbers needs to be adopted. The best model to take account of the aforementioned constraints related to the nature of inequalities, implications and logical conditions is a declarative model. Modeled in a declarative way, the DMRP can be treated as a fuzzy constraint satisfaction problem and solved using both computer simulation and analytical ordered-fuzzy-numbers-driven calculations. It should be noted, that in contrast to standard fuzzy numbers, the support of a fuzzy number does not expand as a result of algebraic operations performed on the ordered fuzzy numbers domain. That is why the algebra of ordered fuzzy numbers (OFN algebra) can be used to increase the competitiveness of the analytical approach in comparison to the time-consuming computer simulation-based calculations of a feasible out-plant milk-run dynamic re-routing scenario. In other words, computer simulation methods, which are commonly used for dynamic route prototyping, can be replaced with an analytical method that employs OFN algebra.

In this context, the goal of our research is to develop an ordered-fuzzy-numbers-driven declarative model of a DMRP subject to fuzzy maintenance time and transportation time constraints, the solution to which provides the possible dynamic rerouting scenarios. Unlike most of the problems discussed in the literature, which concentrate on the search for solutions that optimize the path traveled or the cumulative cost of the mission carried out, in our approach, an answer to the following question is sought: Can the newly reported service requests be accepted or not, and can the completion date of the already submitted requests be corrected or not?

The present study is a continuation of our previous work in which we explored methods for fast prototyping of solutions to problems related to the routing and scheduling of tasks typically performed in batch flow production systems [4, 5, 19, 28, 32], as well as problems related to the planning and control of production flow in departments of automotive companies [2, 3, 30]. Its main contribution is fourfold:

- Out-plant milk-run-driven maintenance services require taking into account disturbances in road traffic and the uncertainty of the timing of remanufacturing services.
- Formulation of the DMRP using OFN algebra allows one to plan vehicle fleet scenarios, taking into account the uncertainty of vehicle travel time and repair time.
- In contrast to standard fuzzy numbers, the support of a fuzzy number does not expand as a result of algebraic operations performed on the OFN domain, which means that analytical methods based on OFN algebra are more efficient than the currently used computer simulation methods.
- The proposed algebraic approach allows to make quick online decisions in situations involving unplanned changes to the current order-servicing conditions which require vehicle re-routing and/or re-scheduling.

The remainder of this paper is organized as follows. Section 2 reviews the literature. Section 3 provides an overview of some basic concepts of the ordered fuzzy numbers theory and constraint programming techniques. The problem statement is provided in Sect. 4, and the methodology used in solving the problem is described in Sect. 5. Computational results are then reported and analyzed in Sect. 6. Conclusions are made in Sect. 7, which also indicates future directions of research.

2 Related Work

2.1 Milk-Run Routing and Scheduling

Most of the problems appearing in milk-run systems are aimed at searching for an optimal periodic distribution policy, i.e., a plan on whom to serve, how much to deliver, and what regularly repeated routes to travel by what fleet of vehicles. Examples of such problems [12] include both simple ones, such as the Mix Fleet VRP, Multi-depot VRP, Split-up Delivery VRP, Pick-up and Delivery VRP, VRP with Time Windows, VRP with Backhauls, and more complex ones, such as the VRP with multi-trip multi-traffic pick-up and delivery problem with time windows and synchronization, which is a combination of variants of the VRP with multiple trips, VRP with a time window, and VRP with pick-up delivery. Since milk-run routing and scheduling problems are NP-hard VRPs, they are solved using heuristic methods. The solutions are derived from the milk-run distribution policy and involve determining, for instance, in what time windows parts can be collected from suppliers and how many logistic trains should run along which routes [11, 16, 17, 20]. Regardless of the class of problems considered, be they problems typical of in-plant or out-plant Milk-run systems [23] or problems that accentuate the dynamic or static character of vehicle routing [11, 12, 17, 18], the goal is always to search for optimal solutions. In focusing on the search for such solutions, these studies implicitly assume that there do exist admissible solutions, e.g., ones that ensure congestion-free flow of concurrently executed transport processes and/or that planned routings and schedules are robust to assumed disturbances [18, 20]. It is worth noting that relatively few studies are devoted to the problems of out-plant milk-run dynamic routing and systems in which services are provided by appointment. In systems of this type, the dynamic multi-period VRP is solved, which boils down to scheduling services in a rolling horizon fashion, in which new requests received, but unfulfilled, during the first period together with the set of customer requests preplanned for the next period constitute a new portfolio of orders to be considered for subsequent scheduling [9, 26, 29]. An exhaustive review of VRP taxonomy-inspired problems formulated for milk-run systems can be found in [6, 21].

In many real-life situations, the uncertainty of DMRP data caused by traffic disturbances (uncertain travel times caused by weather conditions, daily changes in traffic intensity etc.) as well as the degree of difficulty of the service provided (caused by intertwined overhauls, condition monitoring, product repair operations, etc.) cannot be estimated in a precise way. However, the majority of models of the so-called Fuzzy VRP only assume vagueness for fuzzy customer demands to be collected and fuzzy service- or travel times. The literature on these issues is very scarce [7], despite the rapidly growing demand for predictive maintenance oriented service providers [21]. The quickly developing enterprise servitization indicates that there is an increasing need for this type of services [1, 14, 15].

To summarize, the review of the literature shows that there is an urgent need to develop analytical approaches that would replace the labor-intensive and time-consuming computer simulation methods for the assessment of possible maintenance service scenarios. The methods sought should take into account the fact that mobile servicing missions require considering the uncertainty factor resulting from the fuzzy nature of vehicle movement and service period. Our approach, which combines the declarative modeling paradigm with OFN algebra seems to meet the above requirements.

2.2 An Ordered Fuzzy Numbers Framework

The milk-run routing and scheduling problems developed so far have limited use due to the data uncertainty observed in practice. The values describing parameters such as transport time or loading/unloading times, depend on the human factor, which means they cannot be determined precisely. It is difficult to account for data uncertainty by using fuzzy variables due to the imperfections of the classical fuzzy numbers algebra [2]. Equations which describe the relationships between fuzzy variables (variables with fuzzy values) using algebraic operations (in particular, addition and multiplication) do not meet the conditions of the Ring (among others if the condition $\forall_{A \in \mathcal{F}} A + 0 = A$ is met, then condition $\forall_{A \in \mathcal{F}} \exists!_{B \in \mathcal{F}} A + B = 0$ is not met). In addition, algebraic operations based on standard fuzzy numbers follow Zadeh's extension principle. In practice, this means that no matter what algebraic operations are used, the support of the fuzzy number, which is the result of these operations, expands. Consequently, it is impossible to solve algebraic equations with fuzzy variables. In particular, this means that for any fuzzy numbers a, b, c, the following implication $(a + b = c) \Rightarrow [(c - b = a) \wedge (c - a = b)]$ does not hold. This makes it impossible to solve a simple equation $A + X = C$. This fact significantly hinders the use of approaches based on declarative models, in which most of the relationships between decision variables are described as linear/nonlinear equations and/or algebraic inequalities. There are various approaches in the literature that work around the above-mentioned deficiencies [2], but they are quite complex.

We address these issues by proposing a declarative model that implements the formalism of OFN algebra, which assumes that there exists a neutral element (zero) for operations such as addition and multiplication, which makes it possible to solve algebraic equations in the model. The concept of OFNs can be defined as follows [13]:

Definition 1. An OFN is a pair of continuous real functions defined by the interval [0, 1], i.e.:

$$\hat{A} = (f_A, g_A), \text{ where: } f_A, g_A: [0, 1] \rightarrow \mathbb{R}. \tag{1}$$

The functions f_A and g_A are called the up part and the down part of the OFN $\widehat{A}$, respectively. The values of these continuous functions are limited ranges, which can be defined as the following bounded intervals: $UP_A = (l_{A0}, l_{A1})$ and $DOWN_A = (p_{A1}, p_{A0})$. Assuming that: f_A is increasing and g_A is decreasing as well as that $f_A \leq g_A$, the membership function μ_A of the OFN $\widehat{A}$ is as shown in Figs. 1a) and 1b):

$$\mu_A(x) = \begin{cases} f_A^{-1}(x) & when\ x \in UP_A \\ g_A^{-1}(x) & when\ x \in DOWN_A \\ 1 & when\ x \in [l_{A1}, p_{A1}] \\ 0 & \text{otherwise} \end{cases}. \tag{2}$$

A property called orientation (direction) is defined for an OFN. There are two types of orientation: positive, when $\widehat{A} = (f_A, g_A)$ the direction is consistent with the direction of the *OX* axis and negative, when $\widehat{A} = (g_A, f_A)$ the direction is opposite to the direction of the *OX* axis. Assuming that the values of all fuzzy variables may have a different orientation, the definitions of the algebraic operations used are as follows:

Definition 2. Let $\widehat{A} = (f_A, g_A)$ and $\widehat{B} = (f_B, g_B)$ be OFNs. $\widehat{A}$ is a number equal to $\widehat{B}$ ($\widehat{A} = \widehat{B}$), $\widehat{A}$ is a number greater than $\widehat{B}$ or equal to or greater than $\widehat{B}$ ($\widehat{A} > \widehat{B}$; $\widehat{A} \geq \widehat{B}$), $\widehat{A}$ is less than $\widehat{B}$ or equal to or less than $\widehat{B}$ ($\widehat{A} < \widehat{B}, \widehat{A} \leq \widehat{B}$) if: $\forall_{x \in [0,1]} f_A(x) * f_B(x) \wedge g_A(x) * g_B(x)$, where: the symbol $*$ stands for: $=$, $>$, $\geq$, $<$, or $\leq$.

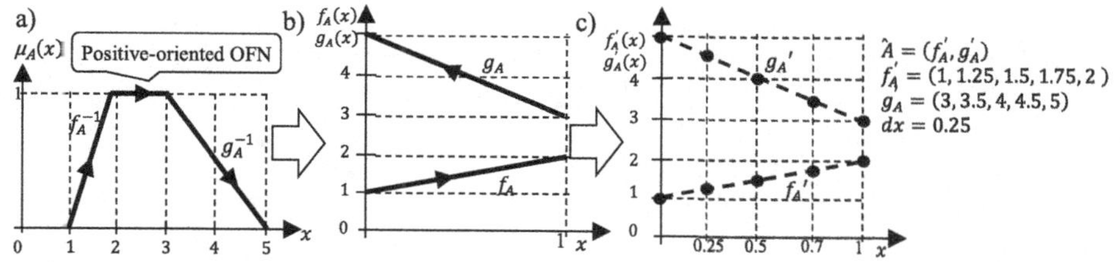

Fig. 1. a) OFN $\widehat{A}$ represented as a convex fuzzy number, b) functions f_A, g_A determining $\widehat{A}$ (positive orientation), c) discrete representation of $\widehat{A}$ ($dx = 0.25$) (based on [13])

Definition 3. Let $\widehat{A} = (f_A, g_A)$, $\widehat{B} = (f_B, g_B)$, and $\widehat{C} = (f_C, g_C)$ be OFNs. The operations of addition $\widehat{C} = \widehat{A} + \widehat{B}$, subtraction $\widehat{C} = \widehat{A} - \widehat{B}$, multiplication $\widehat{C} = \widehat{A} \times \widehat{B}$ and division $\widehat{C} = \widehat{A}/\widehat{B}$ are defined as follows: $\forall_{x \in [0,1]} f_C(x) = f_A(x) * f_B(x) \wedge g_C(x) = g_A(x) * g_B(x)$, where: the symbol $*$ stands for $+$, $-$, $\times$, or $\div$; The operation of division is defined for $\widehat{B}$ such that $|f_B| > 0$ and $|g_B| > 0$ for $x \in [0, 1]$.

In recent years, the concept of OFNs has been continuously developed and used in various practical applications. Many publications have been devoted to the analysis of the OFN model in relation to convex fuzzy sets [8, 10, 22, 24, 25].

3 Preliminaries

Let us consider graph $G = (N, E)$ modelling a transportation network composed of $|N| = \omega = 11$ points (i.e., customers and service), as shown in Fig. 2. The points (hereinafter referred to as nodes) include 1 node representing service N_1 and 10 nodes representing customers $N_2 - N_{11}$. The customers $N_2 - N_{11}$ are cyclically serviced (with period $T = 2000$ u.t.) by vehicles traveling form node N_1. For each node, a variable is assigned which indicates the beginning moments of node N_λ occupation (service) y_λ^k by vehicle U_k. The service is executed in intervals determined by the service deadline $[ld_\lambda; ud_\lambda]$, i.e. $y_\lambda^k \geq ld_\lambda$ and $y_\lambda^k + t_i \leq ud_\lambda$ (see Table 1).

In turn, each edge $\left(N_\beta, N_\lambda\right) \in E$ linking nodes N_β and N_λ is labelled with a fuzzy variable (in the OFN representation) representing the uncertainty of traveling time $d_{\beta,\lambda}$ between nodes N_β and N_λ (see Fig. 3). It is assumed that the set of edges E model the routes traveled by vehicles between nodes N_β and N_λ. Given is a fleet of vehicles $\mathcal{U} = \{U_1, \ldots, U_k, \ldots, U_K\}$ servicing customers spatially dispersed in network G. The routes traveled by vehicle U_k are denoted by sequences of nodes: $\pi_k = \left(N_{k_1}, \ldots, N_{k_i}, N_{k_{i+1}}, \ldots, N_{k_\mu}\right)$, where: $k_i \in \{1, .., K\}$, $\forall_{k_i \neq k_j} N_{k_i} \neq N_{k_j}$, $\left(N_{k_i}, N_{k_{i+1}}\right) \in E$. It is assumed that nodes representing the service base (e.g. N_1) appear on every route, and each node representing a customer (e.g. $N_2 - N_{11}$) occurs on only one route from set Π.

In general, taking into account the assumptions mentioned above (the uncertainty of the transport operations and the uncertainty of the moments of node occupation and release), what is sought is a set of routes and associated schedules guaranteeing that service is provided to customers on time.

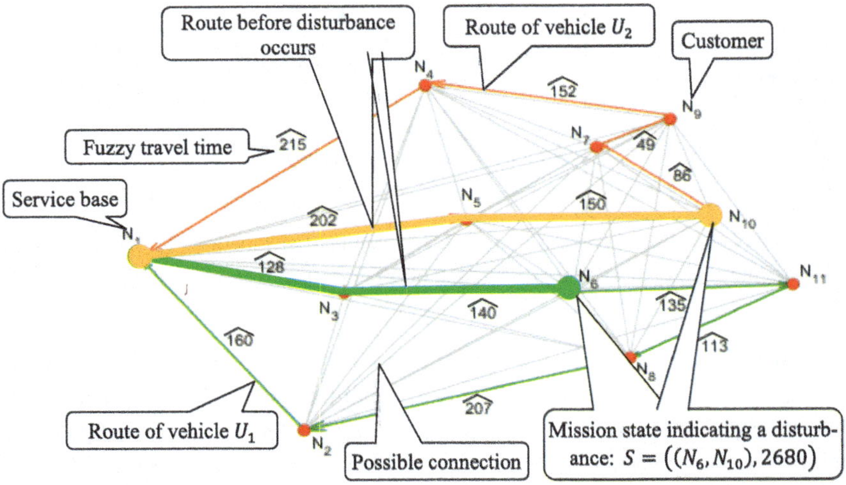

Fig. 2. Graph modeling transportation network G with a mission state indicating a disturbance

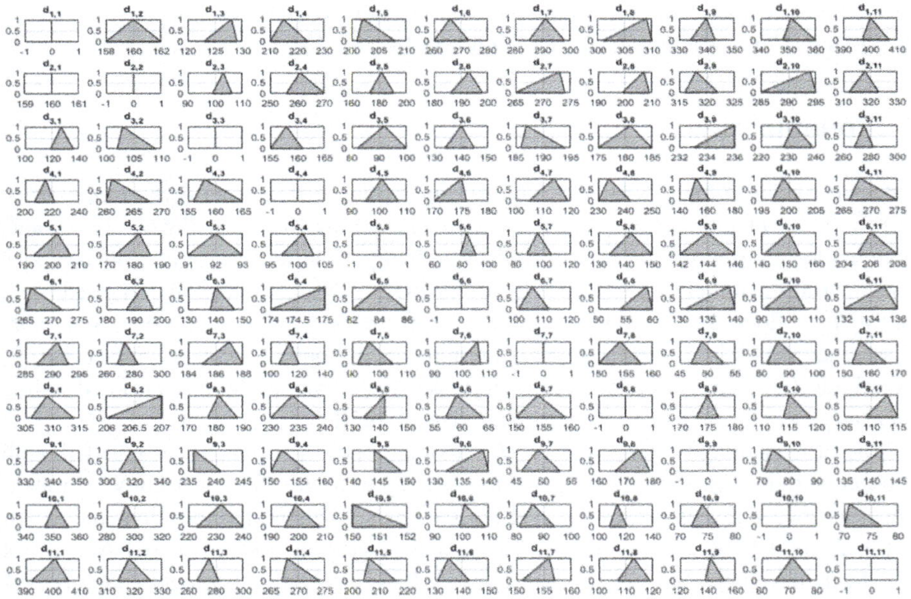

Fig. 3. Traveling times between nodes for the networks from Fig. 2 (an OFN representation)

Examples of such vehicle routings and the associated fuzzy schedule are presented in Figs. 2 and 4. The routes are the following sequences of nodes visited repetitively by U_1 and U_2: $\pi_1 = (N_1, N_3, N_6, N_{11}, N_8, N_2, N_1)$, $\pi_2 = (N_1, N_5, N_{10}, N_7, N_9, N_4)$. In practice, however, changes are quite frequently made to the times at which service is provided to customers.

Table 1. Service deadlines for customers $N_2 - N_{11}$

	N2	N3	N4	N5	N6	N7	N8	N9	N10	N11
ld_λ	1300	300	1100	300	600	700	1000	800	500	800
ud_λ	1900	600	1400	600	900	1000	1500	1200	800	1100

An example of this kind of disturbance is presented in Fig. 4. In the second window shown in Fig. 4 (moment $t^* = 2680$ when U_1 occupies N_6 and U_2 occupies N_{10}), the system receives information that the service deadline on node N_{11} has changed. The following question can be considered: Does there exist a set of routes Π and a fuzzy cyclic schedule that determines how the vehicles in the fleet will travel, in response to ad hoc changes in customer service periods?

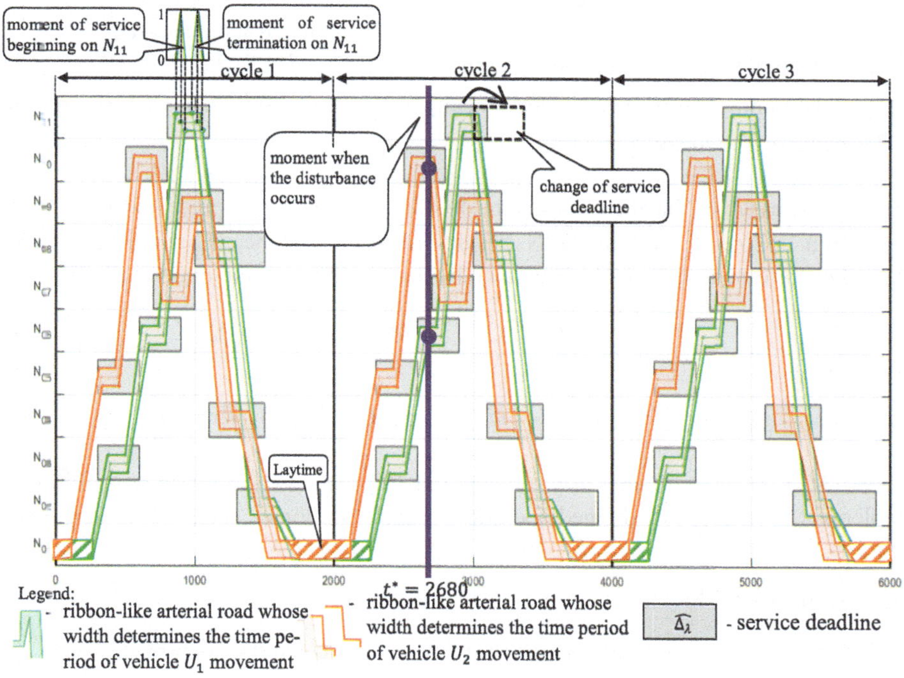

Fig. 4. Cyclic fuzzy schedule

4 Problem Description

4.1 Assumptions

The following assumptions are met:

- given is a network $G = (N, E)$, where N is a set of nodes and E is a set of edges,
- each node N_λ is labelled with a fuzzy value $\widehat{t_\lambda}$ (represented in terms of OFN) denoting the duration of node occupation (service),
- each edge $(N_\beta, N_\lambda) \in E$ is labeled with a fuzzy value $\widehat{d_{\beta,\lambda}}$ (represented in terms of OFN) denoting the travel time between nodes N_β and N_λ,
- given is a fleet of vehicles $\mathcal{U}$, in which each vehicle U_k travels route π_k ($\pi_k \in \Pi$) described by a sequence of successively visited nodes,
- crossings between nodes can only be made along the edges connecting them,
- node N_1 representing a service base occurs exactly on all routes of set Π,
- node N_λ ($\lambda > 1$) representing a customer occurs exactly on one route of set Π,
- customers are serviced cyclically in time windows repeated with period $\widehat{T}$,
- each customer N_λ is serviced by deadline $\Delta_\lambda = [ld_\lambda; ud_\lambda]$,
- fuzzy beginning moments of node N_λ occupation $\widehat{y_\lambda^k}$ (represented as OFN) by vehicle U_k make up the fuzzy cyclic schedule $\widehat{\mathbb{Y}}$.

It is also assumed that the values of the decision variables are represented as OFN, see Definition 1. Consequently the OFN $\widehat{A}$ is described by sequences f_A' and g_A' containing the values of functions f_A and g_A obtained by discretization of the interval [0, 1]:

$$f_A' = (f_A(0), f_A(dx), ..., f_A((M-1)dx), f_A(1)), \tag{3}$$

$$g_A' = (g_A(1), g_A((M-1)dx), ..., g_A(1dx), g_A(0)), dx = \frac{1}{M}, \tag{4}$$

where $(M+1)$ is the number of discrete points (Fig. 1c). This representation allows to implement the previously defined operations (see Definitions 2 and 3).

4.2 Declarative Model

Following the assumptions above, the proposed reference model consists of:
Parameters:
Crisp parameters:

G: graph of a transportation network $G = (N, E)$, where: $N = \{N_1, \ldots, N_\lambda, \ldots, N_n\}$ is a set of nodes and $E = \{(N_i, N_j)|i, j \in N, i \neq j\}$ is a set of edges, n – the number of nodes.
$\mathcal{U}$: fleet of vehicles: $\mathcal{U} = \{U_1, \ldots, U_k, \ldots, U_K\}$, where: U_k is the k-th vehicle.
K: size of the fleet.
Δ: set of service deadlines: $\Delta = \{\Delta_1, \ldots, \Delta_\lambda, \ldots, \Delta_n\}$, where $\Delta_\lambda = [ld_\lambda; ud_\lambda]$ is a deadline for providing service for customer N_λ (see example in Fig. 4).

ts: time interval at which vehicles can take off from the base.
T: window width, understood as a period, repeated at regular intervals, in which all nodes should be serviced (see Fig. 4).

Imprecise parameters: (defined as positive-oriented OFNs and marked by "^"):

$\widehat{d_{\beta,\lambda}}$: traveling time along edge (N_β, N_λ).
$\widehat{t_\lambda}$: time of node N_λ occupation.

Variables:

Crisp variables:

$x^k_{\beta,\lambda}$: binary variable indicating the travel of U_k between nodes N_β, N_λ

$$x^k_{\beta,\lambda} = \begin{cases} 1 \text{ if } U_k \text{ travels from node } N_\beta \text{ to node } N_\lambda \\ 0 \qquad\qquad \text{otherwise} \end{cases},$$

Imprecise variables (positive-/negative-oriented OFNs):

$\widehat{y^k_\lambda}$: fuzzy time at which vehicle U_k arrives at node N_λ.
$\widehat{w^k_\lambda}$: laytime at node N_λ for U_k.
$\widehat{s^k}$: take-off time of vehicle U_k.

Sets and sequences:

π_k: route of vehicle U_k, $\pi_k = (N_{k_1}, \ldots, N_{k_i}, N_{k_{i+1}}, \ldots, N_{k_\mu})$, where: $x^k_{k_i,k_{i+1}} = 1$ for $i = 1, \ldots, \mu - 1$ and $x^k_{k_\mu,k_1} = 1$.
Π: set of routes π_k.
$\widehat{W^k}$: sequence of laytimes $\widehat{w^k_\lambda}$: $\widehat{W^k} = \left(\widehat{w^k_1}, \ldots, \widehat{w^k_\lambda}, \ldots, \widehat{w^k_n}\right)$.
$\widehat{W}$: family of W^k, fuzzy schedule of fleet $\mathcal{U}$.
$\widehat{Y^k}$: sequence of moments $\widehat{y^k_\lambda}$: $\widehat{Y^k} = \left(\widehat{y^k_1}, \ldots, \widehat{y^k_\lambda}, \ldots, \widehat{y^k_n}\right)$.
$\widehat{Y}$: family of Y^k,
$\widehat{\mathbb{Y}}$: fuzzy schedule of fleet $\mathcal{U}$, $\widehat{\mathbb{Y}} = \left(\widehat{Y}, \widehat{W}\right)$.

Constraints:

Routes. Relationships between the variables describing vehicle take-off times/mission start times and task order:

$$\widehat{s^k} \geq 0; k = 1 \ldots K, \tag{5}$$

$$\left(\left|\widehat{s^k} - \widehat{s^q}\right| \geq ts\right); k, q = 1 \ldots K; k \neq q, \tag{6}$$

$$\sum\nolimits_{j=1}^{n} x^k_{1,j} = 1; k = 1 \ldots K, \tag{7}$$

$$\left(x_{1,j}^{k}=1\right) \Rightarrow \left(\widehat{y_j^k}=\widehat{s^k}+\widehat{d_{1,j}}\right); j=1\ldots n; k=1\ldots K, \quad (8)$$

$$\left(\widehat{y_i^k}>0 \wedge \widehat{y_i^q}>0\right) \Rightarrow \left(\left|\widehat{y_i^k}-\widehat{y_i^q}\right| \geq \widehat{t_i}\right); i=1\ldots n; k,q=1\ldots K; k \neq q, \quad (9)$$

$$\left(x_{i,j}^{k}=1\right) \Rightarrow \left(\widehat{y_j^k}=\widehat{y_i^k}+\widehat{d_{i,j}}+\widehat{t_i}+\widehat{w_\lambda^k}\right); j=1\ldots n; i=2\ldots n; k=1\ldots K, \quad (10)$$

$$\widehat{s^k}+T=\widehat{y_1^k}+\widehat{t_1}+\widehat{w_1^k}; k=1\ldots K, \quad (11)$$

$$\widehat{y_i^k} \geq 0; i=1\ldots n; k=1\ldots K, \quad (12)$$

$$\sum\nolimits_{j=1}^{n} x_{i,j}^{k}=\sum\nolimits_{j=1}^{n} x_{j,i}^{k}; i=1\ldots n; k=1\ldots K, \quad (13)$$

$$\widehat{y_i^k} \leq T, i=1\ldots n; k=1\ldots K, \quad (14)$$

$$x_{i,i}^{k}=0; i=1\ldots n; k=1\ldots K. \quad (15)$$

Service deadlines. All customers N_λ should be serviced by the given deadlines $\Delta_\lambda = [ld_\lambda; ud_\lambda]$:

$$\widehat{y_i^k}+\widehat{t_i}+c \times T \leq ud_\lambda, i=1\ldots n; k=1\ldots K, \quad (16)$$

$$\widehat{y_i^k}+c \times T \geq ld_\lambda, i=1\ldots n; k=1\ldots K. \quad (17)$$

4.3 Fuzzy Constraint Satisfaction Problem

The model proposed above allows to define the problem under consideration in the following way:

Given is a fleet $\mathcal{U}$*of maintenance service vehicles servicing customers allocated in a network* G*(customers are serviced by prescheduled deadlines* Δ*). The vehicles move along a given set of routes* Π*according to a cyclic fuzzy schedule* $\widehat{\mathbb{Y}}$*. Assuming that there occurs a disturbance which changes* Δ_λ*to* Δ_λ^**, a feasible way of re-routing and re-scheduling the vehicles is sought, which can guarantee that the vehicles service the customers on time while moving along the newly obtained routes* Π'*according to a corrected fuzzy cyclic schedule* $\widehat{\mathbb{Y}}'$*.*

Re-scheduling and re-routing of the vehicle fleet in response to the signaled disturbance (a change from Δ_λ to Δ_λ^*) requires supplementing the model under consideration with the concept of state S of the mission (18).

$$S=\left((s_1,\ldots,s_k,\ldots,s_K),t^*\right) \quad (18)$$

where: $s_k \in N$ is the node occupied by vehicle U_k (or the node the vehicle U_k is headed to) at time t^* when the information about the disturbance is received. For example, in the situation shown in Fig. 2, information about the disturbance was received at moment $t^* = 2680$ – the mission state at this moment is $S = ((N_6, N_{10}), 2680)$. The concept of mission state makes it possible to view the implementation of the mission of a given fleet of vehicles as a trajectory of states describing the locations of vehicles at the successive discrete moments of time.

Given that the changes relate to service deadlines Δ_λ^*, the disturbance is defined as follows:

$$IS = \left(S, \Delta_\lambda^*\right). \tag{19}$$

Therefore, when disturbance IS occurs, new routes Π' and a new schedule $\widehat{\mathbb{Y}}'$, which guarantee timely servicing of customers, are determined by solving the following Fuzzy Constraint Satisfaction (FCS) Problem (20):

$$\widehat{FCS}(\widehat{\mathbb{Y}}, \Pi, IS) = \left(\left(\widehat{\mathcal{V}}, \widehat{\mathcal{D}}\right), \widehat{\mathcal{C}}(\widehat{\mathbb{Y}}, \Pi, IS)\right), \tag{20}$$

where: $\widehat{\mathcal{V}} = \left\{\widehat{\mathbb{Y}}', \Pi'\right\}$ is a set of decision variables including: $\widehat{\mathbb{Y}}'$ – a fuzzy cyclic schedule guaranteeing timely provision of service to customers in the case of disturbance IS, and Π' – a set of routes determining the fuzzy schedule $\widehat{Y}'$; $\widehat{\mathcal{D}}$ – a finite set of decision variable domains: $\widehat{y_\lambda^k}, \widehat{w_\lambda^k} \in \mathcal{F}$ ($\mathcal{F}$ is a set of OFNs (1)), $x_{\beta,\lambda}^k \in \{0, 1\}$; $\widehat{\mathcal{C}}$ – a set of constraints which take into account the set of routes Π, fuzzy schedule $\widehat{\mathbb{Y}}$ and disturbance IS, while specifying the relationships that link the operations occurring in the milk-run cycles (5)–(17).

To solve $\widehat{FCS}$ (20), it is necessary to determine the values of the decision variables from the adopted set of domains for which the given constraints are satisfied. Implementation of $\widehat{FCS}$ in a constraint programming environment, such as ILOG, enables to find the solution.

5 Solution Methodology

The approach proposed in this paper assumes that traffic disturbances are responded to as they emerge, in a follow-up mode. In other words, it is possible to dynamically change the adopted routes and schedules, adjusting them (if possible) to the changes in service deadlines.

An algorithm that supports dynamic planning, i.e. vehicle fleet re-routing and rescheduling, based on the proposed concept of $\widehat{FCS}$ (20) is shown in Fig. 5. The algorithm is started in the successive customer service cycles. If there is a disturbance ($IS \neq 0$) in a given cycle (at moment t^*), then problem $\widehat{FCS}$ is solved (*solve* function).

In other words, to solve $\widehat{FCS}$, the function *solve* is used, which represents algorithms implemented in declarative programming environments (responsible for the search for admissible solutions to the decision problems considered).

The existence of an admissible solution (i.e. $\left(\widehat{\mathbb{Y}}' \neq \emptyset\right) \wedge \left(\Pi' \neq \emptyset\right)$) means that there are routes which ensure that customers are serviced on time when disturbance IS occurs.

If an admissible solution does not exist, then the currently used routes and the associated vehicle schedule should be modified (*reduce* function) in such a way as to remove the servicing operation at node N_λ at which disturbance *IS* occurs. The *reduce* function is responsible for modifying the routes. The proposed algorithm formulated in the constraints programming framework was implemented in the IBM ILOG environment.

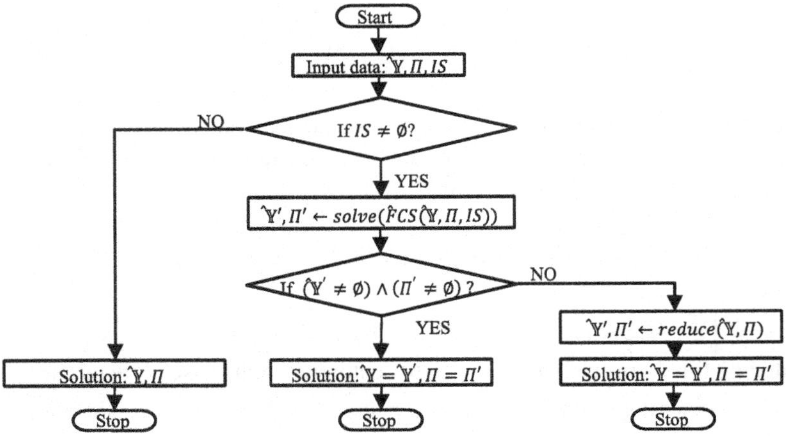

Fig. 5. A dynamic re-routing and re-scheduling algorithm

6 Computational Experiments

Consider the network from Fig. 2, in which the two vehicles $\mathcal{U} = \{U_1, U_2\}$ cyclically service nodes $N_2 - N_{11}$ with period $T = 2000$ [u.t]. The fuzzy traveling times between nodes $\widehat{d_{\lambda,\beta}}$ and the service deadlines Δ are as shown in Table 1 and Fig. 3, respectively. Routes $\pi_1 = (N_1, N_3, N_6, N_{11}, N_8, N_2, N_1)$ and $\pi_2 = (N_1, N_5, N_{10}, N_7, N_9, N_4)$ determine the fuzzy schedule $\widehat{\mathbb{Y}}$, as shown in Fig. 4. It is easy to see (Fig. 4) that in the second cycle of the fuzzy schedule (in state $S = ((N_6, N_{10}), 2680))$, information about unplanned changes in service deadline $\Delta^*_{11} = [1000, 1300]$ (instead $\Delta_{11} = [800, 1100]$) on node N_{11} is announced.

Given this, an answer to the following question is sought: Does there exist a set of routes Π' operated by vehicles U_1 and U_2 for which the fuzzy cyclic schedule $\widehat{\mathbb{Y}}'$ will guarantee that all customers are serviced on time when disturbance $IS = \left(S, \Delta^*_{11}\right)$ occurs? In order to find the answer to this question, we used the algorithm shown in Fig. 5. The problem (20) $\widehat{FCS}$ was then implemented in the constraint programming environment IBM ILOG (Windows 10, Intel Core Duo2 3.00 GHz, 4 GB RAM).

The solution time for problems of this size (with up to 11 nodes) does not exceed 15 s. The results are shown in graphical form in Fig. 6c. The following routes were obtained: $\pi'_1 = (N_1, N_3, N_6, N_9, N_4)$ and $\pi'_2 = (N_1, N_5, N_{10}, N_7, N_{11}, N_8, N_2)$. In fuzzy schedule $\widehat{\mathbb{Y}}'$ (Fig. 6a), transport operations are represented as ribbon-like "arterial roads",

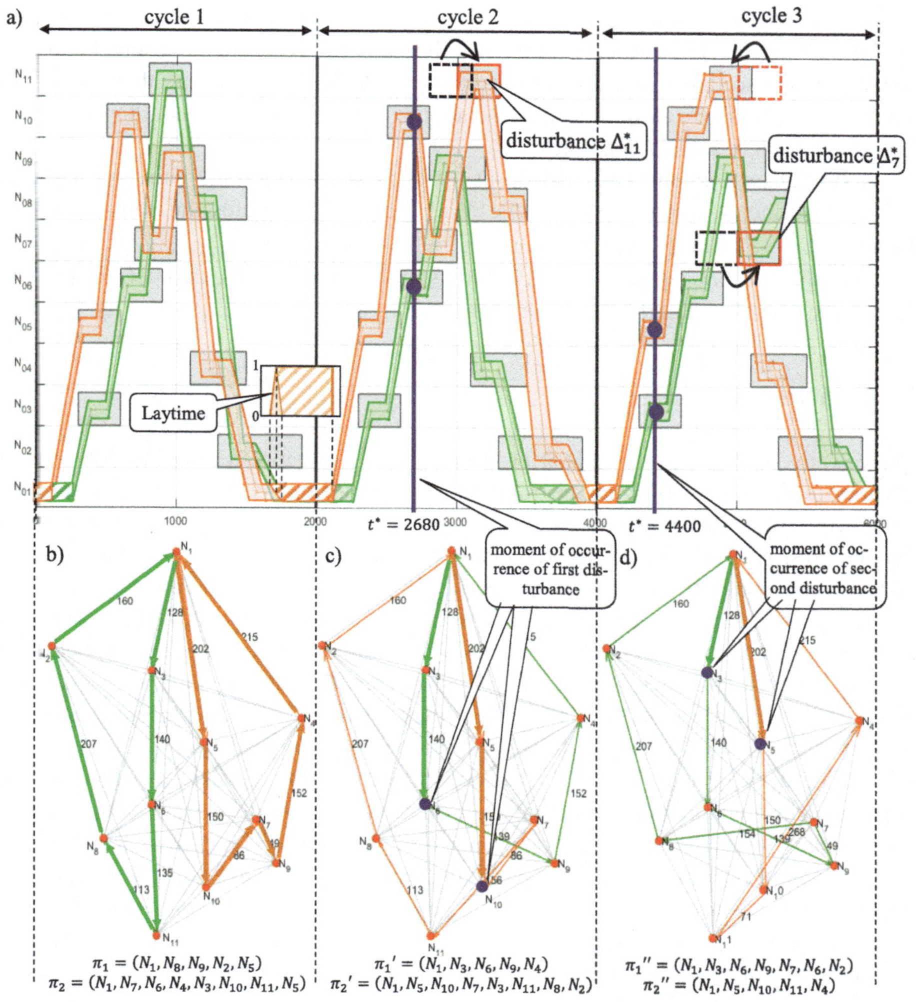

Fig. 6. Cyclic fuzzy schedule a), no disturbance b), occurrence of the first disturbance c), occurrence of the second disturbance d)

whose increasing width represents the time of vehicle movement resulting from the growing uncertainty of the moments of occupation and release of nodes. The width of the ribbon-like arterial roads increases until the next time-window begins.

It is worth noting that the uncertainty of the decision variables is reduced at the end of each time window as a result of the operation of vehicles waiting at node N_1. The increasing uncertainty is not transferred to the subsequent cycles of the system. Uncertainty is reduced as a result of the implementation of the OFN formalism. Fuzzy variables describing the waiting time of trains at nodes N_1 have a negative orientation (laytimes $\widehat{w_1^1}$ and $\widehat{w_1^2}$).

It is also worth noting that routes π_1', π_2' remain unchanged (see routes π_1, π_2 in Fig. 6a) until a disturbance occurs, and then they are re-routed, re-scheduled and re-harmonized so that all customers are serviced on time. This means that the model developed in this study allows to adjust the adopted delivery plans to disturbances which lead to changes in service deadlines.

A similar experiment was carried out for cycle 3, in which the delivery of goods by the deadline was disturbed on node N_7. Figure 6d illustrates the solution to the $\widehat{FCS}$ for $\Delta_7^* = [1000, 1300]$. The routes $\pi_1'' = (N_1, N_3, N_6, N_9, N_7, N_6, N_2)$ and $\pi_2'' = (N_1, N_5, N_{10}, N_{11}, N_4)$ guarantee timely provision of service to all customers despite the disturbance and the uncertain traveling times.

7 Conclusions

The novelty of this study is that it proposes an OFN algebra framework for solving DMRPs formulated as Fuzzy Constraint Satisfaction Problems. The results of the tests demonstrate that the proposed analytical approach allows to cope with problems of dynamic routing and scheduling of fleets of vehicles which service customer requests, taking into account the uncertainty of the travel time and repair/maintenance times.

The results of the experiments reported in this present study indicate that the proposed method should be implemented in systems which support reactive milk-run scheduling of vehicle fleets. In this context the use of commercially available software tools, such as CPLEX, ECLiPSe, Gurobi, etc., which make it possible to tackle practical-scale problems, can be viewed as an attractive solution for problem-oriented DSS.

The model considered in this paper does not take into consideration many of the practical requirements and limitations related to the specificity of the same kind of services and the capabilities of the teams performing those services. We plan to address those issues in our future work. In particular, we want to investigate the question of balancing the needs of the customers with the capabilities of the team which provides the services ordered by those customers. We would also like to extend the DMRP to out-plant dynamic multi-period milk-run routing problems.

Acknowledgments. This research was carried out under the internship: Declarative models of the vehicles fleet mission planning (Aalborg University, 23 November 2020–31 March 2021).

References

1. Baines, T., Bigdeli, Z., Bustinza, O.F., Shi, V.G., Baldwin, J., Ridgway, K.: Servitization: revisiting the state-of-the-art and research priorities. Int. J. Oper. Prod. Manag. **37**(2), 256–278 (2017)
2. Bocewicz, G., Nielsen, I., Banaszak, Z.: Production flows scheduling subject to fuzzy processing time constraints. Int. J. Comput. Integr. Manuf. **29**, 1105–1127 (2016)
3. Bocewicz, G., Bożejko, W., Wójcik, R., Banaszak, Z.: Milk-run routing and scheduling subject to a trade-off between vehicle fleet size and storage capacity. Manag. Prod. Eng. Rev. **10**(3), 41–53 (2019). https://doi.org/10.24425/mper.2019.129597

4. Bozejko, W., Wodecki, M.: Parallel genetic algorithm for minimizing total weighted completion time. In: Lecture Notes in Artificial Intelligence (Subseries of Lecture Notes in Computer Science) (2004). https://doi.org/10.1007/978-3-540-24844-6_58
5. Bożejko, W., Gnatowski, A., Pempera, J., Wodecki, M.: Parallel tabu search for the cyclic job shop scheduling problem. Comput. Ind. Eng. **113**, 512–524 (2017)
6. Braekers, K., Ramaekers, K., Nieuwenhuyse, I.V.: The vehicle routing problem: state of the art classification and review. Comput. Ind. Eng. **99**, 300–313 (2016)
7. Brito, J., Moreno-Pérez, J.A., Verdegay, J.L.: Fuzzy optimization in vehicle routing problems. In: Proceedings of the Joint 2009 International Fuzzy Systems Association, World Congress and European Society of Fuzzy Logic and Technology, pp. 1547–1552 (2009)
8. Chwastyk, A., Kosiński, W.: Fuzzy calculus with applications. Mathematica Applicanda **41**(1), 47–96 (2013)
9. Chen, X.: Multi-period dynamic technician routing and scheduling problems with experience-based service times and stochastic customers. Ph.D. (Doctor of Philosophy) thesis, University of Iowa (2016). https://doi.org/10.17077/etd.g8ozrc3x
10. Czerniak, J.M., Dobrosielski, W.T., Apiecionek, Ł., Ewald, D., Paprzycki, M.: Practical application of OFN arithmetics in a crisis control center monitoring. In: Studies in Computational Intelligence, vol. 655, pp. 51–64 (2016)
11. Franklin, H., Beatrice, O.-B.: Dynamic vehicle routing using genetic algorithms. Appl. Intell. **27**, 89–99 (2007). https://doi.org/10.1007/s10489-006-0033-z
12. Holborn, P.L.: Heuristics for Dynamic Vehicle Routing Problems with Pickups and Deliveries and Time Windows. School of Mathematics, Cardiff University, School of Mathematics, Ph.D. thesis (2013). https://orca.cf.ac.uk/47742/
13. Kosinski, W., Prokopowicz, P., Slezak, D.: On algebraic operations on fuzzy numbers. In: Proceedings of the International Symposium, IIS: IIPWM 2003, Zakopane, Poland, pp. 353–362. Physica Verlag, Heidelberg (2003)
14. Kryvinska, N., Kaczor, S., Strauss, C.: Enterprises' servitization in the first decade—retrospective analysis of back-end and front-end challenges. Appl. Sci. **10**(8), 2957 (2020). https://doi.org/10.3390/app10082957
15. Mastrogiacomo, L., Barravecchia, F., Franceschini, F.: Definition of a conceptual scale of servitization: proposal and preliminary results. CIRP J. Manuf. Sci. Technol. **29**, 141–156 (2020). https://doi.org/10.1016/j.cirpj.2018.11.003
16. Montemanni, R., Gambardella, L.M., Rizzoli, A.E.: Ant colony system for a dynamic vehicle routing problem . J. Comb. Optim. **10**, 327–343 (2005)
17. Okulewicz, M., Mańdziuk, J.: A metaheuristic approach to solve Dynamic Vehicle Routing Problem in continuous search space. Swarm Evol. Comput. **48**, 44–61 (2019). https://doi.org/10.1016/j.swevo.2019.03.008
18. Pavone, M., Bisnik, N., Frazzoli, E., Isler, V.: A stochastic and dynamic vehicle routing problem with time windows and customer impatience. Comput. Sci. Eng. **14**, 350–364 (2009). https://doi.org/10.1007/s11036-008-0101-1
19. Pempera, J., Smutnicki, C.: Open shop cyclic scheduling. Eur. J. Oper. Res. **269**(2), 773–781 (2018)
20. Sáeza, D., Cortésb, C.E., Núñez, A.: Hybrid adaptive predictive control for the multi-vehicle dynamic pick-up and delivery problem based on genetic algorithms and fuzzy clustering. Comput. Oper. Res. **35**, 3412–3438 (2008)
21. Pillac, V., Gendreau, M., Guéret, C., Medaglia, A.L.: A review of dynamic vehicle routing problems. Eur. J. Oper. Res. **225**, 1–11 (2013). https://doi.org/10.1016/j.ejor.2012.08.015
22. Pisz, I., Chwastyk, A., Łapuńka, I.: Assessing the profitability of investment projects using ordered fuzzy numbers. Logforum **15**(3), 377–389 (2019)

23. Purba, H.H., Fitra, A., Nindiani, A.: Control and integration of milk-run operation in Japanese automotive company in Indonesia. Manag. Prod. Eng. Rev. **10**(1), 79–88 (2019). https://doi.org/10.24425/mper.2019.128246
24. Rudnik, K.: Transport trolley control in a manufacturing system using simulation with the FSAW, FWASPAS and FTOPSIS methods. In: Advances in Intelligent Systems and Computing, vol. 637, pp. 440–449 (2018). https://doi.org/10.1007/978-3-319-64465-3_42
25. Rudnik, K., Serafin, R.: Probabilistic fuzzy approach to assessment of supplier based on delivery process. In: Advances in Intelligent Systems and Computing, vol. 835, pp. 254–266 (2019)
26. Wen, M., Cordeau, J.F., Laporte, G., Larsen, J.: The dynamic multi-period vehicle routing problem. Comput. Oper. Res. **37**(9), 1615–1623 (2010)
27. Sitek, P., Wikarek, J.: Capacitated vehicle routing problem with pick-up and alternative delivery (CVRPPAD) – model and implementation using hybrid approach. Ann. Oper. Res. (2017). https://doi.org/10.1007/s10479-017-2722-x
28. Smutnicki, C.: Minimizing cycle time in manufacturing systems with additional technological constraints. In: Proceedings of 22nd International Conference on Methods and Models in Automation & Robotics, pp. 463–470 (2017). https://doi.org/10.1109/MMAR.2017.8046872
29. Subramanyam, A., Gounaris, C.E., Laínez-Aguirre, J.M., Pinto, J.M.: Routing: construction of uncertainty sets and evaluation via Rolling-Horizon simulations. In: Proceedings of the 2018 AlChE Annual Meeting Group: Computing and Systems Technology Division Session (2018)
30. Thibbotuwawa, A., Nielsen, P., Banaszak, Z., Bocewicz, G.: Energy consumption in unmanned aerial vehicles: a review of energy consumption models and their relation to the UAV routing. In: Advances in Intelligent Systems and Computing, vol. 853, pp. 173–184 (2019). https://doi.org/10.1007/978-3-319-99996-8_16
31. Toth, P., Vigo, D.: The Vehicle Routing Problem. SIAM, Philadelphia (2002). https://doi.org/10.1137/1.9780898718515
32. Wójcik, R., Pempera, J.: Designing cyclic schedules for streaming repetitive job-shop manufacturing systems with blocking and no-wait constraints. IFAC-PapersOnLine **52**(10), 73–78 (2019)

Application of the Reference Sliding Variable Profile as an Extension of the Time-Varying Sliding Surface QSM Control

Katarzyna Adamiak(✉)

Institute of Automatic Control, Łódź University of Technology, 18/22 Stefanowskiego Street, 90-924 Łódź, Poland
katarzyna.adamiak@dokt.p.lodz.pl

Abstract. For the last decades the sliding mode control has been considered as one of the most promising continuous-time control methods. The idea assumes switching the structure of the controller, so that the representative point of the system approaches and remains on a predesigned sliding surface. Consequently, in the sliding phase the system becomes insensitive to any external disturbances and model uncertainties. To benefit from these features in the reaching phase as well, time-varying sliding surfaces may be applied. Unique properties of the sliding mode control, have also been widely adapted for discretized systems. However, as in the discrete-time only the quasi-sliding mode exists, the time-varying sliding surfaces have not been popularized for this case. For the quasi-sliding mode design, the reaching law approach has gathered more followers. This paper combines those two control design methods and presents the application of the reference sliding variable profile as an extension of the time-varying sliding surface idea for discrete-time sliding mode control. The study proves, that the application of the time-varying sliding surface guarantees robustness to external disturbances and modelling uncertainties through the whole regulation process.

Keywords: Discrete-time systems · Model reference control · Sliding mode control · Time-varying sliding surface · Variable structure systems

1 Introduction

Originating from the variable structure systems theory, sliding mode control was first described in Russian literature in the mid-20th century [1–3]. The sliding mode controller design begins with the choice of a sliding variable, which describes the distance of the representative point of the system from a predesigned sliding surface in the state space. Next, a nonlinear control law is applied to force the representative point to approach and remain on the sliding plane, which ensures the desired closed-loop system's dynamics. Therefore, the control process may be divided into two stages: the reaching phase, when the value of the sliding variable is monotonously reduced and the sliding phase, when the states of the system become confined to the sliding plane. The main advantage of such a control method is that in the sliding phase the system becomes insensitive to any

R. Szewczyk et al. (Eds.): AUTOMATION 2021, AISC 1390, pp. 37–45, 2021.
https://doi.org/10.1007/978-3-030-74893-7_4

external disturbance or model uncertainties [4]. As the controller is switched every time the representative point deviates from the sliding surface, any disturbance influence is immediately rejected. Due to its computational efficiency and the unique closed-loop system properties, the sliding mode control is nowadays considered as one of the most promising control methods [5, 6].

However, robustness to disturbances and modelling uncertainties is only achieved in the sliding phase. At the first stage of the control process, when the representative point is drawn to the sliding surface, the disturbances still influence the system's trajectories, which may cause undesirable dynamics. On the other hand, attempts to force faster system convergence and reduce the reaching phase require large magnitudes of the control signal. A solution to this problem is application of time-varying sliding surfaces [7–9]. Such surfaces are selected so that at the initial time $t = 0$ they contain the representative point of the system and for $t > 0$ they move inside the state space until they cross the desired state at finite time $t = t_0$. Afterwards the sliding plane remains fixed. Consequently, the reaching phase is completely eliminated and robustness is ensured during the whole control process. Over the years, various authors have discussed different time-varying sliding surfaces for continuous-time sliding mode [9–14].

Recently, progressing computerization has forced the control engineering society to adapt the sliding mode design for discrete-time systems [15]. However, in the discrete-time case, we may only achieve the quasi-sliding mode when the representative point of the system moves in a certain vicinity of the sliding plane [16, 17]. As bounding the system's states strictly to the sliding surface is not possible, the time-varying sliding planes for discrete-time case, although studied [18, 19], have not become so popular. The most common quasi-sliding mode design method is the reaching law approach [20–23], recently often combined with the model reference control [24–26].

This paper considers reference sliding variable based sliding mode control as an extension of the time-varying sliding surface idea for discrete-time systems. The reference sliding variable based approach assumes, first *a priori* defining the demand evolution of the sliding variable. Next, we drive the system's sliding variable to the desired value in each control step. In this study, we show that such control design may be interpreted as the time-varying sliding surface, where the sliding variable describes the distance between the representative point of the system from its reference position. We will prove that such a sliding surface guarantees stability of the closed-loop system and ensures robustness to external disturbance and parameter uncertainties in the entire regulation process.

The reminder of this paper is organized as follows. Section 2 presents the dynamical problem we are considering. Next, Sect. 3 describes the design of the time-varying sliding surface for discrete-time sliding mode control, derives the appropriate control law and discusses the properties of the closed-loop system. Section 4 verifies the proposed control strategy with a simulation example and lastly, Sect. 5 presents some final remarks.

2 Problem Statement

The aim of this study is to design a quasi-sliding mode controller for discrete-time disturbed system with the following dynamics:

$$\boldsymbol{x}[(k+1)T] = \boldsymbol{A}\boldsymbol{x}(kT) + \boldsymbol{\Delta A}[x(kT)] + \boldsymbol{b}u(kT) + \boldsymbol{d}(kT), \tag{1}$$

where $\boldsymbol{x}(kT)$ is an $n \times 1$ state vector, $\boldsymbol{A}$ is the system's state matrix, $\boldsymbol{b}$ denotes the input distribution vector and $u(kT)$ is the control signal. Finally, $\boldsymbol{\Delta A}[\boldsymbol{x}(kT)]$ and $\boldsymbol{d}(kT)$ are column vectors representing the modelling uncertainties and external disturbances, respectively. These disturbance values are unknown but bounded. The system must be driven from its initial state $\boldsymbol{x}_0 = \boldsymbol{x}(0)$ to the desired state $\boldsymbol{x}_d$ in finite time. Therefore, we define the regulation error vector as $\boldsymbol{e}(kT) = \boldsymbol{x}_d - \boldsymbol{x}(kT)$.

In the first step of the quasi-sliding mode control design we should choose the sliding variable as a function of the error vector $s(kT) = f[\boldsymbol{e}(kT)]$, and define the sliding surface as $s(kT) = 0$. Next, the control law may be derived by assigning $s[(k+1)T]$ to zero or with the application of one of many reaching laws presented in the literature [20–23]. However, the reaching law approach provides robustness to external disturbances and model uncertainties in the sliding phase only, when the representative point moves inside the ultimate band. In this study, we attempt to restrict the system's states to the quasi-sliding mode band from the beginning of the control process. For that purpose, in the next section, we introduce a time-varying sliding surface, which completely eliminates the reaching phase.

3 Time-Varying Sliding Surface for QSM Control

In the recent literature, we may find multiple quasi-sliding mode controllers based on a reference sliding variable profile [24–26]. Those strategies assume *a priori* defining the desired profile of the sliding variable, which ensures monotonic convergence from the initial state to the vicinity of the sliding plane. Next, the sliding variable of the real disturbed system is driven to the demand value in each step with a reference sliding variable following reaching law. Such a control scheme guarantees that the disturbance impact on the system is reduced to one control step only. Consequently, $s[(k+1)T]$ is only influenced by $\boldsymbol{d}(kT)$ and all previous disturbance values are rejected.

In this section, we present a different interpretation of the reference sliding variable based control. We propose a discrete-time sliding mode controller with a time-varying sliding surface. Our idea is to design such a sliding plane that it crosses the initial condition of the system at $k = 0$ and for $k > 0$ it moves monotonically till it crosses the desired state at k_0. For any $k > k_0$ the sliding plane remains fixed, containing the steady state of the system. Therefore, we introduce the following sliding variable:

$$s(kT) = \boldsymbol{c}\boldsymbol{e}(kT) - s_r(kT), \tag{2}$$

where $\boldsymbol{c}$ is an $1 \times n$ control vector, chosen so that $\boldsymbol{cb} \neq 0$ and to ensure the desired steady state behavior, and $s_r(kT)$ represents the reference sliding variable. Physically such a sliding variable describes the distance of the representative point of the system from its desired position in each step. The sliding surface becomes:

$$s(kT) = 0. \tag{3}$$

The reference sliding variable profile $s_r(kT)$ may be generated in various ways. However, it must ensure the following properties:

- $s_r(0) = \boldsymbol{ce}(0)$, so that initially the representative point of the system belongs to the sliding plane;
- for $0 < k < k_0$, the absolute value of $s_r(kT)$ is monotonically decreasing;
- for $k \geq k_0$, $s_r(kT) = 0$, so that the sliding plane $s(kT) = \boldsymbol{ce}(kT)$ remains fixed.

In this paper, we propose the following way for the generation of the reference sliding variable profile:

$$s_r(kT) = \begin{cases} \boldsymbol{ce}(0) & \text{for} \quad k = 0 \\ s_r[(k-1)T] - \varepsilon \operatorname{sgn}\{s_r[(k-1)T]\} & \text{for} \quad 0 < k < k_0 \\ 0 & \text{for} \quad k \geq k_0 \end{cases}. \tag{4}$$

The parameter ε is a constant defining the rate of change of the sliding surface in one step. Knowing the initial value $s_r(0)$, we may select ε in order to drive the reference to zero in k_0 number of step. Therefore, ε becomes:

$$\varepsilon = \frac{|s_r(0)|}{k_0}. \tag{5}$$

Having the reference sliding variable, we derive the control law for the real disturbed plant. For this purpose, we neglect the disturbances and model uncertainties. From (1) and (2), by assigning $s[(k+1)T]$ to zero, we obtain:

$$u(kT) = (\boldsymbol{cb})^{-1}\{\boldsymbol{cx}_d - \boldsymbol{cAx}(kT) - s_r[(k+1)T]\}. \tag{6}$$

Finally, we must analyze the stability and the robustness of the closed-loop system. Application of the control signal (6) to the system (1) results in the following state space description:

$$\boldsymbol{x}[(k+1)T] = \\ \left[\boldsymbol{1}_{n\times n} - \boldsymbol{b}(\boldsymbol{cb})^{-1}\boldsymbol{c}\right]\boldsymbol{Ax}(kT) + \boldsymbol{b}(\boldsymbol{cb})^{-1}\boldsymbol{cx}_d - \boldsymbol{b}s_r[(k+1)T] + \boldsymbol{\Delta A}[x(kT)] + \boldsymbol{d}(kT). \tag{7}$$

Consequently, the closed-loop system's state matrix becomes:

$$\boldsymbol{A}_{cl} = \left[\boldsymbol{1}_{n\times n} - \boldsymbol{b}(\boldsymbol{cb})^{-1}\boldsymbol{c}\right]\boldsymbol{A} \tag{8}$$

and its eigenvalues may be found from:

$$M(z) = \det(\boldsymbol{1}_{n\times n}z - \boldsymbol{A}_{cl}). \tag{9}$$

We may conclude that the stability of the closed-loop system directly depends on the selection of the control vector $\boldsymbol{c}$ and must be ensured at the design stage. Taking into account (8), the parameters $c_1, c_2, \ldots, c_{n-1}$ of $\boldsymbol{c} = [1, c_1, c_2, \ldots, c_{n-1}]$ may be chosen with the pole placement method or with the quadratic optimization scheme.

Next, we consider the robustness of the system. We assume that the values of modelling uncertainties and external disturbances are bounded. Therefore, we define their maximum impact on the sliding variable as:

$$\begin{aligned} |\boldsymbol{c}\boldsymbol{\Delta A}[\boldsymbol{x}(kT)]| &\leq \Delta_{\max} \\ |\boldsymbol{c}\boldsymbol{d}(kT)| &\leq D_{\max}. \end{aligned} \tag{10}$$

Now, using (1) and (2), we express $s[(k+1)T]$ for $k > 0$:

$$s[(k+1)T] = \boldsymbol{c}\boldsymbol{x}_d - \boldsymbol{c}\boldsymbol{A}\boldsymbol{x}(kT) - \boldsymbol{c}\boldsymbol{b}u(kT) - \boldsymbol{c}\boldsymbol{\Delta A}[\boldsymbol{x}(kT)] - \boldsymbol{c}\boldsymbol{d}(kT). \tag{11}$$

After application of the control signal (6) we get:

$$s[(k+1)T] = -\boldsymbol{c}\boldsymbol{\Delta A}[\boldsymbol{x}(kT)] - \boldsymbol{c}\boldsymbol{d}(kT). \tag{12}$$

Taking into account (10) we conclude that for $k = 0$

$$s(kT) = 0 \tag{13}$$

and for any $k > 0$ the absolute value of the sliding variable is restricted by:

$$|s(kT)| \leq \Delta_{\max} + D_{\max}. \tag{14}$$

Consequently, the reaching phase has been eliminated and in the whole control process the sliding variable belongs to the band of $2(\Delta_{\max} + D_{\max})$ width.

4 Simulation Example

Let us consider a discrete-time second order system with the following dynamics:

$$x[(k+1)T] = \begin{bmatrix} 2 & -0.5 \\ 1.5 & 0 \end{bmatrix} \boldsymbol{x}(kT) + \boldsymbol{\Delta A}[x(kT)] + \begin{bmatrix} 0 \\ 1 \end{bmatrix} u(kT) + \boldsymbol{d}(kT). \tag{15}$$

The open-loop system is unstable, as its eigenvalues are $z_1 = 1.5$ and $z_2 = 0.5$. Our purpose is to drive the system from the initial state $\boldsymbol{x_0} = [-20\ 0]^{\mathrm{T}}$ to the demand state $\boldsymbol{x}_d = [0\ 0]^{\mathrm{T}}$ and ensure its stable steady state performance. For that purpose, we propose a quasi-sliding mode controller with a time-varying sliding line.

In the first step, we select a control vector $\boldsymbol{c} = [1 - 0.25]$, so the closed-loop system's eigenvalues are placed at the origin of the complex plane. Next, we generate the reference sliding variable profile according to (4). Therefore, $s_r(0) = 20$. We decide to drive the sliding line to the fixed state in $k_0 = 10$ step. Consequently, we set $\varepsilon = 2$, as stated in (5). The evolution of the reference sliding variable is shown in Fig. 1. It is clearly visible that $s_r(kT)$ satisfies all the requirements stated in Sect. 3. Finally, we define the sliding line for the system (15) as $s(kT) = \boldsymbol{c}\boldsymbol{e}(kT) - s_r(kT) = 0$. The system's sliding

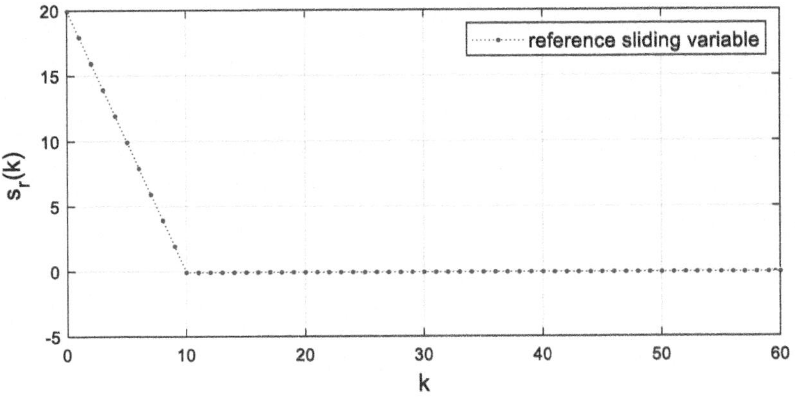

Fig. 1. The reference sliding variable

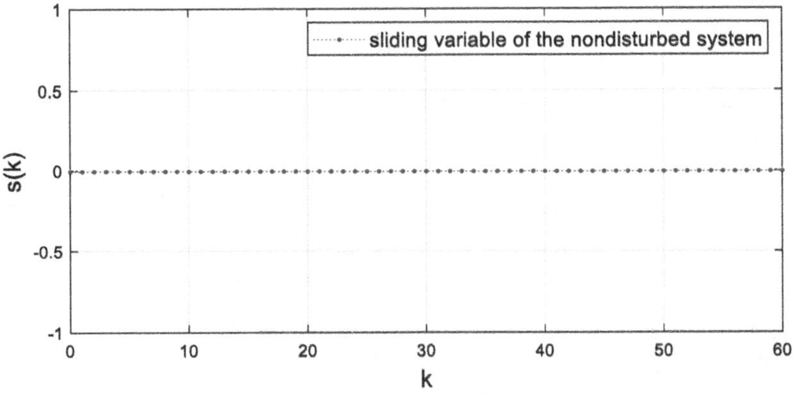

Fig. 2. The sliding variable of the nondisturbed system

variable, without any disturbance influence, is presented in Fig. 2. We may conclude that the application of the time-varying sliding line has completely eliminated the reaching phase.

Next, we analyze the system subject to a sinusoidal disturbance function with amplitude 1. Its maximum impact on the system's representative point is bounded by $\Delta_{\max} + D_{\max} = 0.25$. The simulation's results are presented in Figs. 3, 4 and 5. From the system's sliding variable, we conclude that the representative point remains in the quasi-sliding mode band of $2 \cdot 0.25$ width through the whole regulation process. Therefore, the robustness to external disturbances and modelling uncertainties has been ensured for the entire control time. Finally, Fig. 5 shows that in the quasi-steady state the state variables oscillate around the desired point $\boldsymbol{x}_d$ and the range of these oscillations strictly depends on the disturbance values.

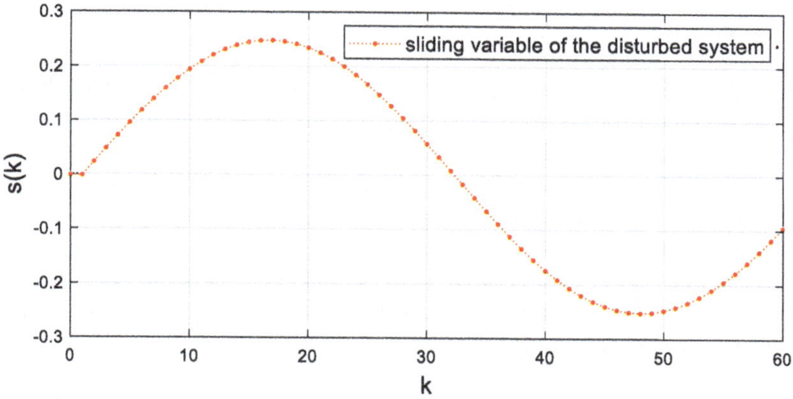

Fig. 3. Sliding variable of the disturbed system

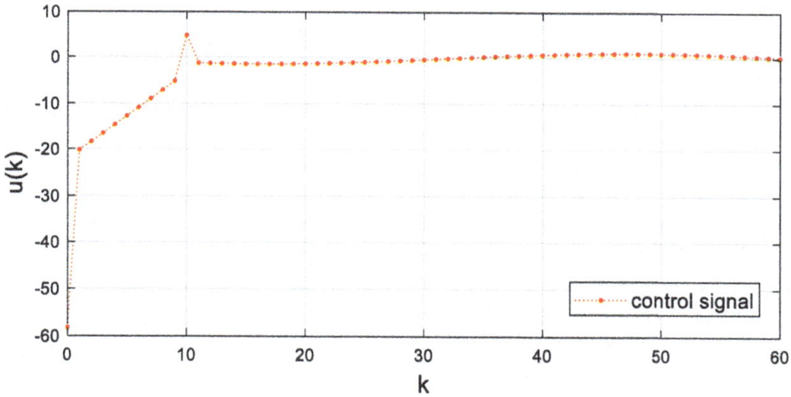

Fig. 4. Control signal of the disturbed system

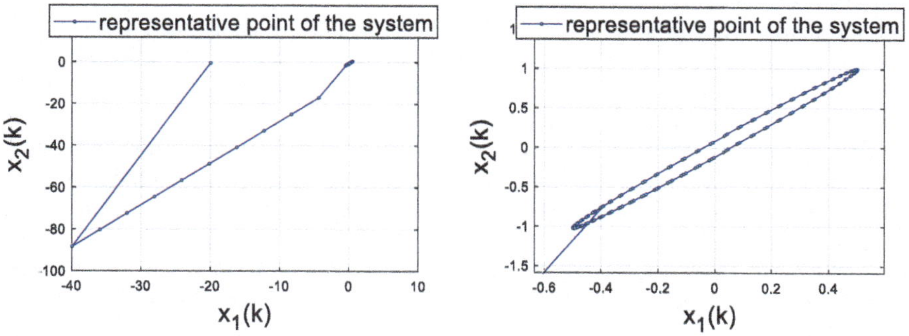

Fig. 5. Evolution of the system's representative point in the whole regulation process and in the quasi-steady state

5 Conclusions

This study presented a new interpretation of the reference sliding variable based quasi-sliding mode control for disturbed uncertain systems. Instead of utilizing the classical reaching law approach, here the reference sliding variable was adopted to define a time-varying sliding plane. Consequently, the system's sliding variable represents the distance between the position of the representative point of the system from its desired state in each step. Once the reference sliding variable is generated so that the sliding plane crosses the initial state of the system at the initial time, then the reaching phase is completely eliminated. As follows the system's robustness to external disturbance and model uncertainties is ensured for the whole control process.

References

1. Emelyanov, S.V.: Variable-Structure Control Systems. Nauka, Moscow (1967). (in Russian)
2. Itkis, U.: Control Systems of Variable Structures. Wiley, New York (1976)
3. Utkin, V.I.: Variable structure systems with sliding modes. IEEE Trans. Autom. Control **22**(2), 212–222 (1977)
4. Draženović, B.: The invariance conditions in variable structure systems. Automatica **5**(3), 287–295 (1969)
5. Hung, J.Y., Gao, W.B., Hung, J.C.: Variable structure control: a survey. IEEE Trans. Ind. Electron. **40**(1), 2–22 (1993)
6. Edwards, C., Spurgeon, S.: Sliding Mode Control: Theory and Applications. Taylor & Francis, London (1998)
7. Choi, S.B., Cheong, C.C., Park, D.W.: Moving switching surfaces for robust control of second-order variable structure systems. Int. J. Control **58**(1), 229–245 (1993)
8. Choi, S.B., Park, D.W., Jayasuriya, S.: A time-varying sliding surface for fast and robust tracking control of second-order uncertain systems. Automatica **30**(5), 899–904 (1994)
9. Bartoszewicz, A.: A Comment on 'A time-varying sliding surface for fast and robust tracking control of second-order uncertain systems.' Automatica **31**(12), 1893–1895 (1995)
10. Bartoszewicz, A.: Time-varying sliding modes for second-order systems. IEE Proc. Control Theory Appl. **143**(5), 455–462 (1996)
11. Bartoszewicz, A., Nowacka-Leverton, A.: SMC without the reaching phase – the switching plane design for the third-order system. IET Control Theory Appl. **1**(5), 1461–1470 (2007)
12. Bartoszewicz, A., Nowacka-Leverton, A.: Time-Varying Sliding Modes for Second and Third Order Systems. Springer, Heidelberg (2009)
13. Bartoszewicz, A., Nowacka, A.: Reaching phase elimination in variable structure control of the third order system with state constraints. Kybernetika **42**(1), 111–126 (2006)
14. Park, D.W., Choi, S.B.: Moving sliding surfaces for high-order variable structure systems. Int. J. Control **72**(11), 960–970 (1999)
15. Milosavljević, D.: General conditions for the existence of a quasi-sliding mode on the switching hyperplane in discrete variable structure systems. Autom. Remote Control **46**(3), 307–314 (1985)
16. Sarpturk, S.Z., Istefanopulos, Y., Kaynak, O.: On the stability of discrete-time sliding mode control systems. IEEE Trans. Autom. Control **22**(10), 930–932 (1987)
17. Utkin, V.I., Drakunov, S.V.: On discrete-time sliding modes. IFAC Proc. Vol. **22**(3), 273–278 (1989)
18. Kanai, Y., Mori, Y.: Discrete time Sliding Mode Control with time varying switching hyper plane. In: 2008 SICE Annual Conference, Tokyo, pp. 2349–2352 (2008)

19. Hu, Q., Du, C., Xie, L., Wang, Y.: Discrete-time sliding mode control with time-varying surface for hard disk drives. IEEE Trans. Control Syst. Technol. **17**(1), 175–183 (2009)
20. Gao, W.B., Wang, Y., Homaifa, A.: Discrete-time variable structure control systems. IEEE Trans. Ind. Electron. **42**(2), 117–122 (1995)
21. Bartoszewicz, A.: Discrete-time quasi-sliding mode control strategies. IEEE Trans. Ind. Electron. **45**(4), 633–637 (1998)
22. Golo, G., Milosavljević, Č: Robust discrete-time chattering free sliding mode control. Syst. Control Lett. **41**, 19–28 (2000)
23. Chakrabarty, S., Bandyopadhyay, B.: A generalized reaching law with different convergence rates. Automatica **63**, 34–37 (2016)
24. Bartoszewicz, A., Adamiak, K.: Model reference discrete-time variable structure control. Int. J. Adapt. Control Signal Process. **32**(10), 1440–1452 (2018)
25. Bartoszewicz, A., Adamiak, K.: Discrete time sliding mode control with a desired switching variable generator. IEEE Trans. Autom. Control **65**(4), 1807–1814 (2020)
26. Leśniewski, P.: Reference trajectory based SMC of DCDC buck converter. In: Bartoszewicz, A., Kabziński, J., Kacprzyk, J. (eds.) Advanced, Contemporary Control. Advances in Intelligent Systems and Computing, vol. 1196, pp. 161–170. Springer, Cham (2020)

A Discrete, Fractional Order, Memory-Effective State Space Model of a RLC Circuit

Krzysztof Oprzędkiewicz(✉)

Department of Automatic Control and Robotics, Faculty of Electrotechnics, Automatic Control, Informatics and Biomedical Engineering, AGH University of Science and Technology, al. A Mickiewicza 30, 30-059 Krakow, Poland
kop@agh.edu.pl

Abstract. The paper presents a new, discrete, memory-effective state-space model of fractional RLC network. For the presented model its stability, positivity and reaction to different inital functions are analysed. Results are illustrated by simulations.

Keywords: Fractional order systems · Fractional order state equation · Fractional order initial problem · CFE approximation

1 Abbreviations

In the paper the following abbreviations are used: MIMO - Multiple Input Mutliple Output, IO - Integer Order, FO - Fractional Order, CFE - Continuous Fraction Expansion, IIR -Infinite Impulse Response, FIR - Finite Impulse Response, PSE - Power Series Expansion.

2 Introduction

The modeling of processes and phenomena hard to descrribe using another mathematical tools is one of the main areas of application of non-integer order calculus. Fractional models of different physical phenomena have been presented by many Authors, for example by [2,4–6,14,20,21].

Fractional RLC circuits have been analyzed by many Authors, for example [7, 9,10]. The use of FO approach allows to precisely describe the behaviour of cirocutis containing supercapacitors or Li-ion capacitors (see for example [11]). Initial problems for FO discrete systems were discussed for example by [13].

This paper presents a proposal of a new, discrete, state-space model of a fractional order RLC circuit. The use of the CFE approximation to solving a state equation allows to keep small size of a model with its good accuracy. The proposed results can be easily adopoted to analysis of other oscillatory systems, for example Spring-Mass-Damper systems, describing a car suspension.

R. Szewczyk et al. (Eds.): AUTOMATION 2021, AISC 1390, pp. 46–57, 2021.
https://doi.org/10.1007/978-3-030-74893-7_5

The paper is organized as follows: at the beginning elementary ideas from fractional calculus as well as the CFE approximation and its implementation to solve FO state equation are recalled. Next the considered system and its proposed, CFE-based state space model are given. Finally illustrative simulations are presented.

3 Preliminaries

3.1 Elementary Ideas

Elementary ideas from fractional calculus can be found in many books, for example: [4,8,17] or [20]. Here only some definitions necessary to explain of main results will be given.

Firstly the fractional-order, integro-differential operator is given (see for example [4,9,20]):

Definition 1 *(The elementary fractional order operator). The fractional-order integro-differential operator is defined as follows:*

$$
{}_aD_t^\alpha f(t) = \begin{cases} \frac{d^\alpha f(t)}{dt^\alpha} & \alpha > 0 \\ f(t) & \alpha = 0 \\ \int\limits_a^t f(\tau)(d\tau)^\alpha & \alpha < 0 \end{cases} . \tag{1}
$$

where a and t denote time limits for operator calculation, $\alpha \in \mathbb{R}$ denotes the non integer order of the operation.

Next remember an idea of Gamma Euler function ([9]):

Definition 2. *The Gamma function*

$$
\Gamma(x) = \int\limits_0^\infty t^{x-1} e^{-t} dt. \tag{2}
$$

Furthermore the Mittag-Leffler function will be recalled. It is a non-integer order generalization of exponential function $e^{\lambda t}$ and it plays crucial rule in solution of FO state equation. It is defined as underneath:

Definition 3. *The one parameter Mittag-Leffler function*

$$
E_\alpha(x) = \sum_{k=0}^{\infty} \frac{x^k}{\Gamma(k\alpha + 1)}. \tag{3}
$$

The fractional-order, integro-differential operator (1) can be described by different definitions, given by Grünwald and Letnikov (GL Definition), Riemann and Liouville (RL Definition) and Caputo (C Definition). Only C definition will be employed in this paper. It is as follows [9]:

Definition 4. *The Caputo Definition of the FO operator*

$$ {}_0^C D_t^\alpha f(t) = \frac{1}{\Gamma(N-\alpha)} \int_0^\infty \frac{f^{(N)}(\tau)}{(t-\tau)^{\alpha+1-N}} d\tau. \tag{4}$$

where $N-1<\alpha<N$ denotes the non-integer order of operation and $\Gamma(..)$ is the complete Gamma function expressed by (2).

For the Caputo operator the Laplace transform can be defined [8]:

Definition 5. *The Laplace transform for Caputo operator*

$$\begin{aligned} &\mathcal{L}({}_0^C D_t^\alpha f(t)) = s^\alpha F(s), \ \ \alpha<0 \\ &\mathcal{L}({}_0^C D_t^\alpha f(t)) = s^\alpha F(s) - \sum_{k=0}^{n-1} s^{\alpha-k-1}{}_0 D_t^k f(0), \\ &\alpha>0, \ \ n-1<\alpha \le n \in N. \end{aligned} \tag{5}$$

A fractional-order linear MIMO state space system, employing C defnition is described as follows:

$$\begin{aligned} &{}_0^C D_t^\alpha x(t) = Ax(t) + Bu(t) \\ &y(t) = Cx(t) \end{aligned} . \tag{6}$$

where $\alpha \in (0,1)$ denotes the fractional order of the state equation, $x(t) \in \mathbb{R}^N$, $u(t) \in \mathbb{R}^L$, $y(t) \in \mathbb{R}^P$ are the state, control and output vectors respectively, A, B, C are the state, control and output matrices respectively.

3.2 The CFE Approximation

An implementation of basic operator s^α at each digital platform (PLC, microcontroller) requires to employ its integer-order, finite-length, discrete-time approximator. The typical approximators are based on PSE and CFE approximations. They allow to express a non integer-order element in the form of digital FIR or IIR filters. The PSE approximator is based directly on discrete-time version of the GL Definition and it takes the form of a FIR filter containg zeros only. However, its digital, high quality implementation requires to apply a long memory buffer (high order of the filter).

The CFE approximator has the form of an IIR filter containing both poles and zeros. It is faster covergent and easier to implement because its order is relatively low, typically not higher than 5. It is obtained via discretization of elementary fractional order element s^α. This can be done using so called generating function $s \approx \omega(z^{-1})$. The new operator raised to the power α has the following form (see for example [3,18]):

$$\begin{aligned}&\left(\omega(z^{-1})\right)^{\alpha} = g_h CFE\{\left(\tfrac{1-z^{-1}}{1+az^{-1}}\right)^{\alpha}\}_{M,M} = \\ &= \frac{P_{\alpha M}(z^{-1})}{Q_{\alpha M}(z^{-1})} = g_h \frac{CFE_N(z^{-1},\alpha)}{CFE_D(z^{-1},\alpha)} = \\ &= g_h \frac{\sum\limits_{m=0}^{M} w_m z^{-m}}{\sum\limits_{m=0}^{M} v_m z^{-m}}\end{aligned} \qquad (7)$$

where M is the order of approximation, g_h is the coefficient depending on sample time and type of approximation:

$$g_h = \left(\frac{1+a}{h}\right)^{\alpha}. \qquad (8)$$

In (8) h is the sample time and a is the coefficient depending on approximation type. For $a = 0$ and $a = 1$ we obtain the Euler and Tustin approximations respectively. For $a \in (0, 1)$ we arrive at the Al-Alaoui-based approximation, which is a linear combination of the Euler and Tustin approaches. Note that in this case the parameter a in Eq. (7) is equal to $a = \frac{1-\gamma}{1+\gamma}$, with γ being the Al-Alaoui weighting coefficient [1,25]. Numerical values of coefficients w_m and v_m and various values of the parameter a can be calculated with the use of the MATLAB function written by I. Petras and available at [19]. If the Tustin approximation is considered (a=1) then $CFE_D(z^{-1}, \alpha) = CFE_N(z^{-1}, -\alpha)$ and the polynomial $CFE_D(z^{-1}, \alpha)$ can be given in the direct form (see [3]).

3.3 The CFE Based Method of Solution a FO State Equation

The method of memory-effective solution a fractional state equation was proposed and analysed by [16] and applied to modeling of heat transfer process by [15]. Its idea consists in replacing the continuous operator s^{α} in the Laplace transform of the FO state equation by its discrete CFE approximant expressed by (7). The CFE approximant is a function of discrete complex variable z^{-1}. This allows to directly pass to the discrete time domain. Then the solution of the state equation takes the following form:

$$\sum_{m=0}^{M} E_m x^+(k-m) = \sum_{m=0}^{M} F_m u^+(k-m) + \sum_{m=-M}^{0} x_0(m). \qquad (9)$$

where matrices E_m and F_m are defined as follows:

$$\begin{cases} E_m = g_h w_m I_{N\times N} - v_m A. \\ F_m = v_m B. \\ m = 0, 1, ..., M. \end{cases} \qquad (10)$$

In (10) g_h is described by (8), w_m and v_m denote coefficients of CFE approximant. From (9) the state vector $x^+(k)$ can be directly calculated as follows:

$$\begin{gathered} x^+(k) = -E_0^{-1} \sum_{m=1}^{M} E_m x^+(k-m) + \\ + E_0^{-1} \sum_{m=0}^{M} F_m u^+(k-m) + E_0^{-1} \sum_{m=-M}^{0} x_0(m). \end{gathered} \tag{11}$$

The Eq. (11) allows to directly solve the discrete-time FO state equation using CFE. The solution has the form of the M-th order difference equation, requiring to know M previous steps of state and control signals.

Notice that the initial condition has the form of a discrete function M steps long, starting M steps before an initial step. This is one of crucial advantages of FO modeling over "classic" approach. It allows to take into account a behaviour of a modeled system not only in an initial moment, but also M steps before. Both initial state and control functions can be considered. An impact of different initial functions on trajectories of the considered system will be shown in simulations. The deep analysis of inital problems for disccete FO systems is presented in [13].

The state Eq. (9) with M delays can also be written in the equivalent extended form without delay:

$$\begin{cases} x_q^+(k+1) = A_q^+ x_q^+(k) + B_q^+ u_q^+(k) \\ y_q^+(k) = C_q^+ x_q^+(k) \end{cases}. \tag{12}$$

where:

$$x_q^+(k) = \begin{bmatrix} x_1^+(k) \\ x_2^+(k) \\ \dots \\ x_M^+(k) \end{bmatrix}_{MN\times 1}. \tag{13}$$

$$\begin{cases} x_1^+(k) = x(k) \\ x_2^+(k) = x(k-1) \\ \dots \\ x_M^+(k) = x(k+1-M) \end{cases}. \tag{14}$$

$$u_q^+(k) = \begin{bmatrix} u_1^+(k) \\ u_2^+(k) \\ \dots \\ u_{M+1}^+(k) \end{bmatrix}_{M+1\times 1}. \tag{15}$$

$$\begin{cases} u_1^+(k) = u(k) \\ u_2^+(k) = u(k-1) \\ \dots \\ u_{M+1}^+(k) = u(k-M) \end{cases}. \tag{16}$$

$$A_q^+ = \begin{bmatrix} -E_0^{-1}E_1, ..., -E_0^{-1}E_M \\ I_{N\times N}, 0, 0, ..., 0 \\ 0, I_{N\times N}, 0, ..., 0 \\ ..., ..., ..., ..., .. \\ 0, ..., I_{N\times N}, 0 \end{bmatrix}_{MN\times MN}. \tag{17}$$

$$B_q^+ = \begin{bmatrix} E_0^{-1}F_0, ..., E_0^{-1}F_M \\ 0, 0, 0, ..., 0 \\ ..., ..., ..., ..., .. \\ 0, 0, 0, ..., 0 \end{bmatrix}_{MN\times M+1}. \tag{18}$$

The output matrix C is as follows:

$$C_q^+ = \left[C^+, 0, ..., 0\right]_{N\times MN}. \tag{19}$$

The initial conditions for state and control (12) also turn to the extended form:

$$x_{q0}^+ = \begin{bmatrix} x^+(0) \\ x^+(-1) \\ ... \\ x^+(-M+1) \end{bmatrix}_{MN\times 1}. \tag{20}$$

$$u_{q0}^+ = \begin{bmatrix} u^+(0) \\ u^+(-1) \\ ... \\ u^+(-M+1) \end{bmatrix}_{M\times 1}. \tag{21}$$

In (20), (21) index $-m$ denotes m-th time instant before initial one. This allows to take in consideration "a history" of process dynamics before start of modeling.

The summarized size of the proposed discrete, FO model equals to NM. This size is significantly smaller than a size of model using PSE approximation. In our case the maximum size of discrete CFE model equals to 10.

Next, notice the extended system (12)–(19) is the IO discrete system due to its "fractionality" is hidden in its parameters. It allows to analyze its properties using IO approach. Typical numerical tools available for example in MATLAB can be employed too.

4 The Considered RLC Circuit and Its FO Models

4.1 The Time-Continuous Model

We consider an elementary, serial RLC circuit shown in Fig. 1, where R is the resistance, C is the capacitance and L is the inductance. The circuit is powered by voltage $u(t)$. Denote the current in the circuit by $i_l(t)$ and the voltage at the capacitor by $v_c(t)$. Then they are described by the following FO differential equation given for example by [7,10]:

$$\frac{LC}{\delta^{2(1-\alpha)}}D^{2\alpha}u_c(t) + \frac{RC}{\delta^{(1-\alpha)}}D^{\alpha}u_c(t) + u_c(t) = u(t). \tag{22}$$

where $0.0 < \alpha \leq 1.0$ is the fractional order and δ is the auxiliary parameter expressed in seconds $[s]$. It allows to keep the dimensionality of the fractional temporal operator consistent to physical system ([7]):

$$\begin{aligned} \frac{d}{dt} &\rightarrow \frac{1}{\delta^{(1-\alpha)}} \frac{d^{\alpha}}{dt^{\alpha}} \\ \frac{d^2}{dt^2} &\rightarrow \frac{1}{\delta^{2(1-\alpha)}} \frac{d^{2\alpha}}{dt^{2\alpha}} \end{aligned}. \tag{23}$$

The free and forced solutions of FO differential Eq. (22) are given in [7] and [10].

The state vector takes the following form:

$$x(t) = \begin{bmatrix} i_l(t) \\ u_c(t) \end{bmatrix}. \tag{24}$$

Then the FO differential Eq. (22) takes the form (6) with:

$$A = \delta^{(1-\alpha)} \begin{bmatrix} -\frac{R}{L} & -\frac{1}{L} \\ \frac{1}{C} & 0 \end{bmatrix}_{2\times 2}. \tag{25}$$

$$B = \delta^{(1-\alpha)} \begin{bmatrix} \frac{1}{L} \\ 0 \end{bmatrix}_{2\times 1}. \tag{26}$$

The output of the system directly equals to its state:

$$C = I_{2\times 2} \tag{27}$$

The characteristic polynomial of state matrix A is as follows:

$$w(\lambda) = LC\lambda^2 + RC\lambda + 1. \tag{28}$$

and the spectrum of the system contains two eigenvalues:

$$\lambda_{1,2} = \frac{-R \pm \sqrt{R^2 - \frac{4L}{C}}}{2L}. \tag{29}$$

For undamped system with $R = 0$ the spectrum is as follows:

$$\lambda_{1,2} = \pm 2j\sqrt{\frac{L}{C}}. \tag{30}$$

The spectrum of the system is located in the left semiplane or directly at imaginary axis for undamped system, if $R = 0$. The system is overdamped when parameters of the circuit meet the known condition (31), otherwise it is underdamped. The border denotes critically damped circuit. The form of forced and free responses of the system depends on its damping.

$$R > 2\sqrt{\frac{L}{C}}. \tag{31}$$

The above elementary FO model is used to construct the proposed discrete CFE model presented beneath.

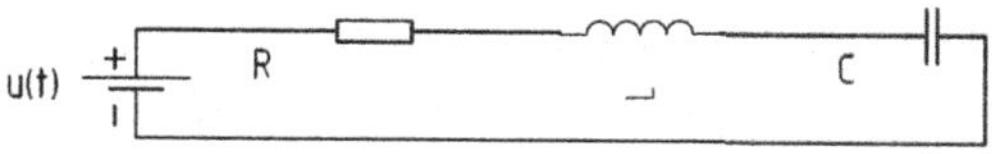

Fig. 1. The considered RLC circuit.

4.2 The Discrete CFE State Space Model

The discrete CFE state space model has the direct form (11) or its equivalent extended form (9)–(19). With respect to (25) the matrices E and F of model (10) take the following form:

$$\begin{cases} E_m = \begin{bmatrix} g_h w_m + \frac{v_m R}{L} & \frac{v_m}{L} \\ -\frac{v_m}{C} & g_h w_m \end{bmatrix}. \\ F_m = \begin{bmatrix} \frac{v_m}{L} \\ 0 \end{bmatrix}. \\ m = 0, 1, ..., M. \end{cases} \tag{32}$$

Matrices E and F are employed both in direct model (9)–(11) and its extended form (12)–(18). The extended form will be applied to simulations presented in the next section.

The continuous model, expressed by the FO state Eq. (6), (24)–(29) is stable for each value of fractional order from range $0.0 < \alpha \leq 1.0$. This is justified by Matignon theorem [12] and form of the system spectrum (29).

Next deal with the stability of the proposed discrete model. In this case the spectrum of the continuous system is also employed. For $0.0 < \alpha \leq 1.0$ it is always located outside the restricted area for stability of CFE model (see [22–24]).

Finally it can be concluded that for parameters of circuit: $R, L, C > 0$ and $0.0, \alpha \leq 1.0$ the considered model is always stable for continuous and discrete case. In addition, stability does not depend on sample time h. In the book [9] it is proved that the considered, elementary RLC ciricuit cannot be positive for any its real parameters.

5 Simulations

Simulations were done with the use of extended model (12)–(21) with matrices E and F defined by (32) and MATLAB. The underdamped and overdamped systems were investigated.

Firstly free responses to inital condition were examined as follows. The initial voltage of the capacitor was described by the functions $f_{l-4}(m)[V]$, defined for time instants $m = -4, ..., 0$. Consequently the initial current was equal: $g_{1,2} = 0[A]$ for functions $f_{1,2}(m)$, $g_3(m) = -0.25[A]$ for function $f_3(m)$ and $g_4(m) = 0.25[A]$ for function $f_4(m)$. Each function has the same value equal 1 in the initial step 0. The function $f_1(m)$ is the impulse equal 1 only for $m = 0$, otherwise it equals 0. The function $f_2(m)$ is the constant equal 1 for all m, the functions $f_3(m)$ and $f_4(m)$ are linearly decreasing and increasing respectively. All these functions are shown in the Fig. 2. Firstly the underdamped system described in the Table 1 was tested. The responses of the system to initial functions $f_{1-4}(m)$ with $u(t) = 0$ are given in Fig. 3.

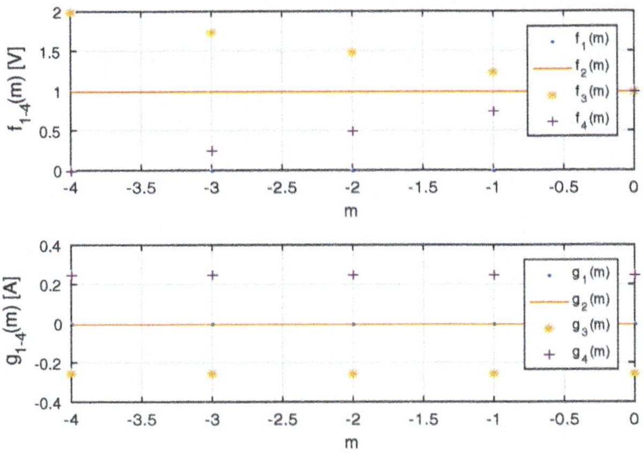

Fig. 2. The initial voltage and current functions.

Table 1. The parameters of simulation for underdamped circuit.

Parameter	α	R $[\Omega]$	L $[H]$	$C[F]$	$h[s]$	a	M
Value	0.9	0.5	1	1	0.05	1	5

Next the overdamped system described by the Table 2 was examined analogically as previously. Its responses to different initial functions are shown in Fig. 4.

Table 2. The parameters of simulation for overdamped circuit.

Parameter	α	R $[\Omega]$	L $[H]$	$C[F]$	$h[s]$	a	M
Value	0.9	2	1	1	0.05	1	5

Next the forced responses of the system for different inputs and homogenous initial conditions were examined. The testing voltage at input is defined as beneath:

$$u(k) = \begin{cases} 1, & k > 0 \\ f_{1-4}(m), & m = -M, .., 0 \end{cases} \tag{33}$$

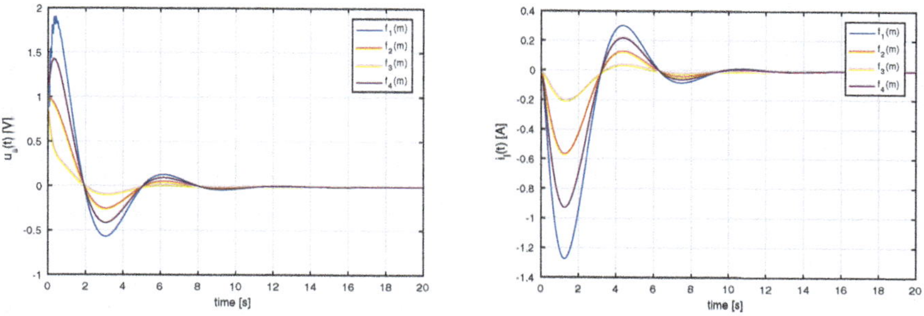

Fig. 3. The voltage $u_c(t)$ and current $i_L(t)$ as functions of $f_{1-4}(m)$, $g_{1-4}(m)$ for the underdamped system.

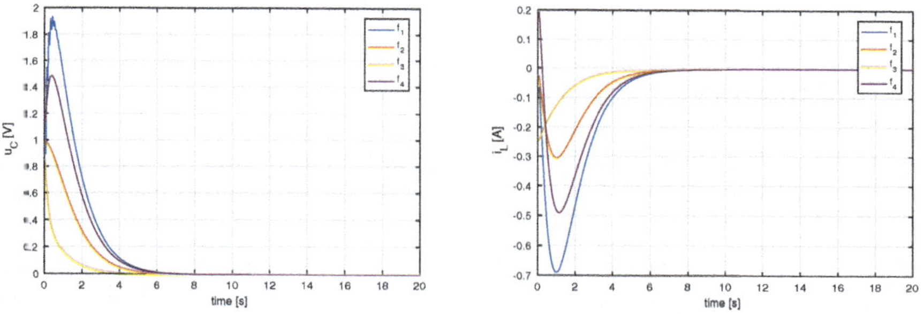

Fig. 4. The $u_c(t)$ and $i_L(t)$ as functions of $f_{1-4}(m)$, $g_{1-4}(m)$ for the overdamped system.

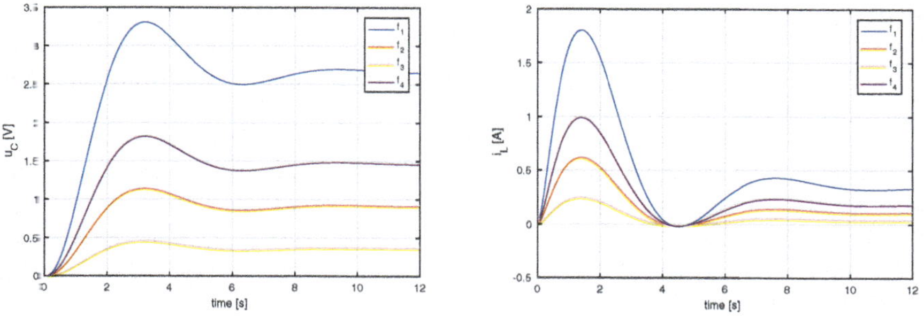

Fig. 5. The $u_c(t)$ and $i_L(t)$ as functions of $f_{1-4}(m)$ for the underdamped system.

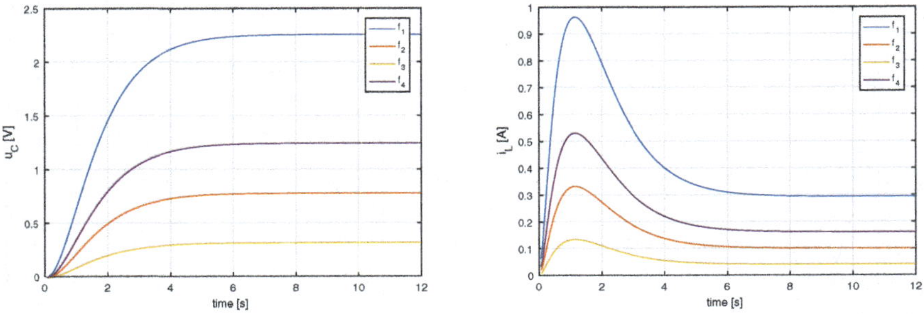

Fig. 6. The $u_c(t)$ and $i_L(t)$ as a function of $f_{1-4}(m)$ for the overdamped system.

where $f_{1-4}(m)$ are shown in Fig. 2. All the testing funtcions have the constant value after initial time instant, but they differ before it. Results of simulations for underdamped and overdamped system are presented in Figs. 5 and 6.

Diagrams 3–4 and 5–6 show that the shape of responses of the considered system strongly depends on the behaviour the system before the initial moment.

6 Conclusions

The main final conclusion from this paper is that the proposed, FO, memory effective, discrete model is able to properly describe an elementary, non positive RLC circuit.

Next, the FO approach allows to illustrate the system trajectory for different initial functions and these trajectories are different for different initial functions.

The further investigations of the discussed problems will cover both better theoretical explaination and experimental validation of the presented results.

Acknowledgments. This paper was sponsored by AGH UST project no 16.16.120.773.

References

1. Al-Alaoui, M.A.: Novel digital integrator and differentiator. Electron. Lett. **29**(4), 376–378 (1993)
2. Caponetto, R., Dongola, G., Fortuna, L., Petras, I.: Fractional order systems: modeling and control applications. In: Leon, O.C. (ed.) World Scientific Series on Nonlinear Science, pp. 1–178. University of California, Berkeley (2010)
3. Chen, Y.Q., Moore, K.L.: Discretization schemes for fractional-order differentiators and integrators. IEEE Trans. Circ. Syst. - I: Fund. Theory Appl. **49**(3), 263–269 (2002)
4. Das, S.: Functional Fractional Calculus for System Identification and Control. Springer, Berlin (2010)
5. Dzieliński, A., Sierociuk, D., Sarwas, G.: Some applications of fractional order calculus. Bull. Pol. Acad. Sci. Tech. Sci. **58**(4), 583–592 (2010)

6. Gal, C.G., Warma, M.: Elliptic and parabolic equations with fractional diffusion and dynamic boundary conditions. Evol. Equn. Control Theory **5**(1), 61–103 (2016)
7. Gómez-Aguilar, J.F., Rosales, J., Guia, M.: RLC electrical circuit of non-integer order. Cent. Eur. J. Phys. **11**(10), 1361–1365 (2013)
8. Kaczorek, T.: Selected Problems of Fractional Systems Theory. Springer, Berlin (2011)
9. Kaczorek, T., Rogowski, K.: Fractional Linear Systems and Electrical Circuits. Bialystok University of Technology, Bialystok (2014)
10. Koksal, M.E.: Time and frequency responses of non-integer order RLC circuits. AIMS Math. **4**(1), 64–78 (2018)
11. Kopka, R.: Minimum energy control of loading of supercapacitors and Li-ion capacitors (in Polish). Opole Univ. Techn. eds., Opole (2019)
12. Matignon, D.: Stability results for fractional differential equations with applications to control processing. In: IMACS-SMC Proceedings, Lille, France, July 1996, pp. 963–968 (1996)
13. Mozyrska, D., Pawluszewicz, E.: Fractional discrete-time linear control systems with initialisation. Int. J. Control **1**(1), 1–7 (2011)
14. Obrączka, A.: Control of heat processes with the use of non-integer models. Ph.D. thesis, AGH University, Krakow, Poland (2014)
15. Oprzędkiewicz, K., Mitkowski, W.: A memory efficient non integer order discrete time state space model of a heat transfer process. Int. J. Appl. Math. Comput. Sci. **28**(4), 649–659 (2018)
16. Oprzędkiewicz, K., Stanisławski, R., Gawin, E., Mitkowski,W.: A new algorithm for a CFE approximated solution of a discrete-time non integer-order state equation. Bull. Pol. Acad. Sci. Tech. Sci. **65**(4), 429–437 (2017)
17. Ostalczyk, P.: Discrete fractional calculus. In: Applications in Control and Image Processing. World Scientific, New Jersey (2016)
18. Petras, I.: Fractional order feedback control of a DC motor. J. Electr. Eng. **60**(3), 117–128 (2009)
19. Petras, I.: http://people.tuke.sk/igor.podlubny/usu/matlab/petras/dfod2.m (2009)
20. Podlubny, I.: Fractional Differential Equations. Academic Press, San Diego (1999)
21. Sierociuk, D., Skovranek, T., Macias, M., Podlubny, I., Petras, I., Dzielinski, A., Ziubinski, P.: Diffusion process modeling by using fractional-order models. Appl. Math. Comput. **257**(1), 2–11 (2015)
22. Stanisławski, R., Latawiec, K.: Normalized finite fractional differences: computational and accuracy breakthroughs. Int. J. Appl. Math. Comput. Sci. **22**(4), 907–919 (2012)
23. Stanisławski, R., Latawiec, K.: Stability analysis for discrete-time fractional-order LTI state-space systems. part i: new necessary and sufficient conditions for asymptotic stability. Bull. Pol. Acad. Sci. Tech. Sci. **61**(2), 353–361 (2013)
24. Stanisławski, R., Latawiec, K.: Stability analysis for discrete-time fractional-order LTI state-space systems. part ii: stability criterion for FD-based systems. Bull. Pol. Acad. Sci. Tech. Sci. **61**(2), 362–370 (2013)
25. Stanisławski, R., Latawiec, K., Łukaniszyn, M.: A comparative analysis of Laguerre-based approximators to the Grünwald-Letnikov fractional-order difference. Hindawi Publ. Corporation Math. Probl. Eng. **2015**(1), 1–10 (2015)

The Comparison of IAE Optimal Time-Varying Sliding Modes for Second Order Systems

Filip Szewczyk(✉) and Katarzyna Adamiak

Institute of Automatic Control, Łódź University of Technology,
18/22 Stefanowskiego Street, 90-924 Łódź, Poland
katarzyna.adamiak@dokt.p.lodz.pl

Abstract. This study presents a comparison of three different sliding mode control strategies for continuous time disturbed systems. Each of the analyzed control methods provides stability and robustness of the considered system from the very beginning of the regulation process by elimination of the reaching phase. This is achieved with application of time-variant switching lines, which cross the system's position at the initial time and then move in the state space until they reach the desired state. Moreover, the switching planes are designed in order to minimize the IAE and provide monotonical convergence of error vector to zero.

Keywords: Sliding mode control · Time-variant sliding mode · IAE minimization

1 Introduction

Sliding mode control is well known and praised for providing insensitivity to external disturbances and model uncertainties. This efficient control scheme drives the representative point of the system from any initial position to a pre-selected sliding plane and confines its dynamics to this plane ever after [1,2]. Therefore, a nonlinear control law is applied, which ensures immediate rejection of matched disturbance effects in the sliding phase. Unfortunately, the benefit of robustness is only guaranteed once the representative point reaches the sliding plane [3]. In the reaching phase the system is still influenced by disturbances and modelling uncertainties, which may cause undesirable dynamics.

An answer to this problem are time-variant sliding planes. With application of a sliding surface, which crosses the initial position of the system at $t = 0$ and for $t > 0$ moves inside the state space, until it reaches the demand state at $t = t_0$, the reaching phase is completely eliminated and the system becomes robust in the whole regulation process. Such sliding lines have been widely introduced for second order systems. The first attempts considered rotating and shifting sliding lines [4,5]. Later on, many other sliding line evolutions were introduced [6–11] and the concept was extended for higher order systems [12–14].

R. Szewczyk et al. (Eds.): AUTOMATION 2021, AISC 1390, pp. 58–67, 2021.
https://doi.org/10.1007/978-3-030-74893-7_6

The aim of this study is to compare three most common designs of time-variant sliding lines: constant-velocity sliding line, constant-acceleration sliding line and a time-variant curve. The strategies are compared on the basis of IAE minimalization and a short simulation example is presented.

2 Problem Description

We consider a second order nonlinear system described with

$$\begin{cases} \dot{x_1} = x_2 \\ \dot{x_2} = f(\mathbf{x}, t) + \Delta f(\mathbf{x}, t) + bu(t) + d(t) \end{cases}, \tag{1}$$

where $\mathbf{x}(t) = [x_1(t)\ x_2(t)]^T$ is the system's state vector, $f(\mathbf{x}, t)$ is a function of state vector and time, $\Delta f(\mathbf{x}, t)$ and $d(t)$ represent system's uncertainty and disturbance respectively, $b \neq 0$ is gain coefficient of the input signal $u(t)$. Next, we assume that $|f(\mathbf{x}, t)| \leq M$ and $|\Delta f(\mathbf{x}, t) + d(t)| \leq \mu$, where M and μ are strictly positive constants. The initial position of the system is $x_1(0) \neq 0$ and its velocity is $x_2(0) = 0$.

In this paper, sliding mode control of system (1) with use of several different switching patterns is taken into consideration. The main purpose is to drive the system to the point $\mathbf{x}_d = [x_{1d}\ x_{2d}]^T$ and ensure its stability and robustness to disturbances and uncertainties. We define the error-state vector as $\mathbf{e}(t) = \mathbf{x}(t) - \mathbf{x}_d(t)$. We also assume that the point $\mathbf{x}_d$ is fixed, meaning the system's desired velocity is zero. Therefore, initial values of $\mathbf{e}(t)$ are $e_1(0) \neq 0$ and $e_2(0) = 0$.

3 Selection of the Sliding Lines

In this section, we analyze the properties of the sliding mode obtained with three different time-variant sliding lines.

3.1 Constant-Velocity Sliding Mode

At first, we consider a straight sliding line moving in the state space with constant velocity. After the line crosses the origin of system coordinates at time t_0, it remains fixed. This approach was presented in [6,7] and [13].

The sliding line changes depending on the current time t, as follows

$$s(\mathbf{e}, t) = \begin{cases} e_2(t) + ce_1(t) + A + Bt, & t \leq t_0 \\ e_2(t) + ce_1(t), & t > t_0 \end{cases}. \tag{2}$$

To ensure stability of the system, constant c must be strictly positive. Since the robustness of the system is required, the point $\mathbf{x}(0) = [x_1(0)\ x_2(0)]^T$ must belong to the sliding line at $t = 0$. Hence, the condition below must be satisfied

$$s[\mathbf{e}(0), 0] = e_2(0) + ce_1(0) + A + B \cdot 0 = 0. \tag{3}$$

Bearing in mind that $e_2(0) = 0$, we may write

$$A = -ce_1(0). \tag{4}$$

Assuming that $\gamma \geq M + \mu$, all sliding mode requirements are ensured with the control signal

$$u(t) = \frac{1}{b}\Big\{\dot{x}_{2d} - ce_2(t) - B - \gamma \text{sgn}\left[s(\mathbf{e}, t)\right]\Big\}. \tag{5}$$

This can be proved by considering the closed-loop system's stability condition

$$s(\mathbf{e}, t)\dot{s}(\mathbf{e}, t) = s(\mathbf{e}, t)\left(\dot{e}_2(t) + c\dot{e}_1(t) + B\right) \leq 0. \tag{6}$$

Using (1) and recalling the definition of error-state vector $\mathbf{e}(t)$ given in Sect. 2, we may write

$$s(\mathbf{e}, t)\big(f(\mathbf{x}, t) + \Delta f(\mathbf{x}, t) + d(t) + bu(t) - \dot{x}_{2d} + ce_2(t) + B\big) \leq 0. \tag{7}$$

Substituting control signal (5), we obtain

$$s(\mathbf{e}, t)\big(f(\mathbf{x}, t) + \Delta f(\mathbf{x}, t) + d(t) - \gamma \text{sgn}\left[s(\mathbf{e}, t)\right]\big) \leq 0. \tag{8}$$

Values of $f(\mathbf{x}, t)$ and $\Delta f(\mathbf{x}, t) + d(t)$ are bounded, as described in Sect. 2. Using the subadditivity, we may write

$$|f(\mathbf{x}, t) + \Delta f(\mathbf{x}, t) + d(t)| \leq M + \mu \leq \gamma. \tag{9}$$

Verification of the signs of both factors in inequality (8) leads directly to the conclusion that the control signal (5) ensures stability of system (1) for any $t > 0$.

Next, we look for the relationship between A, B and c, so that inequality

$$\underset{t \geq 0}{\forall} |u(t)| \leq u_{max}. \tag{10}$$

holds. Substituting (5) into (10), we obtain

$$\Big|\dot{x}_{2d} - ce_2(t) - B - \gamma \text{sgn}\left[s(\mathbf{e}, t)\right]\Big| \leq |b| u_{max}. \tag{11}$$

Using the subadditivity and remembering that $\mathbf{x}_d$ is a fixed point, we can write

$$\Big|ce_2(t) + B\Big| \leq |b| u_{max} - \gamma. \tag{12}$$

The formulas for $e_1(t)$ and $e_2(t)$ can be derived from (2) by equating it to zero. Solving the differential equation with initial condition $e_2(0) = 0$, we get

$$e_1(t) = \begin{cases} -\frac{B}{c^2} e^{-ct} + e_1(0) + \frac{B}{c^2} - \frac{B}{c^1} t, & t \leq t_0 \\ \frac{B}{c^2}\left(-1 + e^m\right) e^{-ct}, & t > t_0 \end{cases}, \tag{13}$$

$$e_2(t) = \begin{cases} \frac{B}{c}\left(e^{-ct} - 1\right), & t \leq t_0 \\ \frac{B}{c}\left(1 - e^m\right) e^{-ct}, & t > t_0 \end{cases}, \tag{14}$$

where

$$m = \frac{c^2 e_1(0)}{B} > 0. \tag{15}$$

At $t = t_0$ the sliding line stops moving and crosses the origin of the system coordinates, which combined with (2) and (4), leads to $A + Bt_0 = 0$. After designation of t_0, we obtain

$$t_0 = -\frac{A}{B} = \frac{ce_1(0)}{B} = \frac{m}{c}. \tag{16}$$

Substituting (14) into (12), we get:

- for $t \leq t_0$:

$$\left|Be^{-ct}\right| \leq |b|u_{max} - \gamma, \tag{17}$$

  the left-hand side of which is maximal for $t = 0$. Therefore, (10) is satisfied when

$$|B| \leq |b|u_{max} - \gamma. \tag{18}$$

- for $t > t_0$:

$$|B|\left|1 + \left(1 - e^m\right) e^{-ct}\right| \leq |b|u_{max} - \gamma, \tag{19}$$

the left-hand side of which is non-monotonous due to the absolute value. However, the expression inside the absolute value equals to zero at time $t < t_0$. Therefore, it does not have to be considered at this point.

Proof. Consider function $g(t) = 1 + \left(1 - e^m\right) e^{-ct}$. We notice monotonicity of $g(t)$, whose derivative is equal to $g'(t) = -c\left(1 - e^m\right) e^{-ct}$, so, as $c > 0$ and $m > 0$, it increases with time. The solution of $|g(t)| = 0$ becomes

$$t_{min} = \frac{1}{c} \ln\left(e^m - 1\right), \tag{20}$$

which determines the point, where the absolute value of $g(t)$ is minimal. To check the difference between t_0 and t_{min}, we write

$$\begin{aligned} t_{min} - t_0 &= \frac{1}{c} \ln\left(e^m - 1\right) - \frac{m}{c} = \frac{1}{c}\left[\ln\left(e^m - 1\right) - \ln\left(e^m\right)\right] \\ &= \frac{1}{c}\left[\ln\left(1 - e^{-m}\right)\right]. \end{aligned} \tag{21}$$

Recalling that $c > 0$ and $m > 0$, we obtain

$$\frac{1}{c}\left[\ln\left(1 - e^{-m}\right)\right] < 0, \tag{22}$$

which is equivalent to

$$t_{min} < t_0. \tag{23}$$

As the value of $g(t)$ increases with time, we calculate the limit

$$\lim_{t \to \infty} g(t) = \lim_{t \to \infty} 1 + \left(1 - e^m\right) e^{-ct} = 1 \tag{24}$$

and we notice that for $t \geq t_0$ condition (19) can be transformed into (18). □

In order to compare the system's performance with the proposed sliding lines, we need an overtime error measure. We choose integral absolute error defined as

$$I = \int_0^\infty |e_1(t)|dt. \tag{25}$$

If the desired point $\mathbf{x_d}$ is reached in finite time, we can equivalently write

$$I = \int_0^{t_f} |e_1(t)|dt, \tag{26}$$

where t_f is the regulation time.

In the considered case, the state vector $\mathbf{x}(t)$ reaches the desired value $\mathbf{x_d}$ asymptotically, which is why we use (25) to calculate IAE. Substituting (13) and performing some calculations, we obtain

$$I_V = \frac{c[e_1(0)]^2}{2|B|} + \frac{|e_1(0)|}{c}. \tag{27}$$

Moreover, using (15), we can write I_V as a function of m and B

$$I_V(m, |B|) = \frac{|e_1(0)|^{\frac{3}{2}}}{\sqrt{|B|}} \left(\frac{\sqrt{m}}{2} + \frac{1}{\sqrt{m}} \right). \tag{28}$$

One can easily notice that the value of I_V is smallest when $|B|$ is maximal. Therefore, due to (18), we assume

$$|B| = |b|u_{max} - \gamma. \tag{29}$$

The minimum of the expression in parentheses on the right-hand side of (28) is obtained for $m = 2$. Hence, the minimal value of I_V is

$$I_{V\ min} = \frac{\sqrt{2}\sqrt{|e_1(0)|^3}}{\sqrt{|b|u_{max} - \gamma}}. \tag{30}$$

3.2 Constant-Acceleration Sliding Mode

The second sliding line we consider starts its movement with non-zero velocity and decelerates from the very beginning to stop at the origin of the system coordinates. Such sliding mode was presented in [6].

The formula for $u(t)$ may be derived similarly as in Sect. 3.1. Sliding line equation is given by

$$s(\mathbf{e}, t) = \begin{cases} e_2(t) + ce_1(t) + A + Bt + Dt^2, & t \le t_0 \\ e_2(t) + ce_1(t), & t > t_0 \end{cases}. \tag{31}$$

Supposing that the sliding line stops moving at time t_0 due to deceleration represented by D (its velocity is continuous), we may write

$$A + Bt_0 + Dt_0^2 = 0, \tag{32}$$

$$(A + Bt + Dt^2)'|_{t=t_0} = B + 2Dt_0 = 0. \tag{33}$$

Moreover, we want to ensure system's robustness. Hence, we write

$$e_2(0) + ce_1(0) + A = 0. \tag{34}$$

Bearing in mind that $e_2(0) = 0$, constants A, B and D can be described as

$$A = -ce_1(0), \quad B = 2\frac{ce_1(0)}{t_0}, \quad D = -\frac{ce_1(0)}{t_0^2}. \tag{35}$$

The stability and robustness are ensured for

$$u(t) = \frac{1}{b}\left(\dot{x}_{2d} - ce_2(t) - \gamma \text{sgn}\left[s(\mathbf{e}, t)\right]\right) - \frac{1}{b}\begin{cases} B + 2Dt, & t \leq t_0 \\ 0, & t > t_0 \end{cases}. \tag{36}$$

Now limitation of the control signal will be considered. From the solution of $s(\mathbf{e}, t) = 0$ for both $t \leq t_0$ and $t > t_0$, we obtain

$$e_1(t) = \begin{cases} e_1(0)\left\{\frac{2}{ct_0}\left(1 + \frac{1}{ct_0}\right)\left[1 - e^{-ct}\right] + \frac{t^2}{t_0^2} - 2\left(1 + \frac{1}{ct_0}\right)\frac{t}{t_0} + 1\right\}, & t \leq t_0 \\ e_1(0)\left\{\frac{2}{c^2 t_0^2}e^{-c(t-t_0)} - \frac{2}{c^2 t_0}\left(c + \frac{1}{t_0}\right)e^{-ct}\right\}, & t > t_0 \end{cases} \tag{37}$$

$$e_2(t) = \begin{cases} e_1(0)\left\{\frac{2}{t_0}\left(1 + \frac{1}{ct_0}\right)e^{-ct} + \frac{2t}{t_0^2} - \frac{2}{t_0}\left(1 + \frac{1}{ct_0}\right)\right\}, & t \leq t_0 \\ e_1(0)\left\{\frac{-2}{ct_0^2}e^{-c(t-t_0)} + \frac{2}{ct_0}\left(c + \frac{1}{t_0}\right)e^{-ct}\right\}, & t > t_0 \end{cases}. \tag{38}$$

We want the control signal $u(t)$ to be upper bounded, so that (10) is satisfied. Assuming $\mathbf{x}_d$ to be a fixed point and substituting (38) to (36), we get:

- for $t \leq t_0$:

$$\left|\frac{2e_1(0)}{t_0^2}\left(e^{-ct} - 1\right) + \frac{2ce_1(0)}{t_0}\right| \leq |b|u_{max} - \gamma, \tag{39}$$

  the left-hand side of which is maximal for $t = 0$. Hence, using (35), we get

$$|B| \leq |b|u_{max} - \gamma. \tag{40}$$

- for $t > t_0$:

$$\left|\frac{-2e_1(0)}{t_0^2}e^{-c(t-t_0)} + \frac{2ce_1(0)}{t_0}e^{-ct} + \frac{2e_1(0)}{t_0^2}e^{-ct}\right| \leq |b|u_{max} - \gamma, \tag{41}$$

the left-hand side of which is a monotonic function reaching its maximum at $t = t_0$. Hence, we transform (41) into

$$\left|\frac{2ce_1(0)}{t_0}e^{-ct_0} + \frac{2e_1(0)}{t_0^2}\left(e^{-ct_0} - 1\right)\right| \leq |b|u_{max} - \gamma. \tag{42}$$

One can notice that, because $e^{-ct_0} < 1$, the left-hand side of (42) is less or equal than $\left|\frac{2ce_1(0)}{t_0}\right|$. Using (35), we conclude that (41) is satisfied as long as (40) holds.

To define the IAE we use (25). After some calculations, we obtain

$$I_A(c, t_0) = |e_1(0)| \left(\frac{t_0}{3} + \frac{1}{c} \right). \tag{43}$$

Deriving c from the formula for B in (35) and substituting into (43), we obtain

$$I_A(t_0) = |e_1(0)| \left(\frac{t_0}{3} + \frac{2|e_1(0)|}{|B|t_0} \right), \tag{44}$$

which is smallest if $|B|$ is the biggest possible. Hence, we assume $|B| = |b|u_{max} - \gamma$. Minimizing (44) with respect to t_0, we obtain

$$I_{A\ min} = \frac{2}{\sqrt{3}} \frac{\sqrt{2}\sqrt{|e_1(0)|^3}}{\sqrt{|b|u_{max} - \gamma}}. \tag{45}$$

3.3 Time-Variant Curve Sliding Mode

The third proposed line is a curve described as

$$s(\mathbf{e}, t) = e_2(t) + c(t)\text{sgn}\left[e_1(t)\right] \sqrt{|e_1(t)|} = 0, \tag{46}$$

where

$$c(t) = \begin{cases} \alpha t, & t \leq t_0 \\ \alpha t_0, & t > t_0 \end{cases}. \tag{47}$$

This approach was presented in [8] and [9]. As $e_2(0) = 0$, the chosen sliding line will always cross the initial point $\mathbf{x}(0)$ at the beginning of the regulation process. We apply the following control signal

$$u = \frac{1}{b}\left\{ \dot{x}_{2d} - \text{sgn}[e_1(t)]\sqrt{|e_1(t)|}\dot{c}(t) - \text{sgn}\left[s(\mathbf{e}, t)\right] \frac{c(t)e_2(t)}{2\sqrt{|e_1(t)|}} - \gamma\text{sgn}[s(\mathbf{e}, t)] \right\}. \tag{48}$$

System's stability with use of (48) was proved in [8]. It was also shown that the control signal constraint (10) is satisfied for

$$t_0 \leq \frac{\sqrt{2}\sqrt{|b|u_{max} - \gamma}}{\alpha}, \tag{49}$$

$$\alpha \leq \frac{|b|u_{max} - \gamma}{\sqrt{|e_1(0)|}}. \tag{50}$$

In order to calculate the IAE, we need to describe $|e_1(t)|$ explicitly. Solving (46) with $y = \sqrt{|e_1(t)|}$, we obtain

$$\sqrt{|e_1(t)|} = \begin{cases} C_1 - \frac{1}{4}\alpha t^2, & t \leq t_0 \\ C_2 - \frac{1}{2}\alpha t_0 t, & t_0 < t < t_f \\ 0, & t \geq t_f \end{cases}. \tag{51}$$

As the function describing $e_1(t)$ must remain continuous, we can derive values of C_1 and C_2. After some calculations, we obtain the following formula for $|e_1(t)|$

$$|e_1(t)| = \begin{cases} \left(\sqrt{|e_1(0)|} - \frac{1}{4}\alpha t^2\right)^2, & t \leq t_0 \\ \left(\sqrt{|e_1(0)|} + \frac{1}{4}\alpha t_0^2 - \frac{1}{2}\alpha t_0 t\right)^2, & t_0 < t < t_f \\ 0, & t \geq t_f \end{cases}. \tag{52}$$

According to (52), the regulation time is finite and equal to t_f. That is why, we may use (26) to calculate the IAE, which in this case is equal to

$$I_C(\alpha, t_0) = \frac{1}{2}|e_1(0)|t_0 - \frac{1}{24}\sqrt{|e_1(0)|}\alpha t_0^3 + \frac{1}{480}\alpha^2 t_0^5 + \frac{2\sqrt{|e_1(0)|^3}}{3\alpha t_0}. \tag{53}$$

The final step is to minimize $I_C(\alpha, t_0)$. However, $\nabla I_C(\alpha, t_0) = 0$ only for $\alpha t_0 = 0$. Both values, α and t_0, are always positive, which is why the value of $I_C(\alpha, t_0)$ will be minimal, if conditions (49) and (50) are satisfied simultaneously. Hence, we obtain

$$\begin{cases} \alpha = \frac{|b|u_{max} - \gamma}{\sqrt{|e_1(0)|}} \\ t_0 = \frac{\sqrt{2}\sqrt{|e_1(0)|}}{\sqrt{|b|u_{max} - \gamma}} \end{cases}. \tag{54}$$

Substituting these values into (53), we finally get

$$I_{C\ min} = \frac{91}{120}\frac{\sqrt{2}\sqrt{|e_1(0)|^3}}{\sqrt{|b|u_{max} - \gamma}}. \tag{55}$$

4 Simulation Example

To verify the considerations of Sect. 3, we take into account a second order nonlinear system

$$\begin{cases} \dot{x}_1 = x_2 \\ \dot{x}_2 = (1+\delta)\sin(x_1) + 0.5 + \frac{x_2^2}{x_2^2+1} + u(t) + d(t) \end{cases}, \tag{56}$$

where $\delta = 0.27$ and $d(t) = 0.7\sin(10t)$. Thus, we assume $M = 2.6$ and $\mu = 0.98$. The desired position of the system is $x_d = [0\ 0]^T$ and its initial state vector is $x_0 = [10\ 0]^T$. The control is limited by $u_{max} = 5$. For each of the sliding modes, we define all needed constants as shown in Sect. 3. We also use formulas (5), (36) and (48) to define the control signal. The simulation's results can be seen in Figs. 1, 2, 3, 4 and 5. First observation to make is the successful reaching phase elimination in each sliding mode. However, only the time-variant curve provides the finite time convergence of the system. The other two sliding modes ensure the stability of the system, but the regulation process is infinite. From the control signals, presented in Figs. 3, 4 and 5, we notice that the control signal constraint was satisfied in all cases. This validates formulas (18), (40), (49) and (50).

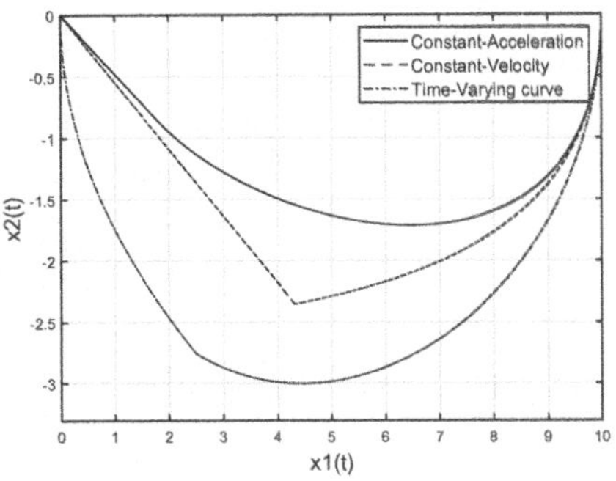

Fig. 1. System's state trajectories.

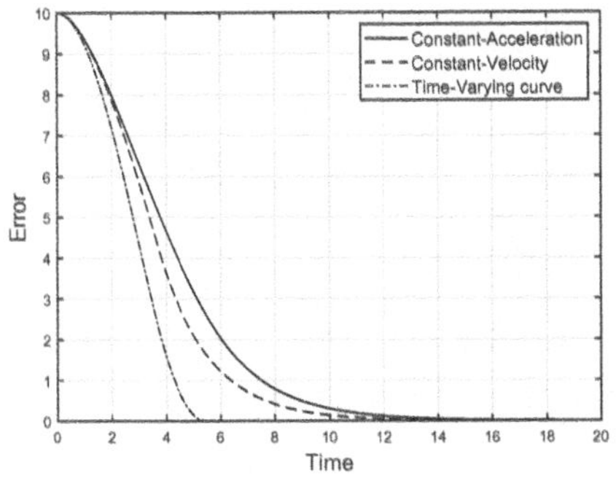

Fig. 2. First state variable error.

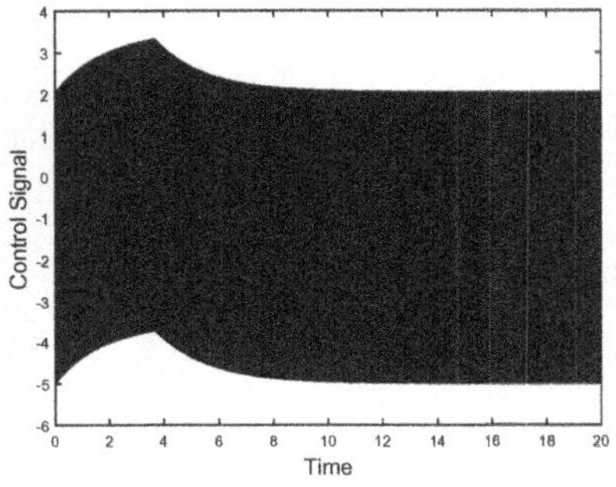

Fig. 3. Constant - velocity sliding mode.

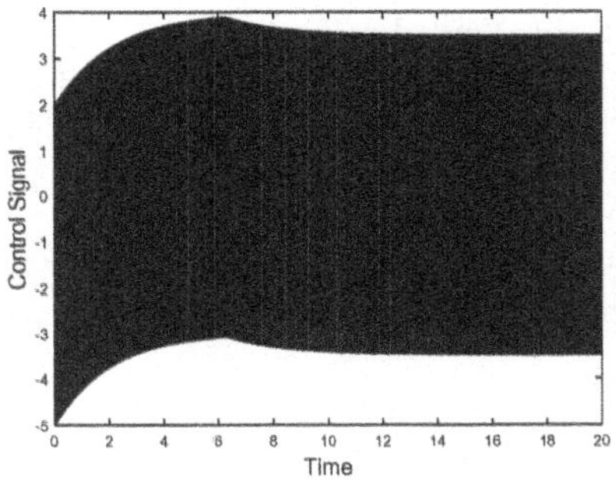

Fig. 4. Constant - acceleration sliding mode.

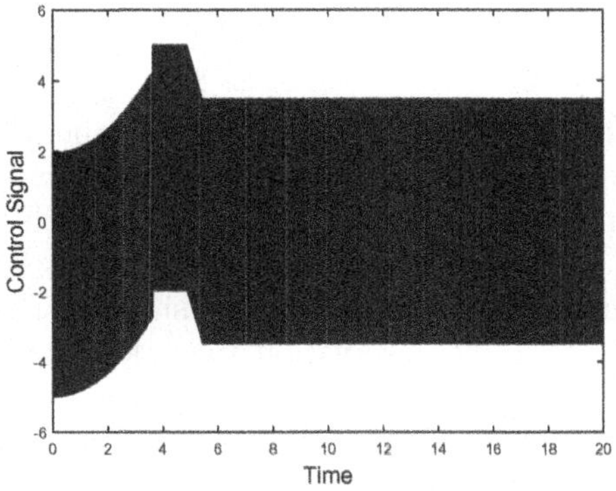

Fig. 5. Time-variant curve sliding mode.

5 Conclusion

This paper compared three popular time-variant sliding line designs. In each case, the sliding line coefficients were chosen to optimize a well known performance quality index IAE and to satisfy a control signal constraint. The results clearly show that the best performance is achieved with a sliding line based on a time-varying curve. Moreover, this choice of the sliding line provides finite-time convergence of the error-state vector to zero. The outcomes of this research were calculated for the second order system. However, similar research could be conducted for third order plants.

References

1. Utkin, V.I.: Variable structure systems with sliding modes. IEEE Trans. Autom. Control **22**(2), 212–222 (1977)
2. Edwards, C., Spurgeon, S.: Sliding Mode Control: Theory and Applications. Taylor and Francis, London (1998)
3. Draženović, B.: The invariance conditions in variable structure systems. Automatica **5**(3), 287–295 (1969)
4. Choi, S.B., Cheong, C.C., Park, D.W.: Moving switching surfaces for robust control of second-order variable structure systems. Int. J. Control **58**(1), 229–245 (1993)
5. Choi, S.B., Park, D.W.: Moving sliding surfaces for fast tracking control of second-order dynamical systems. J. Dyna. Syst. Meas. Control **116**(1), 154–158 (1994)
6. Bartoszewicz, A.: Time-varying sliding modes for second-order systems. In: IEE Proceedings - Control Theory and Application, vol. 143, no. 5, pp. 455–462 (1996)
7. Nowacka-Leverton, A., Bartoszewicz, A.: Variable structure control of second order system with acceleration constraint. Mechanics **28**(3), 96–100 (2009)
8. Bartoszewicz, A., Pietrala, M., Jaskuła, M.: IAE optimal sliding mode control for second order dynamical systems. In: 2018 19th International Carpathian Control Conference (ICCC), Szilvasvarad, Hungary, pp. 169–174 (2018)
9. Bartoszewicz, A., Pietrala, M., Jaskuła, M.: Terminal sliding mode control of second order systems with velocity constraint. In: 2018 19th International Carpathian Control Conference (ICCC), Szilvasvarad, Hungary, pp. 223–227 (2018)
10. Bartoszewicz, A.: Chattering attenuation in sliding mode control systems. Control Cybern. **29**(2), pp. 585–594 (2000)
11. Bartolini, G., Fridman, L., Pisano, A., Usai, E.: Modern Sliding Mode Control Theory. New Perspectives and Applications. Springer-Verlag, Heidelberg (2008)
12. Bartoszewicz, A., Nowacka, A.: Reaching phase elimination in variable structure control of the third order system with state constraints. Kybernetika **42**(1), 111–126 (2006)
13. Bartoszewicz, A., Nowacka-Leverton, A.: Time-Varying Sliding Modes for Second and Third Order Systems. Springer, Heidelberg (2009)
14. Bartoszewicz, A., Nowacka-Leverton, A.: ITAE optimal sliding modes for third order systems with input signal and state constraints. IEEE Trans. Autom. Control **55**(8), 1928–1932 (2010)

Advantages of Using a Reference Model of the Plant in Discrete-Time Sliding Mode Control

Paweł Latosiński(✉)

Institute of Automatic Control, Lodz University of Technology,
B. Stefanowskiego 18/22, 90-924 Lodz, Poland
pawel.latosinski@p.lodz.pl

Abstract. Reaching law based discrete-time sliding mode control strategies are an efficient way of ensuring good robustness of the system with respect to uncertainties. However, since such strategies rely on a recursive formula of the system state, their performance can be gradually deteriorated by the effect of disturbance in each discrete step. This can negatively affect the length of the reaching phase, as well as increase the quasi-sliding mode band width in the sliding phase. In order to remedy this issue, we propose the use of a model reference strategy. In each step, this strategy drives the representative point of the plant to a specific reference point obtained from the model. As a result, residual effect of past uncertainties on sliding motion of the system is eliminated. Furthermore, model reference approach allows one to bypass the assumption about matching conditions, which is otherwise necessary in many cases.

Keywords: Discrete-time systems · Robust control · Sliding mode control

1 Introduction

Discrete-time sliding mode control (DSMC) strategies, first proposed by researchers such as Milosavljevic and Utkin [15,18], are well known for their robustness with respect to uncertainties and their exceptional computational efficiency. Typical design process of such a strategy consists of stating a control signal with variable structure, and then proving that this control signal ensures a stable sliding motion of the system. However, a more recent reaching law approach proposed by Gao and co-authors [8] allows one to bypass the often difficult proof of stability. In this approach the desired sliding mode performance is first stated in the form of a recursive function, then this function is applied to design the control signal.

Many authors have proposed new reaching law based sliding mode control strategies for discrete-time systems with the aim of ensuring particular qualities of system sliding motion. Such strategies can belong to one of two broad categories [6]. The first one consists of switching type strategies [7,10,14,16] which

R. Szewczyk et al. (Eds.): AUTOMATION 2021, AISC 1390, pp. 68–74, 2021.
https://doi.org/10.1007/978-3-030-74893-7_7

follow the classic definition of discrete-time sliding motion introduced by Gao, Wang and Homaifa. The second category includes non-switching type strategies [1,9,17] which follow the updated definition [2].

An interesting recent development in DSMC is the application of a reference model of the plant. This novel approach involves obtaining a desired state trajectory using the deterministic model of the plant and then driving the state of the actual plant alongside that trajectory in each step. Such a control scheme ensures that the desired sliding motion of the system is not distorted by the continued detrimental effect of uncertainties. It has been demonstrated in literature [3–5,13] that the application of a model reference scheme improves robustness of the system compared to conventional reaching laws. Moreover, it has been shown that the model reference approach allows one to bypass the assumption about matching conditions even in strategies that normally require it, such as reaching laws for arbitrary relative degree sliding variables [11,12].

This paper is a brief introduction to the application of a reference model in reaching law based DSMC. The design process of a model reference sliding mode controller is outlined and its key advantages are highlighted. In particular, it is demonstrated that the sliding motion of the system is only affected by disturbance from the single most recent discrete step.

2 Sliding Mode Control of a Discrete-Time System

In this paper we consider a class of discrete-time plants subject to non-matched uncertainties. In particular, in this section we will illustrate the design process of a reaching law based sliding mode control strategy for such systems. Dynamics of the considered plants are expressed in the state space as

$$\boldsymbol{x}_p(k+1) = \boldsymbol{A}\boldsymbol{x}_p(k) + \boldsymbol{b}u_p(k) + \boldsymbol{f}(k), \tag{1}$$

where $\boldsymbol{x}_p \in \mathbb{R}^n$ is the state vector, $u_p \in \mathbb{R}$ is the control signal, $\boldsymbol{f} \in \mathbb{R}^n$ represents uncertainties (that do not have to satisfy the matching conditions) and $\boldsymbol{A}, \boldsymbol{b}$ are of appropriate dimensions. The "p" subscript in this equation refers to the original plant as opposed to its model that will be defined later in this paper.

2.1 Sliding Hyperplane Design

The design process of a sliding mode controller begins with the selection of a particular sliding variable and a sliding hyperplane on which this variable is equal to 0. If the objective of the control process is to drive the system state to a static reference point $\boldsymbol{x}_d \in \mathbb{R}^n$, then the sliding variable and its corresponding sliding hyperplane are typically defined as

$$\sigma_p(k) = \boldsymbol{c}^{\mathrm{T}}\boldsymbol{e}_p(k) = \boldsymbol{c}^{\mathrm{T}}\boldsymbol{x}_d - \boldsymbol{c}^{\mathrm{T}}\boldsymbol{x}_p(k) = 0. \tag{2}$$

Vector $\boldsymbol{c} \in \mathbb{R}^n$ in this equation is selected so that $\boldsymbol{c}^{\mathrm{T}}\boldsymbol{b} \neq 0$. Before any sliding mode control strategy can be successfully designed for system (1), one must first assume that the total effect of uncertainties on the selected sliding variable is limited. Typically, it is assumed that for all time instants k

$$f_{\min} \leq \mathbf{c}^{\mathrm{T}}\boldsymbol{f}(k) \leq f_{\max}, \tag{3}$$

where $f_{\min}, f_{\max} \in \mathbb{R}$ are certain known constants.

2.2 Reaching Law Approach

Conventionally, sliding mode controller design procedure involves stating a control signal that ensures a variable structure of the system and demonstrating that the selected signal ensures stability of the sliding motion. However, in our paper a more recent reaching law approach will be considered. In this approach the desired evolution of the sliding variable is first stated as a recursive function

$$\sigma_p(k+1) = \varphi[\sigma_p(k)] + \mathbf{c}^{\mathrm{T}}\boldsymbol{f}(k), \tag{4}$$

where $\sigma_p(0) = \mathbf{c}^{\mathrm{T}}\boldsymbol{e}(0)$ and $\varphi : \mathbb{R} \to \mathbb{R}$ is selected to guarantee a finite time convergence to a specified vicinity of zero and confine the variable to this vicinity. Thus, φ is typically a monotonical function that ensures at least asymptotic convergence to zero for any initial state. As a result, the representative point of the system is driven to a specific area in the state space

$$\{\boldsymbol{x}_p : |\mathbf{c}^{\mathrm{T}}\boldsymbol{x}_d - \mathbf{c}^{\mathrm{T}}\boldsymbol{x}_p| \leq B\} \tag{5}$$

referred to as a quasi-sliding mode band, where B is the width of this band.

In order to obtain the control signal which ensures the desired evolution specified by the reaching law, one can simply substitute known system dynamics (1) into the left hand side of (4) and solve this equation for $u_p(k)$, which yields

$$u_p(k) = \left(\mathbf{c}^{\mathrm{T}}\boldsymbol{b}\right)^{-1}\left\{\mathbf{c}^{\mathrm{T}}\boldsymbol{x}_d - \mathbf{c}^{\mathrm{T}}\boldsymbol{A}\boldsymbol{x}_p(k) - \varphi[\sigma_p(k)]\right\}. \tag{6}$$

It is important to notice that this control signal only requires information about the system state $\boldsymbol{x}_p(k)$ which can be obtained at time k. Unlike reaching law (4), it does not depend on the unknown disturbance term $\boldsymbol{f}(k)$. However, since this disturbance is present in the recursive function (1), target value of the sliding variable specified by function φ is altered in each individual step. Thus, the desired trajectory of this variable can gradually become distorted during the control process. In order to prevent this phenomenon, one can apply a model reference approach that will be demonstrated in this paper.

3 Sliding Mode Control with a Reference Model

Typically, a reaching law based control strategy is designed directly for the original plant (1), as seen in (4)–(6). However, a more advanced approach using the reference model of the plant can be considered in order to improve sliding mode performance at all stages of the control process. We first consider the model

$$\boldsymbol{x}_m(k+1) = \boldsymbol{A}\boldsymbol{x}_m(k) + \boldsymbol{b}u_m(k), \tag{7}$$

where $\boldsymbol{A}, \boldsymbol{b}$ are the same as in (1) and $\boldsymbol{x}_m(0) = \boldsymbol{x}_p(0)$. The design process of a model reference sliding mode control strategy involves two steps. First, a reaching law for the disturbance-free model (7) is selected with the aim of ensuring favorable properties of its sliding motion. Then, a simple dead-beat controller is designed for the original plant in order to drive its output alongside the trajectory laid out by the model. It will be demonstrated that this approach allows one to eliminate the residual effect of uncertainties on the system and improve its dynamic performance as a result.

3.1 Reaching Law for the Reference Model

Just like in the general case outlined in Sects. 2.1 and 2.2, the design process of a reaching law based strategy for model (7) begins with the selection of an appropriate sliding variable

$$\sigma_m(k) = \boldsymbol{c}^{\mathrm{T}}\boldsymbol{e}_m(k) = \boldsymbol{c}^{\mathrm{T}}\boldsymbol{x}_d - \boldsymbol{c}^{\mathrm{T}}\boldsymbol{x}_m(k) = 0, \tag{8}$$

where $\boldsymbol{e}_m$ is the state error vector and $\boldsymbol{c} \in \mathbb{R}^n$ is selected to ensure $\boldsymbol{c}^{\mathrm{T}}\boldsymbol{b} \neq 0$. Then, a reaching law which defines the desired sliding motion of the system is stated in the form of a recursive formula

$$\sigma_m(k+1) = \psi[\sigma_m(k)]. \tag{9}$$

It can be seen that the reaching law (9) for the reference model does not contain any uncertain terms, which means that the sliding variable will follow the trajectory outlined by function $\psi : \mathbb{R} \rightarrow \mathbb{R}$ exactly. Similarly to relation (6), the reaching law for the reference model can be applied to design the control signal

$$u_m(k) = \left(\boldsymbol{c}^{\mathrm{T}}\boldsymbol{b}\right)^{-1}\left\{\boldsymbol{c}^{\mathrm{T}}\boldsymbol{x}_d - \boldsymbol{c}^{\mathrm{T}}\boldsymbol{A}\boldsymbol{x}_m(k) - \psi[\sigma_m(k)]\right\}. \tag{10}$$

This control signal is purely virtual and is only used to ensure a desired response of model (7). Furthermore, since this control signal is deterministic for all time instants (as it only depends on known model parameters and its initial conditions), desired model dynamics $\boldsymbol{x}_m$ can be calculated in advance and stored in a look-up table. In the next step, the control signal for the original plant (1) will be developed taking these model dynamics into account.

3.2 Model Reference Control Strategy

The objective of the considered strategy is to drive the state of the plant towards that of the model in each step. To that end, we consider the reaching law

$$\sigma_p(k+1) = \sigma_m(k+1) + \boldsymbol{c}^{\mathrm{T}}\boldsymbol{f}(k) = \boldsymbol{c}^{\mathrm{T}}\boldsymbol{x}_d - \boldsymbol{c}^{\mathrm{T}}\boldsymbol{x}_m(k+1) + \boldsymbol{c}^{\mathrm{T}}\boldsymbol{f}(k), \tag{11}$$

where $\boldsymbol{x}_m$ is the state of the model obtained in the previous section. Then, substitution of known system dynamics (1) into reaching law (11) allows one to obtain the following control signal for the original plant

$$u_p(k) = \left(\boldsymbol{c}^{\mathrm{T}}\boldsymbol{b}\right)^{-1}\left[\boldsymbol{c}^{\mathrm{T}}\boldsymbol{x}_m(k+1) - \boldsymbol{c}^{\mathrm{T}}\boldsymbol{A}\boldsymbol{x}_p(k)\right]. \tag{12}$$

It can be demonstrated that this control signal ensures advantageous properties of system sliding motion, which are not necessarily guaranteed by the conventional reaching law approach. It can indeed be seen that, unlike the conventional reaching law (4), the proposed approach does not use a recursive function of sliding variable σ_p, but rather drives this variable to a specific reference point in each step. As a result, the desired trajectory of the sliding variable is not distorted by residual effect of uncertainties on the sliding motion of the system. This property is summarized in the following theorem.

Theorem 1. *If the control signal for system* (1) *is described by* (12) *where* $\boldsymbol{x}_m$ *is the state of reference model* (7), *then for all* $k > 0$

$$|\sigma_p(k) - \sigma_m(k)| \leq \max\left\{|f_{\min}|, |f_{\max}|\right\}, \tag{13}$$

where $f_{\min}$ *and* $f_{\max}$ *are defined in* (3).

Proof. Since reference model (7) is devoid of uncertainties, control signal (12) ensures that sliding variable will evolve exactly according to (9) in each step. Thus, reaching law (11) implies

$$\sigma_p(k+1) - \sigma_m(k+1) = \boldsymbol{c}^{\mathrm{T}}\boldsymbol{f}(k). \tag{14}$$

Then, since the effect of the disturbance on the sliding variable is limited as stated in (3), then for all k one gets

$$f_{\min} \leq \sigma_p(k+1) - \sigma_m(k+1) \leq f_{\max}, \tag{15}$$

which directly implies inequality (13). □

4 Discussion

Conventional reaching law approach outlined in relations (4)-(6) provides an easy way of designing a control strategy that ensures a desired dynamic response of the system. However, since reaching law (4) is subject to unpredictable disturbance in each step, in practical applications this desired trajectory is gradually distorted. This phenomenon can alter the length of the reaching phase or negatively affect the quasi-sliding mode band width.

In order to eliminate the residual effect of uncertainties on the sliding motion of the system, one can apply the model reference strategy described in Sect. 3 of this paper. By first designing reaching law (9) for the deterministic model of the plant and then driving the state of the actual plant towards that of the model, one can ensure that past uncertainties do not drive the state of the plant too far from the reference point. Indeed, Theorem 1 demonstrates that the distance between the system representative point and its target value is only affected by disturbance from the single most recent time instant. As a result, the system will always closely follow the desired trajectory obtained from the reference model.

In formerly published literature, it has been demonstrated that model reference sliding mode control strategies drive the system representative to a narrower quasi-sliding mode band than similar control schemes without the reference model [3–5]. This is an important property, since quasi-sliding mode band width reflects the robustness of the whole system with respect to uncertainties.

Another important advantage of the presented model reference scheme is its ability to bypass the restrictive assumption about matched uncertainties. Indeed, as shown in literature [11, 12] as well as in the analysis conducted in this paper, model reference approach is applicable to any plant (1) even if the disturbance $\boldsymbol{f}$ does not satisfy matching conditions. This is particularly relevant for strategies using arbitrary relative degree sliding variables as they conventionally require matching conditions to ensure good performance of the system.

References

1. Bandyopadhyay, B., Fulwani, D.: High-performance tracking controller for discrete plant using nonlinear sliding surface. IEEE Trans. Ind. Electron. **56**(9), 3628–3637 (2009)
2. Bartoszewicz, A.: Discrete time quasi-sliding mode control strategies. IEEE Trans. Ind. Electron. **45**(4), 633–637 (1998)
3. Bartoszewicz, A., Adamiak, K.: Model reference discrete-time variable structure control. Int. J. Adapt. Control Signal Process. **32**(10), 1440–1452 (2018)
4. Bartoszewicz, A., Adamiak, K.: Reference trajectory based discrete time sliding mode control strategy. Int. J. Appl. Math. Comput. Sci. **29**(3), 87–97 (2019)
5. Bartoszewicz, A., Adamiak, K.: Discrete-time sliding-mode control with a desired switching variable generator. IEEE Trans. Autom. Control **65**(4), 1807–1814 (2020)
6. Bartoszewicz, A., Leśniewski, P.: New switching and nonswitching type reaching laws for SMC of discrete time systems. IEEE Trans. Control Syst. Technol. **24**(2), 670–677 (2016)
7. Chakrabarty, S., Bartoszewicz, A.: Improved robustness and performance of discrete time sliding mode control systems. ISA Trans. **65**, 143–149 (2016)
8. Gao, W., Wang, Y., Homaifa, A.: Discrete-time variable structure control systems. IEEE Trans. Ind. Electron. **42**(2), 117–122 (1995)
9. Golo, G., Milosavljević, Č.: Robust discrete-time chattering free sliding mode control. Syst. Control Lett. **41**(1), 19–28 (2000)
10. Janardhanan, S., Bandyopadhyay, B.: Multirate output feedback based robust quasi-sliding mode control of discrete-time systems. IEEE Trans. Autom. Control **52**(3), 499–503 (2007)
11. Latosinski, P., Bartoszewicz, A.: Model reference DSMC with a relative degree two switching variable. IEEE Trans. Autom. Control **66**(4), 1749–1755 (2021)
12. Latosiński, P., Bartoszewicz, A.: Model reference strategy for mismatched disturbance attenuation in relative degree one and two DSMC. Mech. Syst. Signal Process. **149**, 107188 (2021). https://doi.org/10.1016/j.ymssp.2020.107188

13. Leśniewski, P., Bartoszewicz, A.: Optimal model reference sliding mode control of perishable inventory systems. IEEE Trans. Autom. Sci. Eng. **17**(3), 1647–1656 (2020)
14. Mija, S., Susy, T.: Reaching law based sliding mode control for discrete MIMO systems. In: IEEE International Conference on Control, Automation, Robotics & Vision, pp. 1291–1296 (2010)
15. Milosavljević, Č.: General conditions for the existence of a quasisliding mode on the switching hyperplane in discrete variable structure systems. Autom. Remote Control **46**(3), 307–314 (1985)
16. Milosavljević, Č., Peruničić-Draženović, B., Veselić, B., Mitić, D.: Sampled data quasi-sliding mode control strategies. In: IEEE International Conference on Industrial Technology, pp. 2640–2645 (2006)
17. Niu, Y., Ho, D.W.C., Wang, Z.: Improved sliding mode control for discrete-time systems via reaching law. IET Control Theory Appl. **4**(11), 2245–2251 (2010)
18. Utkin, V., Drakunov, S.V.: On discrete-time sliding mode control. In: IFAC Conference on Nonlinear Control, pp. 484–489 (1989)

Sliding Mode Control with Minimization of ITAE and the Input Signal Limitation

Mateusz Pietrala(✉)

Institute of Automatic Control, Łodź University of Technology,
18/22 Stefanowskiego Street, 90-924 Łodź, Poland
mateusz.pietrala@dokt.p.lodz.pl

Abstract. In this paper a second order, continuous-time system perturbed by external disturbances is presented. The robustness for the whole control process is achieved by applying a time-varying switching line. Moreover, a finite time convergence of the representative point to the demand state is ensured by selecting a nonlinear sliding curve. An absolute value of the control signal is limited from above by a positive parameter. ITAE quality index is applied in order to evaluate the dynamical performance of the system.

Keywords: Sliding mode control · Continuous-time systems · Control signal constraint · Integral Time Absolute Error

1 Introduction

Sliding mode control is an effective regulation strategy. Main advantages of this methods are: robustness to external disturbances and modeling uncertainties and computational efficiency [1–8]. First mentions about sliding mode control can be found in Russian literature. These authors focused mostly on continuous-time systems. In the next years, discrete-time systems became a more interesting area of research due to a wider range of applications. Primary goal of the sliding mode control is to drive a so-called representative point (state vector) to a predefined sliding hyperplane (during the reaching phase) and after that to maintain it in a certain neighborhood of that hyperplane (during the sliding phase). In a consequence a state converges to a demand state. An important issue is to select parameters of a switching manifold in order to achieve a good dynamical performance of our system and a stable sliding mode. It can be ensured by optimal control or pole placement method. Another important aspect is to show that the regulation error decays to zero. If we take into account only a model of the controller, which description is known, then we are able to achieve a perfect trajectory in which the representative point moves exactly on the switching hyperplane. However, in real time systems our object is exposed to some external disturbances such as for example weather conditions. Therefore, the next goal is to dynamically compensate these perturbations. Sliding mode control guarantees

R. Szewczyk et al. (Eds.): AUTOMATION 2021, AISC 1390, pp. 75–84, 2021.
https://doi.org/10.1007/978-3-030-74893-7_8

a robustness to the external disturbances. However, this property is only achieved during the sliding phase. Therefore, in order to obtain the robustness for the whole control process, time-varying sliding hyperplanes can be introduced [9–15]. Moreover, in order to achieve a finite-time convergence of the representative point to the demand state, that hyperplane has to be non-linear [16–19]. It is selected in such a way, that it crosses the initial point and the demand state. After that it starts moving and stops at some time. A dynamical performance of the system can be evaluated using one of a well-known quality indexes such as: regulation time, Integral Absolute Error (IAE), Integral Time Absolute Error (ITAE), Integral Square Error (ISE) etc. From a practical point of view, it is important to take into account some signal constraints such as input or state variables.

This paper is organized as follows. Section 2 comprises sliding mode controller design, selection of the switching curve and an introduction of a stable control signal. In Sect. 3 state equations were derived. Furthermore, the admissible set of the sliding line parameters, which causes that the absolute value of the control signal is constrained, was presented. ITAE quality index was minimized considering two cases: in the first one the sliding line was moving, after that it stopped and remained fixed until the end of the control process and in the second case the sliding line was non-stationary for the whole regulation process. We selected the most beneficial strategy. Section 4 contains a simulation example and Sect. 5 concludes the whole paper.

2 Sliding Mode Controller Design

In the article the following second order, continuous-time system

$$\begin{cases} \frac{d}{dt}x_1(t) = x_2(t) \\ \frac{d}{dt}x_2(t) = f(x_1(t), x_2(t), t) + d(t) + bu(t) \end{cases} \tag{1}$$

is taken into account. Variables $x_1(t)$ and $x_2(t)$ are system's state and they can be considered as system's position and velocity, respectively. The initial position of the representative point is $x_1(0) \neq 0$ and $x_2(0) = 0$. Function f is dependent on the position of the system, its velocity and time. External disturbances are denoted by d. These perturbations are unknown, however their absolute value is bounded from above by a positive parameter. The absolute value of the sum of modeling uncertainties and external disturbances is bounded from above by a positive value denoted by M. Control signal u is connected with a scalar b. As was mentioned before, in Sliding mode control robustness can only be obtained during the sliding phase. Therefore, the main goal of this paper is to eliminate the reaching phase. Furthermore, we would like to achieve a finite-time convergence of the representative point to the demand state $(0,0)$. In order to meet our objectives we propose a time-varying nonlinear sliding curve $s(t) = 0$, where

$$s(t) = x_2(t) + c(t)\,\text{sgn}[x_1(t)]\sqrt{|x_1(t)|}. \tag{2}$$

Function c is connected with the movement of our sliding line. It is given as follows

$$c(t) = \begin{cases} \alpha t & \text{for} \quad t \le t_0 \\ \alpha t_0 & \text{for} \quad t > t_0 \end{cases}. \tag{3}$$

The velocity of the evolution of the sliding curve is described by the parameter $\alpha > 0$. Parameter t_0 is the moment in which the switching line stops and after that it remains fixed until the end of the control process. In the paper two strategies will be considered: when the sliding line stops during the control process and when the sliding line moves for the whole control process. One can observe, that the sliding variable s is equal to zero, when $t = 0$. This results in the elimination of the reaching phase and the robustness to disturbances for the whole control process. We propose the following control signal

$$u(t) = -\frac{1}{b}\left\{ \operatorname{sgn}[x_1(t)]\sqrt{|x_1(t)|}\frac{d}{dt}c(t) + \frac{x_2(t)\,c(t)}{2\sqrt{|x_1(t)|}} + M\operatorname{sgn}[s(t)] \right\} \tag{4}$$

that ensures the stability of the sliding mode.

3 ITAE Minimization with the Control Signal Constraint

This section will begin with deriving the admissible set of the sliding line parameters α and t_0, that causes that the absolute value of the control signal u is constrained by a positive value $u_{\max}$ for the whole regulation process, i.e.

$$|u(t)| \le u_{\max} \tag{5}$$

for any $t \ge 0$. In order to calculate the input we have to derive formulas for the system's velocity and a square root of the system's position. The second state variable can be computed from the fact that the reaching phase is eliminated and the control signal ensures the stable sliding motion. Therefore

$$x_2(t) = -c(t)\operatorname{sgn}[x_1(t)]\sqrt{|x_1(t)|}. \tag{6}$$

If we substitute the above equation to (1) we obtain a Bernoulli differential equation. Solving this problem one can get that the square root of the absolute value of the first state variable is

$$\sqrt{|x_1(t)|} = \begin{cases} C_1 - \frac{1}{4}\alpha t^2 & \text{for} \quad t < t_0 \\ C_2 - \frac{1}{2}\alpha t_0 t & \text{for} \quad t_0 \le t < t_f \\ 0 & \text{for} \quad t \ge t_f \end{cases}, \tag{7}$$

where $C_1 > 0$ and $C_2 > 0$. From the fact that (7) is a continuous function we get the final form of the system's position as follows

$$|x_1(t)| = \begin{cases} \left(\sqrt{|x_1(0)|} - \frac{1}{4}\alpha t^2\right)^2 & \text{for} \quad t < t_0 \\ \left(\sqrt{|x_1(0)|} + \frac{1}{4}\alpha t_0^2 - \frac{1}{2}\alpha t_0 t\right)^2 & \text{for} \quad t_0 \le t < t_f \\ 0 & \text{for} \quad t \ge t_f \end{cases}. \tag{8}$$

From the same fact, we can get a formula for the settling time t_f by substituting this parameter to (8) obtaining

$$t_f = \frac{1}{2}t_0 + \frac{2\sqrt{|x_1(0)|}}{\alpha t_0}. \tag{9}$$

Above equations describe a situation when the sliding line moves from the very beginning of the control process, after that it stops and remains fixed to the end of a control process. When the sliding line moves and does not stop, then the absolute value of the first state variable can be written as follows

$$|x_1(t)| = \begin{cases} \left(\sqrt{|x_1(0)|} - \frac{1}{4}\alpha t^2\right)^2 & \text{for} \quad t \leq t_f \\ 0 & \text{for} \quad t > t_f \end{cases} \tag{10}$$

and the settling time is given by

$$t_f = \frac{2\sqrt[4]{|x_1(0)|}}{\sqrt{\alpha}}. \tag{11}$$

In the next step the admissible set of the sliding line parameters α and t_0 will be derived. Then, the optimal values will be computed and the quality index ITAE will be minimized.

3.1 Admissible Set

In this subsection we will focus on the parameters of the switching curve α and t_0 which cause that the absolute value of the control signal is constrained from above by $u_{\max}$. First, let us focus on the case, when the sliding line stops at some moment and after that it remains fixed. Using (4), (6), (8) and the fact that the absolute value is subadditive and multiplicative we obtain that in order to fulfill (5) the following inequality

$$t_0 \leq \frac{\sqrt{2(|b|\, u_{\max} - M)}}{\alpha} \tag{12}$$

has to be true. However, one can observe that this inequality depends on the value of parameter α. Deriving admissible values of α we obtain

$$\alpha \leq \frac{|b|\, u_{\max} - M}{\sqrt{|x_1(0)|}}. \tag{13}$$

The admissible set is illustrated in the Fig. 1. In the second case, when the sliding line moves for the whole regulation process we obtain that α has to fulfill the following inequality

$$\alpha \leq \frac{|b|\, u_{\max} - M}{2\sqrt{|x_1(0)|}}. \tag{14}$$

One can observe, that (14) is more restrictive than (13).

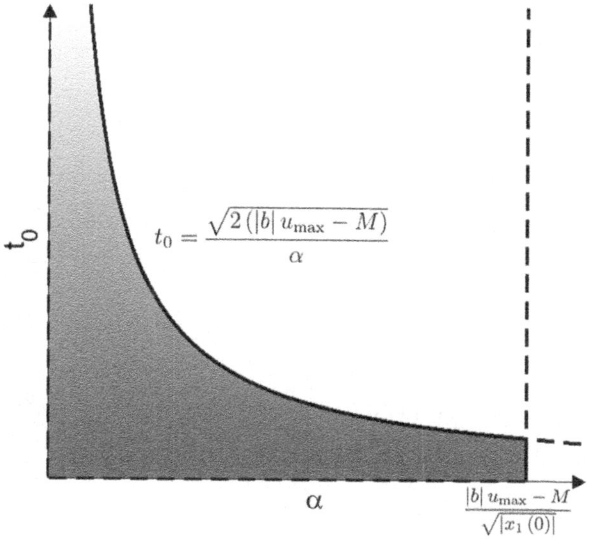

Fig. 1. Admissible set

3.2 ITAE Minimization

In this paper ITAE quality index is selected and minimized in order to evaluate the dynamical performance of the system. It is given by the following equation

$$E = \int_0^{\infty} |x_1(t)|\, t\, dt, \tag{15}$$

where E denotes ITAE. Using the fact that our controller causes that the representative point reaches the demand state in the finite time we can rewrite the above equation as follows

$$E = \int_0^{t_f} |x_1(t)|\, t\, dt. \tag{16}$$

Again, we will consider two cases: in the first one the sliding line will be moving from the very beginning of the control process. After that it will stop and remain fixed. In the second case the demand state will be reached along the moving switching curve.

1. $t_0 < t_f$. Using formula (8) we can rewrite integral (16) in the following manner

$$E = \int_0^{t_0} \left(\sqrt{|x_1(0)|} - \frac{\alpha t^2}{4}\right)^2 t\, dt + \int_{t_0}^{t_f} \left(\sqrt{|x_1(0)|} + \frac{\alpha t_0^2}{4} - \frac{\alpha t_0 t}{2}\right)^2 t\, dt. \tag{17}$$

Both components of the above sum will be computed separately. When the sliding line moves, then

$$E_1 = \int_0^{t_0} \left(\sqrt{|x_1(0)|} - \frac{\alpha t^2}{4} \right)^2 t\, dt$$
$$= |x_1(0)| \frac{t_0^2}{2} - \frac{\alpha t_0^4}{8} \sqrt{|x_1(0)|} + \frac{\alpha^2 t_0^6}{96} \tag{18}$$

and when it stops, then

$$E_2 = \int_{t_0}^{t_f} \left(\sqrt{|x_1(0)|} + \frac{\alpha t_0^2}{4} - \frac{\alpha t_0 t}{2} \right)^2 t\, dt$$
$$= -\frac{3}{8} |x_1(0)| t_0^2 + \frac{5}{48} \alpha t_0^4 \sqrt{|x_1(0)|} - \frac{7}{768} \alpha^2 t_0^6$$
$$+ \frac{\sqrt{|x_1(0)|^3}}{3\alpha} + \frac{|x_1(0)|^2}{3\alpha^2 t_0^2}. \tag{19}$$

Summing up (18) and (19) we present the final formula for ITAE as follows

$$E = \frac{1}{8} |x_1(0)| t_0^2 - \frac{1}{48} \alpha t_0^4 \sqrt{|x_1(0)|} + \frac{1}{768} \alpha^2 t_0^6$$
$$+ \frac{\sqrt{|x_1(0)|^3}}{3\alpha} + \frac{|x_1(0)|^2}{3\alpha^2 t_0^2}. \tag{20}$$

In the next step we will determine optimal parameters α and t_0, for which ITAE is minimized and the absolute value of the control signal is bounded from above by $u_{\max}$. Calculating partial derivatives of (20) with respect to α and t_0 we obtain:

$$\frac{\partial E}{\partial \alpha} = -\frac{1}{48} t_0^4 \sqrt{|x_1(0)|} + \frac{1}{384} \alpha t_0^6 - \frac{\sqrt{|x_1(0)|^3}}{3\alpha^2} - \frac{2|x_1(0)|^2}{3\alpha^3 t_0^2}, \tag{21}$$

$$\frac{\partial E}{\partial t_0} = \frac{1}{4} |x_1(0)| t_0 - \frac{1}{12} \alpha t_0^3 \sqrt{|x_1(0)|} + \frac{1}{128} \alpha^2 t_0^5 - \frac{2|x_1(0)|^2}{3\alpha^2 t_0^3}. \tag{22}$$

Equating (22) to zero we can determine α as a real function of t_0 as follows

$$\alpha = -\frac{4\sqrt{|x_1(0)|}}{3t_0^2} \quad \vee \quad \alpha = \frac{4\sqrt{|x_1(0)|}}{t_0^2}. \tag{23}$$

From the fact that α has to be positive, one can get that the only admissible α is

$$\alpha = \frac{4\sqrt{|x_1(0)|}}{t_0^2}. \tag{24}$$

Substituting (24) to (21) and equating it to zero we get

$$-\frac{1}{24} t_0^4 \sqrt{|x_1(0)|} = 0. \tag{25}$$

From the fact that t_0 and $\sqrt{|x_1(0)|}$ are positive we conclude that ITAE does not have any stationary points and its minimum is obtained on the boundary of the admissible set.

2. $t_c \geq t_f$.
In this case substituting (10) into (16) one can obtain

$$E = \int_0^{t_f} \left(\sqrt{|x_1(0)|} - \frac{\alpha t^2}{4} \right)^2 t\, dt$$
$$= |x_1(0)| \frac{t_f^2}{2} - \frac{\alpha t_f^4}{8} \sqrt{|x_1(0)|} + \frac{\alpha^2 t_f^6}{96}. \tag{26}$$

We substitute regulation time (11) obtaining the final form of ITAE as follows

$$E = \frac{2\sqrt{|x_1(0)|^3}}{3\alpha}. \tag{27}$$

Calculating partial derivative with respect to α one can get

$$\frac{\partial E}{\partial \alpha} = -\frac{2\sqrt{|x_1(0)|^3}}{3\alpha^2}. \tag{28}$$

The absolute value and parameter α are always positive. Therefore, the derivative of ITAE with respect to α is always negative. In consequence, ITAE will be minimized on a boundary of the admissible set.

Theorem 1. *If the absolute value of the control signal is bounded from above by $u_{\max}$ for any $t \geq 0$, then the minimized ITAE is*

$$E = \frac{65\,|x_1(0)|^2}{96\,(|b|\, u_{\max} - M)} \tag{29}$$

and is obtained for

$$\begin{cases} \alpha = \frac{|b|u_{\max} - M}{\sqrt{|x_1(0)|}} \\ t_0 = \sqrt{\frac{2|x_1(0)|}{|b|u_{\max} - M}} \end{cases}. \tag{30}$$

The best strategy is the one in which the switching line moves from the very beginning of the control process, after that it stops and remains fixed.

Proof. In this section it was shown that ITAE is minimized on the boundary of the admissible set. This proof will begin by analyzing the strategy in which the sliding curve stops at some moment. In this case we focus on the lines $\alpha = \frac{|b|u_{\max} - M}{\sqrt{|x_1(0)|}}$ and $t_0 = \frac{\sqrt{2(|b|u_{\max} - M)}}{\alpha}$. On the first line the minimum of ITAE is obtained for the maximum value of t_0 which is

$$t_0 = \sqrt{\frac{2\,|x_1(0)|}{|b|\, u_{\max} - M}} \tag{31}$$

Now we will focus on calculating optimal parameters on the line $t_0 = \frac{\sqrt{2(|b|u_{\max}-M)}}{\alpha}$. Substituting this value to (20) we get

$$E = \frac{|x_1(0)|(|b|u_{\max} - M)}{4\alpha^2} - \frac{\sqrt{|x_1(0)|}(|b|u_{\max} - M)^2}{12\alpha^3} + \frac{(|b|u_{\max} - M)^3}{96\alpha^4} + \frac{\sqrt{|x_1(0)|^3}}{3\alpha} + \frac{|x_1(0)|^2}{6(|b|u_{\max} - M)}. \tag{32}$$

Calculating the derivative of the right-hand side of the above equation and equating it to zero, we obtain that the only real possible stationary point is

$$\alpha = -\frac{(1 + \sqrt[3]{2} + \sqrt[3]{4})(|b|u_{\max} - M)}{2\sqrt{|x_1(0)|}}. \tag{33}$$

Taking into account the fact that $|b|u_{\max} - M$ and $|x_1(0)|$ are positive, we conclude that (33) is negative. Therefore, minimum of the ITAE will be obtained on the boundary of the admissible set, i.e. for α and t_0 given as (30) and

$$E = \frac{65|x_1(0)|^2}{96(|b|u_{\max} - M)}. \tag{34}$$

In the second case, when the switching curve moves for the whole regulation process, then ITAE is equal to the right hand side of (27). One can observe, that ITAE is minimized for the maximum admissible value of α, i.e. for

$$\alpha = \frac{|b|u_{\max} - M}{2\sqrt{|x_1(0)|}}. \tag{35}$$

Substituting (35) to (27) we have

$$E = \frac{4|x_1(0)|^2}{3(|b|u_{\max} - M)}. \tag{36}$$

Value (34) is smaller than (36). Hence, ITAE in the presence of the control signal constraint is minimized, when the demand state is reached along fixed sliding line, which ends the proof.

4 Simulation Example

In this section theoretical considerations will be verified by a simulation example. System's parameters are selected as follows: f is a sine function connected with the first state variable (system's position), function d varies between -2 and 2 twenty times per second. Therefore, the maximum possible absolute value of the sum of f and d is always smaller than or equal to $M = 3$. Parameter b connected with a control signal is equal to 2. The maximum admissible absolute value of the control signal is $u_{\max} = 8$. The system's position at the time 0 is equal to 50. We can calculate optimal parameters of the sliding line obtaining $t_0 \approx 2.7735$ and $\alpha \approx 1.8385$. Settling time is $t_f \approx 4.1603$. Minimized ITAE is given as $E \approx 130.2083$. In the Fig. 2 one can observe a state trajectory. Position of our system decreases monotonically. After time t_0 the switching curve stops and the demand state is reached along the fixed sliding line. The finite-time convergence is achieved. Figure 3 presents the control signal. Starting from the minimal possible value, its average value increases monotonically until the time t_0. Further, it switches, but doesn't exceed its maximum admissible value. After time t_f it varies in order to maintain the representative point exactly on the demand state.

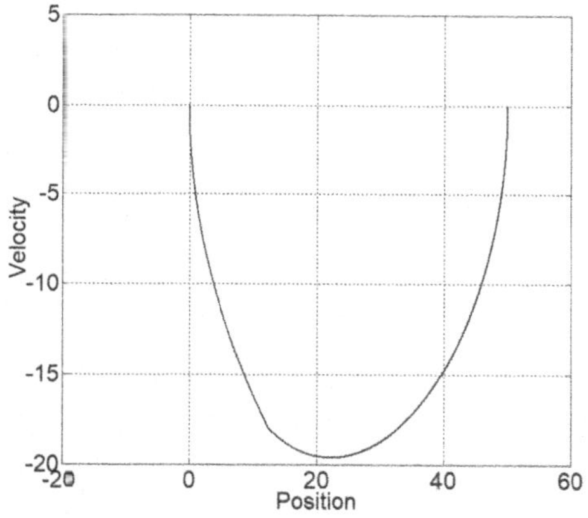

Fig. 2. State trajectory

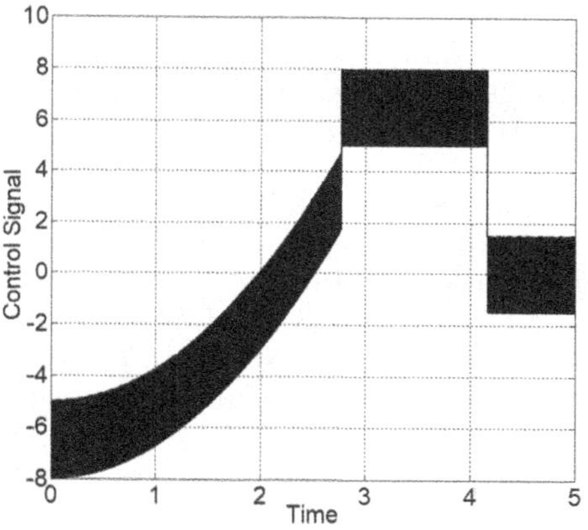

Fig. 3. Control signal

5 Conclusions

In the paper a continuous-time system perturbed by external disturbances was considered. Sliding mode control strategy was applied to obtain a robustness to these perturbations. In order to achieve this goal for the whole regulation process, a non-linear time-varying switching curve was introduced. The absolute value of the control signal was bounded from above by a known, positive parameter. The admissible set of the sliding line parameters was derived. Two strategies in which the representative point reaches the demand state before the switching curve stops and after that this convergence is achieved, were analyzed. ITAE quality index was selected in order to evaluate the dynamical performance of the system. Simulation example verified theoretical considerations.

References

1. Bartoszewicz, A., Kaynak, O., Utkin, V.I. (eds.): Sliding mode control in industrial applications. Special section. IEEE Trans. Ind. Electron. **55**, 3805–4103 (2008)

2. Bartoszewicz, A., Patton, R., (eds.): Sliding mode control. Special Issue. Int. J. Adapt. Control Signal Process. **21**, 635–822 (2007)
3. Basin, M., Fridman, L., Shi, P. (eds.): Optimal sliding mode algorithms for dynamic systems. Special Issue: J. Franklin Inst. **349**, 1317–1616 (2012)
4. Gao, W. (ed): Variable structure control. Special Issue: IEEE Trans. Ind. Electron. **40**, 1–165 (1993)
5. Kaynak, O., Bartoszewicz, A., Utkin, V.I. (eds.): Sliding mode control in industrial applications. Special Issue: IEEE Trans. Ind. Electron. **56**, 3271–3274 (2009)
6. Misawa, A., Utkin, V.I. (eds.): Variable structure systems. Special Issue: Trans. ASME - Jour. Dynamic Syst. Meas. Control **122**, 585–819 (2000)
7. Shtessel, Y., Spurgeon, S., Fridman, L.: Advances in sliding mode observation and estimation. Special Issue: Int. J. Syst. Sci. **38**, 845–942 (2007)
8. Utkin, V.I. (ed.): Sliding mode control. Special Issue: Int. J. Control **57**, 1003–1259 (1993)
9. Choi, S.B., Park, D.W., Jayasuriya, S.: A time-varying sliding surface for fast and robust tracking control of second-order uncertain systems. Automatica **30**(5), 899–904 (1994)
10. Bartoszewicz, A.: A Comment on a time-varying sliding surface for fast and robust tracking control of second-order uncertain systems. Automatica **31**(12), 1893–1895 (1995)
11. Yilmaz, C., Hurmuzlu, Y.: Eliminating the reaching phase from variable structure control. J. Dyn. Syst. Meas. Control **122**(4), 753–757 (2000)
12. Bartoszewicz, A., Nowacka, A.: Reaching phase elimination in variable structure control of the third order system with state constraints. Kybernetika **42**(1), 111–126 (2006)
13. Bartoszewicz, A., Nowacka, A.: Sliding mode control of the third-order system subject to velocity, acceleration and input signal constraints. Int. J. Adaptive Control Signal Process. **21**(8–9), 779–794 (2007)
14. Bartoszewicz, A., Nowacka-Leverton, A.: Time-Varying Sliding Modes for Second and Third Order Systems. Springer, Berlin (2009)
15. Bartoszewicz, A., Nowacka-Leverton, A.: ITAE optimal sliding modes for third order systems with input signal and state constraints. IEEE Trans. Autom. Control **55**(8), 1928–1932 (2010)
16. Bartoszewicz, A.: Time-varying sliding modes for second-order systems. Proc. IEE Part D: Control Theory Appl. **143**(5), 455–462 (1996)
17. Man, Z., Yu, X.H.: Terminal sliding mode control of MIMO linear systems. IEEE Trans. Circ. Syst. - I: Fundam. Theory Appl. **44**(11), 1065–1070 (1997)
18. Man, Z., Paplinski, A., Wu, H.R.: A robust MIMO terminal sliding mode control scheme for rigid robotic manipulators. IEEE Trans. Autom. Control **39**(12), 2464–2469 (1994)
19. Tang, Y.: Terminal sliding mode control for rigid robots. Automatica **34**(1), 51–56 (1998)

Fractional Order vs Integer Order Transfer Function Models of a Pendulum

Krzysztof Oprzędkiewicz(✉), Maciej Rosół, and Jakub Żegleń-Włodarczyk

Department of Automatic Control and Robotics, Faculty of Electrotechnics, Automatic Control, Informatics and Biomedical Engineering, AGH University of Science and Technology, al. A Mickiewicza 30, 30-059 Krakow, Poland
{kop,mr}@agh.edu.pl

Abstract. The paper presents the comparison of fractional order to integer order transfer function models of physical pendulum. The considered transfer functions describe the behaviour of the system around the lower, stable equilibrium point. Fractional order models are built using ORA approximation. Models were verified using real experimental data. Results of tests show that the fractional order transfer function can correctly describe the physical pendulum. Integer order models are a little bit more accurate, but they require more parameters to identify.

Keywords: Fractional order systems · Switchable model · Inverted pendulum · Non linear systems · Fractional order transfer function · Physcial pendulum · 1D crane · ORA approximation

1 Introduction

Models and control of an inverted pendulum has been presented by many Authors, for example in papers: [1,4,7]. This object is an unstable system with nonlinear dynamics. It is an example of the under actuated system – there are less inputs than numbers of degrees of freedom [5]. The switchable linear model of a pendulum was presented i.e. in [2].

The paper is organized as follows. At the beginning some elementary ideas from Fractional Calculus are recalled. Next the considered pendulum and its models are given. Furthermore the Integer Order (IO) and Fractional Order (FO) transfer function models of the considered device are proposed. Finally the experimental verification of results is given and discussed.

2 Preliminaries

2.1 Basic Ideas

Presentation of elementary ideas is started with a definition of a fractional order, integro-differential operator given for example by [8]:

R. Szewczyk et al. (Eds.): AUTOMATION 2021, AISC 1390, pp. 85–95, 2021.
https://doi.org/10.1007/978-3-030-74893-7_9

Definition 1. *(The elementary fractional order operator) The fractional order integro-differential operator is defined as follows:*

$$
{}_aD_t^\gamma f(t) = \begin{cases} \frac{d^\gamma f(t)}{dt^\gamma} & \gamma > 0 \\ 1 & \alpha = 0 \\ \int\limits_a^t f(\tau)(d\tau)^{-\gamma} & \gamma < 0 \end{cases} . \tag{1}
$$

where a and t denote time limits of operator calculation, $\gamma \in \mathbb{R}$ denotes the fractional order of the operation.

Next, an idea of Gamma Euler function is recalled (see for example [9], p.5):

Definition 2. *The Gamma function*

$$
\Gamma(x) = \int\limits_0^\infty t^{x-1} e^{-t} dt, \quad Re(x) > 0. \tag{2}
$$

An idea of Mittag-Leffler function needs to be given next. It is a non-integer order generalization of the exponential function $e^{\lambda t}$ and it plays a crucial role in the solution of fractional order state equation. The one parameter Mittag-Leffler function is defined as follows (see for example [9], p. 7):

Definition 3. *The one parameter Mittag-Leffler function*

$$
E_\gamma(x) = \sum_{k=0}^{\infty} \frac{x^k}{\Gamma(k\gamma + 1)}. \tag{3}
$$

The fractional order operator can be described by different definitions, given by: The Grünwald-Letnikov (the GL definition), Riemann-Liouville (RL definition) and Caputo (C definition). In this paper, only the Caputo definition will be used (e.g. [9], p. 11).

Definition 4. *The Caputo definition of the FO operator*

$$
{}_0^C D_t^\gamma f(t) = \frac{1}{\Gamma(N-\gamma)} \int\limits_0^\infty \frac{f^{(N)}(\tau)}{(t-\tau)^{\gamma+1-N}} d\tau. \tag{4}
$$

For the C definition the Laplace transform is defined as follows (e.g. [9], p. 11):

Definition 5. *(The Laplace transform for Caputo operator)*

$$
\begin{aligned}
&\mathcal{L}({}_0^C D_t^\gamma f(t)) = s^\gamma F(s), \quad \gamma < 0 \\
&\mathcal{L}({}_0^C D_t^\gamma f(t)) = s^\gamma F(s) - \sum_{k=0}^{n-1} s^{\gamma-k-1} {}_0D_t^k f(0), \\
&\gamma > 0, \quad n-1 < \gamma \le n \in N
\end{aligned} \tag{5}
$$

The use of C definition (4) and (5) allows to define the fractional order transfer functions, analogical to known integer order transfer functions:

$$G_1(s) = \frac{b_m s^{\beta_m} + ... + b_1 s^{\beta_1} + b_0}{a_n s^{\alpha_n} + ... + a_1 s^{\alpha_1} + a_0}, \tag{6}$$

where $\beta_m ... \beta_1$ and $\alpha_n ... \alpha_1$ are fractional orders, $b_m ... b_0$ and $a_n, ..., a_0$ are coefficients.

2.2 The Oustaloup Recursive Approximation (ORA Approximation)

The method proposed by [11] allows to approximate an elementary FO transfer function s^α by the finite, integer-order transfer function, close to Pade approximation:

$$s^\alpha \approx k_f \prod_{n=1}^{N} \frac{1 + \frac{s}{\mu_n}}{1 + \frac{s}{\nu_n}} = \frac{L_\alpha(s)}{M_\alpha(s)}. \tag{7}$$

In (7) N is the order of the approximation, k_f is the steady state gain, μ_n and ν_n are calculated as follows:

$$\begin{aligned} \mu_1 &= \omega_l \sqrt{\eta} \\ \nu_n &= \mu_n \gamma,\ n = 1, .., N \\ \mu_{n+1} &= \nu_n \eta,\ n = 1, .., N-1 \end{aligned} \tag{8}$$

where:

$$\begin{aligned} \gamma &= \left(\frac{\omega_h}{\omega_l}\right)^{\frac{\alpha}{N}} \\ \eta &= \left(\frac{\omega_h}{\omega_l}\right)^{\frac{1-\alpha}{N}} \end{aligned} \tag{9}$$

In (9) ω_l and ω_h describe the range of angular frequency, for which parameters should be calculated. The steady state gain k_f is set to assure the convergence the step response of the approximation to step response of the real plant in a steady state.

3 The Pendulum and Its Models

The construction of the inverted pendulum was presented in many papers, for example in [6]. The pendulum employed to tests in this paper is described with details in section "Simulations and experiments". Considered pendulum on a cart can move only in one direction on rails (see Fig. 1). This system has an infinite number of equilibrium points, but in the considered case we are interested only in the one stable equilibrium point when the pendulum is in the lower vertical position. The mathematical model of the pendulum is described by determining the kinetic and potential energy in the system. Define the state vector as follows:

$$x = [\theta \quad \dot{\theta} \quad x \quad \dot{x}]^T = [x_1, x_2, x_3, x_4]^T \tag{10}$$

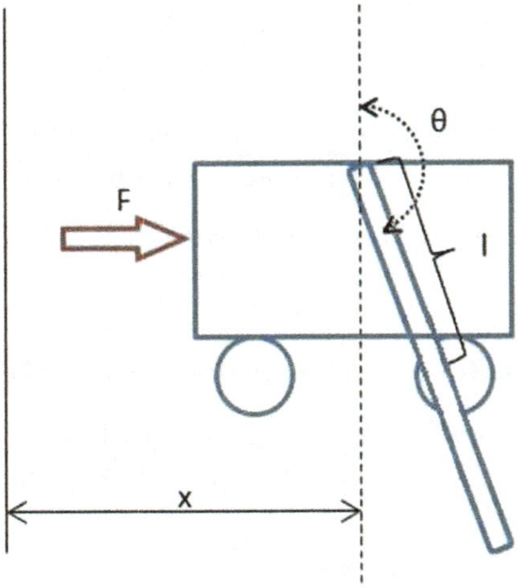

Fig. 1. The pendulum [5].

Then the non-linear state equation (see [6]) takes the following form:

$$\begin{cases} \dot{x}_1 = x_2 \\ \dot{x}_2 = \frac{1}{q}(FmL\cos x_1 - bmL\dot{x}_3\cos x_1 - (M+m)mLg\sin x_1 + m^2\,L^2\dot{x}_1^2\sin x_1\cos x_1 \\ \dot{x}_3 = x_4 \\ \dot{x}_4 = \frac{1}{q}((I+mL^2)(-F+b\dot{x}_3 - mL\dot{x}_1^2\sin x_1) + m^2\,L^2\,g\sin x_1\cos x_2) \end{cases} \tag{11}$$

where:

$$q = m^2L^2\cos^2 x_1 - (M+m)(I+mL^2) \tag{12}$$

In (11) and (12) M denotes mass of the cart, m is the mass of the pendulum, θ is the angular position of the pendulum, g denotes the standard gravity, L is the length of the pendulum, I is moment of inertia of the pendulum, F describes the force applied to the cart and x is cart position.

The above model is non-linear, so it can be linearized around stable equilibrium point $x_0 = [\pi \quad 0 \quad 0 \quad 0]^T$. For values of x_1 close to π it can be assumed that $\sin(\pi + x_1) \approx -x_1$ and finally we obtain the following linear model:

$$\ddot{x}_1 = -\frac{(M+m)g}{ML}x_1 + \frac{b}{ML}\dot{x}_3 - \frac{F}{ML} \tag{13}$$

$$\ddot{x}_3 = \frac{mg}{M}x_1 - \frac{b}{M}\dot{x}_3 + \frac{F}{M} \tag{14}$$

The linear model (13), (14) can be expressed as the following state equation:

$$\begin{cases} \dot{x} = Ax + Bu \\ y = Cx \end{cases} . \tag{15}$$

In (15) $u = F$, A, B and C are defined as follows:

$$A = \begin{bmatrix} 0 & -\frac{(M+m)g}{ML} & \frac{b}{ML} & 0 \\ 1 & 0 & 0 & 0 \\ 0 & \frac{mg}{M} & -\frac{b}{M} & 0 \\ 0 & 0 & 1 & 0 \end{bmatrix}. \tag{16}$$

$$B = \begin{bmatrix} -\frac{1}{ML} \\ 0 \\ \frac{1}{M} \\ 0 \end{bmatrix}. \tag{17}$$

$$C = \begin{bmatrix} 0\,1\,0\,0 \\ 0\,0\,0\,1 \end{bmatrix}. \tag{18}$$

The transfer function of the pendulum takes the following form:

$$G(s) = C(sI - A)^{-1}B = \begin{bmatrix} \frac{\Theta(s)}{F(s)} \\ \frac{X(s)}{F(s)} \end{bmatrix}. \tag{19}$$

In (19) $\Theta(s)$, $X(s)$ and $F(s)$ denote Laplace transforms of angular position θ, cart position x and force F respectively.

4 The Proposed Transfer Function Models of the Pendulum

4.1 Integer Order Model

The integer order models can be easily obtained using experimental data and MATLAB function *tfest*. The general form of IO transfer function well describing the plant under consideration is as beneath:

$$G_{IO} = \frac{b_3 s^3 + b_2 s^2 + b_1 s^2 + b_0}{a_4 s^4 + a_3 s^3 + a_2 s^2 + a_1 s + a_0}. \tag{20}$$

Identification of the above model for the both transfer functions (19) is given in the next section.

4.2 Fractional Order Transfer Functions

The FO transfer functions describing angular position of pendulum θ and cart position x are different. Results of experiments given i.e. in [10] show that the accuracy of the FO model can be significantly improved by adding an IO part. The transfer function describing the angular position is as follows:

$$G_\theta(s) = \frac{\Theta(s)}{F(s)} = \frac{k(1 - T_\beta s^\beta)}{(T_\alpha s^\alpha + 1)(Ts + 1)}. \tag{21}$$

With respect to (7), the transfer function (21) takes the form ready to use at MATLAB:

$$G_\theta(s) \approx \frac{k(M_\alpha(s)M_\beta(s) - T_\beta L_\beta(s)M_\alpha(s))}{(T_\alpha L_\alpha(s) + M_\alpha(s)M_\beta(s))(Ts+1)}. \tag{22}$$

Next the transfer function describing the cart position x was proposed. It is as follows:

$$G_x(s) = \frac{X(s)}{F(s)} = \frac{k(T_\beta s^\beta + 1)}{(T_\alpha s^\alpha + 1)(s^2 + Ts + 1)}. \tag{23}$$

With respect to (7), the transfer function (23) takes the form ready to use at MATLAB:

$$G_x(s) \approx \frac{k(M_\alpha(s)M_\beta(s) + T_\beta L_\beta(s)M_\alpha(s))}{(T_\alpha L_\alpha(s) + M_\alpha(s)M_\beta(s))(s^2 + Ts + 1)}. \tag{24}$$

The both approximated FO transfer functions (22) and (24) were employed to modeling the behaviour of the angular position θ in MATLAB. Results are presented in the next section.

5 Simulations and Experiments

Results were validated experimentally using real experimental system, produced by INTECO company [3] and shown in the Fig. 2. Typically it works as inverted pendulum, but its use as physical pendulum is of course possible too. Trends of the both positions: $\theta(t)$ and $x(t)$ in the system during identification experiment are given in the Fig. 3.

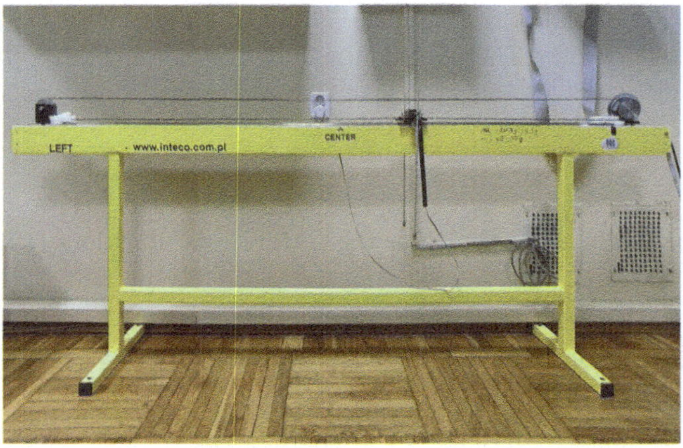

Fig. 2. The experimental system

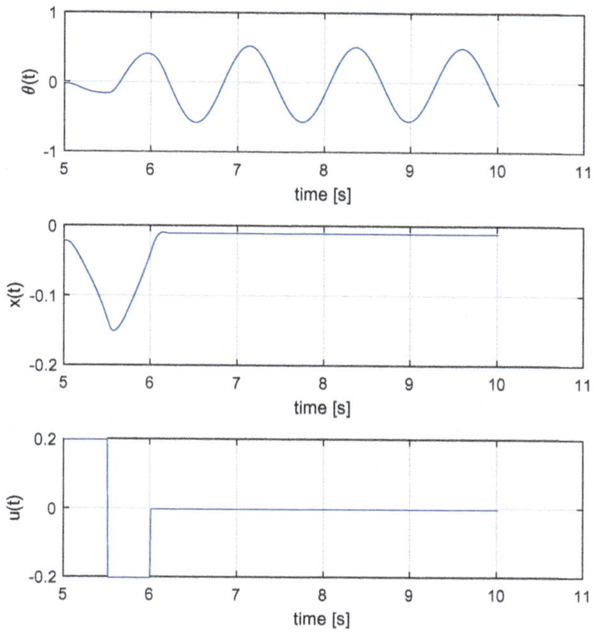

Fig. 3. Trends of angle $\theta(t)$, position $x(t)$ and control $u(t)$ in the experimental system.

The quality of the proposed models was estimated using the Mean Square Error (MSE) and fitting (FIT) cost functions. They are as follows:

$$MSE = \frac{1}{K_s} \sum_{k=1}^{K_s} (y_e(k) - y(k))^2, \tag{25}$$

where $y_e(k)$ and $y(k)$ denote experimental and model responses to the same control signal in time moments k, K_s is the number of all collected samples.

$$FIT = \frac{||y - y_e||}{||y - \overline{y}||}. \tag{26}$$

In (26) $\overline{y_e}$ is the average value from experimental output y_e. The function (26) can be also expressed in %. This function is also employed by the MATLAB functions *compare* and *tfest*.

$$FIT_{100} = (1 - FIT) \cdot 100\%. \tag{27}$$

The elements s^{α} in FO transfer functions were modeled using ORA approximation (7). Parameters of ORA approximation are given in the Table 1.

Table 1. Parameters of the ORA approximation

N	ω_l	ω_h
8	10^{-4}	10^4

Firstly the model of angular position θ was examined. Its IO transfer function (20) is as follows:

$$G_{IO\theta}(s) = \frac{2.388s^3 - 44.94s^2 - 334.2s + 147.7}{s^4 + 19.84s^3 + 113.6s^2 + 523.8s + 2267.0}. \tag{28}$$

Parameters of the transfer function (28) were estimated using MATLAB function *tfest* and experimental data. The comparison of the responses of IO model and real system are given in Fig. 4.

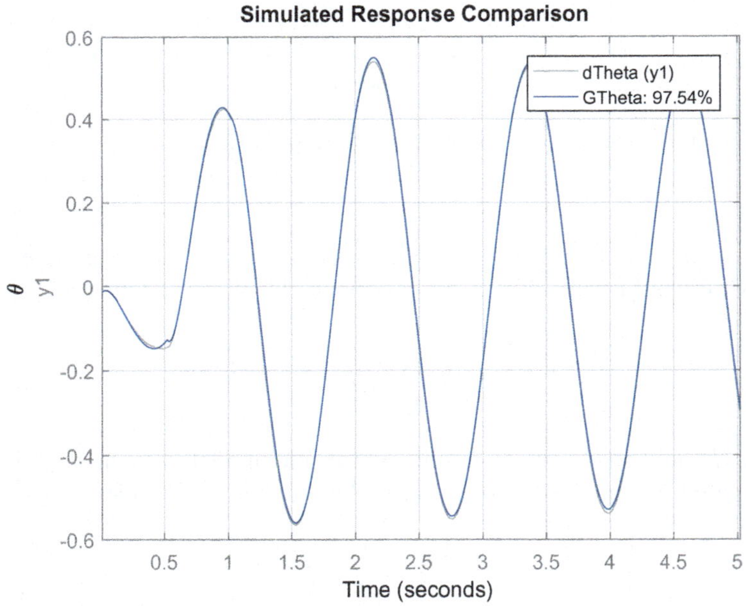

Fig. 4. The IO model of angular position θ.

Next the FO model in the form (21) was tested. The parameters of the FO model were found using experimental data via minimization cost function MSE (25) using MATLAB function *fminsearch*. Then the transfer function (21) takes the following form:

$$G_{\theta}(s) = \frac{\Theta(s)}{F(s)} = \frac{0.3138(1 - 0.2539s^{0.6448})}{(0.0382s^{1.9963} + 1)(0.0608s + 1)}. \tag{29}$$

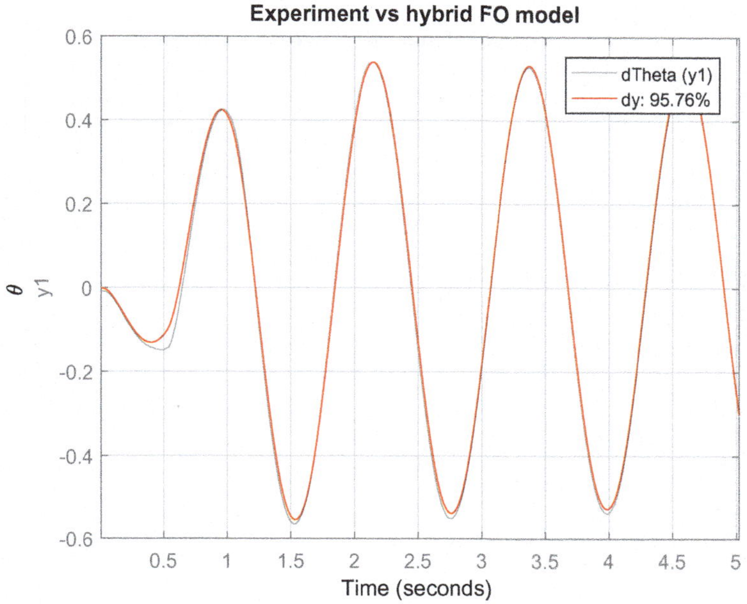

Fig. 5. The FO model of angular position θ.

The comparison of the responses of FO transfer function G_θ and real system are given in the Fig. 5.

The cart position x was examined next. The IO model with respect to (20) was built using MATLAB function *tfest* and experimental data. It is as beneath:

$$G_{xIO} = \frac{0.4227s^3 - 30.63s^2 + 55.79s - 3227}{s^4 + 19.54s^3 + 129.9s^2 + 2086s + 1.891e - 07}. \tag{30}$$

Comparison of responses of transfer function G_{xIO} and real system is given in the Fig. 6.

Next the FO model in the form (23) was tested. The parameters of the FO model were found using experimental data via minimization cost function MSE (25) using MATLAB function *fminsearch*. Then the transfer function (23) is as beneath:

$$G_x(s) = \frac{X(s)}{F(s)} = \frac{-51.2493(9.797s^{1.974} + 1)}{(14.8323s^{2.0527} + 1)(s^2 + 19.5683s + 1)}. \tag{31}$$

The response of the model (31) is compared to experiment in Fig. 7. Results of simulations allow to conclude that the use of IO models allows to obtain a little bit more accurate models, but the analogical FO models require fewer parameters to identify.

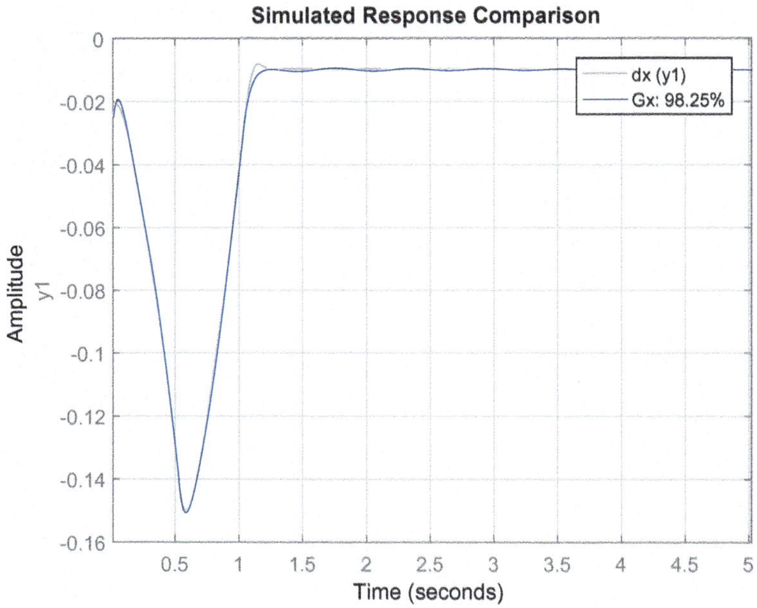

Fig. 6. The IO model of cart position x.

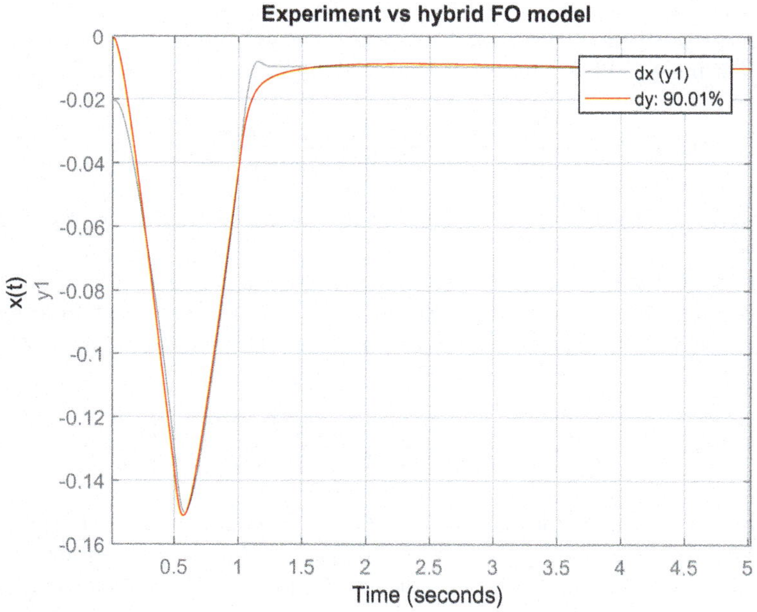

Fig. 7. The FO model of cart position x.

6 Conclusions

The main conclusion from the paper is that the considered physical system can be modeled using fractional order transfer function models. These models are relatively easy to identify and use at the MATLAB platform.

The proposed linear FO model is planned to be employed in construction of model based control algorithms for the pendulum as well as a part of switched linear model of this plant.

The spectrum of further investigations of the considered issue is broad. It covers the construction a FO state space model and tests of impact the parameters of ORA approximation to accuracy of the whole model.

Acknowledgment. This paper was sponsored by AGH project no 16.16.120.773.

References

1. Agarwal, H., Singh, A.P., Srivastava, P.: Fractional order controller design for inverted pendulum on a cart system (POAC). WSEAS Trans. Syst. Control **10** (2015). E-ISSN 2224-2856
2. Domek, S.: Discrete time switched models of non-linear fractional-order systems. In: Bartoszewicz, A., Kabziński, J., Kacprzyk, J. (eds.) Advanced, Contemporary Control, Proceedings of KKA 2020-The 20th Polish Control Conference, Łódź, Poland. Advances in Intelligent Systems and Computing, vol. 1196, pp. 1176–1188 (2020)
3. http://www.inteco.com.pl/products/pendulum-cart-control-system/
4. Mishra, S.K., Chandra, D.: Stabilization and tracking control of inverted pendulum using fractional order PID controllers. J. Eng. **2014**, 9, Article ID 752918 (2014)
5. Paliwal, S., Pathak, V.K.: Analysis & control of inverted pendulum system using PID controller. J. Eng. Res. Appl. **7**(5), 01–04 (2018). ISSN 2248-9622. http://www.ijera.com/papers/Vol7_issue5/Part-4/A0705040104.pdf
6. Żegleń, J.: Advanced control methods of an inverted pendulum on the card implemented in PLC (Zaawansowane metody sterowania modelem laboratoryjnym wahadła na wózku implementowane w sterowniku PLC) Master Thesis at AGH University under supervision M. Rosół (2018)
7. Żegleń, J.: The application of an adaptive controller combined with the LQR controller for the inverted pendulum. Pomiary Automatyka Robotyka **23**(4), 47–54 (2019). https://doi.org/10.14313/PAR_234/47
8. Kaczorek, T.: Selected Problems in Fractional Systems Theory. Springer (2011)
9. Kaczorek, T., Rogowski, K.: Fractional Linear Systems and Electrical Circuits. Bialystok University of Technology (2014)
10. Oprzędkiewicz, K., Mitkowski, W., Gawin, E.: Application of fractional order transfer functions to modeling of high - order systems. In: 20th International Conference on Methods and Models in Automation and Robotics, MMAR 2015, Międzyzdroje, Poland, 24–27 August 2015, pp. 1169–1174 (2015)
11. Oustaloup, A., Levron, F., Mathieu, B., Nanot, F.M.: Frequency-band complex noninteger differentiator: characterization and synthesis. IEEE Trans. Circ. Syst. I: Fundam. Theory Appl. **47**(1), 25–39 (2000)

Biologically Inspired Neural Behavioral Control of the Wheeled Mobile Robot

Paweł Penar(✉) and Zenon Hendzel

Rzeszow University of Technology,
al. Powstańców Warszawy 12, 35-959 Rzeszów, Poland
ppenar@prz.edu.pl

Abstract. In this paper, to solve the task of neural behavioral control of a 2-wheeled mobile robot (WMR), a hierarchical structure is used. At higher levels of the hierarchic generate a desired trajectory of mobile robot motion based on the artificial potential field theory. The generated trajectory is a desired trajectory realized by the neural control algorithm, implemented on the lower level of the hierarchy. Correctness of the solution of the desired trajectory generator and the control system of the elementary robot behavior has been confirmed in numerical simulations.

Keywords: Wheeled robot control · Potential field · Neural network

1 Introduction

The problem of navigation with obstacle avoidance is one of the most important topics in robotics. This decision-making process aims to ensure that the desired transition state and target state of the WMR are achieved. This topic is the domain of artificial intelligence and is implemented by a higher-level hierarchical control [1–3]. The method of synthesis of behavioral control proposed in this article uses the features of animal movement [4], is aimed at extending the existing solutions, with particular emphasis on the application aspect in the field of methods of controlling intelligent WMR. In the paper [4], in reference to the paper [5], three characteristic features of animal movement are described that can be used in planning the WMR movement. Two of them, object proximity-seeking and people approach are shown in Fig. 1. The Fig. 1 shows two types of animal behavior that can be implemented in WMR motion planning:

- **Object proximity-seeking.** A standard approach to generating a trajectory between the starting point and destination in an obstacle environment is shown in Fig. 1-A-1. Another approach, inspired by the behavior of dogs, is shown in Fig. 1-A-2 [5]. In this case, the WMR moves along the obstacles.
- **People approach.** A dog that moves between people, one of which is friendly and one enemy is moving closer to the former. This behavior is accounted for by the trajectory shown in Fig. 1-B-2. If this behavior were not included, the MRK would move as shown in Fig. 1-B-1.

R. Szewczyk et al. (Eds.): AUTOMATION 2021, AISC 1390, pp. 96–105, 2021.
https://doi.org/10.1007/978-3-030-74893-7_10

In Sect. 2, the equations of kinematics and dynamic equations of motion are given. Section 3 provides theory of artificial potential fields in the implementation of elementary robot behavior. In Sect. 4 it was adopted the law of neural control implementing elementary robot behavior. An example is provided in Sect. 5 simulation and simulation results of the adopted solution, which confirm theoretical analysis. The article ends with a summary and bibliography.

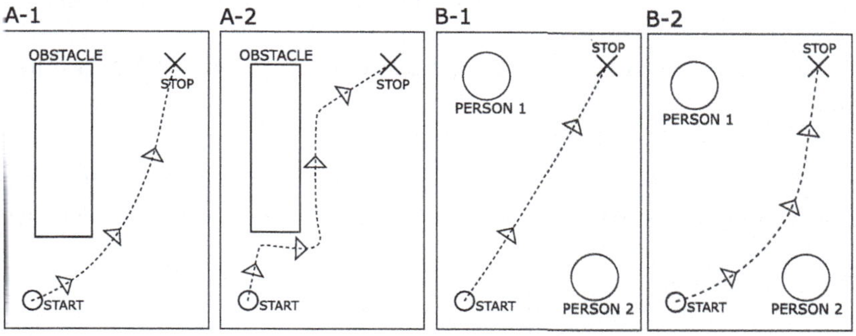

Fig. 1. Features of animal movement [4]

2 Description of Motion a 2-Wheeled Robot

Description of the kinematics of the selected point A of the robot (Fig. 2a) with driving wheels 1 and 2 and a free rolling wheel 3 requires the provision of kinematics equations [6,7]. The position and orientation of the WMR in a fixed coordinate system can be determined from the equations [6,7]

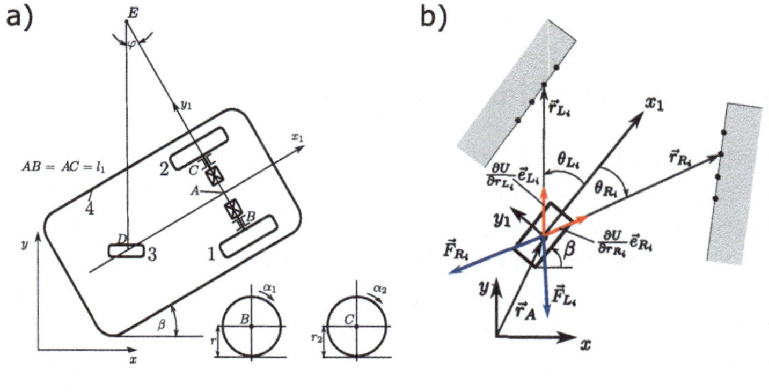

Fig. 2. Scheme of WMR (a) and scheme of the environment with obstacles (b)

$$\begin{bmatrix}\dot{x}_A\\ \dot{y}_A\\ \dot{\beta}\end{bmatrix} = \begin{bmatrix}v_A^* cos\beta & 0\\ v_A^* sin\beta & 0\\ 0 & \omega^*\end{bmatrix}\begin{bmatrix}u_v\\ u_\beta\end{bmatrix} \tag{1}$$

where v_A^* is the maximum speed of point A which is the WMR characteristic point and ω^* is the maximum value of the angular velocity of the WMR frame. Normalized variables $|u_V| \leq u_V^*, |u_\beta| \leq u_\beta^*$ are used to control the linear velocity and orientation of the MRK and their values are generated in the higher control layer.

The velocity vector $\mathbf{v}_A$ of the characteristic point A of the WMR results from the equation

$$\mathbf{v}_A = \dot{x}_A\mathbf{i} + \dot{y}_A\mathbf{j} \tag{2}$$

where $\mathbf{i}, \mathbf{j}$ are the unit vectors of the axes x and y. Assuming the value of point A v_A and orientation MRK $\dot{\beta}$, the inverse of the kinematics problem can be determined on the system of equations [7,8]

$$\begin{cases}\dot{\alpha}_1 = \frac{v_A}{r_1} + \dot{\beta}\frac{l_1}{r_1}\\ \dot{\alpha}_2 = \frac{v_A}{r_2} - \dot{\beta}\frac{l_1}{r_2}.\end{cases} \tag{3}$$

where α_1, α_2 is angle of a self-turn of the drive wheels.

By applying Maggi's mathematical formalism, the dynamic equations of motion of the MWR, as shown in Fig. 2a can be expressed as [3,7]

$$\mathbf{M}\ddot{\alpha} + \mathbf{C}(\dot{\alpha})\dot{\alpha} + \mathbf{F}(\dot{\alpha}) = \mathbf{U} \tag{4}$$

where $\mathbf{a} = [a_1, \ldots, a_6]$ features the parameters derived from the geometry, the distribution of mass, the movement resistances of the WMR, and the dynamic properties of the actuator systems. The matrices and vectors have the following forms

$$\mathbf{M} = \begin{bmatrix}a_1 + a_2 + a_3 & a_1 - a_2\\ a_1 - a_2 & a_1 + a_2 + a_3\end{bmatrix}, \mathbf{C}(\dot{\mathbf{q}}) = \begin{bmatrix}0 & 2a_4(\dot{\alpha}_2 - \dot{\alpha}_1)\\ -2a_4(\dot{\alpha}_2 - \dot{\alpha}_1) & 0\end{bmatrix}, \mathbf{F}(\dot{\mathbf{q}}) = \begin{bmatrix}a_5 sign(\dot{\alpha}_1)\\ a_6 sign(\dot{\alpha}_2)\end{bmatrix} \tag{5}$$

where *sgn* allows for the direction of motion. In application the *sign* function is often approximated by a continuous function. In this case $sign(\dot{\alpha}_i) \approx \frac{1-e^{-cF\dot{\alpha}_i}}{1+e^{-cF\dot{\alpha}_i}}$.

3 Method of Artificial Potential Fields

The method of artificial potential fields [4,9] is a local method because the subsequent positions of the robot that make up the trajectory result from the antigradient search in close proximity to the robot. The value of the artificial potential in the WMR environment is based on the assumption that the robot's movement is the resul of two types of artificial forces acting on it. One type of force pull the robot to the target point and the other pushes the robot away from obstacles. The value of the artificial forces pushing the WMR away from obstacles increases with the approach to the obstacle. The advantage of the method of artificial potential fields is its simplicity and low calculation cost. The disadvantage of this approach is the problem of local minima. Limiting or solving the problem of local minima is widely discussed in the literature.

3.1 Avoidance Obstacles

Figure 2b shows a selected MRK configuration, which results from the location of the characteristic point A in the xy coordinate system [9]. Coordinate system x_1y_1 is related to the point A and the direction of the x_1 has the direction of the velocity vector $\mathbf{v}_A$. The robot has $2n$ sensors that are symmetrically distributed about the x_1 axis. The direction and distance of the sensor R_i, L_i from the obstacle is determined by the radius vector $\mathbf{r}_{R_i}, i = 1, 2, \ldots n$ and $\mathbf{r}_{L_i}, i = 1, 2, \ldots n$. The angles between x_1 axis and $\mathbf{r}_{R_i}, \mathbf{r}_{L_i}$ is $\theta_{R_i}, \theta_{L_i}$, respectively. The vector $\mathbf{e}_{r_{R_i}}, \mathbf{e}_{r_{L_i}}$ are unit vectors in the direction of the respective radius vectors.

In the paper [4] are definitions of force derived from obstacles which take into account the biological behavior of animals, described in Sect. 1. This force are defined as

$$\mathbf{F}_{D_i}(r_{D_i}) = \begin{cases} -\delta(r_{D_i} - d_0)(r_{D_i} - d_1)^2\mathbf{e}_i & r_{D_i} \leq d_1 \\ 0 & r_{D_i} > d_1 \end{cases}, i = 1, 2, \ldots, n, D \in \{R, L\} \tag{6}$$

where d_0 is the constant which is the safe distance of MRK from the obstacle, $d_1 = d_0 + PO$, $\delta > 0$ is the design constant, P, O are scalar parameters related to object proximity-seeking and people approach behavior.

The radius vector $\mathbf{r}_{D_i}$ can be written as $\mathbf{r}_{D_i} = r_{D_i}\mathbf{e}_{D_i}$. Hence obtained

$$\mathbf{F}_{D_i}(r_{D_i}) = \begin{cases} -\frac{\delta}{r_{D_i}}(r_{D_i} - d_0)(r_{D_i} - d_1)^2\mathbf{r}_{D_i} & r_{D_i} \leq d_1 \\ 0 & r_{D_i} > d_1 \end{cases}, i = 1, 2, \ldots, n, D \in \{R, L\} \tag{7}$$

The artificial potential fields U_R for $r_{D_i} \leq d_1$ can be determined from the equation [9]

$$U_R = -\sum_{i=1}^{n}\left[\int F_{R_i} dr_{R_i} + \int F_{L_i} dr_{L_i}\right] \tag{8}$$

The gradient indicates the direction in which the value of the U_R function increases. The direction of the force vectors is opposite to the gradient direction (it indicates the minimum of the U_R function). It is known that the direction of the velocity of the characteristic point A WMR has the direction of the resultant force which is the sum of the forces F_i on the directions $\mathbf{r}_i$. Hence

$$gradU_R \times \mathbf{v}_A = \mathbf{0} \tag{9}$$

The non-zero value of the vector $gradU_R \times \mathbf{v}_A$ is proportional to the value of the angular velocity, therefore in the paper [9] assumed

$$\dot{\beta} = \eta_\beta \left(gradU_R \times \mathbf{v}_A\right), \eta_\beta > 0 \tag{10}$$

where $\eta_\beta > 0$ is design constant. On the xy plane, the equation $gradU_r \times \mathbf{v}_A$ can be written as

$$gradU_R \times \mathbf{v}_A = \begin{vmatrix} \mathbf{i} & \mathbf{j} & \mathbf{k} \\ \frac{\partial U_R}{\partial x_A} & \frac{\partial U_R}{\partial y_A} & 0 \\ \dot{x}_A & \dot{y}_A & 0 \end{vmatrix} \tag{11}$$

where **k** is unit vector of an axis z. Hence

$$\dot{\beta} = \eta_\beta \left(\frac{\partial U_R}{\partial x_A} \dot{y}_A - \frac{\partial U_R}{\partial y_A} \dot{x}_A \right) \mathbf{k} \tag{12}$$

where

$$\frac{\partial U_R}{\partial x_A} = \sum_{i=1}^{n} \left[\frac{\partial U_R}{\partial r_{R_i}} cos(\theta_{R_i} - \beta) + \frac{\partial U_R}{\partial r_{L_i}} cos(\theta_{L_i} + \beta) \right] \tag{13}$$

$$\frac{\partial U_R}{\partial y_A} = \sum_{i=1}^{n} \left[-\frac{\partial U_R}{\partial r_{R_i}} sin(\theta_{R_i} - \beta) + \frac{\partial U_R}{\partial r_{L_i}} sin(\theta_{L_i} + \beta) \right] \tag{14}$$

Projections of the velocity vector v_A on the axis of the xy coordinate system result from the kinematics equations. Hence

$$\dot{\beta} = \eta_\beta \delta \sum_{i=1}^{n} \left[(d_0 - r_{P_i})(d_1 - r_{P_i})^2 sin\theta_{P_i} - (d_0 - r_{L_i})(d_1 - r_{L_i})^2 sin\theta_{L_i} \right] \mathbf{k} = u_{\beta R} \tag{15}$$

Equation (15) is the value of the control variable angular velocity $u_{\beta R}$. The angular velocity of the WMR frame is zero when the right side of (12) is zero. The v_A velocity projections are proportional to the gradient components of the U_R artificial potential function. Hence:

$$\dot{x}_A = \eta_v \frac{\partial U_R}{\partial x_A} = \eta_v \sum_{i=1}^{n} \left[\frac{\partial U_R}{\partial r_{R_i}} cos(\theta_{R_i} - \beta) + \frac{\partial U_R}{\partial r_{L_i}} cos(\theta_{L_i} + \beta) \right] \tag{16}$$

$$\dot{y}_A = \eta_v \frac{\partial U_R}{\partial y_A} = \eta_v \sum_{i=1}^{n} \left[-\frac{\partial U_R}{\partial r_{R_i}} sin(\theta_{R_i} - \beta) + \frac{\partial U_R}{\partial r_{L_i}} sin(\theta_{L_i} + \beta) \right] \tag{17}$$

where $\eta_v > 0$ is a design constant. Hence it follows that

$$\dot{x}_A = \eta_v \delta \sum_{i=1}^{n} \left[(d_0 - r_{R_i})(d_1 - r_{R_i})^2 cos(\theta_{R_i} - \beta) + (d_0 - r_{L_i})(d_1 - r_{L_i})^2 cos(\theta_{L_i} + \beta) \right] \tag{18}$$

$$\dot{y}_A = \eta_v \delta \sum_{i=1}^{n} \left[-(d_0 - r_{R_i})(d_1 - r_{R_i})^2 sin(\theta_{R_i} - \beta) + (d_0 - r_{L_i})(d_1 - r_{L_i})^2 sin(\theta_{L_i} + \beta) \right] \tag{19}$$

Equations (18) and (19) determined control variable linear velocity u_{vR} whose values are $u_{vR} = \sqrt{\dot{x}_A^2 + \dot{y}_A^2}$.

3.2 Goal-Seeking Behaviour

The desired trajectory generated in the higher control layer is designed not only to avoid obstacles, but also to drive WMR from the starting point (A) to the destination point (B). This fact is reflected in the function of the artificial potential.

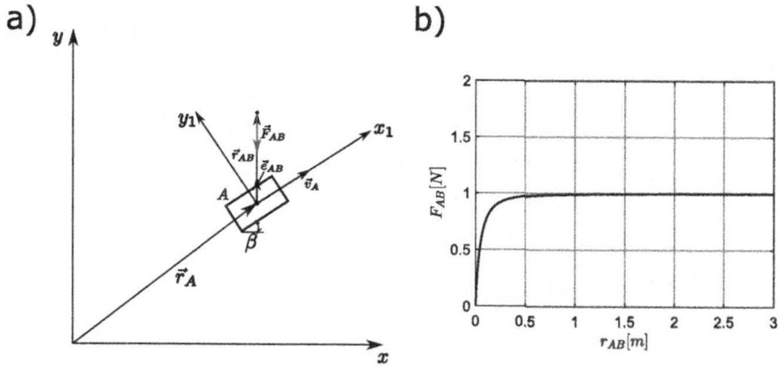

Fig. 3. Goal-seeking behaviour scheme (a) and course of the F_{AB} function (b)

Figure 3a shows the selected MRK configuration in the goal-seeking behaviour. The realization of this behavior is possible through the attractive force $\mathbf{F}_{AB}$, which can be of the form [4]

$$\mathbf{F}_{AB} = \left(\frac{\Delta}{\chi^2} - \frac{\Delta}{(r_{AB} + \chi)^2}\right) \mathbf{e}_{AB} \tag{20}$$

where $\Delta \geq 0$ and $\chi >$ are design parameters. Figure 3b shows the change of the F_{AB} function proposed in [4]. The author points out that the essential feature of the function of the form (20) is its course over long distances from the target. In such cases, it is constant, assuming a fixed level of F^*_{AB}.

The radius vector $\mathbf{r}_{AB}$ can be written as $\mathbf{r}_{AB} = r_{AB}\mathbf{e}_{AB}$. Hence obtained

$$\mathbf{F}_{AB} = \left(\frac{\Delta}{\chi^2} - \frac{\Delta}{(r_{AB} + \chi)^2}\right) \frac{\mathbf{r}_{AB}}{r_{AB}} \tag{21}$$

The artificial potential field in the case goal-seeking behaviour has the form

$$U_{AB} = \frac{\Delta((\chi + r_{AB})^2 + \chi^2)}{\chi^2(\chi + r_{AB})} \tag{22}$$

Assuming that the elements of the velocity vector of the characteristic point A are proportional to the gradient of the artificial potential field, we obtained

$$\dot{x}_A = \eta_{v_{AB}} \frac{dU_{AB}}{dx_A} = \eta_{v_{AB}} \left[\frac{2\Delta(x_A - x_B)}{\chi^2 r}\left(1 - \frac{(\chi + r_{AB})^2 + \chi^2}{2(\chi + r)^2}\right)\right] \tag{23}$$

$$\dot{y}_A = \eta_{v_{AB}} \frac{dU_{AB}}{dy_A} = \eta_{v_{AB}} \left[\frac{2\Delta(y_A - y_B)}{\chi^2 r}\left(1 - \frac{(\chi + r_{AB})^2 + \chi^2}{2(\chi + r)^2}\right)\right] \tag{24}$$

where $r = \sqrt{(x_B - x_A)^2 + (y_B - y_A)^2}$, $\eta_{AB} > 0$ is design constant.

The rotation angle of the MRK, which implements the goal-seeking behaviour, can be determined from the equation

$$\dot{\beta} = arctan \frac{y_B - y_A}{x_B - x_A} - \beta \tag{25}$$

Using the equations (23)–(25) you can find the control variables $u_{vAB}, u_{\beta AB}$ for goal-seeking behaviour task, i.e.

$$u_{vAB} = \sqrt{\dot{x}_A^2 + \dot{y}_A^2}, u_{\beta AB} = \eta_{\beta_{AB}} \dot{\beta} \tag{26}$$

where $\eta_{\beta AB} > 0$ is a design constant.

3.3 Fusion of Behaviour

In order for WMR to drive to the destination with avoiding obstacles, the potential field should be determined as the sum of the potential fields. Hence it follows that the control variables for the complex behavior have the form

$$u_V = -u_{vR} + u_{vAB}, u_\beta = u_{\beta R} + u_{\beta AB} \tag{27}$$

4 Neural Control

From the point of view of the control theory, the most important properties of neural networks are their ability of approximating nonlinear function. These properties will be used in the synthesis of the neural control algorithm to achieve the desired trajectory, which comes from the implementation of the complex behavior of the WMR.

Let the WMR describe Eq. (4). Defining the control error $\mathbf{e} = \boldsymbol{\alpha}_d - \boldsymbol{\alpha}$ and generalised error $\boldsymbol{s} = \dot{\boldsymbol{e}} + \boldsymbol{\Lambda e}$ where $\boldsymbol{\Lambda} \in \mathbb{R}_{2\times 2}$ is the design matrix, the dynamics of WMR can be written in the generalized error space $\mathbf{s}$ [8,10]. Tracking control can take into account the object nonlinearity compensation. Then it has the form:

$$\mathbf{u} = \mathbf{f}_M(\mathbf{v}, \dot{\mathbf{v}}, \dot{\boldsymbol{q}}) + \boldsymbol{K}_D \mathbf{s} \tag{28}$$

where $\mathbf{f}_M(\mathbf{v}, \dot{\mathbf{v}}, \dot{\mathbf{q}}) = \mathbf{M}\dot{\mathbf{v}} + \mathbf{C}(\dot{\boldsymbol{\alpha}})\mathbf{v} + \mathbf{F}(\dot{\boldsymbol{\alpha}})$, $\mathbf{v} = \dot{\boldsymbol{\alpha}}_d + \boldsymbol{\Lambda e}$ In practice, approximations of the nonlinearity compensating control $\mathbf{f}_M$ are used due to the inaccuracy of the mathematical model. For this purpose, we can use random vector functional link (RVFL) neural network (NN) [3,8]. Therefore, the assessment of nonlinear function of $\mathbf{f}_M$ can be written as

$$\hat{\mathbf{f}}_M(\mathbf{x}_{SN}) = \hat{\mathbf{W}}^T \phi(\boldsymbol{V}_S^T \mathbf{x}_{SN}) = \hat{\mathbf{W}}^T \phi(\bar{\boldsymbol{x}}) \tag{29}$$

where $\hat{\mathbf{W}}$ is an assessment of ideal NN weights, ϕ is the activation function vector and $\mathbf{x}_{SN} = \left[\boldsymbol{e}/s_e \; \dot{\boldsymbol{e}}/s_{\dot{e}} \; \boldsymbol{s}/s_s\right]$ is normalized NN input, $s_e, s_{\dot{e}}, s_s$ are normalizing variables and $\boldsymbol{V}_s$, is a randomly generated constant matrix. Assuming (29), the control law of the form (28) we write as

$$\mathbf{u} = \hat{\mathbf{W}}^T \phi(\bar{\mathbf{x}}) + \boldsymbol{K}_D \mathbf{s} \tag{30}$$

For the analyzed system, the law of adaptation of NN weights results from Lapunov's stability [3,8,10]

$$\dot{\hat{\boldsymbol{W}}} = \boldsymbol{F}_{NN} \phi(\mathbf{x}_{SN}) \boldsymbol{s}^T \tag{31}$$

where $\mathbf{F}_{NN} = \mathbf{F}_{NN}^T > 0$ is a design matrix.

5 Numerical Simulation

Generating MRK trajectories using the artificial potential field method, which definition is biologically inspired, was carried out in the simulation. It was implemented using the Matlab/Simulink, Navigation Toolbox and Mobile Robotics Simulation Toolbox packages. Figure 4a shows a diagram of the control system, in which the upper hierarchy layer implementing the complex MRK behavior and the lower layer that implements tracking control. In order to perform the simulation, it was assumed that MRK has 12 distance sensors arranged symmetrically with respect to the x_1 axis every 10°. The range of the sensor is d_1 Kinematics control signals $u_{VR}, u_{\beta R}$ are filtered with the transfer function $G(s) = 1/(0.05s + 1)$. The robot performs the behavior of avoiding obstacles and goal-seeking behaviour from point $(0, 0)$ to the surroundings of ε_G of point $G(x_G, y_G)$. The adaptive control unit is composed of six neurons with bipolar activation functions. By selecting parameters related to the generation of trajectories and the tracking control PD+NN which are summarized in Table 1, simulation were performed for three selected stop points: $G_1(3, 1), G_2(3.5, 3), G_3(3, 1)$ (rys. Fig. 4b).

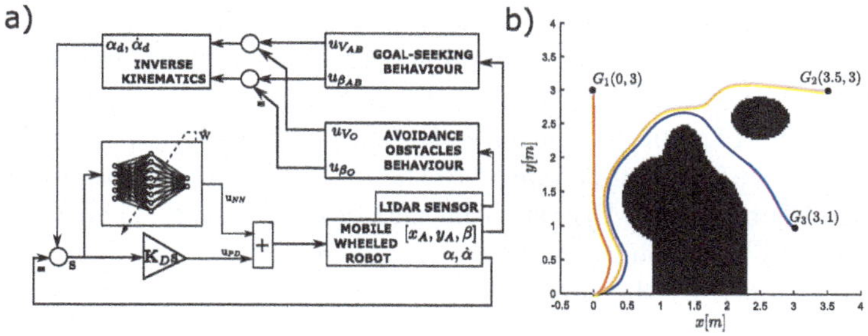

Fig. 4. WMR control scheme (a) and WMR trajectory generated by upper control layer

Table 1. Selected simulation parameters

$[a_1, a_2, a_3, a_4, a_5, a_6, c_F]$	$[0.0825, 0.043, 0.0311, 0.0011, 5.5, 5.5, 0.24]$
P, O, d_0	$1, 0.8, 0.6$
$\delta, \Delta, \chi, F_{NN}$	$3, 7, 0.1, 20$
$\eta_V, \eta_\beta, \eta_{AB}, \eta_{\beta AB}$	$0.05, 0.8, 0.0012, 0.4$
v_{max}, β_{max}	$0.5, 6$
$N, \varepsilon_G, s_e, s_{\dot{e}}, s_s$	$6, 0.1, 0.03, 1, 0.03$

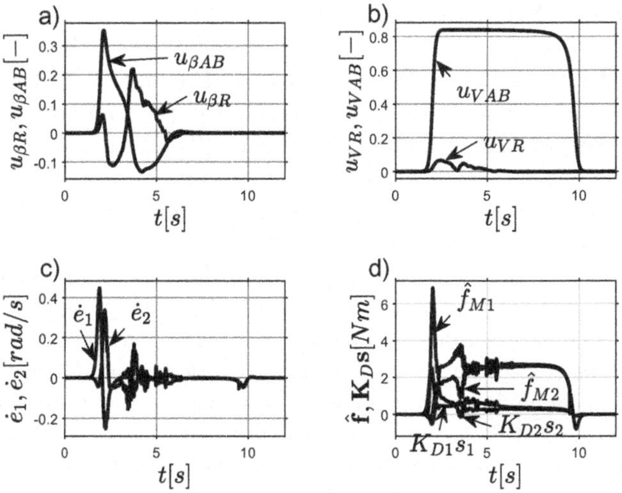

Fig. 5. Kinematics control signals (a, b), angular velocity tracking error (c), adaptive control and PD control (d) when driving to the destination G_1

Figure 5 shows the kinematics control signals (in this case, WMR moves to the G_1), which are generated by the higher control layer. The signals $u_{VAB}, u_{\beta AB}$ correspond to the goal-seeking behaviour, and signals $u_{vR}, u_{\beta R}$ result from the implementation of the obstacle avoidance behavior. For $t < 2[s]$, the kinematics control signals are 0, which is related to the multiplication of the control signals by the term defining the MRK starting and braking phase. Their occurrence is best reflected by the change of the u_{VAB} control signal (Fig. 5b). The braking phase is determined by the time at which the characteristic point WMR was in the vicinity of the G point.

The higher control layer generates the desired trajectory. Tracking control is responsible for its realization. Tracking error for angular velocity $\dot{e}$ WMR is shown in Fig. 5c. Its value is bound, which proves the stability of the adopted control system. The disturbance of the angular velocity error occurring about 4 [s] result from avoiding the obstacle and are caused by a sudden change in the direction of MRK motion and a discrete form of distance sensors arrangement. It is of particular importance in the context of disturbances in kinematics control signals, when the shape of the obstacle is complex.

The Fig. 5d shows the control signals in the lower control layer. The PD control signals is most active in the initial phase of the WMR motion. Then, its influence on the value of total control decreases in favor of neural control, which is performed by the neural network.

6 Summary

The article presents the synthesis of MRK behavioral control in the task of driving to the destination while avoiding obstacles, for which an artificial potential field was used. Proper selection of the forces coming from obstacles, allows for application of the characteristic behavior of animals in the trajectory planning WMR. Further improvement of the presented solution should include an algorithm for leaving the local minimum, and further work may focus on automating the selection of parameters related to the artificial potential field.

References

1. Arkin, R.C.: Behavior-Based Robotics. The MIT Press, Cambridge (1998)
2. Latombe, J.-C.: Robot Motion Planning. Springer, Heidelberg (2012)
3. Giergiel, M.J., Hendzel, Z., Żylski, W.: Modeling and Control of Mobile Wheeled Robots. PWN (2013). (in Polish)
4. Kovács, B., Szayer, G., Tajti, F., Burdelis, M., Korondi, P.: A novel potential field method for path planning of mobile robots by adapting animal motion attributes. Robot. Auton. Syst. **82**, 24–34 (2016)
5. Faragó, T., Miklósi, Á., Korcsok, B., Száraz, J., Gácsi, M.: Social behaviours in dog-owner interactions can serve as a model for designing social robots. Interact. Stud. Soc. Behav. Commun. Biol. Artif. Syst. **15**(2), 143–172 (2014)
6. Giergiel, J., Żylski, W.: Description of motion of a mobile robot by Maggie's equations. J. Theor. Appl. Mech. (Warsaw) **43**(3), 511–521 (2005)
7. Zylski, W.: Kinematics and Dynamics of Mobile Wheeled Robots. Oficyna wydawnicza PRz (1996). (in Polish)
8. Gierlak, P., Hendzel, Z.: Control of Wheeled Mobile Robots and Manipulators. Oficyna Wydawnicza PRz (2011). (in Polish)
9. Żylski, W.: Motion planning for wheeled mobile robot using potential field method. J. Theor. Appl. Mech. **42**(1986), 695–705 (2004)
10. Marcin, S., Hendzel, Z.: Intelligent Optimal Adaptive Control for Mechatronic Systems. Springer, Heidelberg (2018)

Robust Follow-Up Sequencing Algorithm

Sara Bysko[1], Szymon Bysko[2(✉)], Jolanta Krystek[1], and Łukasz Głodek[2]

[1] Silesian University of Technology, Akademicka 16, 44-100 Gliwice, Poland
sara.bysko@polsl.pl
[2] Virtual Commissioning Department,
ProPoint Sp. z o.o. Sp. K., ul. Bojkowska 37R, 44-100 Gliwice, Poland
szymon.bysko@propoint.pl

Abstract. The paper presents the continuation of work carried out over the last few years on development of an effective algorithm for sequencing cars in the paint shop. One of the proposed algorithms is based on the follow-up production control algorithm. The article focuses in particular on the verification of the impact of selected parameters, both production and technological, on the effectiveness of the considered sequencing algorithm. The following parameters were used as part of the research: periodic cleaning period value affecting the quality of the sequencing process, length of the production plan, which is taken into account when determining the priorities of individual body colors by the sequencing algorithm, and distribution of car colors. The aim of the conducted research was to determine whether this algorithm is resistant to the change of the examined parameters. Such research is important from the point of view of possibility of applying following algorithm in a real factory.

Keywords: Follow-up production control · Automotive paint shop · Sequencing · Car Sequencing Problem 4.0

1 Introduction

Currently manufacturers use flexible forms of production organization, typical for multi-version and multi-assortment production, to demand dynamically changing requirements of potential customers. An example of production for which planning in smaller batches, by sequencing (mixing) product variants in the same process is important from the point of view of process optimization is a multi-stage car production.

In the literature, the Car Sequencing Problem (CSP) was presented in several versions: the original problem concerned the assembly line [1], the modified issues also included requirements of paint shop [2] and structures like buffers which enable online

This work was supported by ProPoint Sp. z o.o. Sp. K. under grant POIR.01.01.01-00-0047/15-00.

R. Szewczyk et al. (Eds.): AUTOMATION 2021, AISC 1390, pp. 106–116, 2021.
https://doi.org/10.1007/978-3-030-74893-7_11

sequencing [3–5]. A comprehensive review of the literature presenting the above-mentioned variants of the CSP problem was considered in [6]. This problem is still open, as evidenced by new studies appearing in the literature [7]. However, in most cases the papers concern a simplified problem, which does not take into account the presence of buffers [8], periodic cleanings [9] or the need for online sequencing. Therefore, the developed solutions cannot be applied in industry. The newly defined CSP problem (called CPS 4.0), which takes into account the limitations and requirements of real car production line, is detailed described in [10].

So far, research published in the literature on the CSP problem focus solely on presenting and comparing possible approaches to the problem. The behavior of these solutions in response to changing production or technological parameters was not verified This paper presents the continuation of work carried out over the last few years on development of an effective algorithm for sequencing cars in the paint shop and presents such considerations. The aim of the research was to verify the effectiveness of one of the developed sequencing algorithms in the context of changing selected parameters. The approach, called Follow-up Sequencing Algorithm (FuSA) is based on the idea of a follow-up production control (NSP) [11]. The role of the FuSA algorithm is to generate bodywork sequences targeted for painting that will minimize the number of changes due to color change and synchronize them with periodic cleaning of the paint nozzles. The detailed description of the FuSA is presented in [12, 13]. Previous studies have shown that this algorithm provides the best solutions to the problem compared to other approaches developed by the authors. The FuSA algorithm was compared with such approaches as: priority algorithms [10, 14], game theory-based algorithms [6] or the approach used in the real factory from which the data for research was received [10, 14]. Therefore, this article focuses on the FuSA algorithm, considered to be the best algorithm developed so far.

The article is organized as follows. Section 2 formulates a new sequencing problem that takes into account the current requirements of the automotive industry. Section 3 describes the general concept of proposed sequencing algorithm in the context of follow-up control algorithm. Section 4 contains proposed quality indicators used to evaluate potential solutions. Section 5 presents experimental research and discussion of the results. The final Section concludes the paper and presents further researches.

2 Problem Formulation

If a buffer is used for sequencing, cars transported from the body shop to the paint shop create sequences in real time. A color of a car is not known until the car reaches the buffer input, where the QR code is read. This means that the input sequence to the buffer is not known a priori. Only production plan determined for a certain time horizon is available. Each procedure for the sequencing problem should be based on limited information.

In addition to limited access to information and real-time decision making requirements, the problem of car sequencing in the paint shop is subject to an additional restriction, which results from the applied paint system. This forces the need for cleaning painting guns after a certain number of cars is painted. This limitation determines the optimal length of sequence composed of cars in the same color. This problem is called *Car Sequencing Problem 4.0* (CSP 4.0).

An instance of the considered problem is defined as tuple ($V, C,$ *NRowBuff, NColBuff*, *NPerClean*), based on production plan and technical parameters, where:

- $V = \{v_1, \ldots, v_N\}$ – set of vehicle to be produced,
- $C = \{c_1, \ldots, c_D\}$ – set of available colors and function c: $V \rightarrow C$, that associates color c_i to each vehicle v_i,
- *NRowBuff, NColBuff* – buffer size defined by number of buffer rows and number of buffer columns,
- *NPerClean* – periodic cleaning interval.

The solution of the CSP 4.0 is an order δ in which cars are delivered to the paint station. The sequence is generated in real time. The optimization problem is to find such *NPerClean* – element subsequences, for which the number of color changes requiring gun changeovers is minimal. For economic reasons, the most favorable situation is when the gun changeover coincides with periodic cleaning. In other words, the optimal solution is a sequence composed of single-color *NPerClean*-element subsequences, whereby two consecutive subsequences should be in different color.

3 Follow-Up Sequencing Algorithm

The Follow-up Sequencing Algorithm (FuSA) is designed for the current generation of executive plans – car sequences. Implementation plans are synchronized keep pace with operational plans from the system coordination unit. Decisions of the FuSA algorithm are made on the basis of the algorithm state, which consists of recent decisions and status of backlogs in the implementation of operational plans [12, 13]. The FuSA algorithm integrates two algorithms. The first of them, the loading algorithm, is designed to determine the transport line, on which the body from the loading conveyor is to be loaded. The unloading algorithm generates a current execution plan understood as a set of decisions defining the order of buffer unloading in a defined horizon. In general case, this decision concerns body set, therefore it is also necessary to specify, in which order, the individual bodies from this set should leave the buffer. This approach ensures that the decisions made are the best from the perspective of a current situation on the line, but at the same time, the possibility of their improvement in the planning horizon of *NPerClean* length is not excluded.

4 Quality Indicators

To evaluate the solution of CSP 4.0, four quality indicators are proposed (1–4):

- Number of Changeovers (*NCs*) – defines the number of color changes excluding the changes occurring during periodic cleaning.

$$NCs(\delta) = \sum_{\substack{1 \leq n \leq N-1 \\ n \bmod NPerClean \neq 0}} \mathrm{isCC}_{\delta n, \delta n+1}, \tag{1}$$

where:

N – number of vehicles to be produced,
$\text{isCC}_{\delta n,\delta n+1} = \begin{cases} 0, & c(v_n) = c(v_{n+1}) \\ 1, & else \end{cases}$ – the variable determining whether there is a color change between painting of the next two cars,
$c(v_n)$ – the color of n-th vehicle.

- Effectiveness of Synchronization (ES) – determines the total number of color changes occurring between two subsequences in relation to the number of all color changes.

$$ES = \frac{\sum_{\substack{1 \le n \le N-1 \\ n\, mod NPerClean = 0}} \text{isCC}_{\delta n,\delta n+1}}{\left[\frac{N-1}{NPerClean}\right]} \cdot 100\% \tag{2}$$

- Number of Color Changes (NCC) – defines the number of all color changes.

$$NCC(\delta) = \sum_{n=1}^{N-1} \text{isCC}_{\delta n,\delta n+1} \tag{3}$$

- Synchronization Ratio (SR) – determines the sum of the number of color changes occurring between two sub-sequences in relation to the number of color changes.

$$SR(\delta) = \frac{\sum_{n=1}^{N-1} \text{isCC}_{\delta n,\delta n+1}}{NCC(\delta)} \cdot 100\% \tag{4}$$

5 Numerical Experiments

The simulation tests carried out consisted of verifying the impact of three selected parameters on the effectiveness of the FuSA algorithm, i.e.: periodic cleaning period value (*NPerClean*), length of the production plan, and color distribution.

5.1 Testing the Effect of the Periodic Cleaning Period Value

One of the parameters assessing the effectiveness of the sequencing algorithm is the interval of periodic cleaning imposed by the factory and related to technological limitations. This value determines the number of car bodies to be painted between consecutive periodic cleanings. The performed tests assumed the verification of the following values of the *NPerClean* parameter in comparison with length of the transport line *NColBuff*:

- $NPerClean = 3$ (N3) – $NPerClean < NColBuff$,
- $NPerClean = 5$ (N5) – $NPerClean = NColBuff$,
- $NPerClean = 7$ (N7) – the default value of the periodic cleaning period,
- $NPerClean = 9$ (N9) – $NPerClean > NColBuff$.

For the purposes of the research, 5 instances were generated, differing in the order of car bodies transported to the buffer, with the following characteristics:

- $V = \{v_1, \ldots, v_{1000}\}$,
- $C = \{c_1, \ldots, c_6\}$,
- $NRowBuff = 5$, $NColBuff = 5$,
- *NPerClean* – tested parameter,
- color distribution: c_1: 6%, c_2: 38%, c_3: 29%, c_4: 14%, c_5: 10%, c_6: 3%.

The results obtained during the simulation tests are presented in Tables 1, 2, 3 and 4.

Table 1. NCs indicator values for different values of the *NPerClean*

Dataset no.	*NCs*			
	N3	N5	N7	N9
Data_01	34	70	137	196
Data_02	52	74	128	182
Data_03	53	80	131	208
Data_04	32	79	148	183
Data_05	42	72	151	194

Table 2. NCC indicator values for different values of the *NPerClean*

Dataset no.	*NCC*			
	N3	N5	N7	N9
Data_01	312	249	249	258
Data_02	326	246	254	278
Data_03	320	257	260	272
Data_04	303	251	261	278
Data_05	310	243	256	283

Based on the results presented in Tables 1, 2, 3 and 4, it can be concluded that the value of the *NPerClean* parameter has a significant impact on the number of color changes (*NCC*). The FuSA algorithm provides the best solutions when the value of this parameter is 5 car bodies, which corresponds to the number of columns (*NColBuff*) of the analyzed buffer. If the value of the *NPerClean* parameter is different from the *NColBuff* value, the *NCC* indicator is higher. If $NPerClean < NColBuff$, the number of color changes is greater compared to $NPerClean > NColBuff$. As expected, the value of the *NCs* indicator increases with the increase of the *NPerClean* parameter value, while the values of the *ES* indicator do not change significantly. In turn, *SR* values decrease with increasing *NPerClean* period. This is because the difficulty of creating single-color sub-sequences increases as the number of cars that these sub-sequences are composed of increases.

Table 3. ES indicator values for different values of the *NPerClean*

Dataset no.	*ES*			
	N3	N5	N7	N9
Data_01	83%	90%	79%	56%
Data_02	82%	86%	89%	86%
Data_03	80%	89%	91%	58%
Data_04	81%	86%	80%	86%
Data_05	80%	86%	74%	80%

Table 4. SR indicator values for different values of the *NPerClean*

Dataset no.	*SR*			
	N3	N5	N7	N9
Data_01	89%	72%	45%	24%
Data_02	84%	70%	50%	35%
Data_03	83%	69%	50%	24%
Data_04	89%	69%	43%	34%
Data_05	86%	70%	41%	31%

5.2 Testing the Effect of the Plan Production Length

The next stage of the research was to verify whether the length of the production plan affects the effectiveness of the FuSA algorithm. According to the concept of the proposed sequencing algorithm, the production plan is used to prioritize individual colors. When making decisions in the subsequent steps of the algorithm, these priorities are used to select a car body color variant that will be sent to the buffer output.

For the tests, 1000-element instances generated in the first stage of the research were used. It was assumed that the value of the *NPerClean* parameter is 7. For the purposes of testing, an additional parameter was defined – the production plan update period. The following nomenclature was used: HX_Y – means production plan with length X and update period Y. For example, H10_5 means that the production plan includes 10 cars and is refreshed after 5 car bodies are manufactured.

Tables 5, 6, 7 and 8 present the values of quality indicators, respectively *NCs* and *NCC* as well as *ES* and *SR* for individual production plans.

Based on the results presented in Tables 5, 6, 7 and 8, it can be concluded that the FuSA algorithm is robust to the length of the production plan. This is due to the fact that this plan is only used to calculate the color priorities that are used in the final steps of the algorithm (in most cases the decision is made in the previous steps).

Table 5. *NCs* indicator values for different lengths of production plans

Dataset no.	*NCs*				
	H10_5	H25_10	H50_25	H100_50	H200_100
Data_01	143	148	122	134	131
Data_02	147	148	15	141	121
Data_03	150	141	157	147	146
Data_04	136	135	133	115	144
Data_05	150	138	128	146	167

Table 6. *NCC* indicator values for different lengths of production plans

Dataset no.	*NCC*				
	H10_5	H25_10	H50_25	H100_50	H200_100
Data_01	266	268	243	256	251
Data_02	264	271	273	257	241
Data_03	269	265	279	275	271
Data_04	259	261	257	224	268
Data_05	271	256	256	266	264

Table 7. *ES* indicator values for different lengths of production plans

Dataset no.	*ES*				
	H10_5	H25_10	H50_25	H100_50	H200_100
Data_01	87%	85%	85%	86%	85%
Data_02	82%	87%	87%	82%	85%
Data_03	84%	87%	86%	90%	88%
Data_04	87%	89%	87%	77%	87%
Data_05	85%	83%	90%	85%	68%

5.3 Study of the Influence of the Color Distribution of Instances

As part of the tests, the impact of the color distribution of the instances on the effectiveness of the FuSA algorithm was verified. Following distributions were tested:

- distribution consistent with the real (factory) distribution (R),
- distribution with two dominant colors (2C),
- distribution with one dominant color (1C),
- distribution with equal proportion of all colors (E).

Table 8. *SR* indicator values for different lengths of production plans

Dataset no.	*SR*				
	H10_5	H25_10	H50_25	H100_50	H200_100
Data_01	46%	45%	50%	48%	48%
Data_02	44%	45%	45%	45%	50%
Data_03	44%	47%	47%	47%	46%
Data_04	47%	48%	49%	49%	46%
Data_05	45%	46%	45%	45%	37%

Table 9 shows the percentage of individual colors for the proposed distributions.

Table 9. Characteristics of the tested color distributions

Distrib.	Color					
	C1	C2	C3	C4	C5	C6
R	38%	29%	14%	10%	6%	3%
2C	34.1%	34.1%	13.6%	13.6%	2.3%	2.3%
1C	70%	12%	12%	2%	2%	2%
E	16.7%	16.7%	16.7%	16.7%	16.6%	16.6%

For the purposes of the tests, for each type of the analyzed distribution, with the exception of the primary distribution, 5 instances were generated, differing in the order of car bodies transported to the buffer, with the following characteristics:

- $V = \{v_1, \ldots, v_{1000}\}$,
- $C = \{c_1, \ldots, c_6\}$,
- $NRowBuff = 5, NColBuff = 5$,
- $NPerClean = 7$,
- color distribution – tested parameter.

The experimental data proposed in the first stage of the research were used for the primary distribution tests.

The results of the conducted simulation tests are presented in Tables 10, 11, 12 and 13.

For the tests of instances representing the Gaussian distribution with one dominant color (G2), the obtained results are the best in terms of the number of *NCC* color changes and the number of changeovers of *NCs* paint nozzles. The highest values of these indicators were obtained for the uniform distribution. The values of the *ES* and *SR* indicators are the best for the Gaussian distribution with two dominant colors.

Table 10. Bill of quality indicators for R distribution

Dataset no.	Quality indicator			
	NCs	*NCC*	*ES*	*SR*
Data_01	137	249	79%	45%
Data_02	128	254	89%	50%
Data_03	131	260	91%	50%
Data_04	148	261	80%	43%
Data_05	151	256	74%	41%

Table 11. Bill of quality indicators for 2C distribution

Dataset no.	Quality indicator			
	NCs	*NCC*	*ES*	*SR*
Data_01	120	242	86%	50%
Data_02	114	233	84%	51%
Data_03	128	253	88%	49%
Data_04	110	235	88%	53%
Data_05	108	232	87%	53%

Table 12. Bill of quality indicators for 1C distribution

Dataset no.	Quality indicator			
	NCs	*NCC*	*ES*	*SR*
Data_01	76	161	60%	53%
Data_02	83	176	65%	53%
Data_03	85	166	57%	49%
Data_04	88	171	58%	49%
Data_05	85	178	65%	52%

Table 13. Bill of quality indicators for E distribution.

Dataset no.	Quality indicator			
	NCs	*NCC*	*ES*	*SR*
Data_01	170	296	89%	43%
Data_02	232	294	44%	21%
Data_03	179	301	86%	41%
Data_04	226	293	47%	23%
Data_05	267	301	24%	11%

6 Conclusions

Sequencing of production at the stage of painting cars is important especially for economic reasons. Minimizing the number of changeovers of the paint nozzles reduces in turn production costs. It is necessary to include both the changeovers of the paint nozzles, resulting from changes in paint colors, as well as periodic cleaning of the nozzle, ensuring good varnish. In addition, proper management of the car flow through the buffer ensures the continuity of production. The efficient sequencing algorithm allows to automate the decision-making process, which in many cases is carried out by operators.

The paper considered the influence of selected parameters on the operation of the follow-up algorithm. The conducted research showed that, the efficiency of the algorithm deteriorated with increasing period of batch cleaning. The study of changes in the production plan length showed that this parameter did not affect the algorithm's operation. As expected, the type of color distribution in the tested data is decisive for the quality of the solutions obtained.

Due to the fact that the cleaning period can be set by the manufacturer, future works will focus mainly on refining the proposed algorithm so as to make it independent of the influence of color distribution. As part of the next stages of work, the research will also cover the game theory based approach, which gives slightly worse solutions compared to the FuSA algorithm.

References

1. Parello, B.D., Kabat, W.C.: Job-shop scheduling using automated reasoning: a case study of the car sequencing problem. J. Autom. Reason. **2**(1), 1–42 (1986)
2. Solnon, C., et al.: The car sequencing problem: overview of state-of-the-art methods and industrial case-study of the ROADEF 2005 challenge problem. Eur. J. Oper. Res. **191**(3), 912–927 (2008). https://doi.org/10.1016/j.ejor.2007.04.033
3. Sun, H., et al.: A colour-batching problem using selectivity banks in automobile paint shops. Int. J. Prod. Res. **53**, 1124–1142 (2015)
4. Ko, S.S., Han, Y.H., Choi, J.Y.: Paint batching problem on M-to-1 conveyor systems. Comput. Oper. Res. **74**, 118–126 (2016)
5. Hartmann, S.A., Runkler, T.A.: Online optimization of a color sorting assembly buffer using ant colony optimization. In: Kalcsics, J., Nickel, S. (eds.) Operations Research Proceedings, vol. 2007, pp. 415–420. Springer, Heidelberg (2008)
6. Bysko, S., Krystek, J.: A game theory approach for solving the new concept of car sequencing problem. In: Proceedings of 2019 5th International Conference on Control, Automation and Robotics (ICCAR), pp. 531–535 (2019). https://doi.org/10.1109/ICCAR.2019.8813372
7. Moya, I., Chica, M., Bautista, J.: Constructive metaheuristics for solving the car sequencing problem under uncertain partial demand. Comput. Ind. Eng. **137**, 106048 (2019). https://doi.org/10.1016/j.cie.2019.106048
8. Bysko, S., Krystek, J., Bysko, S., Lenort, R.: Buffer management in solving a real se-quencing problem in the automotive industry – paint shop 4.0 concept. Arch. Control Sci. **29**(3), 413–433 (2019). https://doi.org/10.24425/acs.2019.130203
9. Krystek, J., Alszer, S.: Współczesne aspekty problemu sekwencjonowania samochodów na wydziale lakierni. Mechanik **7**, 527–529 (2017). https://doi.org/10.17814/mechanik.2017.7.67

10. Bysko, S., Krystek, J., Bysko, S.: Automotive paint shop 4.0. Comput. Ind. Eng. **139**, 1 (2020). https://doi.org/10.1016/j.cie.2018.11.056
11. Krystek, J.: Nadążne sterowanie produkcją jako metoda sekwencjonowania wariantów produktów. In: Karkula, M., Matuszek, J., Skołud, B. (eds.) Wokół Inżynierii Produkcji, pp. 121–128. Wydawnictwa AGH, Kraków (2017)
12. Krystek, J., Bysko, S.: The follow-up control of the body sequencing process at the paint shop. Mechanik **92**(7), 462–464 (2019). https://doi.org/10.17814/mechanik.2019.7.57
13. Bysko, S., Krystek, J.: Follow-up sequencing algorithm for car sequencing problem 4.0. In: Szewczyk, R., Zieliński, C., Kaliczyńska, M. (eds.) Automation 2019. Advances in Intelligent Systems and Computing, vol. 920, pp. 145–154. Springer, Cham (2019). https://doi.org/10.1007/978-3-030-13273-6_15
14. Alszer, S., Krystek, J., Bysko, S.: Complex approach to the control of car body buffering in the paint shop. In: Szewczyk, R., Zieliński, C., Kaliczyńska, M. (eds.) Automation 2018. Advances in Automation, Robotics and Measurement Techniques, AISC, vol. 743, pp. 152–161. Springer, Berlin (2018). https://doi.org/10.1007/978-3-319-77179-3_14

State and Control Signal Limitations in Discrete Time Sliding Mode Control with Time Varying System's Uncertainties

Marek Jaskuła(✉) and Piotr Leśniewski

Institute of Automatic Control, Łodź University of Technology, 18/22 Stefanowskiego Street, 90-924 Łodź, Poland
marek.jaskula@dokt.p.lodz.pl, piotr.lesniewski@p.lodz.pl

Abstract. This paper studies the problem of bounding the control signal, state variables and state variables variations in the discrete time sliding mode control approach with external disturbances and time varying system's uncertainties. Presented results provide sufficient conditions for the fastest, monotonic, finite time convergence of the representative point to the predefined switching hyperplane, at the same time preserving the mentioned limitations. These conditions have the linear forms, therefore they can be easily verified, using the linear optimization methods. The results extend our previous research on this issue. The comparison is depicted in the simulation example.

Keywords: Sliding mode control · Discrete time object · Bounding state variable variation · Bounding control signal · Reaching law method

1 Introduction

In real world applications, systems are influenced by variety of uncertainties, caused by measurement, parameter identification and environmental problems. What is more, even the dynamics of the plant can vary during the regulation process. Therefore, there is a need for control strategies that could handle such difficulties, especially in cases were safety is of utmost importance. Nowadays, the sliding mode control approach has been found to be an effective method due to its robustness and computational efficiency [2,8]. This methodology can deal with many perturbed systems, however there is still research space to be explored when it comes to extending its applicability to uncertain systems and when considering control signal or state variables constraints issues.

The overall idea of the sliding mode control firstly appeared in the works of Utkin [14] and Emelyanov [7], where continuous time plants were researched. Soon after, the methodology was extended onto the discrete time systems [1,9,

R. Szewczyk et al. (Eds.): AUTOMATION 2021, AISC 1390, pp. 117–127, 2021.
https://doi.org/10.1007/978-3-030-74893-7_12

15]. In order to design the sliding mode controller, at first the switching hyperplane must be selected. Then, the control law can be derived using one of two methods: by demonstrating the stability of the sliding motion or a priori stating the evolution of the sliding variable, i.e. implementing the reaching law [3–6,10–12]. In this article we apply the latter approach, and extend the previous results [13], by taking into account the time varying system's uncertainties and by weakening the sufficient conditions for the fastest, monotonic, finite time convergence of the representative point to the switching hyperplane, simultaneously limiting the control input, state variables and state variables variations.

2 Control Law Design

First of all, let us take into consideration the discrete time plant of arbitrary order with time varying system's uncertainties and perturbed by external disturbances. The object dynamics is given by

$$\boldsymbol{x}(k+1) = \left[\boldsymbol{A} + \widetilde{\boldsymbol{A}}(k)\right]\boldsymbol{x}(k) + \boldsymbol{b}u(k) + \boldsymbol{d}(k), \tag{1}$$

where $\boldsymbol{x}(k)$ is $n \times 1$ state vector, $\boldsymbol{A}$ is $n \times n$ state matrix, $\widetilde{\boldsymbol{A}}(k)$ represents the time varying system's uncertainties and the control signal $u(k)$ is a scalar. Let us outline that perturbations $\boldsymbol{d}(k) = [d_1(k), \ldots, d_n(k)]^T$ are limited in the following manner: $\max|d_i(k)| = D_i$ for every $i \in \{1, \ldots, n\}$. We also assume that $\widetilde{a}_{ij}(k)$, elements of matrix $\widetilde{\boldsymbol{A}}(k)$, are bounded: $|\widetilde{a}_{ij}(k)| \le v_{ij}$ and $\boldsymbol{V} = [v_{ij}]$. We define the switching hyperplane $\boldsymbol{c}^T\boldsymbol{x}(k) = 0$, where $\boldsymbol{c}^T = [c_1, \ldots, c_{n-1}, 1]$ fulfills the condition $\boldsymbol{c}^T\boldsymbol{b} \neq 0$. As a consequence, the sliding variable $s(k) = \boldsymbol{c}^T\boldsymbol{x}(k)$. What is more, in the sliding phase we do not demand the representative point to cross the switching hyperplane in each consecutive step, but to remain in the certain neighborhood of that plane. Now, let us formulate the reaching law

$$s(k+1) = s(k) - K\mathrm{sgn}[s(k)] + \boldsymbol{c}^T\boldsymbol{d}(k) + \boldsymbol{c}^T\widetilde{\boldsymbol{A}}(k)\boldsymbol{x}(k), \tag{2}$$

where $K > 0$ is the convergence rate. From (1), $s(k) = 0$ and the above reaching law we compute the control input getting

$$u(k) = -\left(\boldsymbol{c}^T\boldsymbol{b}\right)^{-1}\left\{\boldsymbol{c}^T\boldsymbol{A}\boldsymbol{x}(k) - s(k) + K\mathrm{sgn}[s(k)]\right\}. \tag{3}$$

This control signal guarantees a stable quasi-sliding motion in the sliding phase. However, for the purpose of obtaining the terminal reaching phase, we must ensure $|s(k)| - |s(k+1)| \ge \lambda > 0$, up to the moment when the neighborhood of the switching hyperplane is reached. This condition is true when $K - D_{\max} - \boldsymbol{c}^T\widetilde{\boldsymbol{A}}(t)\boldsymbol{x}(t) \ge \lambda > 0$, where $D_{\max} = |c_1|D_1 + \ldots |c_{n-1}|D_{n-1} + D_n$. Moreover, by $V_{\max}$, let us denote the maximal possible impact of system's uncertainties on the sliding variable in case when the state belongs to the admissible set i.e. to $\mathbf{X} = \{\boldsymbol{x} \in \Re^n : \forall_{i\in\{1,\ldots,n\}}|x_i| \le r_i\}$. Therefore, $V_{\max} = \max_{\boldsymbol{x}\in X}\{\boldsymbol{c}^T\boldsymbol{V}\boldsymbol{x}(k)\}$. Our control methodology will result in keeping the representative point in $\mathbf{X}$.

Taking into account the fact that the initial point belongs to the admissible set, we obtain that $|s(k)| - |s(k+1)| \geq \lambda > 0$ is true when

$$K - D_{\max} - V_{\max} \geq \lambda > 0. \tag{4}$$

Although, we cannot evaluate the moment when the representative point reaches the neighborhood of the switching hyperplane, we can approximate it using (4).

3 Bounding the Control Signal and the State Variables

In this section we compute maximal possible convergence rates that ensure the control signal and the state variables restrictions, extending the previous results in [13], by including the influence of the time varying system's uncertainties. Firstly, we take into consideration the control input constraint $-r_u \leq u(k) \leq r_u$, where $k \geq 0, r_u > 0$. Using (3) we rewrite this inequality, getting

$$-r_u \leq -\left(\boldsymbol{c}^T\boldsymbol{b}\right)^{-1}\left\{\boldsymbol{c}^T\boldsymbol{A}\boldsymbol{x}(k) - s(k) + K\text{sgn}\left[s(k)\right]\right\} \leq r_u. \tag{5}$$

Noticing that the impact of the system's uncertainties is not present in (5), we obtain the same convergence rate as in [13]. Therefore,

$$K_u = \left|\boldsymbol{c}^T\boldsymbol{b}\right| r_u - \text{sgn}\left[s(k)\right]\boldsymbol{c}^T\boldsymbol{A}\boldsymbol{x}(k) + |s(k)|. \tag{6}$$

Applying K_u into the control signal will result in its maximal or minimal admissible value. Let us also derive the lowest convergence rate that still guarantees the control input limitation. Similar calculations lead us to

$$K_{\min_u} = -\left|\boldsymbol{c}^T\boldsymbol{b}\right| r_u - \text{sgn}\left[s(k)\right]\boldsymbol{c}^T\boldsymbol{A}\boldsymbol{x}(k) + |s(k)|. \tag{7}$$

Next, we will focus on the state variables restrictions $-r_i \leq x_i(k) \leq r_i$, where $i \in \{1, \ldots, n\}, k \geq 0$ and $D_i + V_i < \frac{r_i}{2}$. We start from substituting (3) into the state equation. Then, we left multiply by versor $\boldsymbol{e}_i$ obtaining

$$\begin{aligned} -r_i \leq &\boldsymbol{e}_i\boldsymbol{A}\boldsymbol{x}(k) + \boldsymbol{e}_i\widetilde{\boldsymbol{A}}(k)\boldsymbol{x}(k) - b_i\left(\boldsymbol{c}^T\boldsymbol{b}\right)^{-1}\boldsymbol{c}^T\boldsymbol{A}\boldsymbol{x}(k) + b_i\left(\boldsymbol{c}^T\boldsymbol{b}\right)^{-1}s(k) \\ &- b_i\left(\boldsymbol{c}^T\boldsymbol{b}\right)^{-1}K\text{sgn}\left[s(k)\right] + d_i(k) \leq r_i. \end{aligned} \tag{8}$$

Let us denote the maximal possible influence of the system's uncertainties on the i-th state variable, for states in $\mathbf{X}$, by $V_i = \max_{\boldsymbol{x}\in X}\{\boldsymbol{e}_i\boldsymbol{V}\boldsymbol{x}(t)\}$. Taking into account that $|d_i(k)| \leq D_i$ and $|\boldsymbol{e}_i\boldsymbol{V}\boldsymbol{x}(t)| \leq V_i$ we have that (8) is true when

$$\begin{aligned} &r_i + \boldsymbol{e}_i\boldsymbol{A}\boldsymbol{x}(k) - b_i\left(\boldsymbol{c}^T\boldsymbol{b}\right)^{-1}\boldsymbol{c}^T\boldsymbol{A}\boldsymbol{x}(k) + b_i\left(\boldsymbol{c}^T\boldsymbol{b}\right)^{-1}s(k) - D_i - V_i \geq \\ &b_i\left(\boldsymbol{c}^T\boldsymbol{b}\right)^{-1}K\text{sgn}\left[s(k)\right] \geq \\ &-r_i + \boldsymbol{e}_i\boldsymbol{A}\boldsymbol{x}(k) - b_i\left(\boldsymbol{c}^T\boldsymbol{b}\right)^{-1}\boldsymbol{c}^T\boldsymbol{A}\boldsymbol{x}(k) + b_i\left(\boldsymbol{c}^T\boldsymbol{b}\right)^{-1}s(k) + D_i + V_i. \end{aligned} \tag{9}$$

By deriving the biggest convergence rate $K = K_i$ that satisfies (9) we obtain

$$\begin{aligned} K_i = & \left|b_i^{-1}\boldsymbol{c}^T\boldsymbol{b}\right|(r_i - D_i - V_i) + \operatorname{sgn}[s(k)]\, b_i^{-1}\boldsymbol{c}^T\boldsymbol{b}\boldsymbol{e}_i\boldsymbol{A}\boldsymbol{x}(k) \\ & - \operatorname{sgn}[s(k)]\,\boldsymbol{c}^T\boldsymbol{A}\boldsymbol{x}(k) + |s(k)|. \end{aligned} \quad (10)$$

Moreover, the lowest convergence rate $K = K_{\min_i}$ that fulfills (9) is of the form

$$\begin{aligned} K_{\min_i} = & -\left|b_i^{-1}\boldsymbol{c}^T\boldsymbol{b}\right|(r_i - D_i - V_i) + \operatorname{sgn}[s(k)]\, b_i^{-1}\boldsymbol{c}^T\boldsymbol{b}\boldsymbol{e}_i\boldsymbol{A}\boldsymbol{x}(k) \\ & - \operatorname{sgn}[s(k)]\,\boldsymbol{c}^T\boldsymbol{A}\boldsymbol{x}(k) + |s(k)|. \end{aligned} \quad (11)$$

4 Bounding the State Variables Variations

This section consists the design of the biggest convergence rate that keeps the i-th state variable variation in the admissible range, i.e. satisfies $-\delta_i \le x_i(k+1) - x_i(k) \le \delta_i$, for the whole reaching phase, for $\delta_i > 0$. In order to ensure this property in the sliding phase it is necessary to select a proper switching hyperplane. The process of computing the mentioned convergence rate we start from substituting the state equation and the control law into $-\delta_i \le x_i(k+1) - x_i(k) \le \delta_i$. Then, after calculation, we obtain

$$\begin{aligned} & \boldsymbol{e}_i\boldsymbol{A}\boldsymbol{x}(k) - b_i\left(\boldsymbol{c}^T\boldsymbol{b}\right)^{-1}\boldsymbol{c}^T\boldsymbol{A}\boldsymbol{x}(k) + b_i\left(\boldsymbol{c}^T\boldsymbol{b}\right)^{-1}s(k) + d_i(k) + \boldsymbol{e}_i\widetilde{\boldsymbol{A}}(k)\boldsymbol{x}(k) \\ & - \boldsymbol{e}_i\boldsymbol{x}(k) + \delta_i \ge b_i\left(\boldsymbol{c}^T\boldsymbol{b}\right)^{-1}K\operatorname{sgn}[s(k)] \ge \boldsymbol{e}_i\boldsymbol{A}\boldsymbol{x}(k) - b_i\left(\boldsymbol{c}^T\boldsymbol{b}\right)^{-1}\boldsymbol{c}^T\boldsymbol{A}\boldsymbol{x}(k) \\ & + b_i\left(\boldsymbol{c}^T\boldsymbol{b}\right)^{-1}s(k) + d_i(k) + \boldsymbol{e}_i\widetilde{\boldsymbol{A}}(k)\boldsymbol{x}(k) - \boldsymbol{e}_i\boldsymbol{x}(k) - \delta_i. \end{aligned} \quad (12)$$

Taking into account the extreme values of disturbances and uncertainties we get

$$\begin{aligned} & \boldsymbol{e}_i\boldsymbol{A}\boldsymbol{x}(k) - b_i\left(\boldsymbol{c}^T\boldsymbol{b}\right)^{-1}\boldsymbol{c}^T\boldsymbol{A}\boldsymbol{x}(k) + b_i\left(\boldsymbol{c}^T\boldsymbol{b}\right)^{-1}s(k) - D_i - V_i \\ & - \boldsymbol{e}_i\boldsymbol{x}(k) + \delta_i \ge b_i\left(\boldsymbol{c}^T\boldsymbol{b}\right)^{-1}K\operatorname{sgn}[s(k)] \ge \boldsymbol{e}_i\boldsymbol{A}\boldsymbol{x}(k) - b_i\left(\boldsymbol{c}^T\boldsymbol{b}\right)^{-1}\boldsymbol{c}^T\boldsymbol{A}\boldsymbol{x}(k) \\ & + b_i\left(\boldsymbol{c}^T\boldsymbol{b}\right)^{-1}s(k) + D_i + V_i - \boldsymbol{e}_i\boldsymbol{x}(k) - \delta_i. \end{aligned} \quad (13)$$

From (13), we derive the biggest convergence rate, denoted by K_{δ_i}, obtaining

$$\begin{aligned} K_{\delta_i} = & \left|b_i^{-1}\boldsymbol{c}^T\boldsymbol{b}\right|(\delta_i - D_i - V_i) + \operatorname{sgn}[s(k)]\, b_i^{-1}\boldsymbol{c}^T\boldsymbol{b}\boldsymbol{e}_i(\boldsymbol{A} - \boldsymbol{I})\boldsymbol{x}(k) \\ & - \operatorname{sgn}[s(k)]\,\boldsymbol{c}^T\boldsymbol{A}\boldsymbol{x}(k) + |s(k)|. \end{aligned} \quad (14)$$

Moreover, the lowest possible convergence rate that ensures $-\delta_i \le x_i(k+1) - x_i(k) \le \delta_i$, denoted by $K_{\min_{\delta_i}}$, is of the form

$$\begin{aligned} K_{\min_{\delta_i}} = & -\left|b_i^{-1}\boldsymbol{c}^T\boldsymbol{b}\right|(\delta_i - D_i - V_i) + \operatorname{sgn}[s(k)]\, b_i^{-1}\boldsymbol{c}^T\boldsymbol{b}\boldsymbol{e}_i(\boldsymbol{A} - \boldsymbol{I})\boldsymbol{x}(k) \\ & - \operatorname{sgn}[s(k)]\,\boldsymbol{c}^T\boldsymbol{A}\boldsymbol{x}(k) + |s(k)|. \end{aligned} \quad (15)$$

5 Control Strategy and Sufficient Conditions

Now we present the methodology of selecting the convergence rate K during the control process. Further we will formulate and prove the theorems that will a priori guarantee the fastest, monotonic, finite time convergence of the sliding variable to zero, at the same time preserving the control signal, state variables and state variables variations from violating their admissible ranges. To begin with, convergence rate is chosen as $K = \min\{K_u, K_{\delta_i}, \ldots, K_{\delta_n}, K_1, \ldots, K_n, |s(k)|\}$. Before we move on, let us notice that the initial point is chosen in the admissible set and if the monotonic convergence is obtained, then the state will belong to

$$\{\mathbf{X}_0 = \boldsymbol{x} \in \Re^n : \forall_{i\in\{1,\ldots,n\}} |x_i| \leq r_i, \left|\boldsymbol{c}^T\boldsymbol{x}\right| \leq \left|\boldsymbol{c}^T\boldsymbol{x}_0\right|\}. \tag{16}$$

We present the conditions that system must satisfy in order to make the mentioned properties hold. To obtain the monotonic convergence of the sliding variable to the neighborhood of zero we must verify if the following inequalities are true for the whole reaching phase:

$$K_u - D_{\max} - V_{\max} \geq \lambda > 0, \quad K_{\min_u} \leq |s(k)|, \tag{17}$$

$$K_i - D_{\max} - V_{\max} \geq \lambda > 0, \quad K_{\min_i} \leq |s(k)|, \tag{18}$$

$$K_{\delta_i} - D_{\max} - V_{\max} \geq \lambda > 0, \quad K_{\min_{\delta_i}} \leq |s(k)|. \tag{19}$$

Theorem 1. *The sufficient condition for (17) is that for $\boldsymbol{x}(k) \in \mathbf{X}_0$*

$$\left|\boldsymbol{c}^T\boldsymbol{b}\right| r_u \geq sgn[s(k)]\, \boldsymbol{c}^T\boldsymbol{A}\boldsymbol{x}(k) - |s(k)| + D_{\max} + V_{\max} + \lambda, \tag{20}$$

$$\left|\boldsymbol{c}^T\boldsymbol{b}\right| r_u \geq -sgn[s(k)]\, \boldsymbol{c}^T\boldsymbol{A}\boldsymbol{x}(k). \tag{21}$$

Proof. After transforming above inequalities and using formulas for $K_u, K_{\min_u}$, respectively, we get conditions (17).

Theorem 2. *The sufficient condition for (18) is that for $\boldsymbol{x}(k) \in \mathbf{X}_0$*

$$\begin{aligned}\left|b_i^{-1}\boldsymbol{c}^T\boldsymbol{b}\right|(r_i - D_i - V_i) \geq\, & sgn[s(k)]\left(\boldsymbol{c}^T - b_i^{-1}\boldsymbol{c}^T\boldsymbol{b}\boldsymbol{e}_i\right)\boldsymbol{A}\boldsymbol{x}(k) - |s(k)| \\ & + D_{\max} + V_{\max} + \lambda,\end{aligned} \tag{22}$$

$$\left|b_i^{-1}\boldsymbol{c}^T\boldsymbol{b}\right|(r_i - D_i - V_i) \geq -\, sgn[s(k)]\left(\boldsymbol{c}^T - b_i^{-1}\boldsymbol{c}^T\boldsymbol{b}\boldsymbol{e}_i\right)\boldsymbol{A}\boldsymbol{x}(k). \tag{23}$$

Proof. By implementing formulas for $K_i, K_{\min_i}$ into (22) and (23), respectively, we obtain conditions (18).

Theorem 3. *The sufficient condition for (19) is that for $\boldsymbol{x}(k) \in \mathbf{X}_0$*

$$\begin{aligned}\left|b_i^{-1}\boldsymbol{c}^T\boldsymbol{b}\right|(\delta_i - D_i - V_i) \geq\, & sgn[s(k)]\left[\boldsymbol{c}^T\boldsymbol{A} - b_i^{-1}\boldsymbol{c}^T\boldsymbol{b}\boldsymbol{e}_i(\boldsymbol{A} - \boldsymbol{I})\right]\boldsymbol{x}(k) - |s(k)| \\ & + D_{\max} + V_{\max} + \lambda,\end{aligned} \tag{24}$$

$$\left|b_i^{-1}\boldsymbol{c}^T\boldsymbol{b}\right|(\delta_i - D_i - V_i) \geq -\, sgn[s(k)]\left[\boldsymbol{c}^T\boldsymbol{A} - b_i^{-1}\boldsymbol{c}^T\boldsymbol{b}\boldsymbol{e}_i(\boldsymbol{A} - \boldsymbol{I})\right]\boldsymbol{x}(k). \tag{25}$$

Proof. Substituting (14), (15) to above inequalities, respectively, results in (19).

Now, it is crucial to verify if the convergence rates enable us to bound every state variable, state variable variation and control signal. For this purpose, we demand that convergence rates with subscript "min" are lower then convergence without "min". Otherwise, there will not exist a value of K that allows to satisfy all limits. For the sake of clarity, we will start from conditions

$$K_{\min_u} \leq K_i, K_{\min_i} \leq K_u. \tag{26}$$

Theorem 4. *The sufficient condition for (26) is that for $\boldsymbol{x}(k) \in \boldsymbol{X}_0$*

$$r_u + \left|b_i^{-1}\right| (r_i - D_i - V_i) \geq \left|b_i^{-1} \boldsymbol{e}_i \boldsymbol{A} \boldsymbol{x}(k)\right|. \tag{27}$$

Proof. We will begin the proof from showing that $K_{\min_i} \leq K_u$. Multiplying both sides of above inequality by $\left|\boldsymbol{c}^T \boldsymbol{b}\right|$, and then subtracting $\left|\boldsymbol{c}^T \boldsymbol{b} b_i^{-1}\right| (r_i - D_i - V_i) + \operatorname{sgn}\left[s\left(k\right)\right] \boldsymbol{c}^T \boldsymbol{A} \boldsymbol{x}\left(k\right) - \left|s\left(k\right)\right|$ we get

$$\begin{aligned} &\left|\boldsymbol{c}^T \boldsymbol{b}\right| r_u - \operatorname{sgn}\left[s\left(k\right)\right] \boldsymbol{c}^T \boldsymbol{A} \boldsymbol{x}\left(k\right) + \left|s\left(k\right)\right| \geq \\ &- \left|\boldsymbol{c}^T \boldsymbol{b} b_i^{-1}\right| (r_i - D_i - V_i) + \left|\boldsymbol{c}^T \boldsymbol{b} b_i^{-1} \boldsymbol{e}_i \boldsymbol{A} \boldsymbol{x}\left(k\right)\right| - \operatorname{sgn}\left[s\left(k\right)\right] \boldsymbol{c}^T \boldsymbol{A} \boldsymbol{x}\left(k\right) + \left|s\left(k\right)\right|. \end{aligned} \tag{28}$$

Let us notice that the left side of this inequality is equivalent to the formula for K_u, while the right side is greater or equal to (11). Thus, we obtain $K_{\min_i} \leq K_u$. Now, we will prove $K_{\min_u} \leq K_i$. Firstly, we multiply both sides of (27) by $-\left|\boldsymbol{c}^T \boldsymbol{b}\right|$ and then we add $\left|\boldsymbol{c}^T \boldsymbol{b} b_i^{-1}\right| (r_i - D_i - V_i) - \operatorname{sgn}\left[s\left(k\right)\right] \boldsymbol{c}^T \boldsymbol{A} \boldsymbol{x}\left(k\right) + \left|s\left(k\right)\right|$. As a consequence,

$$\begin{aligned} &- \left|\boldsymbol{c}^T \boldsymbol{b}\right| r_u - \operatorname{sgn}\left[s\left(k\right)\right] \boldsymbol{c}^T \boldsymbol{A} \boldsymbol{x}\left(k\right) + \left|s\left(k\right)\right| \leq \\ &\left|\boldsymbol{c}^T \boldsymbol{b} b_i^{-1}\right| (r_i - D_i - V_i) - \left|\boldsymbol{c}^T \boldsymbol{b} b_i^{-1} \boldsymbol{e}_i \boldsymbol{A} \boldsymbol{x}\left(k\right)\right| - \operatorname{sgn}\left[s\left(k\right)\right] \boldsymbol{c}^T \boldsymbol{A} \boldsymbol{x}\left(k\right) + \left|s\left(k\right)\right|. \end{aligned} \tag{29}$$

Noticing that the left hand side is equivalent to the formula for $K_{\min_u}$ and the right hand side is lower or equal to (10) we get $K_{\min_u} \leq K_i$.

Now, because $K_{\min_i} \leq K_i$ is always true, we will demand that for $i \neq j$

$$K_{\min_i} \leq K_j. \tag{30}$$

Theorem 5. *The sufficient condition for (30) is that for $\boldsymbol{x}(k) \in \boldsymbol{X}_0$*

$$\begin{aligned} &\left|b_i^{-1}\right| (r_i - D_i - V_i) + \left|b_j^{-1}\right| (r_j - D_j - V_j) \geq \\ &sgn\left[s\left(k\right) \boldsymbol{c}^T \boldsymbol{b}\right] \left(b_i^{-1} \boldsymbol{e}_i - b_j^{-1} \boldsymbol{e}_j\right) \boldsymbol{A} \boldsymbol{x}(k). \end{aligned} \tag{31}$$

Proof. At first, let us move to the left hand side all terms connected with j-th index and to the right hand side all terms with i-th index and then multiply both sides by $\left|\boldsymbol{c}^T \boldsymbol{b}\right|$. As a consequence

$$\begin{aligned} &\left|b_j^{-1} \boldsymbol{c}^T \boldsymbol{b}\right| (r_j - D_j - V_j) + \operatorname{sgn}\left[s\left(k\right)\right] b_j^{-1} \boldsymbol{c}^T \boldsymbol{b} \boldsymbol{e}_j \boldsymbol{A} \boldsymbol{x}\left(k\right) \geq \\ &- \left|b_i^{-1} \boldsymbol{c}^T \boldsymbol{b}\right| (r_i - D_i - V_i) + \operatorname{sgn}\left[s\left(k\right)\right] b_i^{-1} \boldsymbol{c}^T \boldsymbol{b} \boldsymbol{e}_i \boldsymbol{A} \boldsymbol{x}\left(k\right). \end{aligned} \tag{32}$$

Let us observe that after subtracting $\operatorname{sgn}[s(k)]\, \boldsymbol{c}^T \boldsymbol{A}\boldsymbol{x}(k) - |s(k)|$ the left hand side is equivalent to K_j, while the right hand side is equivalent to $K_{\min_i}$. Therefore, (31) implies (30).

Furthermore, let us move to the conditions

$$K_{\min_u} \leq K_{\delta_i}, K_{\min_{\delta_i}} \leq K_u. \tag{33}$$

Theorem 6. *The sufficient condition for (33) is that for $\boldsymbol{x}(k) \in \boldsymbol{X}_0$*

$$r_u + \left|b_i^{-1}\right| (\delta_i - D_i - V_i) \geq \left|b_i^{-1} \boldsymbol{e}_i (\boldsymbol{A} - \boldsymbol{I})\, \boldsymbol{x}(k)\right|. \tag{34}$$

Proof. We shall begin with proving $K_{\min_{\delta_i}} \leq K_u$. After multiplying both sides of the above inequality by $\left|\boldsymbol{c}^T \boldsymbol{b}\right|$ and then subtracting $\left|\boldsymbol{c}^T \boldsymbol{b} b_i^{-1}\right| (\delta_i - D_i - V_i) +$ $\operatorname{sgn}[s(k)]\, \boldsymbol{c}^T \boldsymbol{A}\boldsymbol{x}(k) - |s(k)|$ we get

$$\begin{aligned} &\left|\boldsymbol{c}^T \boldsymbol{b}\right| r_u - \operatorname{sgn}[s(k)]\, \boldsymbol{c}^T \boldsymbol{A}\boldsymbol{x}(k) + |s(k)| \geq |s(k)| \\ &- \left|\boldsymbol{c}^T \boldsymbol{b} b_i^{-1}\right| (\delta_i - D_i - V_i) + \left|\boldsymbol{c}^T \boldsymbol{b} b_i^{-1} \boldsymbol{e}_i (\boldsymbol{A} - \mathbf{I})\, \boldsymbol{x}(k)\right| - \operatorname{sgn}[s(k)]\, \boldsymbol{c}^T \boldsymbol{A}\boldsymbol{x}(k). \end{aligned} \tag{35}$$

The left hand side of inequality (35) is equal to K_u. Moreover, the right hand side is greater or equal to $K_{\min_{\delta_i}}$. Hence, we get $K_{\min_{\delta_i}} \leq K_u$. Next, we move to the proof for $K_{\min_u} \leq K_{\delta_i}$. Firstly, let us multiply both sides of (34) by $-\left|\boldsymbol{c}^T \boldsymbol{b}\right|$. Then, we subtract $\left|\boldsymbol{c}^T \boldsymbol{b} b_i^{-1}\right| (\delta_i - D_i - V_i) - \operatorname{sgn}[s(k)]\, \boldsymbol{c}^T \boldsymbol{A}\boldsymbol{x}(k) + |s(k)|$. Thus,

$$\begin{aligned} &- \left|\boldsymbol{c}^T \boldsymbol{b}\right| r_u - \operatorname{sgn}[s(k)]\, \boldsymbol{c}^T \boldsymbol{A}\boldsymbol{x}(k) + |s(k)| \leq |s(k)| \\ &+ \left|\boldsymbol{c}^T \boldsymbol{b} b_i^{-1}\right| (\delta_i - D_i - V_i) - \left|\boldsymbol{c}^T \boldsymbol{b} b_i^{-1} \boldsymbol{e}_i (\boldsymbol{A} - \mathbf{I})\, \boldsymbol{x}(k)\right| - \operatorname{sgn}[s(k)]\, \boldsymbol{c}^T \boldsymbol{A}\boldsymbol{x}(k). \end{aligned} \tag{36}$$

Similarly to the previous case, left hand side is equal to $K_{\min_u}$, while the right hand side is lower or equal to K_{δ_i}. Therefore, we have $K_{\min_u} \leq K_{\delta_i}$.

Before we move on, let us notice that conditions $K_{\min_i} \leq K_{\delta_i}$, $K_{\min_{\delta_i}} \leq K_i$ are always true, because $D_i + V_i < \frac{r_i}{2}$. Thus, we just need to verify if for $i \neq j$

$$K_{\min_j} \leq K_{\delta_i}, K_{\min_{\delta_i}} \leq K_j. \tag{37}$$

Theorem 7. *The sufficient condition for (37) is that for $\boldsymbol{x}(k) \in \boldsymbol{X}_0$*

$$\left|b_i^{-1}\right| (\delta_i - D_i - V_i) + \left|b_j^{-1}\right| (r_j - D_j - V_j) \geq \left|\left(b_i^{-1} \boldsymbol{e}_i (\boldsymbol{A} - \boldsymbol{I}) - b_j^{-1} \boldsymbol{e}_j \boldsymbol{A}\right) \boldsymbol{x}(k)\right|. \tag{38}$$

Proof. First of all, we will focus on $K_{\min_j} \leq K_{\delta_i}$. From (38) one can get

$$\begin{aligned} &\left|b_i^{-1}\right| (\delta_i - D_i - V_i) + \left|b_j^{-1}\right| (r_j - D_j - V_j) \geq \\ &- \operatorname{sgn}\left[s(k)\, \boldsymbol{c}^T \boldsymbol{b}\right] \left(b_i^{-1} \boldsymbol{e}_i (\boldsymbol{A} - \boldsymbol{I}) - b_j^{-1} \boldsymbol{e}_j \boldsymbol{A}\right) \boldsymbol{x}(k). \end{aligned} \tag{39}$$

Now, we move elements with i-th indexes to the left hand side of inequality and elements with j-th to the right hand side. Hence, we have

$$\begin{aligned}&\left|b_i^{-1}\right|(\delta_i - D_i - V_i) + \operatorname{sgn}\left[s(k)\,\boldsymbol{c}^T\boldsymbol{b}\right] b_i^{-1}\boldsymbol{e}_i(\boldsymbol{A}-\boldsymbol{I})\,\boldsymbol{x}(k) \geq \\ &-\left|b_j^{-1}\right|(r_j - D_j - V_j) + \operatorname{sgn}\left[s(k)\,\boldsymbol{c}^T\boldsymbol{b}\right] b_j^{-1}\boldsymbol{e}_j\boldsymbol{A}\boldsymbol{x}(k).\end{aligned} \tag{40}$$

Next, we multiply both sides of above inequality by $\left|\boldsymbol{c}^T\boldsymbol{b}\right|$ and subtract term $\operatorname{sgn}[s(k)]\,\boldsymbol{c}^T\boldsymbol{A}\boldsymbol{x}(k) + |s(k)|$. Taking into account the formulas for K_{δ_i}, $K_{\min_j}$ we get $K_{\min_j} \leq K_{\delta_i}$. Now, we will prove $K_{\min_{\delta_i}} \leq K_j$. Firstly, from (38) we get

$$\begin{aligned}&\left|b_i^{-1}\right|(\delta_i - D_i - V_i) + \left|b_j^{-1}\right|(r_j - D_j - V_j) \geq \\ &\operatorname{sgn}\left[s(k)\,\boldsymbol{c}^T\boldsymbol{b}\right]\left(b_i^{-1}\boldsymbol{e}_i(\boldsymbol{A}-\boldsymbol{I}) - b_j^{-1}\boldsymbol{e}_j\boldsymbol{A}\right)\boldsymbol{x}(k).\end{aligned} \tag{41}$$

Similarly to the previous case, we combine like terms, obtaining

$$\begin{aligned}&\left|b_j^{-1}\right|(r_j - D_j - V_j) + \operatorname{sgn}\left[s(k)\,\boldsymbol{c}^T\boldsymbol{b}\right] b_j^{-1}\boldsymbol{e}_j\boldsymbol{A}\boldsymbol{x}(k) \geq \\ &-\left|b_i^{-1}\right|(\delta_i - D_i - V_i) + \operatorname{sgn}\left[s(k)\,\boldsymbol{c}^T\boldsymbol{b}\right] b_i^{-1}\boldsymbol{e}_i(\boldsymbol{A}-\boldsymbol{I})\,\boldsymbol{x}(k).\end{aligned} \tag{42}$$

Further, we multiply both sides of (42) by term $\left|\boldsymbol{c}^T\boldsymbol{b}\right|$ and then subtract $|s(k)| + \operatorname{sgn}[s(k)]\,\boldsymbol{c}^T\boldsymbol{A}\boldsymbol{x}(k)$. After using formulas for $K_{\min_{\delta_i}}$, K_j we get $K_{\min_{\delta_i}} \leq K_j$.

The last condition is that for $i \neq j$

$$K_{\min_{\delta_i}} \leq K_{\delta_j}. \tag{43}$$

Theorem 8. *The sufficient condition for (43) is that for* $\boldsymbol{x}(k) \in \boldsymbol{X}_0$

$$\begin{aligned}&\left|b_i^{-1}\right|(\delta_i - D_i - V_i) + \left|b_j^{-1}\right|(\delta_j - D_j - V_j) \geq \\ &sgn\left[s(k)\,\boldsymbol{c}^T\boldsymbol{b}\right]\left(b_i^{-1}\boldsymbol{e}_i - b_j^{-1}\boldsymbol{e}_j\right)(\boldsymbol{A}-\boldsymbol{I})\,\boldsymbol{x}(k).\end{aligned} \tag{44}$$

Proof. Let us multiply both sides of (44) by $\left|\boldsymbol{c}^T\boldsymbol{b}\right|$. Then after combining terms connected with i-th index and j-th index we get

$$\begin{aligned}&\left|b_j^{-1}\boldsymbol{c}^T\boldsymbol{b}\right|(\delta_j - D_j - V_j) + \operatorname{sgn}[s(k)]\, b_j^{-1}\boldsymbol{c}^T\boldsymbol{b}\boldsymbol{e}_j(\boldsymbol{A}-\boldsymbol{I})\,\boldsymbol{x}(k) \geq \\ &-\left|b_i^{-1}\boldsymbol{c}^T\boldsymbol{b}\right|(\delta_i - D_i - V_i) + \operatorname{sgn}[s(k)]\, b_i^{-1}\boldsymbol{c}^T\boldsymbol{b}\boldsymbol{e}_i(\boldsymbol{A}-\boldsymbol{I})\,\boldsymbol{x}(k).\end{aligned} \tag{45}$$

After subtracting term $\operatorname{sgn}[s(k)]\,\boldsymbol{c}^T\boldsymbol{A}\boldsymbol{x}(k) - |s(k)|$ and substituting formulas for $K_{\min_{\delta_i}}$ and K_{δ_j} we obtain (43).

6 Numerical Example

The simulation example will present the advantage over the analysis done in [13]. First of all, let us take into account the discrete time object with time varying uncertainties, described by the Eq. (1), where

$$\boldsymbol{A} = \begin{bmatrix} 1 & 0.5 \\ 0.5 & -1 \end{bmatrix}, \qquad \boldsymbol{b} = \begin{bmatrix} 1/2 \\ 1 \end{bmatrix}, \qquad \widetilde{\boldsymbol{A}}(k) = \begin{bmatrix} 0.01\sin(k) & 0 \\ 0 & 0.1\sin(k) \end{bmatrix} \tag{46}$$

and $d_1(k) = (-1)^{\lfloor k/10 \rfloor}, d_2(k) = 0.1\,(-1)^k$. What is more, the parameters of the switching line are $c_1 = c_2 = 1$. During the regulation process we require $|x_1(k)| \leq 50$, $|x_2(k)| \leq 5$, $|x_1(k) - x_1(k-1)| \leq 25$, $|x_2(k) - x_2(k-1)| \leq 3$, $|u(k)| \leq 35$. The initial state $\boldsymbol{x}_0 = [50, 5]^T$ is selected at the corner of the admissible set. One can observe that the conditions from Theorems (1–8) are true. Although, the condition for the second state variable constraint presented in [13] is not satisfied (at the initial point). Hence, the improvement has been made, even apart from taking into account the influence of the system's uncertainties.

From Fig. 1 we can see that both state variables constraints hold for the whole regulation process. The representative point, starting from the corner of the admissible set, is driven onto the neighborhood of the second state limitation and then moves along it until the switching line is reached. Figure 2 illustrates the control input. We can notice that its admissible range is not fully utilized. It is caused by the need of bounding the state and the state variation. Figure 3 and Fig. 4 depict the first state variable variation and the second state variable variation, respectively. We can observe that both state variables variations are limited from the very beginning of the control process. Moreover, the admissible range of the second state variable variation is utilized (with accuracy to the external disturbances and the system's uncertainties) for the first few steps. This is the reason why the control signal could not be set at its maximum magnitude.

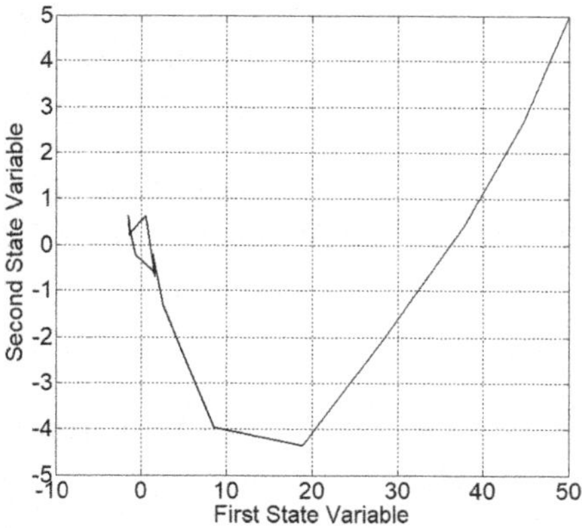

Fig. 1. State trajectory

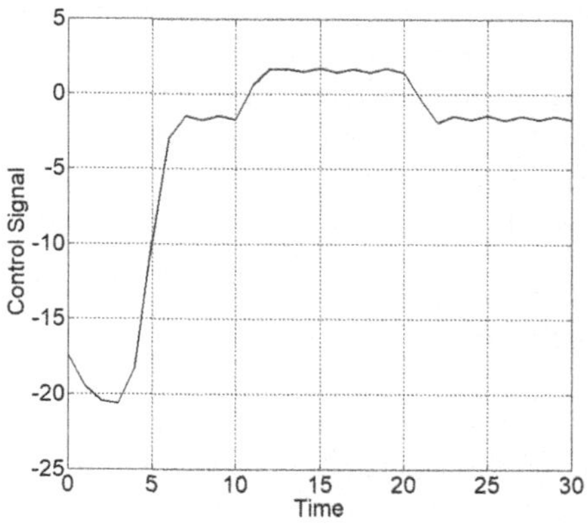

Fig. 2. Control signal

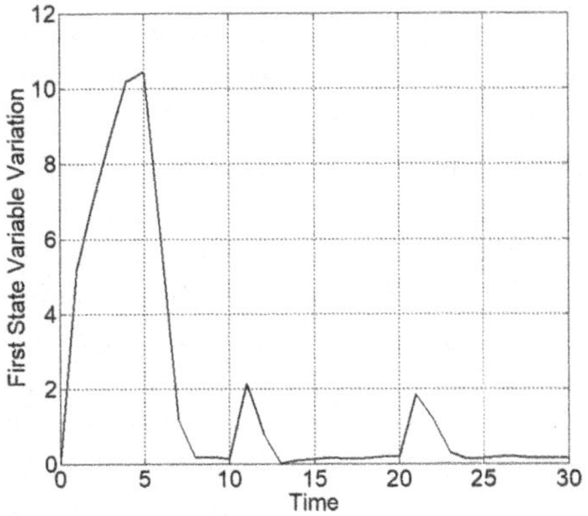

Fig. 3. First state variable variation

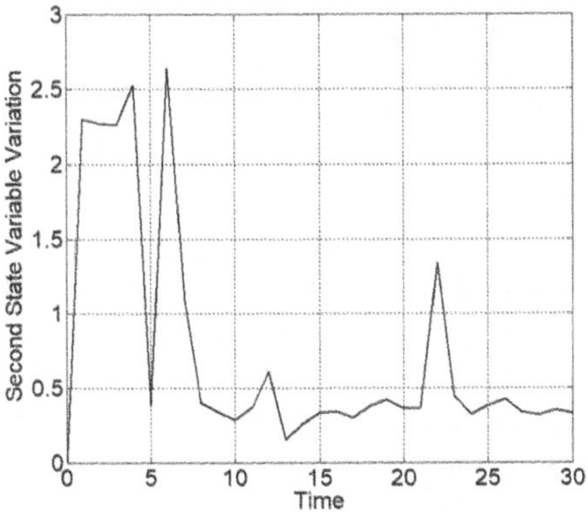

Fig. 4. Second state variable variation

7 Final Remarks

This paper is devoted to the problem of bounding the control signal, state variables and state variables variations in the discrete time domain. Moreover, the system was not only perturbed by the unknown, external disturbances but also by the time varying system's uncertainties. By the use of the presented reaching law, we have achieved the fastest, monotonic, finite time convergence of the representative point to the predefined switching hyperplane, simultaneously preserving the mentioned limitations for the whole regulation process. The sufficient condition presented in this paper was weaker than the one in the [13], which was verified also by the numerical analysis. Our future aim it to consider the time delayed system and still guarantee all of the considered properties.

References

1. Bartoszewicz, A.: Discrete-time quasi-sliding-mode control strategies. IEEE Trans. Ind. Electron. **45**, 633–637 (1998)
2. Bartoszewicz, A., Kaynak, O., Utkin, V.I. (eds.): Sliding mode control in industrial applications. Spec. Sect. IEEE Trans. Industr. Electron. **55**, 3805–4103 (2008)
3. Bartoszewicz, A., Latosiński, P.: Discrete time sliding mode control with reduced switching - a new reaching law approach. Int. J. Robust Nonlinear Control **26**, 47–68 (2016)
4. Bartoszewicz, A., Leśniewski, P.: Reaching law approach to the sliding mode control of periodic review inventory systems. IEEE Trans. Autom. Sci. Eng. **11**, 810–817 (2014)
5. Bartoszewicz, A., Leśniewski, P.: New switching and nonswitching type reaching laws for SMC of discrete time systems. IEEE Trans. Control Syst. Technol. **24**, 670–677 (2016)
6. Chakrabarty, S., Bandyopadhyay, B.: A generalized reaching law for discrete sliding mode control. Automatica **52**, 83–86 (2015)
7. Emelyanov, S.V.: Variable Structure Control Systems. Nauka, Moscow (1967)

8. Fallaha, C.J., Saad, M., Kanaan, H.Y., Al-Haddad, K.: Sliding-mode robot control with exponential reaching law. IEEE Trans. Ind. Electron. **58**, 600–610 (2011)
9. Furuta, K.: Sliding mode control of a discrete system. Syst. Control Lett. **14**, 145–152 (1990)
10. Gao, W., Hung, J.C.: Variable structure control of nonlinear systems: a new approach. IEEE Trans. Ind. Electron. **40**, 45–55 (1993)
11. Gao, W., Wang, Y., Homaifa, A.: Discrete-time variable structure control systems. IEEE Trans. Ind. Electron. **42**, 117–122 (1995)
12. Golo, G., Milosavljević, Č.: Robust discrete-time chattering free sliding mode control. Syst. Control Lett. **41**, 19–28 (2000)
13. Jaskuła, M.: Constraining the state and control signal variation using the sliding mode control. In: Automation 2020: Towards Industry of the Future, pp. 24–33. Springer, Cham (2020)
14. Utkin, V.I.: Variable structure systems with sliding modes. IEEE Trans. Autom. Control **22**, 212–222 (1977)
15. Utkin, V.I., Drakunow, S.V.: On discrete-time sliding mode control. In: IFAC Conference Nonlinear Control, pp. 484–489 (1989)

Modeling of Systems of Automated Auxiliary Processes in Pharmaceutical Industry

Igor Korobiichuk[1(✉)], Nataliia Shybetska[2], Vladyslav Shybetskyi[3], and Sergii Kostyk[3]

[1] Łukasiewicz Research Network – Industrial Research Institute for Automation and Measurements PIAP, Warsaw, Poland
igor.korobiichuk@piap.lukasiewicz.gov.pl
[2] Joint Stock Company «Farmak», Kyiv, Ukraine
n.shibetska@farmak.ua
[3] National Technical University of Ukraine "Igor Sikorsky Kyiv Polytechnic Institute", Kyiv, Ukraine
kostyksergey@ukr.net

Abstract. The computer model of the device with the CIP (clean in place) system allows at the design stage to reduce the cost of implementation and commissioning. A computer model of the bioreactor washing process using the CIP system was built. Using the ANSYS finite element analysis system, diagrams of the distribution of fluid flows for different supply systems are constructed. The influence of the design of the spray device on the quality of the washing process of the bioreactor is analyzed.

Keywords: Bioreactor · CIP system · ANSYS · Modeling · Hydrodynamics · Sprayball

1 Introduction

In modern practice, the implementation of pharmaceutical manufacturing automation processes is increasingly focused on supporting processes. The duration and cost of conducting which usually very strongly affect the cost of the final product. On the one hand, auxiliary processes should provide the proper conditions for the main technological operations, on the other hand, be as short and cheap as possible.

Optimization issues, designed to reduce production costs, have recently increasingly affected at all stages of production [1, 2]. Including issues of cleaning equipment and facilities, which for sterile production can take up to 40% of working time. Which is significantly reduces line performance.

In the case of combining the production of several drugs on the same production line, equipment washing becomes a bottleneck, as well as the expectation of an analysis confirming the quality of washing. The introduction of CIP (clean in place) automatic washing systems can help avoid the need for analysis by validating the washing process [3]. Since manual operations cannot be validated.

R. Szewczyk et al. (Eds.): AUTOMATION 2021, AISC 1390, pp. 128–135, 2021.
https://doi.org/10.1007/978-3-030-74893-7_13

CIP systems can use not only water for washing, but also detergents, which in turn ensures a reduction in water costs. Thus, it significantly reduces the cost of production and reduces the negative impact on the environment [4].

The CIP system for capacitive equipment can use both rotating and stationary spray elements [5]. In general, it is a perforated atomizer located in the cavity of the apparatus through which detergent or water is supplied (Fig. 1).

Fig. 1. CIP system sprayer

The implementation of CIP systems requires detailed study at the design stage to reduce the risk of dead zones and determine the optimal parameters with minimal water consumption.

The purpose of this article is to simulate the process of the distribution of water flows within a bioreactor during washing, for different spraying elements.

To achieve this goal, it is necessary to solve the following tasks:

1. Construction of the geometry of the internal cavity of the bioreactor.
2. Building a computer model of the process of the CIP system for different designs of the spray element.
3. Analysis of the influence of the geometry of the spray element on the efficiency of the washing system.

2 Materials and Methods of Research

Modeling of the process is implemented in a 2-D. For modeling, the built-in "Fluent" solver of the ANSYS universal element analysis software system is used [6–9].

A geometric model of the internal space of the bioreactor with a nominal volume of 0.1 m^3 was constructed (Fig. 2). The internal diameter of the apparatus is 0.5 m, and the height is 0.545 m. In the upper part there is a inlet of the CIP system. Modeling is carried out for two types of spraying: without nozzle, hole with diameter 5 mm (Fig. 2, a); with ball nozzle, 8 holes of 1 mm each (Fig. 2, b).

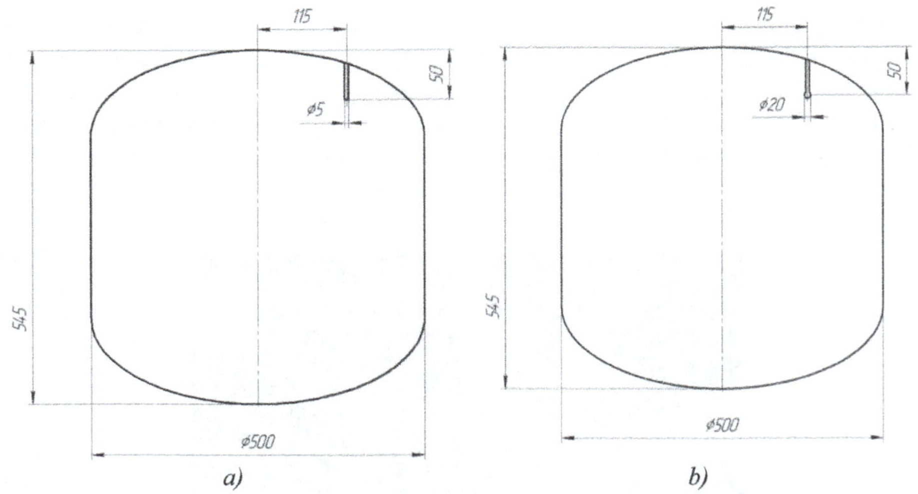

Fig. 2. The geometry of the inner space of the bioreactor: a - spraying without nozzle; b - with a spray stationary ball

For computer-aided design of the washing process of the internal cavity of the apparatus, the finite element method is used. When implemented, the simulated area is divided into a set of smaller volumes. Using the built-in Mesh module, the internal cavity of the device was divided into a mesh with the following parameters: the number of nodes 6504, elements - 10785, the minimum element size - 74 microns (Fig. 3). For a more accurate analysis of fluid movement in the near-wall region, where it is necessary to carry out the washing process, a finer mesh is created using the "Inflation" function. In addition, 8 layers were created, with an increase coefficient of 1.1 and an initial thickness of 0.4 of the size of the element.

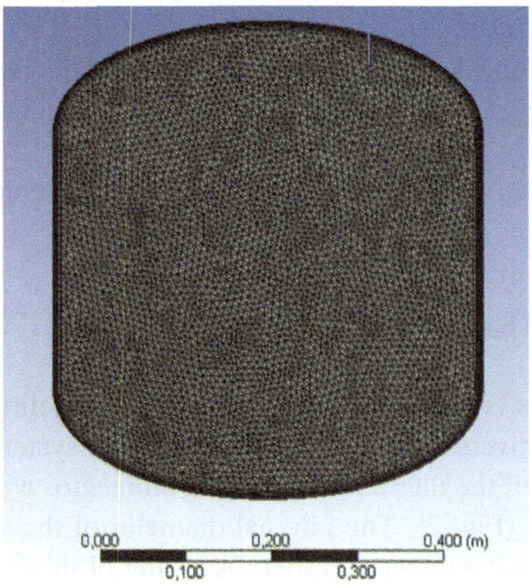

Fig. 3. Mesh

Definition of uniqueness conditions. Two substances take part in the modeling process: water and air. Therefore, the "Volume of fluid" model is used for a multiphase system.

The VOF model is used to model the processes of interaction of two or more immiscible fluids. In this case, a single set of momentum equations is solved and the volume fraction of each fluid is monitored throughout the simulation area. Therefore, this model is ideally suited for modeling the processes of distribution of water flows in the middle of the cavity of the apparatus and for constant monitoring of the phase separation boundary.

The momentum transfer equations are solved over the entire computational domain. The resulting velocity field is distributed between air and water. The momentum equations take into account the volume fractions of substances through their rheological properties: density ρ and viscosity μ [10].

$$\frac{\partial}{\partial t}(\rho\vec{\upsilon}) + \nabla\cdot(\rho\vec{\upsilon}\vec{\upsilon}) = \\ = -\nabla p + \nabla\cdot\left[\mu\left(\nabla\vec{\upsilon} + \nabla\vec{\upsilon}^T\right)\right] + \rho\vec{g} + \vec{F} \tag{1}$$

When calculating the general velocity fields, there is one feature: if the flow rates of substances differ significantly, then at the boundary of the separation of phases would be a significant effect of one quantity on another.

Variables and properties in any cell can be represented both for a specific liquid and for a mixture of phases, taking into account the volume fraction. For each liquid, three states are possible for the volume fraction value α:

- $\alpha_q = 0$ - the cell is empty (there is no qth phase in it);
- $\alpha_q = 1$ - the cell is full (there is only the qth phase in it);
- $0 < \alpha_q < 1$ - there are several phases in the cell.

To track the volume fraction, continuity conditions are used, which takes the following form [11]:

$$\frac{1}{\rho_q}\cdot\left[\begin{array}{c}\frac{\partial}{\partial t}(\alpha_q\rho_q) + \nabla\cdot(\alpha_q\rho_q\vec{\upsilon}_q) \\ = S_{\alpha_q} + \sum_{p=1}^{n}(\dot{m}_{pq} - \dot{m}_{qp})\end{array}\right] \tag{2}$$

Where $\dot{m}_{pq}$ is mass transfer from substance p to q and $\dot{m}_{qp}$ is mass transfer from substance q to p. In our case, the coefficient S_{α_q} is zero and there is no mass transfer between the phases [11].

For the primary phase, in our case of air, the equation will not be calculated. The volume fraction will be calculated from the constraint

$$\sum_{q=1}^{n}\alpha_q = 1. \tag{3}$$

The parameters of substances are set using standard materials of the "Fluent" library. At the initial moment of time, the entire cavity of the apparatus is filled with air, water is supplied through the nozzle and sprayed in the middle of the apparatus. According to this, the initial conditions set the following: volume fraction of air at the initial time in domain 1; volume fraction of water - 0; all three speed components are 0, the pressure inside the apparatus is 1 atm.

The following conditions are used as boundary:

- "Inlet" for entering water through the nozzle of the CIP system with a pressure of 4 atm.
- "Wall" for other surfaces, which corresponds to a smooth wall relative to which the fluid moves without slipping
- "Outlet" - bottom fitting for draining water.

3 Results and Discussion

As a result of modeling, plots of the distribution of the volume fraction of water (Fig. 4) and the fluid velocity vector (Fig. 5) were obtained.

Based on the obtained diagrams, it can be seen that the absence of a spray structure does not provide a proper distribution of water during the washing of equipment (Fig. 4, a). And when using a spray ball, water is redirected to the walls of the apparatus and is used more economically (Fig. 4, b).

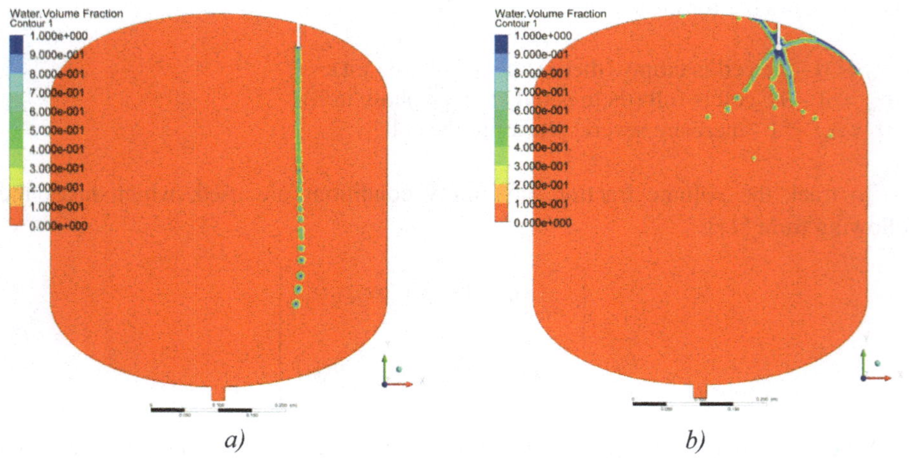

Fig. 4. Contours of the volume fraction of water: a - spraying without nozzle; b - with a spray stationary ball

When using a nebulizer in the form of a stationary perforated ball (Fig. 5, b), the water velocity vectors are directed to the walls and, when interacting with stationary air, at the phase separation boundary, they change their directions. This creates a turbulization of the flow, improves the mass transfer process, which leads to better washing of the equipment. In the case of using a nozzle without a nozzle, the main direction of the water flow does not change (Fig. 5, a).

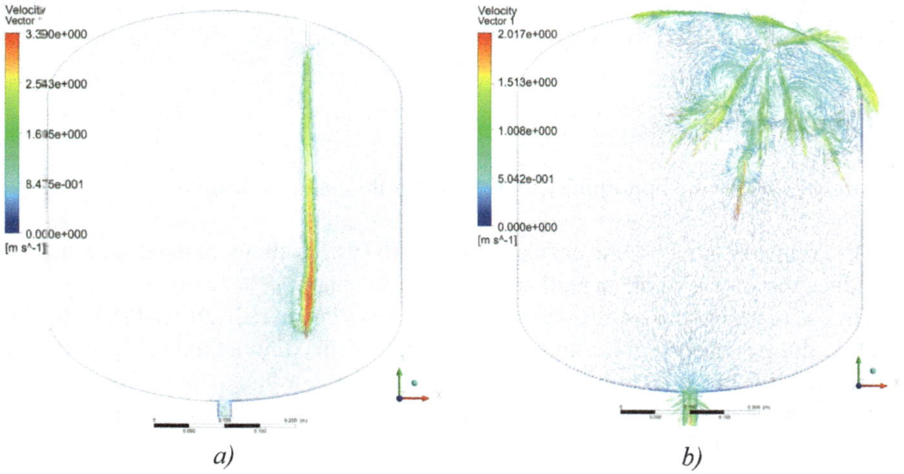

Fig. 5. Vector of water velocity: a - spraying without nozzle; b - with a spray stationary ball

In addition, the maximum values of shear stresses in the near-wall region were obtained for the CIP system with a stationary perforated ball (Fig. 6).

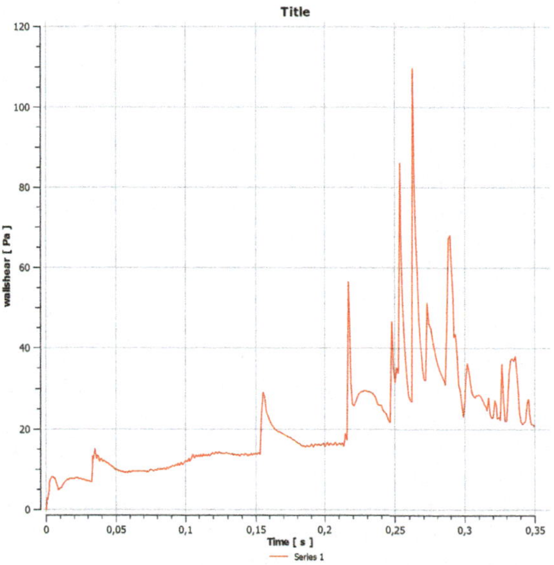

Fig. 6. Chart of shear stresses over time

These data can serve as a basis for analyzing the influence of the water supply rate into the system on the quality of washing of highly adhered objects. As we can see on the chart, the maximum value of the shear stresses is 100 Pa, and the average is 40 Pa. At the initial moments of time, the magnitude of the stresses is not significant, but then it grows and become variable, when the maximum value increases and decreases. This indicates good prospects for using this design for washing highly adhering dirt.

4 Conclusions

The study provides the opportunity to make the following conclusions:

1. The computer model of the device with the CIP system allows at the design stage to reduce the cost of implementation and commissioning.
2. The use of a stationary spray perforated ball allows you to radically redistribute fluid flows during equipment washing. And although the maximum fluid velocity remains within 2 m/s, the direction of flow indicates the best washing parameters.
3. When changing the design of the spray device, the hydraulic resistance of the system increases to 100 Pa.

References

1. Korobiichuk, I., Lobok, A., Goncharenko, B., Savitska, N., Sych, M., Vihrova, L.: The problem of the optimal strategy of minimax control by objects with distributed parameters. In: Szewczyk, R., Zieliński, C., Kaliczyńska, M. (eds.) Automation 2019. AUTOMATION 2019. Advances in Intelligent Systems and Computing, vol. 920, pp. 77–85. Springer, Cham (2020). https://doi.org/10.1007/978-3-030-13273-6_8
2. Rokochinskiy, A., Korobiichuk, I., Kuzmych, L., Volk, P., Kuzmych, A.: The system optimization of technical, technological and construction parameters of polder systems. In: Szewczyk, R., Zieliński, C., Kaliczyńska, M. (eds.) Automation 2020: Towards Industry of the Future. AUTOMATION 2020. Advances in Intelligent Systems and Computing, vol. 1140, pp. 78–86. Springer, Cham (2020). https://doi.org/10.1007/978-3-030-40971-5_8
3. García, L.A., Díaz, M.: Cleaning in place. Comprehensive Biotechnology, pp. 960–973 (2019). https://doi.org/10.1016/B978-0-444-64046-8.00120-8
4. Medina-Rodríguez, A.C., Ávila-Sierra, A., Ariza, J.J., Guillamón, E., Baños-Arjona, A., Vicaria, J.M., Jurado, E.: Clean-in-place disinfection of dual-species biofilm (Listeria and Pseudomonas) by a green antibacterial product made from citrus extract (2020). Food Control **118**, art. no. 107422 (2019). https://doi.org/10.1016/j.foodcont.2020.107422
5. Khan, I.A., Lee, Y.-S., Kim, J.-O.: Identification of scaling during clean-in-place (CIP) in membrane water treatment process. Chemosphere **237**, art. no. 124398 (2019). https://doi.org/10.1016/j.chemosphere.2019.124398
6. Kostyk, S., Shybetskyy, V., Povodzinsy, V., Fesenko, S.: Revealing special features of hydrodynamics in a rotor-disk film vaporizing plant. Eastern Eur. J. Enterp. Technol. **1/6**(97), 28–33 (2019). https://doi.org/10.15587/1729-4061.2019.156649
7. Shybetskiy, V., Semeniuk, S., Kostyk, S.: Design of construction and hydrodynamic modeling in a roller bioreactor with surface cultivation of cell cultures. ScienceRise **7**(36), 53–59 (2017). https://doi.org/10.15587/2313-8416.2017.107176
8. Karachun, V.V., Mel'nick, V.M.: Acoustic radiation energy focus in a shell with liquid. In: Recent Advances in Systems, Control and Information Technology: Proceedings of the International Conference, SCIT 2016, pp. 459–463 (2016). https://doi.org/10.1007/978-3-319-48923-0_48

9. Plashikhin, S.V.: Computer simulation of the hydrodynamic processes of cyclone dust collectors. J. Eng. Phys. Thermophys. **89**, 1093–1102 (2016). https://doi.org/10.1007/s10891-016-1472-y
10. Momentum Equation. www.afs.enea.it/project/neptunius/docs/fluent/html/th/node301.htm
11. Volume Fraction Equation. www.afs.enea.it/project/neptunius/docs/fluent/html/th/node299.htm

Analysis of the 3D Object Reconstruction Accuracy in an Mi-17 Mixed Reality Simulator

Piotr Golański(✉), Marek Szczekala, and Michał Roguszewski

Air Force Institute of Technology, ul. Ks. Bolesława 6, 01-494 Warsaw, Poland
piotr.golanski@itwl.pl

Abstract. The article is devoted to the interaction of a human with mixed reality systems. Mixed Reality (MR) allows for the integration of real world with a virtual world – real objects with computer-generated objects. A human who interacts with such a system receives both visual and tactile sensations. Thus, high reconstruction analysis of virtual objects and the corresponding real objects is necessary to ensure proper interaction of a human with this type of systems. The article aims to illustrate the results of the accuracy analysis of reconstructing flight controls into their virtual 3D models in a prototype of Mi-17 helicopter made in MR technology.

Keywords: Mixed Reality (MR) · Mi-17 helicopter simulator · Unity 3D · IMU Sensor · HTC vive tracker · 3D tracking · Indoor positioning

1 Introduction

Mixed Reality (MR) is a broader reality-virtuality continuum containing both Augmented and Virtual Reality [4]. MR is a combination of the physical and virtual world, which includes real and computer-generated objects. Two worlds are 'combined' together in a way to create a realistic environment. A human can move in this environment and interact with real and virtual objects. An example of the system built based on the above assumptions is a simulator prototype of Mi-17 helicopter designed in Air Force Institute of Technology (AFIT) [2], which was illustrated in Fig. 1.

Human-machine interaction takes place with the use of real flight controls: control stick, collective-pitch lever and rudder pedals, which enable a human to control the simulated object. In response, he receives imagery of surroundings: view of one's own hands, cockpit and terrain view, which are rendered on Head Mounted Display in the form of virtual 3D objects (Fig. 2). The imagery also includes 3D models of real flight controls.

The simulation of flight controls was used with Helicopter Flight Control Sets – Puma [8]. Flight controls that are certain bodies denoted by $\tilde{B}_n \subset \Re^3$ are reconstructed to depict its virtual 3D models - $\widehat{\boldsymbol{B}}_n \subset \Re^3$, which are the objects of the Unity 3D graphics engine

R. Szewczyk et al. (Eds.): AUTOMATION 2021, AISC 1390, pp. 136–145, 2021.
https://doi.org/10.1007/978-3-030-74893-7_14

[10]. To express a spatial description of these bodies, a homogeneous transformation matrix is used, which can be defined as follows:

$$A = \begin{bmatrix} \boldsymbol{R} & \boldsymbol{T} \\ \boldsymbol{0} & \boldsymbol{1} \end{bmatrix} \tag{1}$$

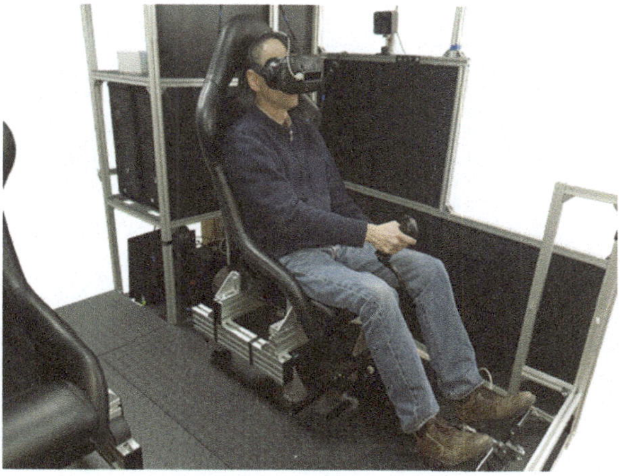

Fig. 1. Cockpit simulator in MR technology

Fig. 2. Virtual world and virtual view of the real world (hand and joystick on the light background)

where $\boldsymbol{R}$ is a 3×3 orthogonal matrix and vector $\boldsymbol{T}_n \in \Re^3$ [7].

Data on the position of flight controls are collected from their potentiometer sensors, which are then transferred to the flight dynamics module via the USB port (Fig. 3). Then, the data are sent to the Visual Display System via Ethernet port (UDP protocol). Based on the obtained data, in the VDS module are updated transformation matrices (1) of

Unity 3D graphical objects (Control Stick Transform), which estimate the transformation matrices of the real (simulated) flight controls $\widehat{\boldsymbol{A}}_n$ and clearly define the position of the flight controls $\widetilde{\boldsymbol{x}}_n \in \widetilde{\boldsymbol{B}}_n$.

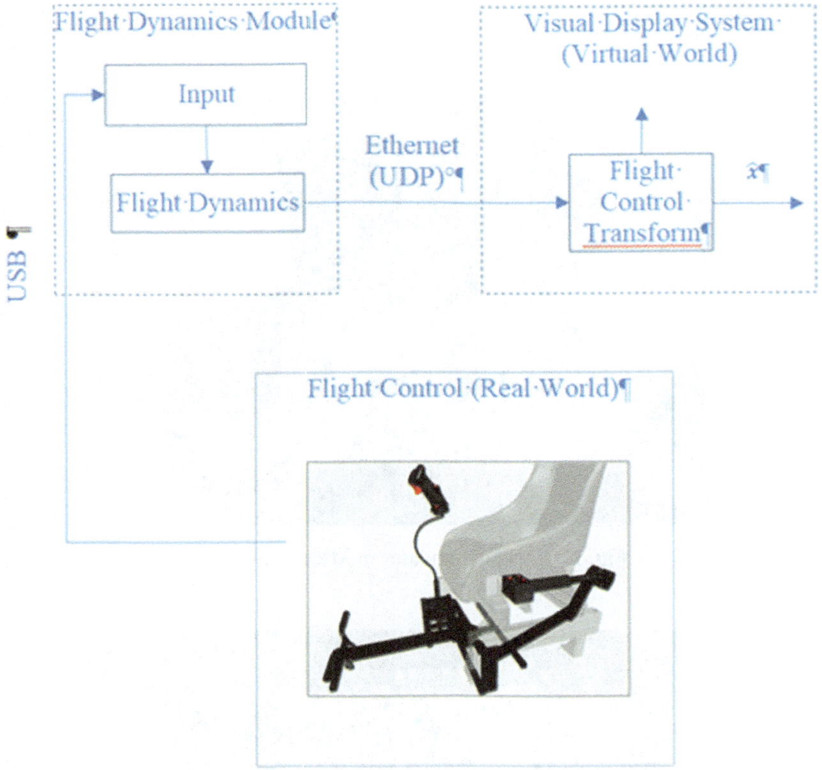

Fig. 3. Mi-17 Mixed Reality simulator architecture

A prerequisite for a proper human-machine interaction is to ensure high accuracy of reconstructing the real flight controls in the virtual world. In the next part of this article, test results of reality reconstruction accuracy for the above-mentioned helicopter simulator prototype will be elaborated, and the conclusions on the improvement of this accuracy will be provided.

2 Method Description

The 3D reconstruction accuracy of flight controls was tested on the example of the control stick. To estimate the reconstruction accuracy, it is necessary to know the real and estimated position of the control stick. The estimation of the position of the control stick $\hat{\boldsymbol{x}}_{stick}$ is done in the simulator, as it was described in the previous point.

In order to determine the real position of the control stick, just like at work [3], the IMU Sensor was used. In the presented manuscript, the IMU Sensor is the VIVE

Tracker Sensor [11]. This sensor was mounted to the control stick so that its parameters correspond to the position of the hand on the control stick (Fig. 4).

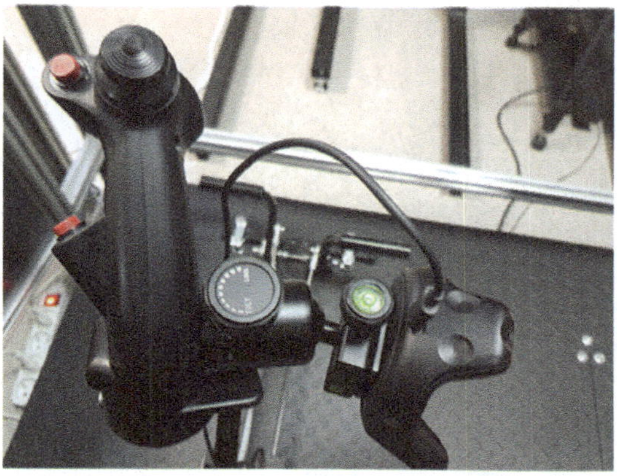

Fig. 4. Mounting position of the IMU sensor on the real control stick

Due to this, an additional measurement path appeared in the simulator, as it was illustrated in Fig. 5. As it results from the figure, data from the Vive Tracker sensor were transferred via USB port directly to the visualization module. In this module, transformation matrix of the object of the IMU Sensor Transform was updated, which is a real homogeneous transformation matrix of the control stick $\widetilde{\boldsymbol{A}}_{stick}$ that is defined the same as in (1), which describes the real position of the control stick $\widetilde{\boldsymbol{x}}_{stick} \in \widetilde{B}_{stick}$.

The discrepancy of the real and estimated position of the control stick was adopted as a quality measure of reconstruction. This discrepancy should be understood as a particular vector of error $\boldsymbol{e}$, the start of which is attached at a place when the IMU sensor is mounted on the real control stick, and the end is attached at a place when the sensor is mounted on the virtual control stick:

$$error = \widetilde{\boldsymbol{x}}_{stick} - \hat{\boldsymbol{x}}_{stick}, \tag{2}$$

where:

$\widetilde{\boldsymbol{x}}_{stick}$ – coordinates vector of the real position of the control stick,
$\hat{\boldsymbol{x}}_{stick}$ – coordinates vector of the estimated position of the control stick.
Error vector (2) has two components – dynamic and static:

$$error = \boldsymbol{e}_{stat} + \boldsymbol{e}_{dyn}. \tag{3}$$

Next section will address the measurement results of both components of the error.

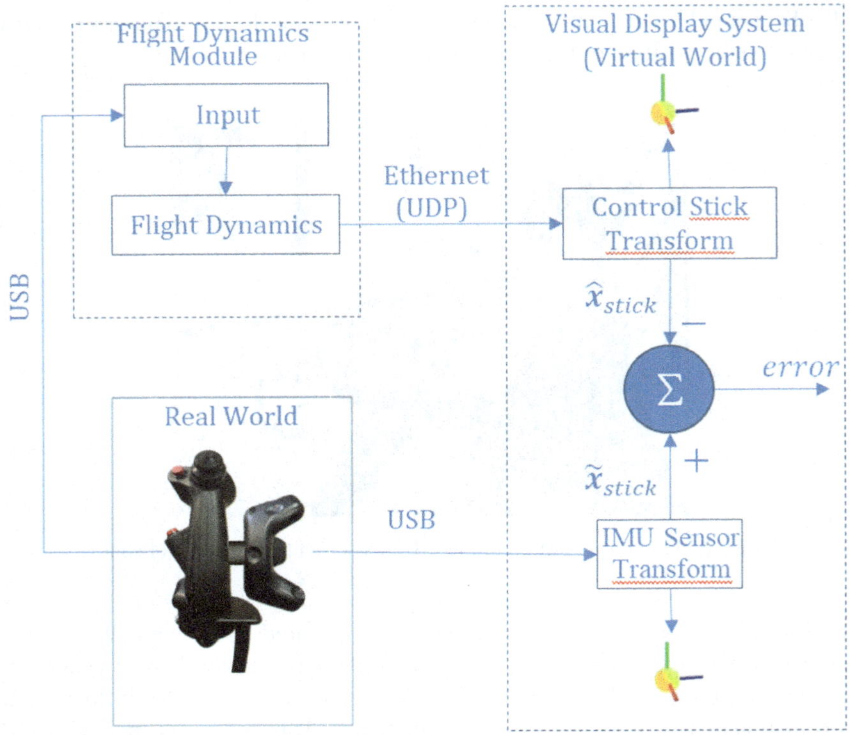

Fig. 5. Diagram of determining the projection accuracy of the control stick in mixed reality

3 Results Analysis

3.1 Static Error Analysis

To verify the statical discrepancy, the position of the real control stick and its virtual counterpart, the averaged processed data were recorded. Every parameter value was calculated as a mean of five subsequent measurements. The measurements were made for both the movement of the stick in the pitch and roll channel. Each measurement was based on the initial setting of the real control stick in the so-called middle position and performing calibration of the position of spheres, which in the virtual world reflect the position of the sensor's attachment point for the real control stick and the corresponding sensor position on the virtual control stick. Then, the real control stick is positioned at a preset angle of pitch or bank (assuming that in the second channel angle value was zero) in the range from $-14°$ to $+14°$, every two degrees, and the data was registered.

Figure 6 illustrated the observed distance changes between the mounting position of the IMU sensor on the real control stick and the mounting position of its counterpart on the virtual control stick as a function of the change in the angle of the control stick position. The discrepancies in the pitch channel were marked in blue, but the discrepancies in the bank channel were marked in red. The value of vector length is given in the scheme in millimeters. The negative value of the vector means that the control stick is situated on the left/away from the middle position. The diagram shows that the farther the control stick is located from the middle position, the bigger the difference of position between a real-world object and its virtual counterpart.

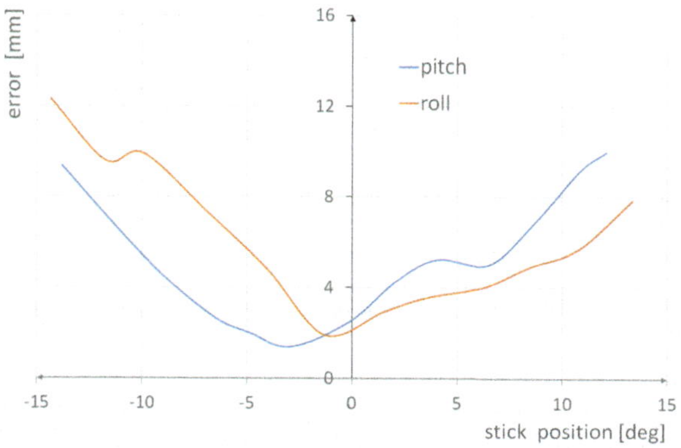

Fig. 6. Values of a static error for the deflection of a control stick

Inertia sensor was installed near the control stick handle. Thus, it can be assumed that the observed errors are the maximum possible. The error changes range from 2 to 12 mm. Such discrepancies are visible also while using the LeapMotion module [9]. In our case, the error increase is due to the mismatch of the real control stick with its virtual counterpart. A virtual control stick, although it can be extended, it doesn't enable to change the position of its attachment point in relation to the object-parent. Thus it is not possible to change the location of its pivot point in relation to the entire object, i.e. flight control. It results in the fact that either the frame of the control system or the attachment point of the control stick will be matched to the aircraft model.

The graphs shown in Fig. 7 illustrate non-overlapping corresponding points on the real and virtual control stick while moving the real control stick. The graphs demonstrate the distribution of vector components in three planes: OXY, OXZ and OYZ, where according to the coordinate system used in Unity 3D, the axes orientations are as follows:

- OX - left,
- OY - up,
- OZ - forward.

The displacement of the real control stick shows that the increase in the difference of inertia sensor's positions changes only in two planes. In the third plane, along which the control stick is moved, the error is much smaller and does not exceed 4 mm in the pitch channel and 6 mm in the roll channel.

3.2 Dynamic Error Analysis

To determine the size and character of dynamic errors, tests were conducted on rapid movements of the control stick, and the differences between the position of the real

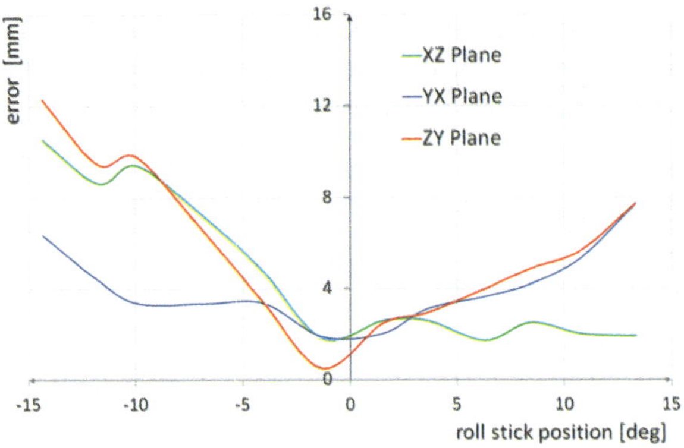

Fig. 7. Distribution of the vector components of the static error

control stick and its virtual counterpart were measured. Each measurement was based on initial setting the control stick in the so-called middle position and performing calibration of the position of virtual spheres, as it was described in Sect. 3.1. After that, the control stick was positioned at a maximum negative angle in the pitch/roll channel (assuming that in the second channel the angle value was zero), i.e. approx. −14° and it was dynamically moved to its extreme position, i.e. to the value of 14°.

Figure 8 outlines discrepancy changes between the mounting position of the IMU sensor on the real control stick and its corresponding position on the virtual control stick in time function. The change in the position angle of the control stick was marked in grey, and the discrepancy between the virtual control stick and the real control stick during the rapid change in the position of the real control stick was marked in orange. The inertia sensor was installed near the control stick handle, as it was done in the previous test. Rapid changes of the control stick position result in very high, instantaneous increase in discrepancies in both channels. In the pitch channel, the difference in the distance is up to 190 mm. In the bank channel, these values are two times smaller and amount up to 90 mm. If instantaneous increase in the discrepancy higher, the changes in the angular position of the control stick are faster.

Generation of dynamic errors can be caused by the long propagation time of the measuring signal. It is the result of the long distance that this signal has to travel from real flight controls, through the USB port to the flight dynamics module and then via the Ethernet port to the visualization module, where the position of the virtual models of flight controls is updated (Fig. 5).

Figure 9 shows the impact of the vector components, which constitute a determinant of the discrepancy between the real object and virtual object during the dynamic change in the position of the control stick in the pitch channel. It can be seen in the diagram that the delay in the reaction of the virtual object in relation to its real-world counterpart is very similar irrespective of the direction.

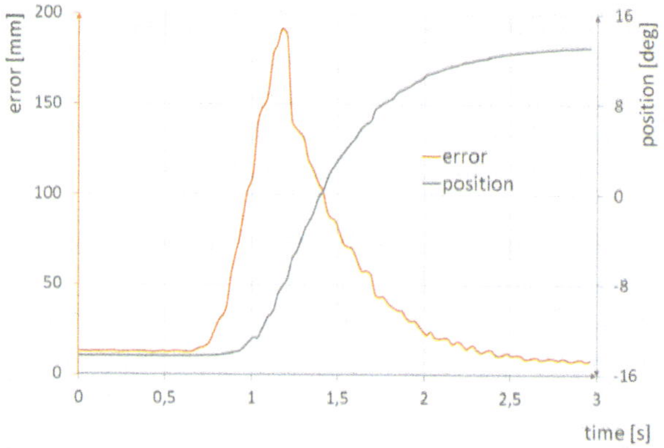

Fig. 8. Dynamic error values for the examplary test control stick deflection

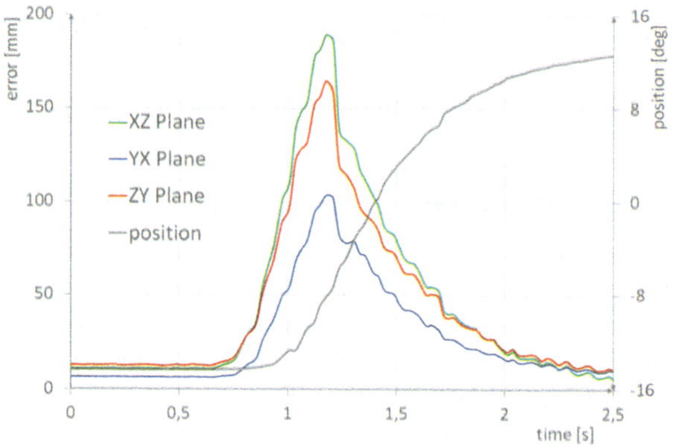

Fig. 9. Components of the dynamic error of discrepancy for the test of dynamic change in the position of the control stick in the roll channel

4 Summary

The article describes the results of the accuracy analysis of real-world objects and their corresponding virtual 3D objects in MR systems. This accuracy has a significant impact on the possibility of interaction of a human with this type of systems.

The investigation into the possibility of precise reconstruction of the real flight controls in the virtual world and the possibility of using them was conducted on the Mi-17 helicopter made in MR technology and designed in AFIT.

The investigation was based on the verification of the position discrepancy of the real control stick and its 3D model in the virtual world in static and dynamic conditions.

In the case of static conditions, several series of measurements were taken. These measurements were obtained to record the positions of the control stick at different angles in the pitch and bank channel. While changing the position of the control stick from the middle position, irrespective of the channel, the increase in the difference of inertia sensor positions was observed. The reason for these errors is a mismatch between the virtual control stick and its real-world counterpart. Although the virtual control stick can be extended, it is not possible to change the position of its attachment point in relation to the object-parent. Thus it is not possible to change the location of its pivot point in relation to the entire object, i.e. flight control. It seems that the elimination of the mismatch error of flight controls is possible by changing the design of the virtual 3D object.

To determine the size and the character of errors under dynamic conditions, discrepancy measurement of the position of the real control stick and its virtual counterpart was performed. Rapid changes in the control stick position resulted in the very high, instantaneous increase in the discrepancy. The instantaneous increase in the discrepancy was the higher, the faster took place the angular position of the control stick.

The reason for dynamic errors is the distance of the path that the data has to travel from the real flight controls to their virtual 3D models. It arises out from the suggested design of the system, which induces considerable delay during operation. This problem can be solved thanks to reducing the measurement path by rebuilding the system: aircraft flight controls would be connected to the visualization module. Moreover, it appears that these changes would prove beneficial not only in terms of the significant reduction of existing delays between worlds, but also the simplification of the structure of dynamics module, and thus, the whole system.

Another solution could be the application of a miniaturized sensor presented in [5], which would allow to determine its location without the necessity to create a measurement path that estimates the position of $\hat{\boldsymbol{x}}_{stick}$ (Fig. 3).

The above solutions are hardware based. In the work [1], it was proposed to increase the accuracy by using tracking algorithms and calibration procedures for the Vive sensor, allowing to achieve submillimeter accuracy for static errors. In a dynamic case, the errors vary from millimeters to centimeters.

Additionally the work [6] shows that the varying offset between the virtual and the physical tracking space makes the HTC Vive at present unsuitable for scientific experiments that require accurate visual stimulation of self-motion through a virtual world.

Taking the above assumptions into consideration, based on the conducted research it can be concluded that Vive HTC sensor can be used to determine the position of real objects for the needs of NUI in a Mi-17 helicopter simulator devised in MR technology. This is evidenced by the obtaining of the characteristics appropriate for the Mi-17 helicopter, determined from the tests performed on the basis of this simulator [2]. However, it is important to find solutions mentioned in this point, which would reduce the reconstruction errors of objects, and it is:

1) to improve the design of virtual 3D objects that reconstruct real objects to decrease static errors,
2) to shorten the measurement path by rebuilding the system, which includes connecting the aircraft flight controls to the visualization module.

An optional solution, though it is a more difficult one, is to determine the location of objects proposed in [5] by the hardware.

References

1. Borges, M., Symington, A., Coltin, B., Smith, T., Ventura, R.: HTC vive: analysis and accuracy improvement. In: 2018 IEEE/RSJ International Conference on Intelligent Robots and Systems (IROS), pp. 2610–2615 (2018)
2. Golański, P., Szczekala, M., Manerowski, J., Roguszewski, M.: Parameterized model of flight dynamics of the transport helicopter. J. KONBiN **50**(3), 237–246 (2020)
3. Guo, H., Uradziński, M., Yin, H., Yu, M.: Indoor positioning based on foot-mounted IMU. In: 2018 IEEE/RSJ International Conference on Intelligent Robots and Systems (IROS), Bull. Pol. Ac.: Tech., vol. 63, no. 3, pp. 629–634 (2015)
4. Haller, M., Billinghurst, M., Thomas, B.: Emerging Technologies of Augmented Reality – Interfaces and Design. Idea Group Publishing, Hershley, London, Melbourne, Singapore (2007)
5. Honnet, C., Lopes, G.: Demo: HiveTracker - 3D positioning for ubiquitous embedded systems. In: UbiComp/ISWC 2019 Adjunct: Adjunct Proceedings of the 2019 ACM International Joint Conference on Pervasive and Ubiquitous Computing and Proceedings of the 2019 ACM International Symposium on Wearable Computers, September 2019, pp 288–291 (2019)
6. Niehorster, D.C., Li, L., Lappe, M.: The accuracy and precision of position and orientation tracking in the HTC vive virtual reality system for scientific research. Iperception **8**, 2041669517708205 (2017)
7. Tchon, K., Muszynski, R., Mazur, A., Duleba, I., Hossa, R.: Manipulators and mobile robots – models, traffic planning, control. AOW PLJ, Warszawa (2000)
8. Pro Flight Trainer. https://pro-flight-trainer-com.myshopify.com. Accessed 10 Nov2018
9. Ultraleap Ltd. LEAP Motion – Documentation. developer.leapmotion.com/documentation. Accessed 10 Oct 2019
10. Unity. unity.com. Accessed 01 Nov 2018
11. HTC Corporation. vive.com/us/support/vive. Accessed 28 May 2019

Extraction of Information from a PSD for the Control of Vehicle Suspension

Tadas Lenkutis(✉), Aurimas Čerškus, Nikita Edgar Sitiajev, Kęstutis Dumbrava, Ieva Staugaitė, Nikolaj Šešok, Andrius Dzedzickis, and Vytautas Bučinskas

Vilnius Gediminas Technical University, Saulėtekio al. 11, 10223 Vilnius, Lithuania
{tadas.lenkutis,vytautas.bucinskas}@vgtu.lt

Abstract. Road irregularities are the main effect influencing driving quality, due to its unpredictability driver can not find the best driving parameters easily and fast. Identification of road profile lets us control the vehicle's damping characteristics and increase driving quality as fast as possible. Power Spectral Density (PSD) could be easily found from the road profile. Deduction of road waviness and amplitudes from PSD could be used to provide the mandatory information required for controllable damping system. Using our developed dynamic model of the car, we performed a simulation of the vehicle passing artificially generated road profiles with different waviness. Obtained results show that the straight-line approximation of PSD of the road profile as recommended by ISO standard doesn't provide suitable information required for the control of car suspension. We define that, two split approximations of PSD spectra is a better solution for this purpose. It provides reliable results and allows to choose optimal damping coefficient value with a small delay for long distances.

Keywords: Road profile · Power spectra density · Waviness · Damping

1 Introduction

A road surface is never perfectly flat, even rarely smooth. Potholes cause damages to vehicles [1]. Road irregularities have a direct impact on ride quality, fuel consumption, tire wear, maintenance costs, etc. Knowledge of road profiles is vital for transport authorities maintaining adequate ride quality for vehicles [2]. The suspension system is being developed from passive to semi-active and active control in automotive technology [3]. Thus, knowledge of road profiles provides information needed to adjust control parameters to improve ride comfort and vehicle handling [4].

Road profile estimation algorithms can be divided into three distinct types, namely contact measurement (as profilograph or longitudinal profile analyzer), non-contact measurement (as inertial profiler), and system response-based estimation [5]. Response-based methods give possibilities to change from dedicated vehicles with special sensors to general-purpose sensors mounted on each vehicle. So finally, it is necessary to identify road profile, then count off parameters and relate them with required control parameters.

R. Szewczyk et al. (Eds.): AUTOMATION 2021, AISC 1390, pp. 146–153, 2021.
https://doi.org/10.1007/978-3-030-74893-7_15

Mostly two methods are used in semi-active suspension constructions: control from sensors readings [6–9] or self-control systems, that uses shock absorbers movement value [10–12]. Despite differences in control structures and control types, the main goal of such systems is to adjust suspension to achieve the best vehicle performance and efficiency. Road profile recognition can be performed by reading signals from sensors (for example linear variable differential transformer (LVDT)) mounted on the shock absorber [6] or measuring acceleration in the z-axis as shown in [7–9]. Some authors [9, 13] use PSD of road displacement to control suspension. Information about road irregularities, road waviness or road class is critical to determine right parameters for damping control.

Here we show that one road waviness value and road class are insufficient parameters to control suspension properly. Also, we propose a new principle to maintain ride comfort.

2 PSD Approximation and Road Profile Generation

The ISO 8608 standard [14] regulates that the PSD first should be smoothed (from the lowest calculated frequency up to the centre frequency of 0.0312 cycles/m in octave bands; then to a centre frequency of 0.25 cycles/m in third-octave bands; and up to the highest calculated frequency in twelfth-octave bands) and approximated with a straight line in log-log scale (1).

$$G_d(\Omega) = G_d(\Omega_0) \cdot (\Omega/\Omega_0)^{-w} \tag{1}$$

where $G_\mathrm{d}(\Omega_0)$ is the degree of roughness, Ω ($\Omega_0 = 1$ rad/m) is the angular spatial frequency, and w is waviness. According to the review of approximation functions made by Andrén [15], the two split approximation has the smallest error.

$$G_\mathrm{d}(\Omega) = \begin{cases} G_\mathrm{d}(\Omega_0)\Omega^{-w_1} \text{ for } \quad \Omega \le \Omega_1 \\ G_\mathrm{d}(\Omega_0)\Omega^{-w_2} \text{ for } \Omega_1 \le \Omega \le \Omega_2 \\ G_\mathrm{d}(\Omega_0)\Omega^{-w_3} \text{ for } \quad \Omega_2 \le \Omega \end{cases} \tag{2}$$

where $\Omega_0 = 1$ rad/m, $\Omega_1 = 0.21 \times 2\pi$ rad/m and $\Omega_2 = 1.22 \times 2\pi$ rad/m. We used this principle for PSD approximation in the range of spatial frequency n from 0.0442 cycles/m up to 1.2241 cycles/m.

Longitudinal road profiles were generated using a sinusoidal approximation in the time domain:

$$z(t) = \sum\nolimits_{i=1}^{N} Z_i \cos(\Omega_i vt + \varphi_i) \text{ with} \tag{3}$$

$$Z_i = \sqrt{2G_\mathrm{d}(\Omega_i)\Delta\Omega}. \tag{4}$$

Where Z_i is the amplitude of ith harmonic, $N = (\Omega_\mathrm{U} - \Omega_\mathrm{L})/\Delta\Omega$ is number of frequency bands into which the total PSD spectrum is divided, Ω_U, Ω_L are upper and lower spatial frequencies in the PSD spectrum [rad/m], $\Delta\Omega$ is the width of each frequency band,

$\Omega_i = \Omega_L + (i-1)\Delta\Omega$ is the spatial frequency of ith harmonic or i^{th} frequency band. The phase angle of ith harmonic φ_i uniformly distributed in the interval $[0, 2\pi]$.

There is a suggestion that the lower spatial frequency limit has no reason to measured lower than $n_L = 0.01$ cycles/m for on-road vehicles and 0.05 cycles/m for off-road vehicles [14]. For vibration analysis of suspension on-road vehicles an upper spatial frequency limit of $n_U = 10$ cycles/m has been suggested by considering the low pass filtering effect due to tire enveloping. For spatial angular frequency, this means the range $2\pi \times 0.01 \leq \Omega \leq 2\pi \times 10$ rad/m. The frequency band was divided into $N = 1000$ intervals. Time step is $dt = 0.0015$ s. The degree of roughness corresponds to the ISO road class B ($G_d(\Omega_0) = 4 \times 10^{-6}$ m^3 [14]). To more coincide measured PSD spectra of real road, we modified Eq. (2) by reducing the amplitude of frequencies lower than 0.04 $\times\ 2\pi$ rad/m and higher than Ω_2:

$$G_d(\Omega) = \begin{cases} G_d(\Omega_0)\Omega^{-1} & \text{for} \quad \Omega \leq 0.04 \times 2\pi\,\text{rad/m} \\ G_d(\Omega_0)\Omega^{-w_1} & \text{for}\ 0.04 \times 2\pi\,\text{rad/m} \leq \Omega \leq \Omega_1 \\ G_d(\Omega_0)\Omega^{-w_2} & \text{for} \quad \Omega_1 \leq \Omega \leq \Omega_2 \\ G_d(\Omega_0)\Omega^{-5} & \text{for} \quad \Omega_2 \leq \Omega \end{cases} \tag{5}$$

The road profile repeats at each $1/n_L$ distance. To avoid repeatability of profile segments, we divided profile length to blocks and for each block, generate a new random set of initial phases. Such a technique can develop the appearance of abrupt shift at the connections between blocks. To minimize this effect, we added Welch type windowing function with the value of exponent equal 10 to get flatter surface:

$$\varpi(i) = 1 - \left(\frac{i - \frac{1}{2}(n-1)}{\frac{1}{2}(n+1)} \right)^{10}, \tag{6}$$

where n is the number of blocks, and i is the number of samples in the block.

3 Results and Discussion

Generated road profiles using Eqs. (3)–(6) with different w_1 and w_2 values are shown in Fig. 1. Each profile is generated with the same sequences of random numbers for comparison. The profiles are named as "prw_1w_2". Root mean square (RMS) values of displacement of each profile, expressed in mm are presented in brackets.

Our developed dynamic model of the full car [16] was used to simulate vehicle passing these profiles with speed $v = 90$ km/h. Damping coefficients that define suspension behaviour were optimized to reach minimal RMS value of driver acceleration; in other words, the suspension has been adjusted for maximum comfort level. Optimized suspension's damping coefficient values for generated profiles are shown in Table 1. For example, profiles pr22C and pr22D, which match ISO class C and D (4 and 16 times higher degree of roughness than pr22 and the same waviness $w = 2$), have higher RMS values of displacement than pr22 profile, their values are equal to 8.63 mm and 17.26 mm, respectively. Nevertheless, their optimized damping coefficient remains the same as for profile pr22. We can conclude that the required information for control parameters one can obtain from waviness w, but not from the degree of roughness $G_d(\Omega_0)$.

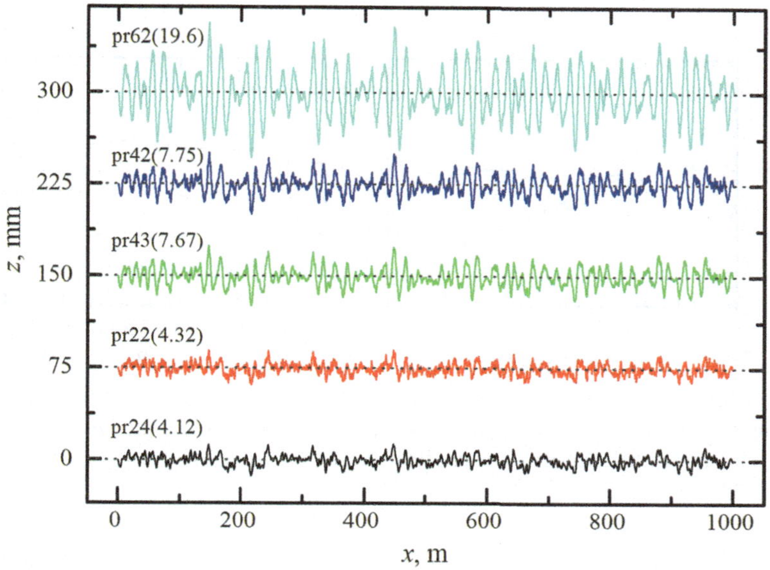

Fig. 1. Generated road profiles with different w_1 and w_2 values. Profiles are shifted vertically for clarity

Table 1. Optimized suspension's damping coefficient h_s values

Profile	pr22	pr22C	pr22D	pr42	pr43	pr62
h_s, Ns/m	4210	4210	4210	7820	8060	11920

Displacement PSD spectra of four profiles were calculated using Fast Fourier Transformation (FFT) with averaging and are presented in Fig. 2(a). FFT was calculated for 100 m distance (from $N = 100\,[\mathrm{m}]/\mathrm{v}[\mathrm{m/s}]/\Delta t[\mathrm{s}]$ points and recalculated after $N/20$ points, where time step $\Delta t = 0.005$ s). Each subsequent FFT result was added to previous and averaged. Finally, the result was smoothed using a method provided by ISO [14], however not in twelfth octave bands above spatial frequency 0.25 cycles/m. We can clearly distinguish PSD spectra related to different waviness w_1 and w_2 values in Fig. 2(a). Points indicate the centres of bands.

Waviness w values of straight-line approximation of averaged PSD spectra are shown in Fig. 2(b). As can be seen, profiles pr43 and pr62 have nearly same waviness w. However, the optimized damping coefficient of profile pr43 is much closer to pr42 case than pr62 profile. Such distribution indicates that linear approximation is insufficient for proper selection of the required damping coefficient and confirms our choice to use two split approximation. Results of two split approximation of averaged PSD spectra are shown in Fig. 3. Here in Fig. 3 (a), w_1 is the slope in the range (0.044; 0.223) cycles/m and w_2 is the slope in the range (0.223; 1.224) cycles/m (Fig. 3(b)). Estimated values of w_1 and w_2 closely match waviness values used for generating a profile.

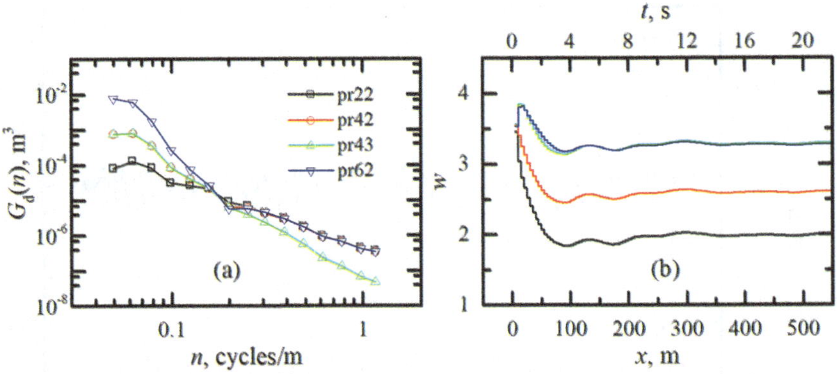

Fig. 2. (a) Smoothed and averaged displacement PSD spectra of profiles for 1000 m; (b) waviness of straight-line approximation of averaged displacement PSD spectra of profiles. The colours of the curves in (a) and (b) corresponds to profile name, as shown in (a)

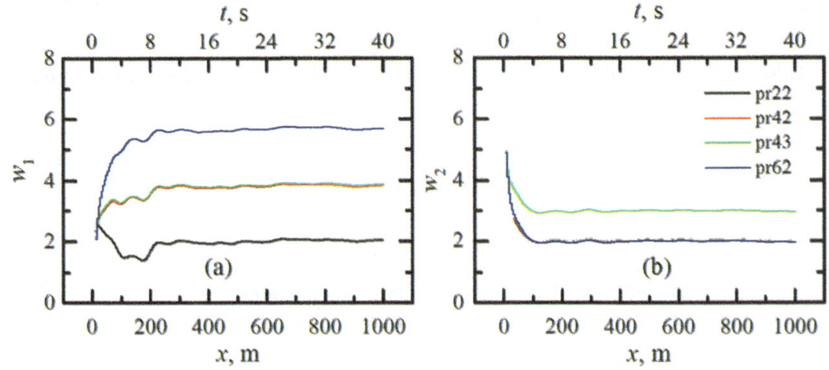

Fig. 3. Waviness (a) w_1 and (b) w_2 of two split approximation of averaged displacement PSD spectra of profiles. The colours of the curves in (a) and (b) corresponds to profile name, as shown in (b)

Waviness values vary mostly for first 100 m (see Fig. 2(b), Fig. 3(a, b)) because in the beginning FFT does not have all needed points for calculation, and instead of missing points FFT algorithm adds zeros. Also, this distance is related to the lowest frequency used for generating profiles. To accommodate the lowest frequency harmonic n_L, the required length will be $x \geq \frac{1}{n_L} = 100$ m. Taking into account that and results of Fig. 2(b) and Fig. 3(a, b)) we increased block length for FFT up to 200 m.

Results obtained by applying a straight line and two split approximations to PSD spectrum of the merged profile are shown in Fig. 4. It should be noted that here FFT calculation was done without averaging. Vertical dotted lines show the boundary between different profiles. Using two split approximation, the changed profile is defined well enough with the delay of FFT block length. The decrease of this length minimizes the delay but increases deviations and vice versa.

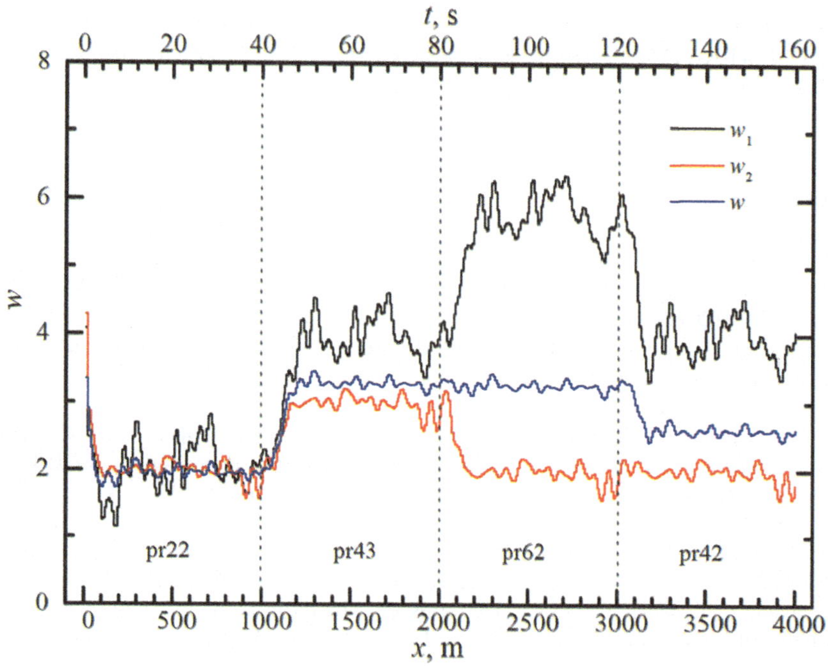

Fig. 4. Results of straight line and two split approximations of PSD spectra of merged profile

This principle we can use to control ride comfort depending on road profile. Knowledge of transfer function between measured acceleration and road displacement and relations between waviness parameters and optimized damping coefficients gives a possibility to control vehicle suspension using the accelerometers. The drawback of the proposed method is that it has some delay and can control averaged parameters for some distances.

4 Conclusions

Our simulation of full car dynamics, and definition of optimized damping coefficients for each artificially generated road profile with different waviness, showed that ISO standard recommended straight-line approximation with two parameters (degree of roughness $G_d(\Omega_0)$ and waviness w) is not the best way to relate road profile with parameters used for suspension control. From obtained results, it is seen that the damping coefficient is independent of the roughness degree $G_d(\Omega_0)$. Moreover, road profiles with identical waviness value could require to apply different damping coefficient to ensure maximum driver's comfort level.

Our obtained results allow us to state that two split approximation of PSD is a promising solution for road profile definition and adjustment of optimal suspension parameters. Delay after changed profile directly depends on FFT block length. However, the decrease of this length increases the deviation of waviness.

Acknowledgement. This project has received funding from the European Social Fund (project No 09.3.3-LMT-K-712–22-0328) under a grant agreement with the Research Council of Lithuania (LMTLT).

References

1. van der Horst, B.B., Lindenbergh, R.C., and Puister, S.W.J.: Mobile laser scan data for road surface damage detection. Int. Arch. Photogr. Remote Sens. Spat. Inf. Sci. **XLII-2/W13**, 1141–1148 (2019). https://doi.org/10.5194/isprs-archives-XLII-2-W13-1141-2019
2. Loprencipe, G., Zoccali, P.: Ride quality due to road surface irregularities: comparison of different methods applied on a set of real road profiles. Coatings **7**(5), 59 (2017). https://doi.org/10.3390/coatings7050059
3. Bhise, A.R., Desai, R.G., Yerrawar, R.N., Mitra, A.C., Arakerimath, R.R.: Comparison between passive and semi-active suspension system using matlab/simulink. IOSR J. Mech. Civ. Eng. **13**(4), 1–6 (2016). https://doi.org/10.9790/1684-1304010106
4. Qin, Y., Rath, J.J., Hu, C., Sentouh, C., Wang, R.: Adaptive nonlinear active suspension control based on a robust road classifier with a modified super-twisting algorithm. Nonlinear Dyn. **97**(4), 2425–2442 (2019). https://doi.org/10.1007/s11071-019-05138-8
5. Nguyen, T., Lechner, B., Wong, Y.D.: Response-based methods to measure road surface irregularity: a state-of-the-art review. Eur. Transp. Res. Rev. **11**(1), 43 (2019). https://doi.org/10.1186/s12544-019-0380-6
6. Wang, D.H., Wang, T., Bai, X.X., Yuan, G., Liao, W.H.: A self-sensing magnetorheological shock absorber for motorcycles. In: 19th International Conference on Adaptive Structures and Technologies 2008 (ICAST 2008), pp. 630–649. Curran Associates, Red Hook, October 2008
7. Harikrishnan, P.M., Gopi, V.P.: Vehicle vibration signal processing for road surface monitoring. IEEE Sens. J. **17**(16), 5192–5197 (2017). https://doi.org/10.1109/JSEN.2017.2719865
8. Choi, S.B., Han, Y.M., Song, H.J., Sohn, J.W., Choi, H.J.: Field test on vibration control of vehicle suspension system featuring ER shock absorbers. In: Electrorheological Fluids And Magnetorheological Suspensions, pp. 496–503. World scientific (2007). https://doi.org/10.1142/9789812771209_0069
9. Uys, P.E., Els, P.S., Thoresson, M.: Suspension settings for optimal ride comfort of off-road vehicles travelling on roads with different roughness and speeds. J. Terramech. **44**(2), 163–175 (2007). https://doi.org/10.1016/j.jterra.2006.05.002
10. Ebrahimi, B., Bolandhemmat, H., Khamesee, M.B., Golnaraghi, F.: A hybrid electromagnetic shock absorber for active vehicle suspension systems. Veh. Syst. Dyn. **49**(1–2), 311–332 (2011). https://doi.org/10.1080/00423111003602400
11. Choi, S.B., Lee, H.K., Chang, E.G.: Field test results of a semi-active ER suspension system associated with skyhook controller. Mechatronics **11**(3), 345–353 (2001). https://doi.org/10.1016/S0957-4158(00)00022-2
12. Ebrahimi, B., Khamesee, M.B., Golnaraghi, M.F.: Feasibility study of an electromagnetic shock absorber with position sensing capability. In: 2008 34th Annual Conference of IEEE Industrial Electronics, pp. 2988–2991. IEEE (2008). https://doi.org/10.1109/IECON.2008.4758436
13. Liu, W., Wang, R., Ding, R., Meng, X., Yang, L.: On-line estimation of road profile in semi-active suspension based on unsprung mass acceleration. Mech. Syst. Signal Process. **135**(1), 106370 (2020). https://doi.org/10.1016/j.ymssp.2019.106370

14. ISO 8606:2016. https://www.iso.org/cms/render/live/en/sites/isoorg/contents/data/standard/07/12/71202.html. Accessed 11 May 2020
15. Andren, P.: Power spectral density approximations of longitudinal road profiles. Int. J. Veh. Design **40**(1–3), 2–14 (2006). https://doi.org/10.1504/IJVD.2006.008450
16. Bucinskas, V., Mitrouchev, P., Sutinys, E., Sesok, N., Iljin, I., Morkvenaite-Vilkonciene, I.: Evaluation of comfort level and harvested energy in the vehicle using controlled damping. Energies **10**(11), 1742 (2017). https://doi.org/10.3390/en10111742

Genetic Algorithm for Parameters Tuning of Two Stage Switching Controller for UAV Autonomous Formation Flight

Arkadiusz Bożko(✉), Leszek Ambroziak, and Ewa Pawluszewicz

Department of Robotics and Mechatronics,
Bialystok University of Technology, Białystok, Poland
a.bozko@doktoranci.pb.edu.pl,
{l.ambroziak,e.pawluszewicz}@pb.edu.pl

Abstract. The idea of adapting genetic algorithms for tuning of the formation flight multi stage control system parameters is presented. The results were conducted on the simulation model with the switching control of the leader-follower. The different configurations of the parameters selection were tested. The fitness function based on position error and with non constant coefficient parameters was introduced. The obtained results were compared with those of the classically tuned system.

Keywords: Genetic algorithms · Machine learning · Multi-stage switching control · Autonomous flight · Aerial navigation

1 Introduction

The autonomous formation flight problem has been an absorbing research topic for last decades. The fixed wing and rotorcraft formations flight algorithms have been proposed in many publications. See for example [1–3], where different approaches to the problem of group flights were developed for various applications. One of them was distributed control approach for swarm algorithms based on potential fields [4] or non-potential factors [5]. The other set of algorithms, based on the formation leader, use defined formation topology, see for example [6,7]. Among them the simple two MAV leader-follower algorithms might be indicated [8–10]. All of the approaches have their advantages and disadvantages.

The group flight systems and their implementations consist of multiple parameters and variables. They can also be affected by the environmental conditions and UAV dynamics changes in unpredictable ways. The high complexity often results in difficulties in the tuning of systems operating parameters. This problem has been investigated using different approaches like treating UAVs as particles of a deformable body [11]. The rapid development of Artificial Intelligence (AI), accelerated the progress in robotics with the new solutions for trajectory generation problem and control issues. Techniques like deep and reinforced learning are used in walking robots for gait learning [12,13] and in many

R. Szewczyk et al. (Eds.): AUTOMATION 2021, AISC 1390, pp. 154–165, 2021.
https://doi.org/10.1007/978-3-030-74893-7_16

UAV applications [14–18]. The group flight algorithms are also beneficial from the application of the machine learning [19–21]. Particularly, the deep reinforcement learning was used for leader follower formation control in [22]. The aim of the presented research is optimisation of the switching control parameters of the leader-follower system presented in [23] with the application of the genetic algorithm.

2 Research Object

Presented results are based on simulation with two UAVs in leader-follower configuration with two stage flight: approaching from afar and keeping prescribed distance in the vicinity. The simulation environment has been created in Matlab & Simulink. The model dynamics of small unmanned aircrafts has been based on mathematical model presented in [23–25]. The Euler angles roll (ϕ), pitch (θ) and yaw (ψ) describe the MAV state in body coordinate system (Fig. 1).

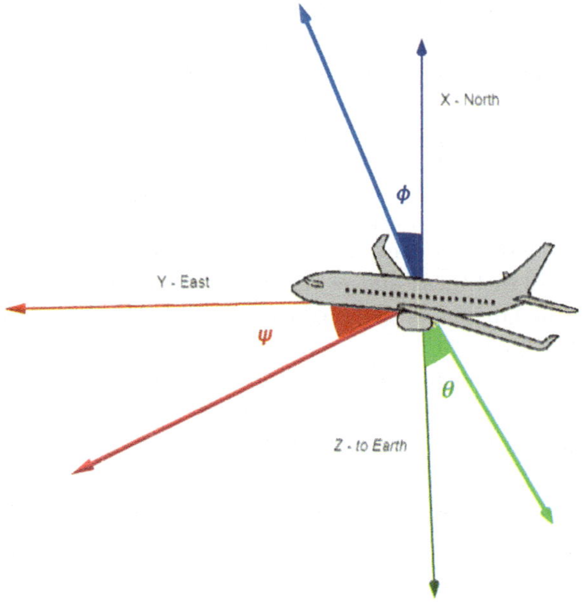

Fig. 1. Rotation between global NED coordinate frame and the body frame

The aerodynamical coefficients for this model were calculated using the Vortex Lattice Method (VLM) and Computational Fluid Dynamics (CFD) [26] for the delta fixed-wing craft "Bullit 60", which is made of balsa wood and is covered in heat-shrinkable foil. It generates thrust with brushless direct current motor powering 14×8". foldable propeller. The control surfaces are two elevons substituting ailerons and elevator for pitch and roll control. This configuration is

Table 1. Main parameters of the simulated UAV

Parameter	Value	Unit
Wing span	1.095	[m]
Wing area	0.497	[m^2]
Mass	1.8	[kg]
Mean chord	0.66	[m]
Propeller	14×8	[in]
Angle of attack	3	[°]
Min airspeed	11	[m/s]
Max airspeed	34	[m/s]

incapable of yaw angle control/correction as there is no rudder or active winglets. The models primary parameters are presented in Table 1.

The simulation model incorporates nested control loop with layers: angular velocity control, attitude control and formation flight control. The additional modules provide the capability of sensor bias and uncertainty simulation. The simulation environment was expanded with the genetic algorithm module.

3 Formation Flight Algorithm

The analysed algorithm from [23] uses the hysteresis for switching control between two ancillary algorithms. The first one called initial guidance on a leader is activated on long distances and as an output it produces the desired heading directly toward the current position of the leader UAV. The desired airspeed is calculated by multiplying the commanded leader airspeed by constant proportional gain k. Its main purpose is to minimize the distance between UAVs in formation without saturating the outputs of PID on long distances. The second algorithm is called precision formation flight control. It uses two PID controllers, one for the airspeed correction and one for heading maintenance. This approach is meant to improve tracking of the position in formation. It secures smooth response for the trajectory variations of the leader UAV.

3.1 Initial Guidance Algorithm

The distance between the UAV's is measured in the NED (North-East-Down) global coordinate system [27]. As the surface of the sphere can be locally approximated by the flat plane, the distances are measured in Cartesian frame. Positions are calculated relatively to the common 'home' position (0,0,0). Then positions of the leader and follower, denoted respectively as L_{pos} and F_{pos} can be expressed in the form:

$$L_{pos} = [X_L; Y_L; Z_L] \tag{1}$$

$$F_{pos} = [X_F; Y_F; Z_F] \tag{2}$$

The error e between the leader and follower positions:

$$e = [e_x; e_y; e_z] = L_{pos} - F_{pos} \tag{3}$$

is denoted as lateral, longitudinal and altitudinal compounds calculated from global leader and follower positions (L_{pos}, F_{pos}) in NED coordinate system. Since altitude is holded by the lower level controler, the hysteresis input $|R| = \sqrt{e_x^2 + e_y^2}$ is the distance between the MAV's on the vertical projection on the surface.

The visual representation of the leader-follower formation is shown in Fig. 2. The heading of the leader ϕ_L and follower ϕ_F are values in range <0;2π) with zero heading North.

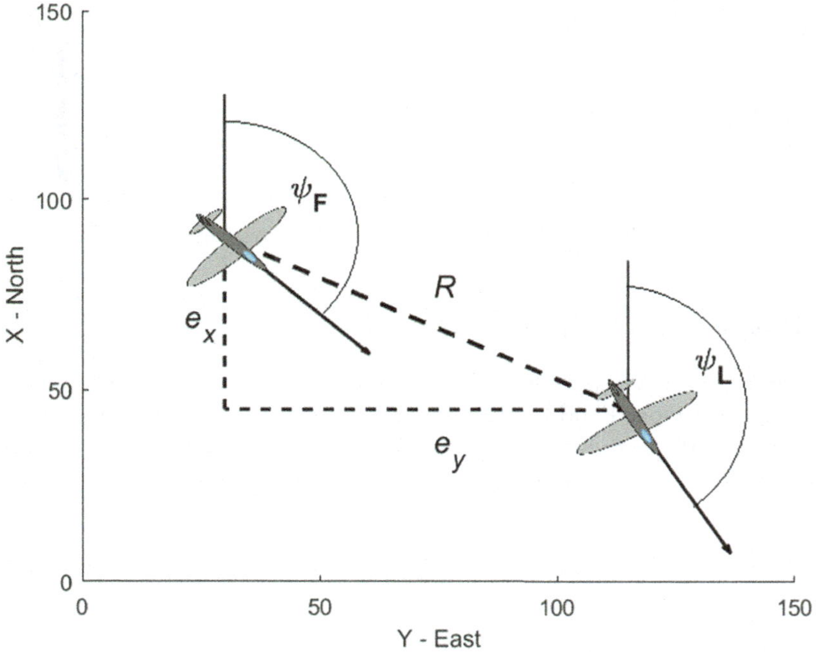

Fig. 2. MAV relative positions in the NED coordinate frame

If the initial guidance requirements are fulfilled the follower can be located in one of the four quarters of the coordinate system relatively to the leader. Each possibility has the different formula for the desired follower angle ϕ_F^c. The desired follower airspeed is an actual leader airspeed multiplied by the gain k. The detailed flow chart of the algorithm is presented in the Fig. 3.

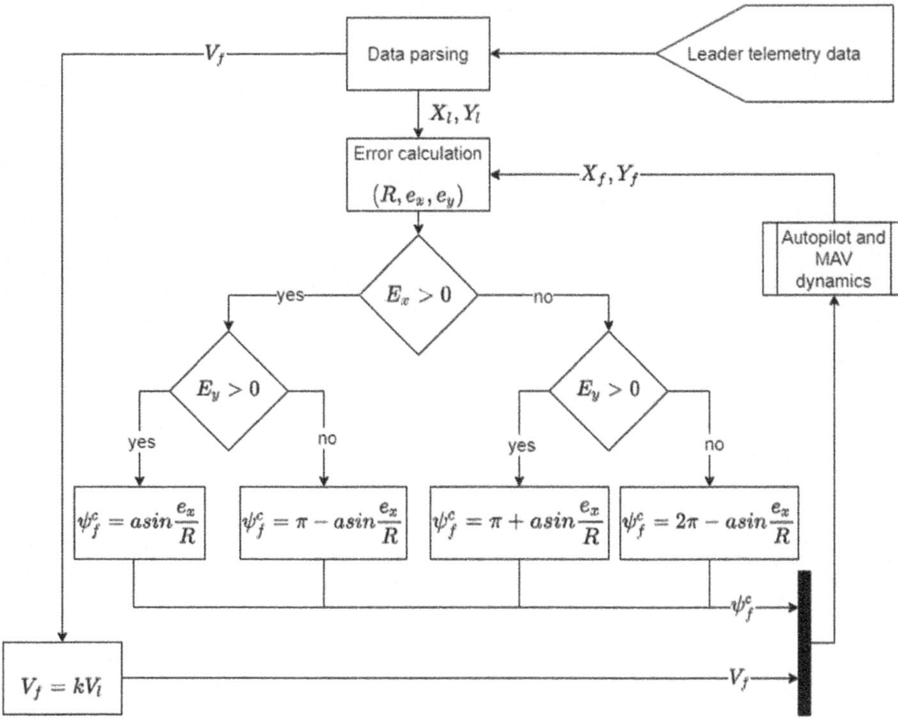

Fig. 3. Initial guidance algorithm flowchart

3.2 Precise Formation Flight Control

In the second step two parallel PID controllers minimizes the tracking error by correcting the airspeed and the heading of the follower. The input values for the controllers are respectively errors e_v and e_Ψ measured in formation coordinate system (relative to the leader position and heading Ψ_L, see Fig. 2):

$$\begin{bmatrix} e_v \\ e_\psi \end{bmatrix} = \begin{bmatrix} e_{x_{rot}} \\ e_{x_{rot}} \end{bmatrix} + \begin{bmatrix} x_d \\ y_d \end{bmatrix} \tag{4}$$

where

$$\begin{bmatrix} e_{x_{rot}} \\ e_{x_{rot}} \end{bmatrix} = \begin{bmatrix} \sin\Psi_L & \cos\Psi_L \\ \cos\Psi_L & -\sin\Psi_L \end{bmatrix} \begin{bmatrix} e_x \\ e_y \end{bmatrix} \tag{5}$$

The output of the controllers are desired airspeed and heading of the follower. To prevent the eventual collision the positioning offset $[x_d, y_d]$ is applied.

The algorithm flow chart is presented in Fig. 4. The errors $e_{x_{rot}}$ and $e_{y_{rot}}$ in formation reference frame are calculated by the multiplication of the vector of global errors by the rotation matrix:

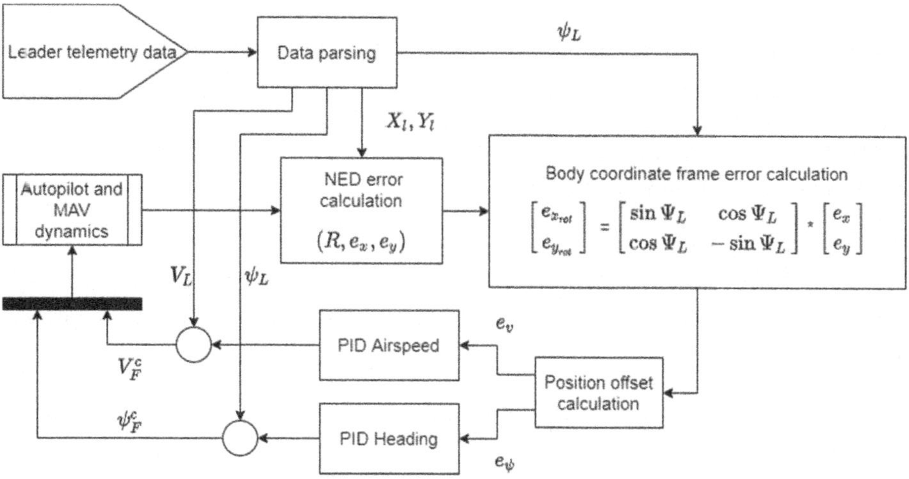

Fig. 4. Precise formation flight algorithm flowchart

4 Genetic Algorithm Implementation

The genetic algorithm used in this research is capable of optimising up to nine parameters of the system. In the precise formation flight control they are proportional, integral and derivative gains of PID controllers. In the initial guidance algorithm the hysteresis parameters and airspeed gain can be refined. Each parameter has one gene assigned, whose value is normalised to the range $(0; 1)$. Genes are transformed to parameters target values on the initialisation of given simulation. From each generation best fitting half of the population is selected in mating pool process for breeding. The number of generated children is equal to the number of parents.

4.1 Crossing-Over

Each object from the mating pool is breeding with another randomly paired parent. As the descendant/ancestor ratio is constant and is equal to one, every parent selected for mating has at least one descendant. It is possible that one parent will be ancestor of all newly generated objects. For the purpose of crossing over, the genome is divided into four sections: PID airspeed (3 gens), PID heading (3 gens), hysteresis (2 gens) and initial guidance KP (1 gene). Each section is cohesively inherited from one random parent. Algorithm forbids descendant to inherit all sections from one parent.

4.2 Mutation

The mutations are performed randomly after the crossing over process. The chance to mutate for every single gene in set by parameter and mutations are regardless of themselves. If the condition is fulfilled then the gene value is modified by addition of normally distributed random number from range with standard deviation $\sigma = 0.16$ and mean value $\bar{x} = 0$. This ensures the advantage over the uniform distribution in late generations, when the small steps in parameters variation are more promising. The mutation ratio was adjusted to the population size based on the results from [28].

4.3 Fitness Function

The chosen fitness function is defined as an integral of the tracking error over time:

$$F_{fitness} = \int_{t=0}^{t_n} \sqrt{e_x e_y}\, T_{f_i}\, dt\, E_f\, S_f \tag{6}$$

The formula (6) is expanded with three additional factors: T_{f_i}, E_f and S_f. Time penalty is the first of them. It enables to diminish the overwhelming influence of the initial guidance algorithm due to the high values of the error in the first stage. It prevents the algorithm from stacking in local minimum, when easier to tune initial stage is used for exaggerated time with frequent sharp algorithm switching in narrow distance between drones. The error samples are multiplied by the linearly increasing factor T_{f_i}:

$$T_{f_i} = 1 + \frac{t_i}{t_n}, \quad \frac{t_i}{t_n} \in <1;2> \tag{7}$$

The second factor E_f is based on the commanded throttle signal as it is meant to optimise energy efficiency of the controller. The estimation is made under the assumption that the ratio of the integral of throttle signals approximates the ratio of the energy cost. Assuming that the follower reaches desired position in formation, the trajectories of the UAV's are nearing and the energy expenses are similar. Thus the theoretical ratio of thrust command integrals will be close to one. The desired smooth trajectory tracking uses less thrust than sharp and oscillating movement. The factor is calculated as in (8), where THR_F and THR_L are throttle command signals:

$$E_f = \frac{\int_{t=0}^{t_n} THR_F(t)\, dt}{\int_{t=0}^{t_n} THR_L(t)\, dt} \qquad \lim_{t \to t_n} E_f = 1 \tag{8}$$

The third factor S_f objective is to minimise the number of algorithm switches SW_{num}. That approach counteracts to multiple algorithm switching on close nearing, what might be an effect of the poor PID tuning or hysteresis parameters inefficiency. The formula (9) is used to calculate its value:

$$S_f = \sqrt{\frac{SW_{num}}{1 + SW_{num}}} \tag{9}$$

4.4 Hysteresis Generation

The hysteresis for algorithm switching uses two parameters R_1 and R_2. When the distance between leader and follower R is greater than R_2, the initial guidance is used. If it is less than R_1 the precision formation control is used. To prevent the situation, when $R_1 > R_2$ the following dependency is applied:

$$R_1 = G_1, \qquad R_2 = G_1 + G_2 \tag{10}$$

where G_1 and G_2 are gens values scaled and shifted to fit the simulation environment. The switching function is presented in Fig. 5.

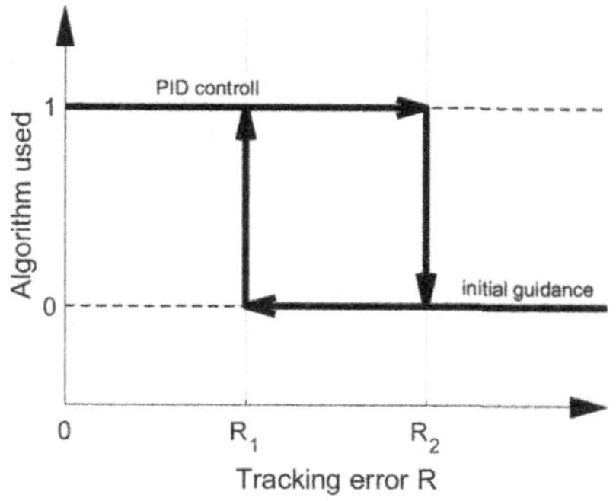

Fig. 5. Hysteresis switching schematic

5 Simulations Performed

The tests were demerged into two phases. The first was focused on evolutioning six gens that corresponds to the tuning of PID parameters in the precision formation flight. The initial distance between UAVs was 100 m in north-east direction and the simulation duration was set for 50 s. In the second phase both algorithms were tuned simultaneously with all nine available parameters active. The initial follower translation in relation to leader was 350 m to the south direction and 100 m to the west. The simulation time was 100 s. In both scenarios the leader UAV was approaching the circular path from the center of the orbit with the radius equal to 500 m. The leader commanded airspeed was 15 m/s.

5.1 Precise Formation Flight Tuning

In this scenario the initial guidance algorithm was playing minor role due to the small distance between UAVs and its evolution parameters were deactivated. Five series of learning with 40 generations and 20 solutions per generations are presented in Fig. 6 (a). The learning process in all cases is overally similar, regardless of the initial quality of the randomly generated system in the generation zero. After five generations the quality of the most fitted solution is approaching the average value between all learning series. Despite the low number of parameters tuned the visible tendency to fell into local minimum is negligible, only in series 4 there is more than eight generations interval between next step forward in solution quality. All the other trends minimize on quasi regular rate with the significant downturn after 30–35 generation. The final results in all learning sessions were better than in the intuitively tuned system.

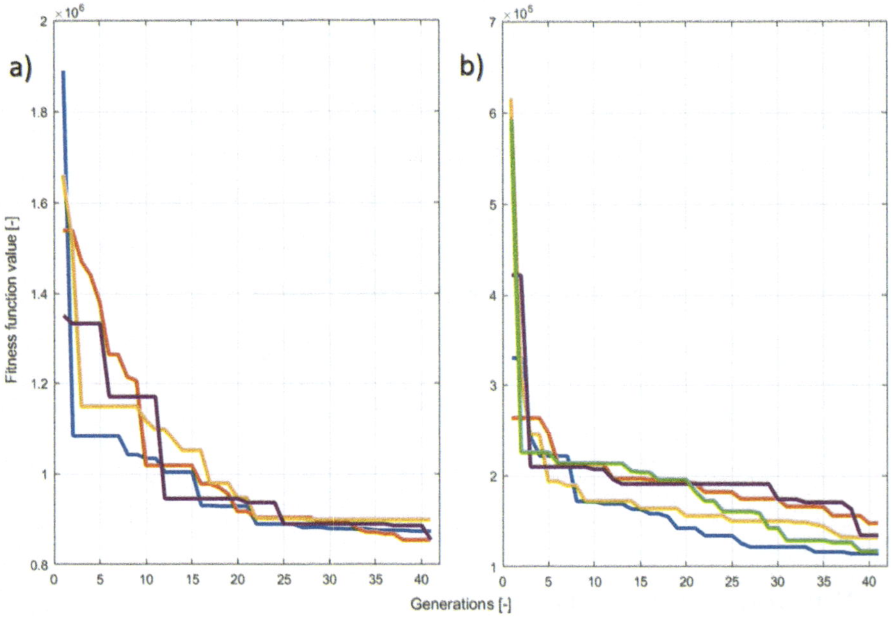

Fig. 6. Comparison of learning sessions for 9 (a) and 6 (b) parameters

5.2 Full System Tuning

Despite the higher complexity and flexibility of the system in this research stage the trend of fitness function value over generations approached the reached minimum in fewer steps. The initial growth of the quality was slower in comparison to the first phase and was relatively smaller. It was caused by the strong impact of the initial guidance algorithm on the fitness function value due to its higher predictability and lesser sensitivity to parameters fluctuations. All the training sessions reached the higher functionality that the intuitively tuned system. The data is presented in Fig. 6 (b).

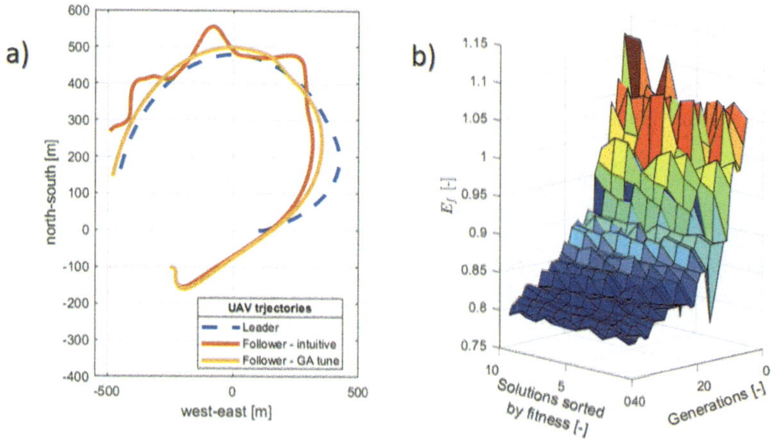

Fig. 7. Comparison of trajectories of the systems tuned with different methods (a). Evolution of the E_f factor over generations (b).

5.3 Fitness Function Evaluation

In Fig. 7 (a) the comparison of trajectories is presented. The classically tuned system is oscillating around the desired position while the tuned system is maintaining commanded offset to the leader. The E_f parameter of the fitness function evolution over generations can be pointed in Fig. 7 (b). Despite the low value and limited influence on the evolution process the parameter value trend and its influence on the performance is significant. The 3-dimensional chart presents values of the E_f for genomes sorted by the fitness function values. In the first generations average value is high and the deviation is noticeable. With the learning process progress the slow minimalisation of E_f value is explicit. The low final value of the factor is the result of learning to intercept the leader trajectory on the turn. The correlation between the fitness function value for genomes and the E_f measured on five learning series is varying from 0.7 to 0.9.

6 Conclusions

The genetic algorithm parameters tuning method for formation flight proposed in this paper is showing promising results. The general performance of the tuned system exceeds the intuitively tuned one. The results obtained during the simulations are repeatable. The proposed fitness function has the advantage of the lack of weighted coefficients that have to be set and may vary between the situations and environmental conditions. The tuning method will be further developed and tested on the physical objects in real conditions. The eventual objective is to create the system capable of live learning from the in-flight collected data.

Acknowledgment. The work is supported with University Grants No. WZ/WM-IIM/1/2019 and WI/WM-IIM/6/2020, Faculty of Mechanical Engineering, Bialystok University of Technology.

References

1. Paul, T., Krogstad, R., Gravdahl, J.: Modelling of UAV formation flight using 3D potential field. In: Simulation Modelling Practice and Theory, pp. 1453–1462 (2008)
2. Jiang Z., Hui, C., Zefeng, Z.: Autonomous formation flight of UAVs: control algorithms and field experiments. In: Proceedings of the 35th Chinese Control Conference (CCC), Chengdu, pp. 7585–7591 (2016)
3. Hafez, A.T., Givigi, S.N., Schwartz, H.M., Yousefi, S., Iskandarani, M.: Real time tactic switching for multiple cooperative UAVs via model predictive control. In: Proceedings of the Annual IEEE Systems Conference (SysCon), Vancouver, pp. 432–438 (2015)
4. Sun, F., Han, S.: A flight path planning method based on improved artificial potential field. In: Proceedings of the 2016 International Conference on Computer, Information and Telecommunication Systems (CITS), Kunming, pp. 1–5 (2016)
5. Glazok, O.M.: A non-potential target function for controlling the UAVs group flight in presence of concave obstacles. In: Proceedings of the IEEE 5th International Conference Actual Problems of Unmanned Aerial Vehicles Developments, Kiev, pp. 238–241 (2019)
6. Zhang, J., Yan, J., Kong, X.: UAV formation flight cooperative tracking controller design. In: Proceedings of the 15th International Conference on Control, Automation, Robotics and Vision (ICARCV), Singapore, pp. 856–861 (2018)
7. Yu, Z., Qu, Y., Su, C.-Y., Zhang, Y.: Distributed fractional-order finite-time control for multiple unmanned aerial vehicles. In: Proceedings of the IEEE Conference on Control Technology and Applications (CCTA), Copenhagen, pp. 1058–1063 (2018)
8. Yun, B., Chen, B.M., Lum, K.Y., et al.: Design and implementation of a leader-follower cooperative control system for unmanned helicopters. J. Control Theory Appl. **8**, 61–68 (2010)
9. Yun, B., Chen, B.M., Lum, K.Y., Lee, T.H.: A leader-follower formation flight control scheme for UAV helicopters. In: Proceedings of the 2008 IEEE International Conference on Automation and Logistics, Qingdao, pp. 39–44 (2008)
10. de Souza Neto, A.M., Romero, R.A.F.: A decentralized approach to drone formation based on leader-follower technique. In: Proceedings of the 2019 Latin American Robotics Symposium (LARS), Rio Grande, pp. 358–362 (2019)
11. Wang, J., Jia, G., et al.: Cooperative task allocation for heterogeneous multi-UAV using multi-objective optimization algorithm. J. Central South Univ. **27**, 432–448 (2020)
12. Kolathaya, S., et al.: Trajectory based deep policy search for quadrupedal walking. In: Proceedings of the 2019 28th IEEE International Conference on Robot and Human Interactive Communication (RO-MAN), New Delhi, pp. 1–6 (2019)
13. Gu, S., Holly, E., Lillicrap, T., Levine, S.: Deep reinforcement learning for robotic manipulation with asynchronous off-policy updates. In: Proceedings of the IEEE International Conference on Robotics and Automation (ICRA), Piscataway, pp. 3389–3396 (2017)
14. Hwangbo, J., Sa, I., Siegwart, R., Hutter, M.: Control of a quadrotor with reinforcement learning. IEEE Robot. Autom. Lett. **2**, 2096–2103 (2017)
15. Liu, C.H., Chen, Z., Tang, J., Xu, J., Piao, C.: Energy-efficient UAV control for effective and fair communication coverage: a deep reinforcement learning approach. IEEE J. Sel. Areas Commun. **36**(9), 2059–2070 (2018)

16. Lin, J., Wang, L., Gao, F., Shen, S., Zhang, F.: Flying through a narrow gap using neural network: an end-to-end planning and control approach. In: Proceedings of the IEEE/RSJ International Conference on Intelligent Robots and Systems (IROS), Macau, pp. 3526–3533 (2019)
17. Faust, A., Palunko, I., Cruz, P., Fierro, R., Tapia, L.: Learning swing-free trajectories for UAVs with a suspended load. In: Proceedings of the 2013 IEEE International Conference on Robotics and Automation, Karlsruhe, pp. 4902–4909 (2013)
18. Sampedro, C., Bavle, H., Rodriguez-Ramos, A., de la Puente, P., Campoy, P.: Laser-based reactive navigation for multirotor aerial robots using deep reinforcement learning. In: Proceedings of the 2018 IEEE/RSJ International Conference on Intelligent Robots and Systems (IROS), Madrid, pp. 1024–1031 (2018)
19. Lin, Y., Wang, M., Zhou, X., Ding, G., Mao, S.: Dynamic spectrum interaction of UAV flight formation communication with priority: a deep reinforcement learning approach. IEEE Trans. Cogn. Commun. Netw. **6**(3), 892–903 (2020)
20. Izzo, D., Martens, M., Pan, B.: A survey on artificial intelligence trends in spacecraft guidance dynamics and control. Astrodynamics **3**, 287–299 (2019)
21. Duan, H., Luo, Q., Shi, Y., Ma, G.: Hybrid particle swarm optimization and genetic algorithm for Multi-UAV formation reconfiguration. IEEE Comput. Intell. Mag. **8**(3), 16–22 (2013)
22. Zhou, Y., et al.: Adaptive leader-follower formation control and obstacle avoidance via deep reinforcement learning. In: Proceedings of the 2019 IEEE/RSJ International Conference on Intelligent Robots and Systems (IROS), Macau, pp. 4273–4280 (2019)
23. Ambroziak, L., Gosiewski, Z.: Two stage switching control for autonomous formation flight of unmanned aerial vehicles. Aerosp. Sci. Technol. **46**, 221–226 (2015)
24. Roskam, J.: Airplane Flight Dynamics and Automatic Flight Controls, Parts I & II. DARcorporation, Lawrence, Kansas (1998)
25. Stevens, B.L., Lewis, F.L.: Aircraft Control and Simulation, 2nd edn. Wiley, Hoboken (2003)
26. Ambroziak, L., Gosiewski, Z., Kondratiuk, M.: Aerodynamics characteristics identification of micro air vehicle. Trans. Inst. Aviat. **216**, 17–29 (2011). (in Polish)
27. Beard, R., McLain, T.: Small Unmanned Aircraft: Theory and Practice. Princeton University Press, Princeton (2012)
28. Haupt, R.L.: Optimum population size and mutation rate for a simple real genetic algorithm that optimizes array factors. In: Proceedings of the IEEE Antennas and Propagation Society International Symposium, Salt Lake City, vol. 2, pp. 1034–1037 (2000)

Robustness Analysis of a Distributed MPC Control System of a Turbo-Generator Set of a Nuclear Plant – Disturbance Issues

Paweł Sokólski[1(✉)], Tomasz A. Rutkowski[2], Bartosz Ceran[3], and Dariusz Horla[1]

[1] Faculty of Automatic Control, Robotics and Electrical Engineering, Poznań University of Technology, Poznań, Poland
dariusz.horla@put.poznan.pl
[2] Faculty of Electrical and Control Engineering, Gdańsk University of Technology, Gdańsk, Poland
tomasz.adam.rutkowski@pg.gda.pl
[3] Faculty of Environmental Engineering and Energy, Poznań University of Technology, Poznań, Poland
bartosz.ceran@put.poznan.pl

Abstract. Typically, there are two main control loops with PI controllers operating at each turbo-generator set. In this paper, a distributed model predictive controller with local quadratic model predictive controllers for the turbine generator is proposed instead of a set of classical PI controllers. The local quadratic predictive controllers utilize step-response models for the controlled system components. The parameters of these models are determined based on the proposed black-box models of the turbine and synchronous generator, which parameters are identified on-line with the recursive least-squares algorithm. A robustness analysis of the control system with respect to different disturbances is presented in the paper. There are various configurations considered, such as change in disturbance levels from the side of electrical and thermal systems, or changes in prediction horizons.

Keywords: Model predictive control · PI controller · Power system · Heat system · Power plant · Robustness

1 Introduction

The quality of electrical energy plays a unique and significant role in development of a modern society. The demand for the generation of electric power grows in accordance with the speed of economic development of societies. Accordingly, there is a growing need to increase the power plants' efficiency and improve the electrical energy quality. The conventional power plants, as well as nuclear

R. Szewczyk et al. (Eds.): AUTOMATION 2021, AISC 1390, pp. 166–184, 2021.
https://doi.org/10.1007/978-3-030-74893-7_17

ones, utilize turbine-generator sets with steam turbines cooperating with the synchronous generators to produce electrical energy. The major novelty of the results presented in the paper, is to show the means to improve the quality of the turbine generator control and, in consequence, the electrical energy quality delivered to the power system network.

Control systems currently used in the power industry are based on classic proportional/integral/derivative control system extended with additional modules and feedback loops, such as, e.g., system stabilizers. Despite the fact that they are often complex systems and take into account a number of phenomena occurring in a regulated system, they do not fully use the knowledge about the object. Currently, it is possible to use more complex algorithms, e.g., predictive control, which fully take advantage of the availability of the information about the structure of the object (mathematical models) and take the constraints imposed on the control and controlled values or state into account. Therefore, in the paper, it is proposed to use the QDMC predictive control, which allows to use the knowledge of the turbine-generator set model. Additionally, the article proposes using the algorithm to estimating parameters of the model in an on-line fashion, with the use of the recursive least squares method (RLS). This is to ensure that the discussed control system works with the current version of the plant model at all times.

Nuclear power plants generate a large amount of power and can significantly affect the stability of the entire electricity system (around 3/4 of electricity in France comes from nuclear energy). Due to the requirements for the quality of electricity and the safety of nuclear power plant operation, it seems important to analyze the behavior of the proposed solution in the event of various disturbances.

The steam turbine and the synchronous generator are complex objects with a non-linear character. Currently, the control methods for the turbine generator are typically based on the Proportional-Integral (PI) controllers [1–3]. With the current state of the control theory and access to the modern computing units, with high computing power, it is possible to use more complex and sophisticated control algorithms for control purposes [4–12]. In this paper, the distributed model predictive control (DMPC) methodology is discussed [13, 14] for the turbine generator control purposes. With the model predictive control (MPC) technology one can design a truly multi-variable optimizing control system that can handle the process constraints and accommodate the model-based knowledge combined with the hard measurements [15–17]. To achieve better closed-loop control performance, some level of communication should be established between local MPC controllers. The local controllers proposed in the paper are designed and implemented in the form of quadratic dynamic matrix control (QDMC) algorithm [5, 16–18], which exchanges **with each other** information about their control signal values. The QDMC consists of the on-line solution to a quadratic programming problem (QP) where a sum of squared deviations of controlled variable predictions from their set-points to maintaining predictions of constrained variables within bounds is minimized. In contrast to the DMC controller, where constraints are enforced via least squares method, the use of a QP provides rigorous handling of constraint violations by formulating them as inequalities linear with respect to the decision variables, and allowing tighter constraint violation control (comparison of DMC and QDMC for a turbine MPC controller can be found in [5]). The distributed model predictive controller for the nuclear power plant turbo-generator set based on the mentioned QDMC controllers and which is a basis for the research described in the paper is described in [19].

Potentially, there might exist configurations, in which a QP problem is infeasible as per constraints imposed on input and output signals simultaneously. There are several ways to relax the constraints that allow to restore feasibility, i.e. minimal-time approach, soft-constraint approach, or hard constraint relaxation with prioritization [20,21]. In the discussed case, the MPC turbine predictive control system is subject to hard constraints imposed on the control valve opening level. This particular constraint cannot be relaxed in any way, and the only acceptable solution is a change of horizons, which increases the number of degrees of freedom in optimization [20].

QDMC controllers are based on the step-response model of the components of the considered system. Taking into account a potentially wide range of operating point changes in the system the step-response model parameters are calculated, based on a simplified linear or black-box models of the turbo-generator set components, which parameters are identified on-line with the recursive least squares algorithm (RLS) [22,23]. The RLS method can include robustness modifications that will cope with estimation errors [23].

Two different types of turbine-generator set models are considered in the paper. In one group are simplified models used for the control system synthesis (as models mentioned above) and education or training [1,3,24–31]. In the second group, there are complex nonlinear models that recreate the object in a detailed way [11,12,32,33]. A complex nonlinear model of the nuclear power plant's turbine-generator set is used as a reference model for the calculations performed for the research and all the discussed control systems are tested with such a plant simulation [4,33].

The paper is organized as follows. In Sect. 2, the turbo-generator set DMPC control structure with local QDMC controllers is described. Section 3 describes the disturbances that may occur in the system. The results of simulation tests with varying disturbances are presented in Sect. 4. Finally, a brief summary of the obtained results is given in Sect. 5.

2 The Turbo-Generator Set DMPC Control Structure with Local QDMC Controllers

Typical turbo-generator set control system consists of two control loops with the PI controllers. In this paper, instead of typical PI controllers the control structure with distributed model predictive controller (DMPC), in the form of two local QDMC controllers for turbine and synchronous generator control, is proposed. To achieve better closed-loop performance, the turbo-generator set control performance, some level of communication is established between the local QDMC controllers – assuming that the information about their control signal values is exchanged.

The diagram in Fig. 1 presents the proposed paper solution's main concept with the linear MPC - QDMC controller. A turbo-generator set is a typical non-linear object. Hence the on-line identification algorithm, in the recursive least-squares (RLS) algorithm structure, is introduced for the linear model parameters used for the object output prediction purposes.

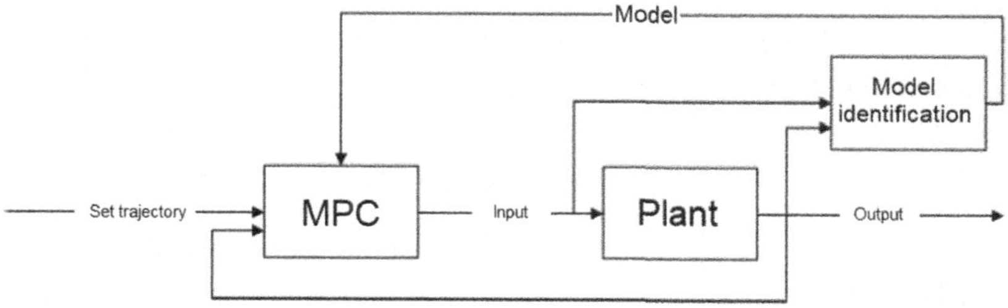

Fig. 1. Control structure with MPC - QDMC controller and with the RLS algorithm module for on-line model parameters identification for plant output prediction purposes.

In the paper, this solution will be compared with the well known from literature classical turbo-generator set control structure which consists of the typical PI controller and a simple power system stabilizer (Fig. 2).

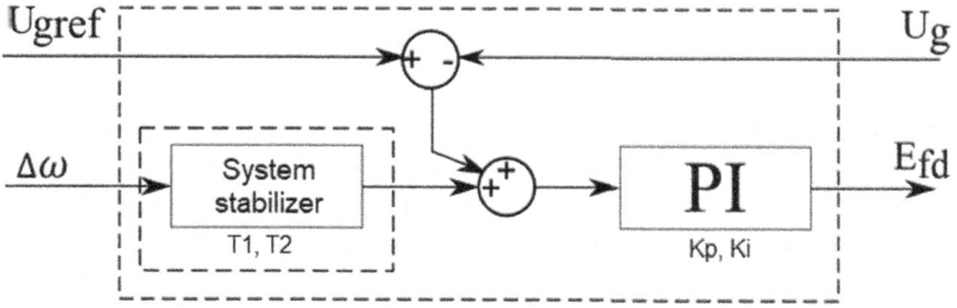

Fig. 2. The classical turbo-generator set control structure with the PI controller and power system stabilizer, where: U_{gref} - reference for the generator voltage, dw - deviation in the generator shaft rotational speed, U_g - generator voltage, E_{fd} - generator excitation voltage, T_1 and T_2 are lead-lag time constant of the system stabilizer, K_p and K_i are proportional and integral gains of PI controller.

The standard QDMC algorithm uses the linear process model in the form of a step response model for prediction purposes [15–18]. For the multiple-input, multiple-output (MIMO) systems, with s controller outputs (manipulated variables) and r measured variables, it may be presented as:

$$\underline{y}_{k+1|k} = \underline{y}_{k+1|k-1} + \boldsymbol{A}\Delta\underline{u}_k + \underline{y}^d_{k+1|k}, \tag{1}$$

where k is the current sampling instant, $\underline{y}_{k+1|k}$ is a value in instant $k+1$ based on information from instant k, $\underline{y}_{k+1|k}$ is a $(r \cdot p) \times 1$ vector representing the prediction of future output trajectory at $t = k$ on the prediction horizon p

$$\underline{y}_{k+1|k} = [(y_{1(k+1|k)}, \ldots, y_{r(k+1|k)}), \ldots, (y_{1(k+p|k)}, \ldots, y_{r(k+p|k)})]^T, \tag{2}$$

the $\underline{y}_{k+1|k-1}$ is a $r \cdot p \times 1$ vector representing the unforced output trajectory, which means the open-loop prediction while the controller output remains constant at the previous value

$$\underline{y}_{k+1|k-1} = [(y_{1(k+1|k-1)}, \ldots, y_{r(k+1|k-1)}), \ldots, (y_{1(k+p|k-1)}, \ldots, y_{r(k+p|k-1)})]^T, \tag{3}$$

$\Delta\underline{u}_k$ is a $(s \cdot m) \times 1$ model manipulated variables adjustments vector defined on the control horizon m

$$\Delta\underline{u}_k = [(\Delta u_{1(k)}, \ldots, \Delta u_{s(k)}), \ldots, (\Delta u_{1(k+m-1)}, \ldots, \Delta u_{s(k+m-1)})]^T, \tag{4}$$

$\underline{y}^d_{k+1|k}$ is a $(r \cdot p) \times 1$ vector representing the unmeasured disturbance estimates, assumed to be the difference between the actual measurements and the unforced output model trajectory components, and $\boldsymbol{A}$ is a $(r \cdot p) \times (s \cdot m)$ dynamic matrix containing the MIMO system model step-response coefficients in the following form:

$$\boldsymbol{A} = \begin{bmatrix} \bar{a}_1 & 0 & 0 & \ldots & 0 \\ \bar{a}_2 & \bar{a}_1 & 0 & \ldots & 0 \\ \vdots & \vdots & \vdots & \ddots & \vdots \\ \bar{a}_p & \bar{a}_{p-1} & \bar{a}_{p-3} & \ldots & \bar{a}_{p-m+1} \end{bmatrix} \tag{5}$$

where every matrix element $\bar{a}_i$ is a $r \times s$ matrix containing the $r \cdot s$ coefficients in the form of samples of the unit step response model, of each controller output to measured process variable pair, at each considered sampling period.

To compute the optimal changes in the manipulated variables vector $\Delta\underline{u}_k$, the multi-variable QDMC quadratic optimization problem (QP) is solved at each sampling instant k [15–18]

$$\begin{aligned} \min_{\Delta\underline{u}_k} \quad & J = [\underline{y}^{\text{ref}}_k - \underline{y}_{k+1|k}]^T \boldsymbol{\Gamma} [\underline{y}^{\text{ref}}_k - \underline{y}_{k+1|k}] + [\Delta\underline{u}_k]^T \boldsymbol{\Lambda} [\Delta\underline{u}_k], \\ \text{s.t.} \quad & \underline{y}_{\min} \leq \underline{y}_{k-1|k} \leq \underline{y}_{\max}, \\ & \Delta\underline{u}_{\min} \leq \Delta\underline{u}_k \leq \Delta\underline{u}_{\max}, \\ & \underline{u}_{\min} \leq \underline{u}_k \leq \underline{u}_{\max}, \end{aligned} \tag{6}$$

where $\underline{u}_k$ is a $(s \cdot m) \times 1$ model manipulated variables vector on the horizon m

$$\underline{u}_k = [(u_{1(k)}, \ldots, u_{s(k)}), \ldots, (u_{1(k+m-1)}, \ldots, u_{s(k+m-1)})]^T, \tag{7}$$

and the $\boldsymbol{\Gamma} > 0$ is a square-diagonal matrix of controlled variable weights, and $\boldsymbol{\Lambda} \geq 0$ is a square-diagonal matrix to penalize control updates. The first matrix is positive definite, the second - positive semi-definite, because there is no need to take explicit soft control limits into consideration.

In case of the described model, the following notation holds:

- outputs - $\underline{y} = [P_g, U_g, \omega_g]$,
- set values - $\underline{y}^{\text{ref}} = [P_{g,\text{ref}}, U_{g,\text{ref}}, \omega_{g,\text{ref}}]$,
- control signals - $\underline{u} = [\alpha, E_{fd}]$
- constraints - $\alpha \in [0, 100]$, $E_{fd} \in [-0.1, 0.1]$.

Finally, the optimal vector of changes in manipulated variables is obtained based on the solution of above-mentioned QP (6) with constraints. Only the first elements from the optimal solution are applied as the control signal to the plant. In the next time instant, the optimization task is solved again, according to the receding horizon rule.

The sampling period has been selected on the basis of the GTHW-600 generator's model description [24], where a dominating time constant of the generator has been identified as $\hat{T} = 0.0017\,\text{s}$ (all turbine's time constraints are larger than those of the generator). The resulting sampling period - $T = 0.00001\,\text{s}$ - abides a rule of thumb to fit at least approximately 10 sampling periods per dominating time constant and avoiding simulation errors occurring for larger time constants. This does not give rise to any problems with the excessive output-averaging properties of the predictive controller, as the considered prediction horizons in the span $40 \div 50$ refer to time ranges $0.0004 \div 0.0005\,\text{s}$.

The functional structures of turbine QDMC and synchronous generator QDMC are presented in Figs. 3–4, respectively. Both QDMC controllers require the unit step response models. It is proposed that their parameters are determined based on the appropriate black-box models (Eqs. 8–10), which parameters are subsequently identified on-line according to the wide range of turbo-generator set operating point changes. In the paper, the RLS identification algorithms for that purposes is proposed [22]. The unknown black-box models parameters are estimated on-line based using the RLS, with respect to the set of the input and output measurements data.

The proposed black-box model for the turbine QDMC controller purposes, in the form of a discrete-time model, is presented with following structure:

$$P_g(k) = \sum_{j=1}^{n} a(j)P_g(k-1-j) + \sum_{j=1}^{n} b(j)\alpha(k-j) + \sum_{j=1}^{n} c(j)E_{fd}(k-j), \quad (8)$$

$$U_g(k) = \sum_{j=1}^{n} d(j)U_g(k-1-j) + \sum_{j=1}^{n} e(j)\alpha(k-j) + \sum_{j=1}^{n} f(j)E_{fd}(k-j), \quad (9)$$

$$\omega_g(k) = \sum_{j=1}^{n} g(j)\omega_g(k-1-j) + \sum_{j=1}^{n} h(j)\alpha(k-j) + \sum_{j=1}^{n} i(j)E_{fd}(k-j), \quad (10)$$

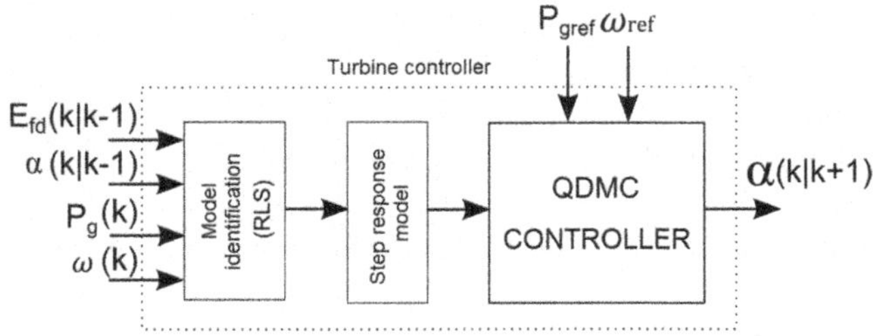

Fig. 3. Turbine QDMC controller's structure

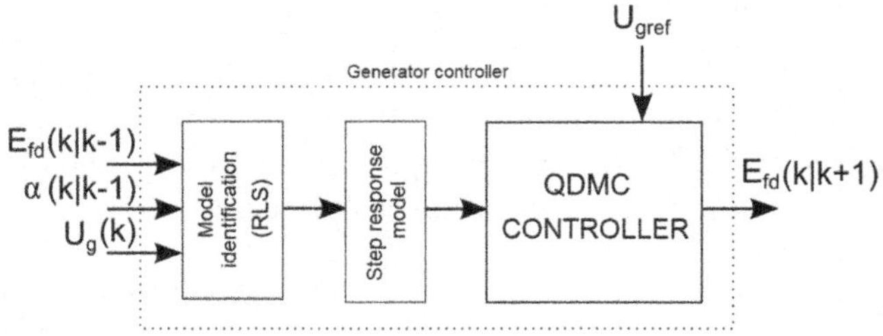

Fig. 4. Generator QDMC controller's structure

where a, b, c, d, e, f, g, h and i are turbine black-box model parameters that are calculated on-line using a recursive least squares estimation method.

The synchronous generator MPC controller uses the same model structure, but the generators' MPC has a different set of parameters estimated in parallel (turbine's and generator's models are independent).

The number of successive samples n considered in black-box models equations (Eqs. 8–10), was chosen arbitrary ($n = 7$) taking into consideration the order of turbine-generator model in the most complex model path ($\alpha \rightarrow U_g$). The number 7 was selected as a compromise between a simple model (e.g., $n = 2$) and an overparametrization. **The order of the model has been calculated on the basis of the sum of the orders of the consecutive components taken into consideration, namely: order of a steam turbine (equal 2)** [34] **and the order of the synchronous generator (equal 5)** [32], **which sums up to 7.**

3 Parameters' Tuning

Before performing experiments with disturbances parameters of compared control systems were tuned in an optimization process.

The PI controller parameters were tuned using the turbine-generator set simulation and assessed using the ISE (Integral of Squared Error) criterion in such a form:

$$f_{\text{ISE}} = \int_{t_0}^{t_f} \left(a(P_{g,\,\text{ref}} - P_g)^2 + b(U_{g,\,\text{ref}} - U_g)^2 + c(\omega_{g,\,\text{ref}} - \omega_g)^2 \right) dt \qquad (11)$$

where:
P_g, $P_{g,\text{ref}}$ – active power and active power set point.
U_g, $U_{g,\text{ref}}$ – voltage and voltage set point.
ω_g, $\omega_{g,\text{ref}}$ – speed and speed set point.
$a, b\ c$ – weights (for P_g, U_g, ω_g, respectively),
t_0, t_f denote the ranges for which the ISE is calculated.

A continuous-time representation of the performance index has been selected to capture all possible rapid changes in values of the signals in the system, which would potentially remain unidentified if a discrete-time domain were selected.

Individual parameters of the controllers were changed and used in the simulation of a the control system with a turbine-generator set model and the quality of the control was assessed. A gradient optimisation method was used with a Simulink simulation output for the objective function calculation purposes (equations describing the turbo-generator set cooperating with the infinite-bus system used by the optimizer in implicit form as a simulation model) - Fig. 5. To find the best parameters of the PI controllers (K_p, K_i, T_i - both for the turbine and the generator controllers, and T_1 and T_2 parameters for a simple power system stabilizer) 100 iterations of gradient algorithm were performed starting from the random initial point.

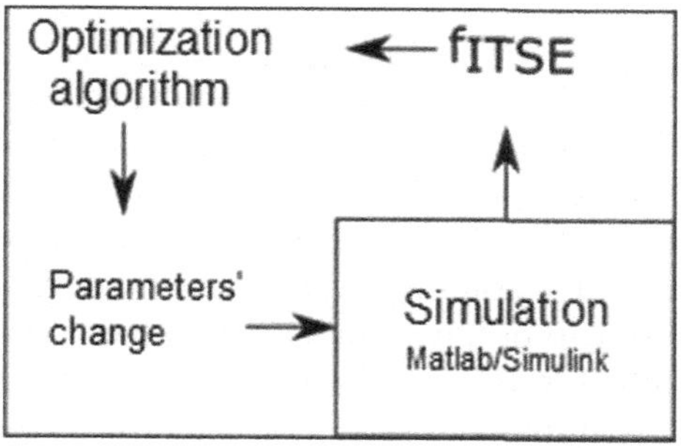

Fig. 5. Tuning of controllers' parameters

Parameters for the turbine MPC controller were selected using a similar method. Two different sets of weights were taken to distinguish between system responses subject to a pair different controller actions: following set trajectory of power P_g and stabilization of voltage U_g and angular speed ω_g. These two actions stand in contrast, i.e. when the system reacts faster to the power set point, it introduces more disturbances to U_g and ω_g. On the other hand, if the system is configured to cope with the disturbances better, it reacts slower after a set point change. To expose this difference one, of the ISE criteria was selected in such a way as to react more to the P_g error (all weights a, b and c are equal to 1) and the second one was chosen to emphasis the U_g component of criterion (weight $b = 100,000$). Additionally, due to the stability issues only the scope of the prediction horizon $40 \leq p_T \leq 50$ was analysed (in some test cases the system with the prediction horizon less than 40, and bigger than 50 became unstable). Using a Simulink simulation plots of the ISE quality indicator ware created for the prediction horizon (p_T) for two different sets of weights in the (Fig. 6, Fig. 7).

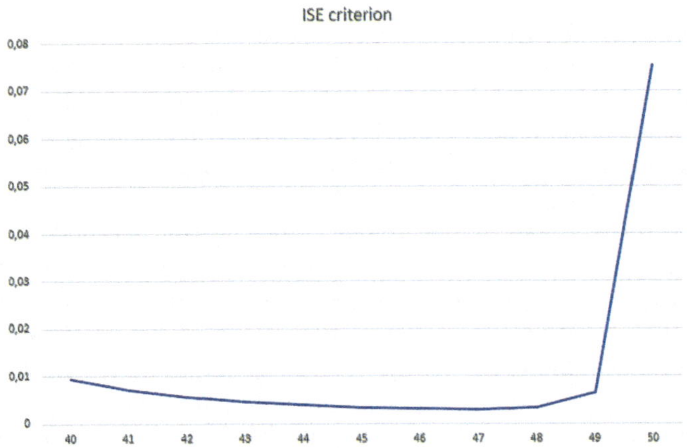

Fig. 6. ISE criterion in function of the prediction horizon p_T - weights $a = 1$, $b = 1$, $c = 1$

Table 1 shows the results of the ISE criterion calculation for different prediction horizons and different ISE criterion's weights. As in the case of the PI controller, the MPC parameters taken as the minimizers were selected. In this case, 10 iterations were performed as per integer value of the prediction horizon p_T. During each of the calculations, power, voltage and speed components were calculated. That allowed to obtain value of a complete ISE criterion with different weights without the need of repeating the simulation. Based on this two different ISE criteria with different sets of weights were used for quality assessment. The parameter values obtained in the optimization process are presented in Table 2.

4 Disturbances

The synchronous generator works nominally with a useful power of 470 MW (1 in relative units), a voltage at the generator terminals of 21 kV (1 r.u.) and an angular speed of 314 $\frac{\text{rad}}{\text{s}}$ (1 r.u.). However, this system works in an environment (Fig. 8), which strongly influences it. This leads to a significant change in the operation condition of the system. Non-linear nature of the object and changes in the environment alter the properties of the model of the plant. On the other hand, to ensure the speed of calculations needed for a real-life implementation, it was decided to use a linear control system with a linear model of the turbine-generator set. Taking this two facts into consideration - frequent changes in the non-linear model and linear control solutions - it was decided to use the on-line

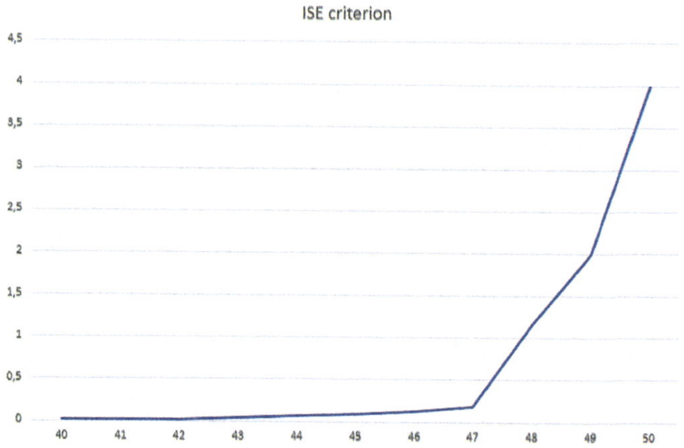

Fig. 7. ISE criterion in function of the prediction horizon p_T - weights $a = 1$, $b = 100.00$, $c = 1$

Table 1. ISE criterion value for the step of active power set point in function of prediction horizon p_T

p_T	ISE_1 - $a = 1$, $b = 1$, $c = 1$	ISE_2 - $a = 1$, $b = 100, 000$, $c = 1$
40	0.0095	**0.0146**
41	0.0071	0.0170
42	0.0057	0.0240
43	0.0046	0.0388
44	0.0039	0.0640
45	0.0034	0.0920
46	0.0030	0.1302
47	**0.0029**	0.1879
48	0.0032	1.1652
49	0.0064	1.9984
50	0.0752	3.9861

Table 2. Controllers' parameters

	K_p	K_i	K_d	T_1	T_2		
PI+PSS	12.82	29.03	0	0.65	1.74		
	p_T	p_G	s_T	s_G	T	$diag(\Gamma_T)$	Λ
$DMPC_1$	**40**	17	1	1	0.01	1;1;1	0
$DMPC_2$	**47**	17	1	1	0.01	1;1;1	0

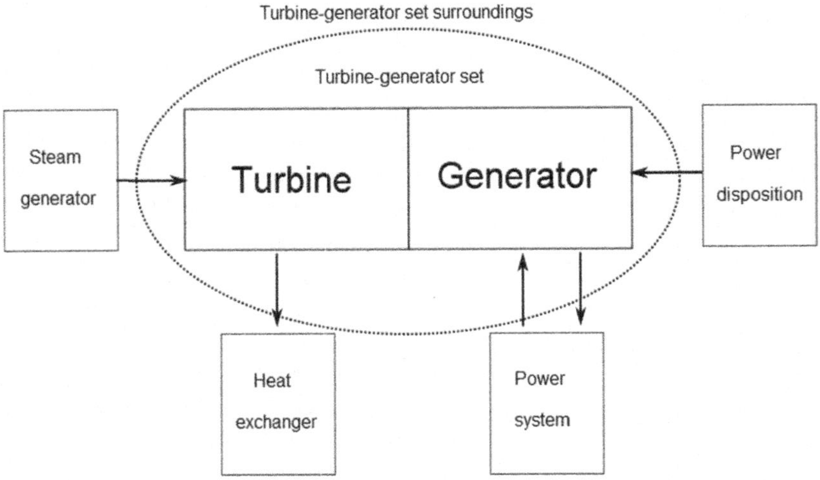

Fig. 8. Turbine-generator set surroundings

model identification, which allows the system to follow up the changes in the plant and to use the control horizon $s = 1$. This is to ensure that decisions about the control signal are made only at the current instant. As due to the change of the operation point of the plant or some disturbances the model may change every instant, also the control signal is calculated every instant using the constantly updated model to follow this changes.

Apart from the impact of the set value of the power through the power disposition, there may also be disturbances from the primary circuit of the power plant, namely, the heat consumption or the power system. After analyzing the operation of the turbo-set predictive control system [5,19], an attempt was made to analyze the impact of external influences on the system operation and to compare the quality of the proposed solutions with classical solutions in the presence of disturbances. As in previous research, only influence from the power disposition was analysed (a change in set points) further research - described in this paper - focused on the rest of the turbine-generator set i.e. the power system, the heating system and a steam generator (primary circuit of the power plant). This concludes the analysis of the system behaviour influenced by its environment.

The main experiment which the parameters of the control system were determined for was the simulation in the change of power set point value by $\pm 10\%$, in the 15th second of simulation (from the level 85%). The power of the turbine is regulated by the turbine control system and the control valve, and the voltage by the generator control system and the excitation system. A similar experiment is described in more details in [19] - in this case it was used to tune up predictive controllers' parameters for further experiments.

The influence of the change in heat consumption load is the first analysed external influence on the operation of the turbine (with co-generation of heat).

The aim of the experiment was to introduce abrupt changes into steam consumption for the needs of co-generation while following the set active power trajectory. This corresponds to a jump in temperature in the heating system, which, through the heat exchanger, leads to a jump in steam intake from the turbine passage.

In the second experiment, in order to visualize the influence of disturbances from the side of the primary circuit of the nuclear power plant on the operation of the turbine set, a simulation was carried out, in which, with a constant electric load 85% of the active power of the generator), the parameters of the steam generator (steam pressure at the input of the control valve) were changed every 30 s of the simulation (−5%, −5%, +10%).

The final experiment was carried out to analyze the impact of disturbances originating from the power system on the operation of the turbine set. The simulation was performed in which the voltage value was changed (−10%, +5%, +5%) every 30 s of the simulation.

The next section describes these experiments in details with a summary and conclusions.

5 Simulation Test Results

The proposed control system and the reference process model were simulated with Matlab/Simulink environment. The results obtained with three different controller sets for turbine-generator were compared, including:

- a typical PI controller with a simple power system stabilizer [4,6],
- two distributed MPC controllers with different parameters marked as $DMPC_1$ and $DMPC_2$, respectively.

It was assumed that the turbine-generator set operated in the power system via the simple infinite-bus [4,6].

The synchronous generator PI controller and power system stabilizer were characterized by the following set of the parameters [6]: K_P, K_I, T_1, T_2 relating to the proportional, integral, derivative gains and integral/derivative time constants, respectively. Their numeric values may be found in Table 1. The QDMC controllers are characterized by following parameters: the control step T, the prediction horizon p, the control horizon s, and the weights on the diagonal of the weights matrix $\boldsymbol{\Gamma}_T$ in the objective function solved by the turbine QDMC controller. Theirs numeric values may be either found in Table 1.

The constraints considered in the simulation test with the QDMC controllers include the lower and upper bounds on the manipulated variables turbine control valve opening degree $\alpha \in [0, 100]$, and synchronous generator excitation voltage $E_{fd} \in [-0.1, 0.1]$.

To make some of the details more visible, along with the results of the whole experiment, also zoomed plot fragments for the short period after the 60th second will be shown (each experiment includes a different disturbance at this time) (Tables 3, 4 and 5).

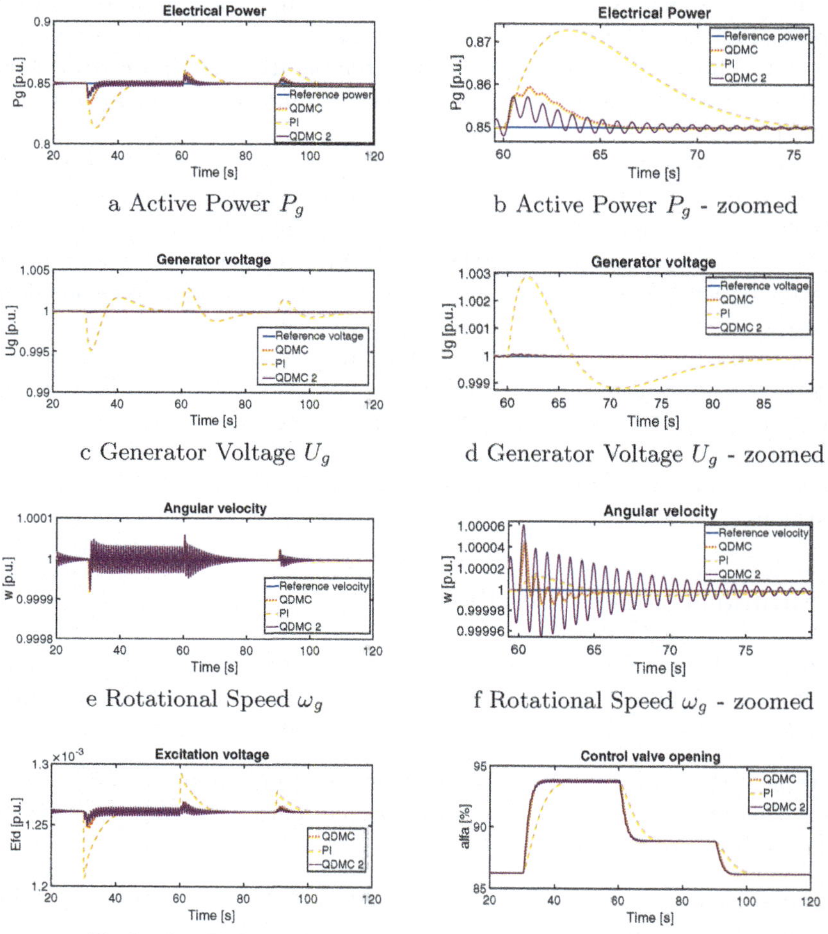

a Active Power P_g

b Active Power P_g - zoomed

c Generator Voltage U_g

d Generator Voltage U_g - zoomed

e Rotational Speed ω_g

f Rotational Speed ω_g - zoomed

g Excitation Voltage E_{fd}

h Control Valve Opening α

Fig. 9. System response to a step change (+90%, −60%, −30%) of the heat demand in 30, 60 and 90th second of the simulation (with 60th second zoomed)

Table 3. ISE criterion value for the heat demand step experiment (ISE_1 - $a = 1$, $b = 1$, $c = 1$ and ISE_2 - $a = 1$, $b = 100{,}000$, $c = 1$)

	PI	DMPC_1	DMPC_2
ISE P_g	$1.03\text{E}-02$	$6.90\text{E}-04$	$2.60\text{E}-04$
ISE U_g	$1.27\text{E}-04$	$1.94\text{E}-08$	$3.37\text{E}-08$
ISE ω_g	$1.90\text{E}-09$	$3.52\text{E}-09$	$2.40\text{E}-08$
ISE_1	$1.04\text{E}-02$	$6.81\text{E}-04$	$\mathbf{2.63E-04}$
ISE_2	$1.27\text{E}+01$	$\mathbf{2.62E-03}$	$3.63\text{E}-03$

Table 4. ISE criterion value for the power system voltage step experiment (ISE_1 - $a = 1$, $b = 1$, $c = 1$ and ISE_2 - $a = 1$, $b = 100{,}000$, $c = 1$)

	PI	$DMPC_1$	$DMPC_2$
ISE P_g	1.88E − 03	7.18E − 02	3.30E − 01
ISE U_g	2.63E − 03	2.82E − 04	3.48E − 04
ISE ω_g	5.99E − 08	2.36E − 05	1.08E − 04
ISE_1	**4.51E − 03**	7.21E − 02	3.30E − 01
ISE_2	2.63E + 02	**2.83E + 01**	3.51E + 01

Table 5. ISE criterion value for the heat demand step experiment (ISE_1 - $a = 1$, $b = 1$, $c = 1$ and ISE_2 - $a = 1$, $b = 100{,}000$, $c = 1$)

	PI	$DMPC_1$	$DMPC_2$
ISE P_g	9.17E − 01	6.41E − 01	6.57E − 01
ISE U_g	2.92E − 02	3.89E − 07	3.37E − 06
ISE ω_g	5.04E − 08	2.12E − 06	6.30E − 06
ISE_1	9.20E − 01	**6.41E − 01**	6.57E − 01
ISE_2	2.93E + 02	**6.80E − 01**	1.01E 00

The results show that due to different controller parameters (prediction horizon) different robustness properties can be achieved. The QDMC controller with $p_T = 40$ follows the set power value, slower, though due to this property, it is capable of rejecting disturbances better, originating both from the side of the heat system, as well as from the power system. In both cases, the $QDMC_2$ controller - due to the stronger power stabilizing action - introduces much larger voltage and rotational speed oscillations.

6 Summary

As part of the research, the impact of disturbances from the heating system and the power system on the operation of the control system of the nuclear power plant turbine-generator set was analyzed. This research is the follow up of the previous analysis of the model predictive control of the turbine-generator set of the nuclear power plant [5,19] extending previous conclusions by the analysis of the system behaviour in the presence of external disturbances. Due to the requirements for the quality of electrical power and the operational safety of a nuclear power plant, it is necessary to analyze the system's robustness. The previous research performed by the authors on the distributed DMPC control of the turbine set required to be supplemented with tests and analysis of the proposed solution in the presence of various disturbances.

With respect to the reference-like PI performance for the heat load change experiment (Fig. 9), the maximum overshoot identified during transients in the

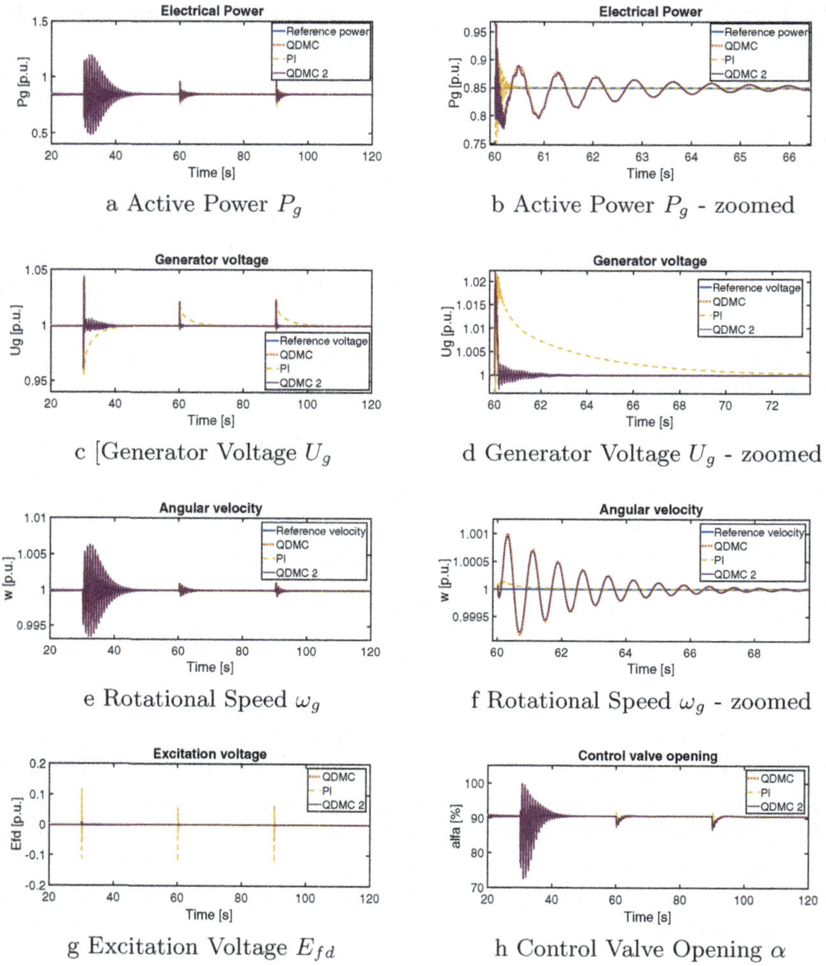

a Active Power P_g

b Active Power P_g - zoomed

c [Generator Voltage U_g

d Generator Voltage U_g - zoomed

e Rotational Speed ω_g

f Rotational Speed ω_g - zoomed

g Excitation Voltage E_{fd}

h Control Valve Opening α

Fig. 10. System response to a step change (−10%, +5%, +5%) of the power system voltage in 30, 60 and 90th second of the simulation (with 60th second zoomed)

case of QDMC1 control is reduced from 2.4% to 0.9%, and in the case of QDMC2 controller – to 0.55% as per electrical power measurements, but it comes with the cost of appearing power oscillations. It is related to high angular velocity oscillations with long settling time. For both QDMC1 and QDMC2, in the case of generator voltage, the overshot is almost two orders of magnitude smaller. In this case - as the overshoots are similar for QDMC1 and QDMC2 and better in both electric power and voltage stabilisation than the PI solution and because the slightly faster controller QDMC2 introduces unwanted oscillations it can be indicated that the QDMC1 shows the best performance in this experiment.

In the second experiment (Fig. 10) – step changes in the power system voltage - the overshoots identified during transients in the case of QDMC1, QDMC2 and

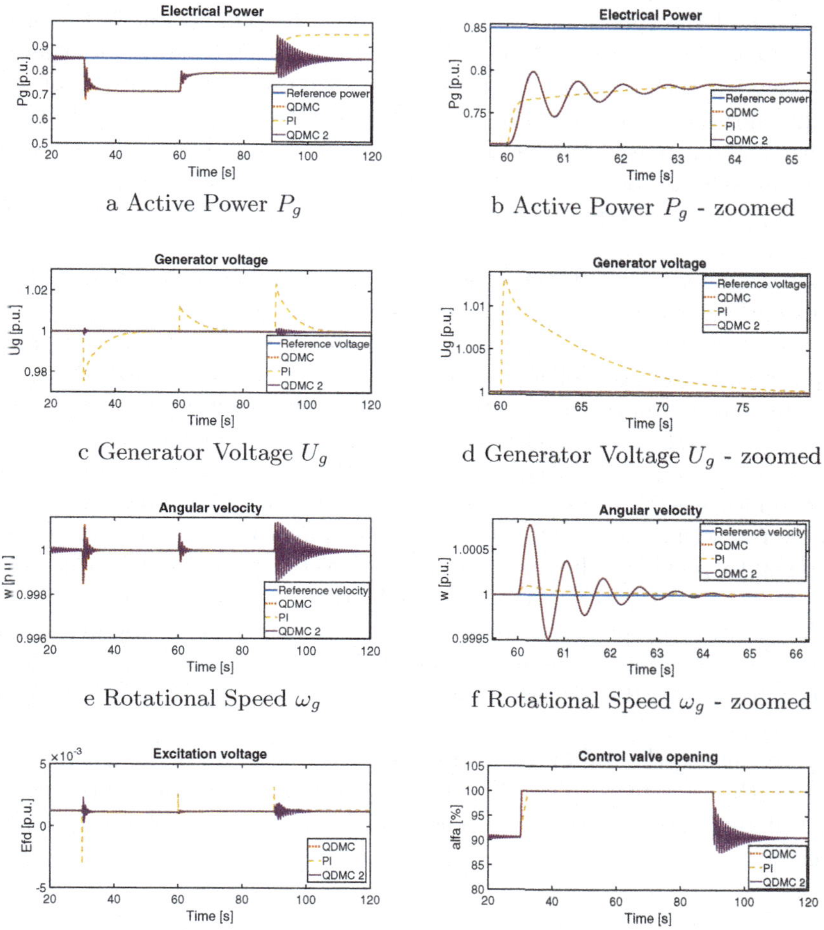

a Active Power P_g b Active Power P_g - zoomed

c Generator Voltage U_g d Generator Voltage U_g - zoomed

e Rotational Speed ω_g f Rotational Speed ω_g - zoomed

g Excitation Voltage E_{fd} h Control Valve Opening α

Fig. 11. System response to a step change (−10%, +5%, +5%) of the steam pressure before the turbine's control valve in 30, 60 and 90th second of the simulation (with 60th second zoomed)

PI control are similar, but PI and QDMC controllers show completely different behaviour. The system with the PI controller restores the set power faster than the QDMCs, but with the cost of slow setting of the generator voltage. On the other hand the QDMCs are much faster in stabilizing the voltage while introducing longer oscillations to the active power. In this case, QDMC2 – unlike in the previous experiment - shows worse performance as both QDMC1 and PI controllers.

The third experiment (Fig. 11) included the disturbance of the steam pressure at the control valve of the turbine. Because of the hard limitations of the valve neither of the controllers was able to keep the requested electrical power level.

In the case of the generator voltage however both QDMCs allowed to achieve smaller disturbances (2.3% PI's overshoot to 0.1% for QDMCs) or to avoid them completely (in 60th second). In this experiment lack of the anti-windup in the PI controller also influences the results. Both QDMC controllers include the knowledge about plant model and it's limits. Due to the lack of the anti-windup the PI controller is not able to set the power at 85% after the nominal pressure is restored. Also in this experiment the QDMC2 introduced the biggest oscillations to the system.

The presented results together with the general analysis of the MPC control system of the turbine and the generator of a nuclear power plant [5,19] confirm the possibility of using predictive control in the control of this type of facility. Adequate choice of controllers' parameters can change the systems behaviour from a faster one (quickly acting to the changes of the set active power steps or disturbances) to the more robust one (minimising all the oscillations of important process values after a disturbance). Even though the purpose of the power plant is to produce electrical energy and it is determined by the power provided to the power system, the quality of the energy is also very important. It is very important to provide voltage and frequency stabilization, and in this context $DMPC_1$ shows better performance as the $DMPC_2$ and PI controllers.

The value of a proper choice of prediction horizons is visible also in the newest research, see for example the paper [35]. This might be a potentially attractive direction to couple MPC optimization with parallel tuning of prediction horizons, using, e.g. [36].

Subsequent works will focus on analysing the behaviour of the generator controller (with constant turbine MPC parameters) and a precise multi-parameter tuning of the whole distributed control system in presence of external disturbances.

References

1. Lipo, T.: Analysis of Synchronous Machines. CRC Press, Boca Raton (2012)
2. Kundur, P.: Dynamic models for steam and hydro turbines in power system studies. IEEE Trans. Power Appar. Syst. **PAS-92**(6), 1904–1915 (1973)
3. Kundur, P.: Analysis of synchronous machines. In: Power System Stability and Control (1994)
4. Sokolski, P., Rutkowski, T., Duzinkiewicz, K.: The excitation controller with gain scheduling mechanism for synchronous generator control. In: 2015 20th International Conference on Methods and Models in Automation and Robotics (MMAR), pp. 23–28 (2015)
5. Sokolski, P., Rutkowski, T., Duzinkiewicz, K.: The QDMC model predictive controller for the nuclear power plant steam turbine control. In: Trends in Advanced Intelligent Control, Optimization and Automation, KKA 2017. Advances in Intelligent Systems and Computing, vol. 577, pp. 241–250 (2017)
6. Sokólski, P., Kobylarz, A., Kulkowski, K., Duzinkiewicz, K., Rutkowski, T., Grochowski, M.: Advanced control structures of turbo generator system of nuclear power plant. Acta Energetica **3**, 83–96 (2015)

7. Grochowski, M., Rutkowski, T.: Supervised model predictive control of wastewater treatment plant. In: 21st International Conference on Methods and Models in Automation and Robotics (MMAR), pp. 613–618 (2016)
8. Ciminski, A., Duzinkiewicz, K.: Optimized robust model predictive control - application to drinking water distribution systems hydraulics. IFAC Proc. Vol. **43**(8), 395–400 (2010)
9. Drewa, M., Brdys, M., Cimiñski, A.: Model predictive control of integrated quantity and quality in drinking water distribution systems. IFAC Proc. Vol. **40**(5), 95–100 (2007)
10. Puchalski, B., Duzinkiewicz, K., Rutkowski, T.: Multi-region fuzzy logic controller with local PID controllers for u-tube steam generator in nuclear power plant. Arch. Control Sci. **25**(4), 429–444 (2015)
11. Czapliński, M., Sokólski, P., Duzinkiewicz, K., Piotrowski, R., Rutkowski, T.: Comparison of state feedback and PID control of pressurizer water level in nuclear power plant. Arch. Control Sci. **23**(4), 381–398 (2013)
12. Puchalski, B., Rutkowski, T., Tarnawski, J., Duzinkiewicz, K.: Comparison of tuning procedures based on evolutionary algorithm for multi-region fuzzy-logic PID controller for non-linear plant. In: 20th International Conference on Methods and Models in Automation and Robotics, pp. 897–903 (2015)
13. Camponogara, E., Jia, D., Krogh, B., Talukdar, S.: Distributed model predictive control. IEEE Control Syst. **22**(1), 44–52 (2002)
14. Cristofides, P., Scattolini, R., de la Pena, D., Jinfeng, L.: Distributed model predictive control: a tutorial review and future research directions. Comput. Chem. Eng. **51**, 21–41 (2013)
15. Camacho, E., Alba, C.: Model Predictive Control. Springer, London (2013)
16. Rossiter, J.: Model-Based Predictive Control: A Practical Approach (2013)
17. Tatjewski, P.: Advanced control of industrial processes (2007)
18. Carlos, E., Morshedi, A.: Quadratic programming solution of dynamic matrix control (QDMC). Chem. Eng. Commun. **46**(1–3), 73–87 (1986)
19. Sokólski, P., Rutkowski, T., Duzinkiewicz, K.: The distributed model predictive controller for the nuclear power plant turbo-generator set. In: Trends in Advanced Intelligent Control, Optimization and Automation, KKA 2017. Advances in Intelligent Systems and Computing, vol. 577, pp. 682–687 (2017)
20. Afonso, R., Galvão, R.: Infeasibility handling in constrained MPC. In: Frontiers of Model Predictive Control, pp. 47–64 (2012)
21. Maciejowski, J.M.: Predictive Control: With Constraints (2002)
22. Eakvoort, R.: System identification for robust process control: nominal models and error bounds. Technische Universitet Delft (1994)
23. Sokolski, P.: Implementation of robust estimation methods for regulatory purposes predictive MPC type. Innov. Solut. Autom. Area **22**(1), 44–52 (2011)
24. Imieliński, A.: Mathematical model of synchronous generator for full-scope simulator. Gdañsk University of Technology, Faculty of Electrical and Control Engineering (1987, unpublished)
25. Perycz, S., Próchnicki, W.: The mathematical model of a nuclear power plant VVER block steam turbine allowing to study transient processes with $w = var$. Gdañsk University of Technology, Faculty of Electrical and Control Engineering (1989, unpublished)
26. Puchalski, B., Rutkowski, T., Duzinkiewicz, K.: Multinodal PWR reactor model-methodology proposition for power distribution coefficients calculation. In: 21st International Conference on Methods and Models in Automation and Robotics, pp. 385–390 (2016)

27. Kulkowski, K., Kobylarz, A., Grochowski, M., Duzinkiewicz, K.: Dynamic model of nuclear power plant steam turbine. Arch. Control Sci. **25**(1), 65–86 (2015)
28. Kulkowski, K., Grochowski, M., Duzinkiewicz, K., Kobylarz, A.: Nuclear power plant steam turbine - modeling for model based control purposes. Appl. Math. Model. **48**, 491–515 (2017)
29. Juszczuk, D., Tarnawski, J., Karla, T., Duzinkiewicz, K.: Real-time basic principles nuclear reactor simulator based on client-server network architecture with WebBrowser as user interface. Adv. Intell. Syst. Comput. **577**, 344–353 (2017)
30. Tarnawski, J., Karla, T.: Real-time simulation in non real-time environment. In: 21st International Conference on Methods and Models in Automation and Robotics (MMAR), pp. 577–582 (2016)
31. Karla, T., Tarnawski, J., Duzinkiewicz, K.: Cross-platform real-time nuclear reactor basic principle simulator. In: 20th International Conference on Methods and Models in Automation and Robotics (MMAR), pp. 1074–1079 (2015)
32. Loo, C., Vanfretti, L., Liceaga-Castro, E., Acha, E.E.: Synchronous generators modeling and control using the framework of individual channel analysis and design: part 1. Int. J. Emerg. Electr. Power Syst. **8**(5), 1–28 (2007)
33. Sokolski, P., Rutkowski, T., Duzinkiewicz, K.: Simplified, multiregional fuzzy model of a nuclear power plant steam turbine. In: 21st International Conference on Methods and Models in Automation and Robotics (MMAR), pp. 379–384 (2016)
34. Power System Dynamic Performance Committee: Dynamic models for turbine–governors in power system studies. IEEE Power & Energy Society (2013)
35. Nebeluk, M.R.: Tuning of multivariable model predictive control for industrial tasks. Model Predictive Control: Algorithms and Applications, 3 January 2021
36. Giernacki, W., Horla, D., Báča, T., Saska, M.: Real-time model-free minimum-seeking autotuning method for unmanned aerial vehicle controllers based on Fibonacci-search algorithm. Sensors **19**, 312 (2019)

Simulation and Testing of Flight Stabilisation System Using Trimmers

Mariusz Krawczyk, Albert Zajdel(✉), and Cezary Szczepański

Łukasiewicz Research Network – Institute of Aviation, Space Technologies Center, Avionics Division, Al. Krakowska 110/114, 02-256 Warsaw, Poland
{mariusz.krawczyk,albert.zajdel,
cezary.szczepanski}@ilot.lukasiewicz.gov.pl

Abstract. The main goal of the research presented in this paper is testing the possibility of introducing the automatic trimming system for the aeroplane flight stabilisation purpose. If successful, it can perform the stabilisation and even automatic control of the aeroplane flight, in the certain flight stages. Model-based attitude towards the automatic control system design has been adopted. Initial models of the aeroplane, flight control system with the control laws and models of servomechanisms have been developed and tested in the MATLAB/Simulink environment. Preliminary designed flight stabilisation system elements have been validated at the HIL testing stand in the two modes: simulation and HIL. Performed research has shown that flight stabilisation system based on the trimming tabs deflection is capable of automatic control of the aeroplane flight even in scenarios typical for autopilots. Proposed stabilisation system assures the safety flight conditions with the use of limits imposed by the control software. Results presented in the paper are original ones achieved during the research project (ISSLOT – Innovative system of flight stabilisation with use of trimmers). The next stage of the project will be the stabilisation system validation during the flight tests.

Keywords: Flight trimming · Automatic flight stabilisation · Mechanics of flight · Hardware in the loop simulation

1 Introduction

In most cases, vehicles are underactuated dynamic systems, as a result of the applied technical solutions. Indeed, there are some exceptions where the number of degrees of freedom and the number of control channels is equal (e.g. train or cableway). However, such situations are relatively uncommon. In many systems, a deficit of control channels may be noticed. An extreme example of such an underactuated system is the aeroplane [8]. If considered as a dynamic system, it has six degrees of freedom with four control channels only – conventionally: rudder, elevator, ailerons and thrust controller [9, 12, 13, 15]. Due to the control channels shortage, the control rearrangement in case of failure in one channel is practically impossible. The only exception here is the rudder – its malfunction in most cases can be compensated with coordinated deflection of ailerons

R. Szewczyk et al. (Eds.): AUTOMATION 2021, AISC 1390, pp. 185–196, 2021.
https://doi.org/10.1007/978-3-030-74893-7_18

and elevator. For that reason, the total reliability of an aircraft control system is a product of three components, i.e. the reliability values of each of the three control channels. Because of the high safety standards in aviation [4, 6], the requirements imposed on the aircraft control system are among the most stringent compared to the other vehicles.

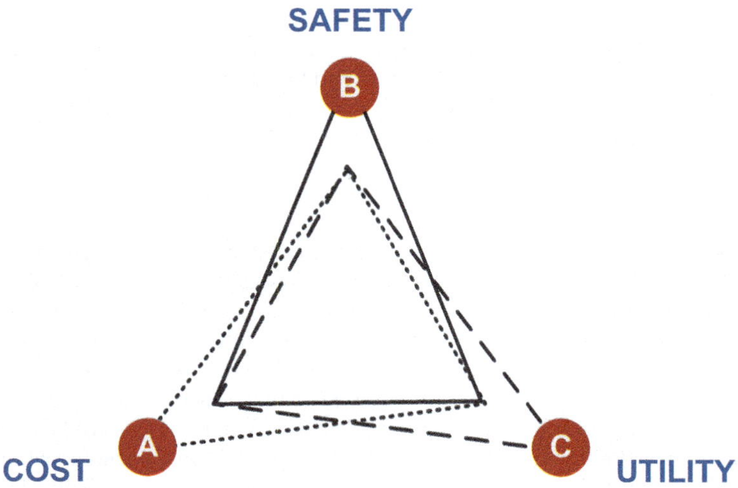

Fig. 1. Factors determining solutions for aircraft control systems

Several factors are determining the structure and the technical solutions of the existing aircraft control systems. In case of small aeroplanes, like ultralights with MTOW not exceeding 450 kg, the pilot directly operates the control surfaces that are physically connected with a control stick and pedals using pulleys, cranks or tension cables. In bigger aeroplanes, weighing even 1000 times more than the lightest ultralights increased control surface area and high airspeed lead to increased forces required from the pilot to move them. At some cases, those forces are too big to be accepted for the operational aeroplanes. Then the reliable hydraulically moved aeroplane's primary control surfaces are used.

In general, the architecture of a flight control system is determined by functional features of the aircraft, system implementation and maintenance cost, and system reliability. For different aircraft groups, the emphasis may be put on a different factor. As shown in Fig. 1, in case of cargo aeroplanes (group A) the dominant factor is "Cost", for passenger aeroplanes (group B) most important is "Safety", while for military aeroplanes (group C) – "Utility". Especially for fighter aeroplane, high manoeuvrability is crucial, and the introduction of fly-by-wire systems (F-16, JAS 39 Gripen) in the 70' allowed for significant progress in this field. Then, owing to the development of electronics that could fulfil the reliability requirements recommended in DO-178 and DO-254 standards, fly-by-wire flight control systems were gradually introduced to civil aviation.

The next trend noticed in the evolution of control systems is the replacement of energy-consuming and massive hydraulic systems with electric ones. It is not always feasible due to enhanced power requirements regarding actuators designed for large aeroplanes. In this case, the high-current wiring and high power electric motors are needed what results in an unfavourable mass balance, even compared to the hydraulic systems. The solution may be a new generation control system where the primary control surfaces are moved by aerodynamic moments generated on the smaller tabs hinged to them, working similarly to the currently used trimmers.

In this paper, a flight stabilisation system using trimmers is presented. It was designed for a turboprop trainer aeroplane, PZL-130 Orlik. In particular, the architecture of altitude and heading stabilisation channels is described. The designed system underwent the Hardware In the Loop (HIL) tests. The results suggest that this solution in certain circumstances may be treated as an autopilot (AP) or fly-by-wire control system. More research is needed to confirm its utility.

2 Aeroplane Stabilisation System Using Trimming Surfaces

The controlled object, which is shown in Fig. 2 is a PZL-130 Orlik turboprop training aeroplane. To perform a synthesis of the 3D flight stabilisation system controller shown in Fig. 2, the non-linear model was linearised using the small perturbation theory.

The essential role of trimmers is to balance the hinge moments and moments of aerodynamic forces, thus causing a reduction of the forces that the pilot has to exert in the process of operation of the control elements [10]. The concept presented in this paper assumes that trimming surfaces can also be used as active controlling elements in the pitch, yaw, and roll control channels – elements of the flight stabilisation system. Such an idea of the flight control system adds new functionality to the typical aircraft trimming system.

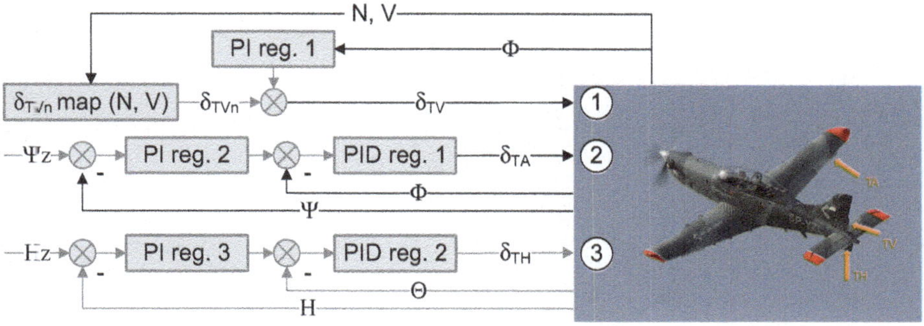

Fig. 2. The automatic 3D aircraft flight stabilisation system

An analysis of the results obtained during the in-flight tests [14] made it possible to determine the transition functions that described the relationship between the values of the ailerons, elevator and rudder deflections and the deflections of the appropriate trimmers surfaces:

$$\delta_H = \frac{-0,6}{0,25s+1}\delta_{TH} \quad \delta_A = \frac{-0,42}{0,25s+1}\delta_{TA} \quad \delta_V = \frac{-0,75}{0,3s+1}\delta_{TV}$$

where: δ_H – elevator angle, δ_{TH} – elevator trim tab angle, δ_A – aileron angle, δ_{TA} – aileron trim tab angle, δ_V – rudder angle, δ_{TV} – rudder trim tab angle, s – Laplace operator.

The stabilisation system presented in Fig. 2 has a cascade structure [1, 2] that is separable in altitude and heading channels. It means that the altitude control can be performed entirely by deflection of elevator trimmer δ_{TH}, whereas heading control is provided by deflection of aileron trimmer δ_{TA}. Both pitch angle Θ and roll angle Φ stabilisation channels are realised in the inner loops. In those channels, the PID controllers were used for eliminating static error and introduce damping of pitch Q and roll P rates. In outer loops, the PI controllers were implemented. They ensure the elimination of static errors of altitude H and heading Ψ.

The third channel, which controls rudder trimmer deflection δ_{TV}, aims at compensating the slipstream effect automatically. It is undesirable yawing moment stemming from the spiralling slipstream generated by the running propeller and altering the airflow around the aeroplane [3]. Figure 3 illustrates the nominal deflection of rudder trimmer δ_{TV} that compensates the undesirable yawing moment depending on the most important parameters – the aircraft velocity V and the thrust N. According to Fig. 3, the biggest yawing moment corresponds to flight with minimum velocity and maximum surplus thrust, that appears during take-off and go-around manoeuvres [3].

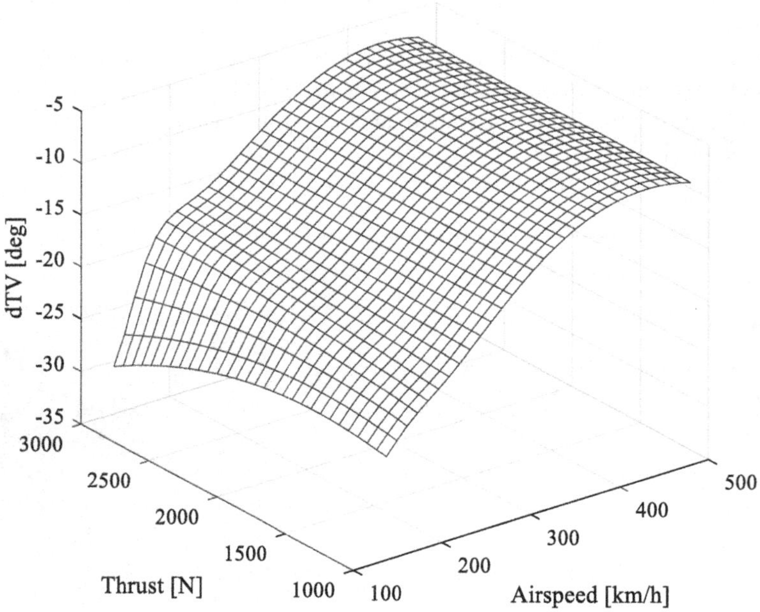

Fig. 3. The relation between rudder trim tab deflection compensating slipstream effect, as a function of thrust and airspeed

Accuracy of the results presented in Fig. 3 is limited due to measurement errors and simplifications that exclude other effects potentially affecting yawing moment. For that reason, the PI controller was chosen for the rudder trimmer control channel. This type of regulator is capable of compensating the errors fully and providing accurate trimming.

3 Results of the HIL Simulation

Usually, the HIL tests of real-time control systems [5] take place after completion of simulation tests in a virtual testing environment. They are directly followed by the final flight tests [7]. The HIL tests are beneficial from an economic point of view, as they effectively allow for a reduction of the number of necessary flight test hours – which are often budget consuming. In Fig. 4 and Fig. 5, the simplified architecture of the stabilisation system for the PZL-130 Orlik aeroplane as well as the HIL stand layout are presented.

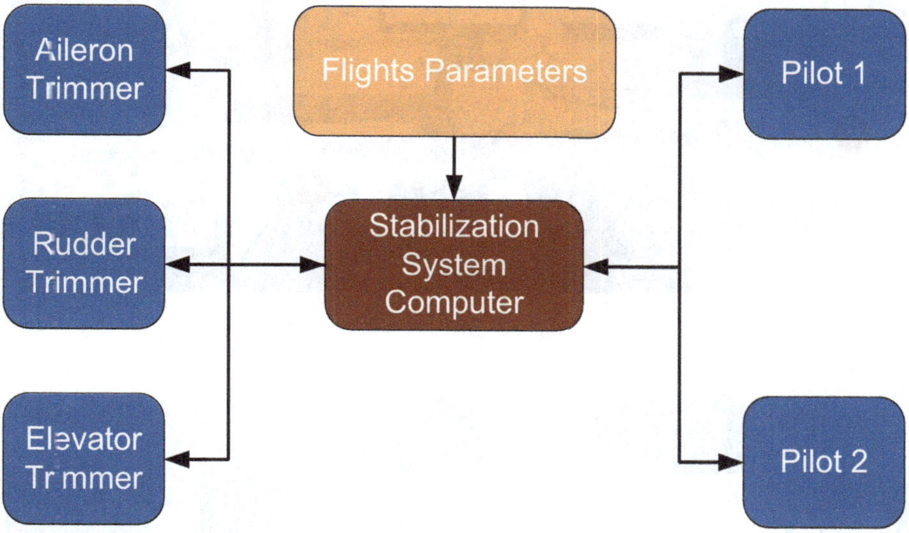

Fig. 4. Trimmer Stabilisation System (TSS) scheme

The following components are installed on the HIL test stand, operating in MATLAB/Simulink plus Speedgoat environment: 1 – TSS computer, 2 – three servomotors, 3 – Real-time computer, and 4 – two pilot-aircraft interfaces. The simulation results are displayed in the real-time on the two screens: current values of the main parameters on the first one, and flight visualisation on the other. The HIL test stand configuration shown in Fig. 6 allows for carrying out the tests in two different modes: Sim_Loop and HIL_Loop.

While operating in the first mode, the MATLAB/Simulink simulation in an exclusively virtual environment is running. Therefore the TSS computer source code, servomotors' models and aeroplane's (A/C) model are used.

In the second mode, the TSS computer and servomotors' models are substituted with real hardware connected to the Speedgoat computer with the use of the ARINC-429 data bus.

The HIL tests were divided into three stages. At first, the real servomechanisms' dynamics were investigated. The results were crucial for validation of servos speed and accuracy. Additionally, the quality of servomotors' and gearbox' models used in

Fig. 5. HIL test stand

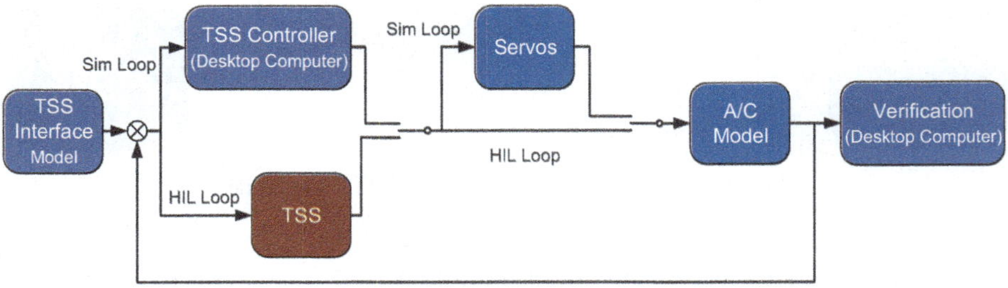

Fig. 6. Configuration of HIL test stand – two options Sim_Loop and HIL_Loop

simulations was assessed. Secondly, the performance of stabilisation in longitudinal motion was analysed, and eventually, the operation of TSS in lateral motion as well as rudder trimming.

Servomechanisms' dynamics assessment

The response of a real servomechanism to the input signal – the unit step function with the value of – 15° – is presented in Fig. 7.

The obtained results proved the assumed control effectiveness and accuracy. The selected servomotor is fast enough to stabilise the desired object and needs just 0,5 s to reach the desired value. The apparent slight noise in the measured signal is a consequence of analogue measurement using a rotary potentiometer integrated with the servo.

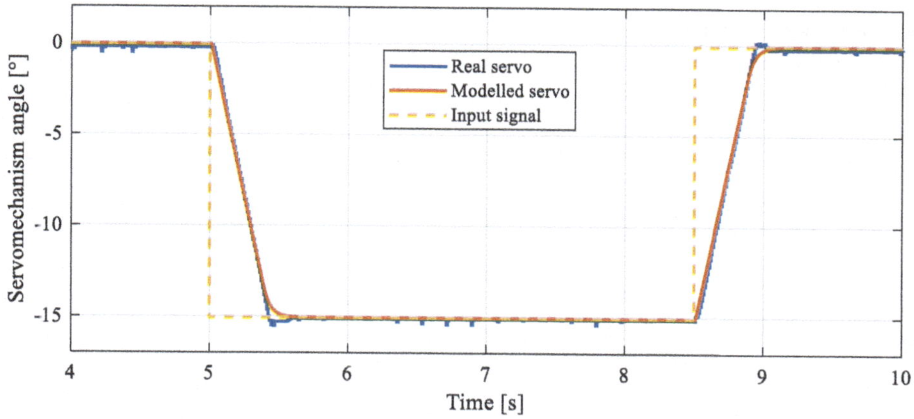

Fig. 7. Real servomechanism response compared to modelled servo response

Longitudinal motion studies

Flight stabilisation in the longitudinal control channel performed by the TSS consists in maintaining flight altitude occupied at the moment of system activation. During the HIL tests, the desired altitude was 2950 m. The disturbance value was intentionally set at 225 m, so that it might have been representative for the more complex device – the autopilot (AP) – or manual control of the aeroplane. Figure 8 illustrates the altitude response for changing the desired altitude from 3175 m to 2950 m.

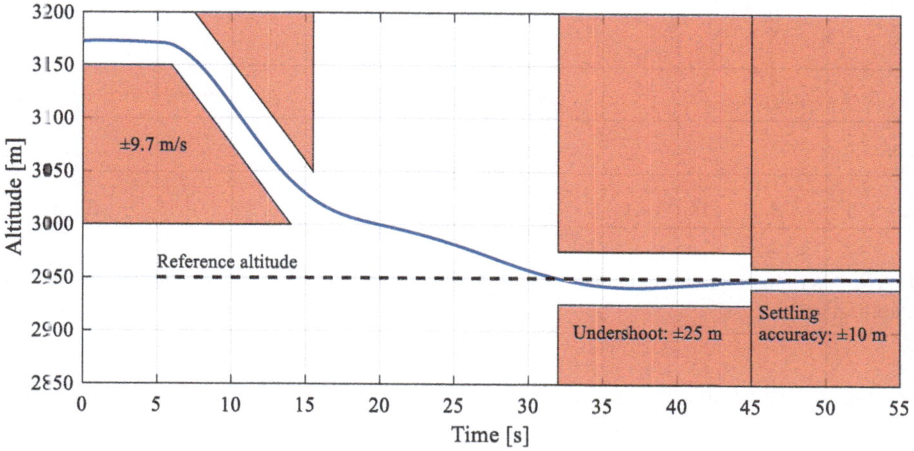

Fig. 8. Altitude response: from 3175 m to 2950 m

Analysis of the graph shown in Fig. 8 leads to the conclusion that the assumed values of the TSS operating parameters were reached:

- Altitude settling accuracy of ±10 m,

- Undershoot within the range of ±25 m,
- Descent/climbing rate not exceeding ±10 m/s.

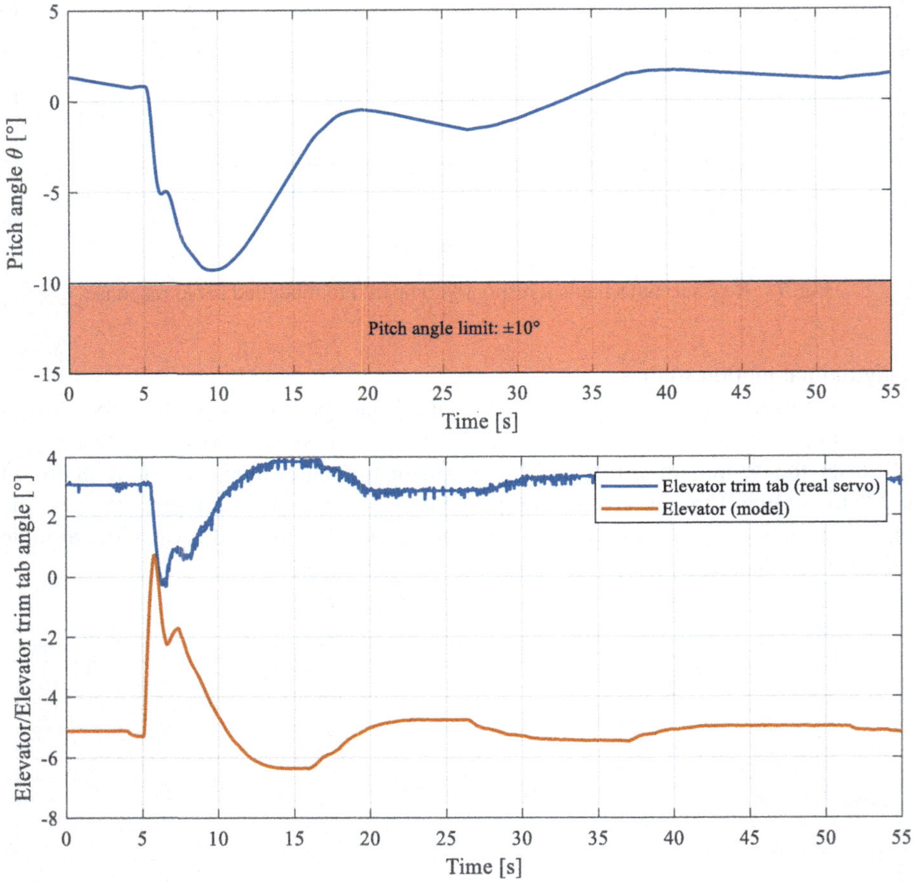

Fig. 9. Aeroplane pitch angle and elevator and trim tab deflections angles during the altitude stabilisation process

Figure 9 illustrates changes in aeroplane pitch angle θ during the altitude stabilisation process compared with the matching deflections of the elevator induced by the elevator trim tab movements. The pitch angle stays within the imposed limit of $\pm 10°$ (Fig. 10).

During the simulated manoeuvre of lowering the flight altitude, the airspeed rises, what directly affects the value of the undesirable yawing moment stemming from the spiralling slipstream, according to Fig. 3. As a result, a non-zero value of the sideslip angle appears. The sideslip angle is effectively compensated with rudder trim tab deflection.

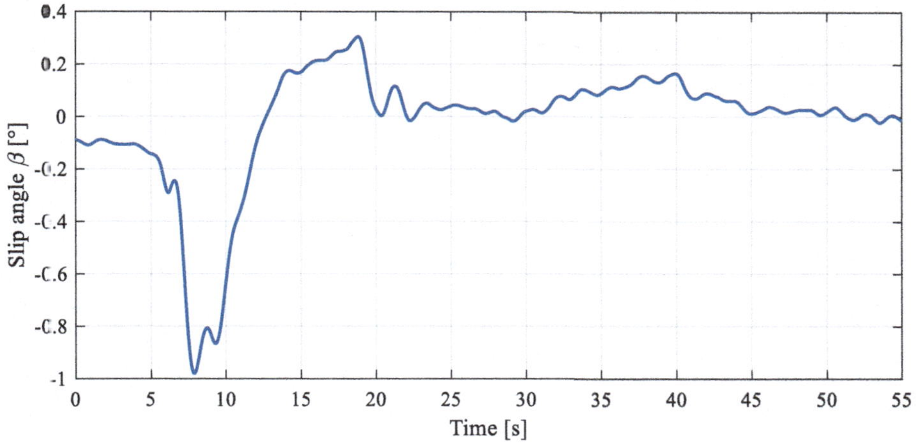

Fig. 10. Aeroplane sideslip angle generated by the slipstream effect in altitude stabilisation

Lateral motion studies

The aim of the aeroplane lateral motion stabilisation is maintaining the heading that existed at the instant of the system activation. While simulating the heading response, it was assumed that the desired heading was 30°. Due to disturbances at the beginning of the stabilisation process, the aeroplane was flying with the heading of 0°. Similarly to the longitudinal motion studies, the disturbance of 30° is more representative for AP systems than TSS. The purpose of choosing such a relatively high value was to evaluate if the designed TSS system may have any AP features.

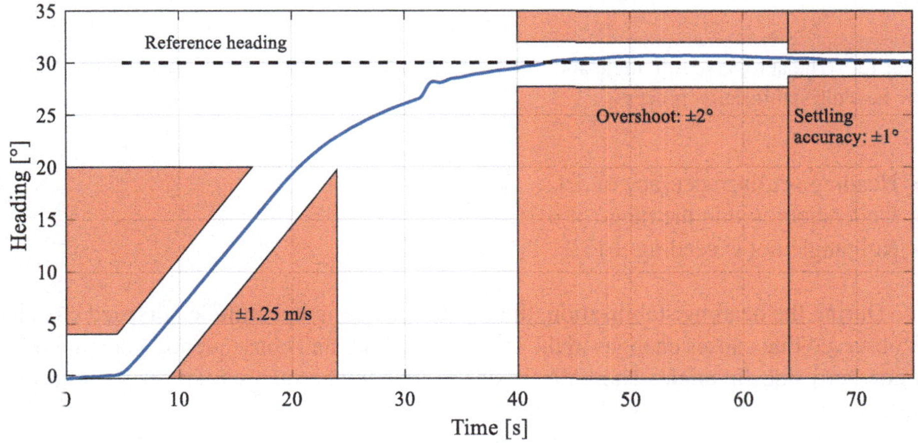

Fig. 11. Aeroplane heading response: from 0° to 30 heading input

Simulation results shown in Fig. 11 and Fig. 12, confirm fulfilling the requirements specified by the following parameters:

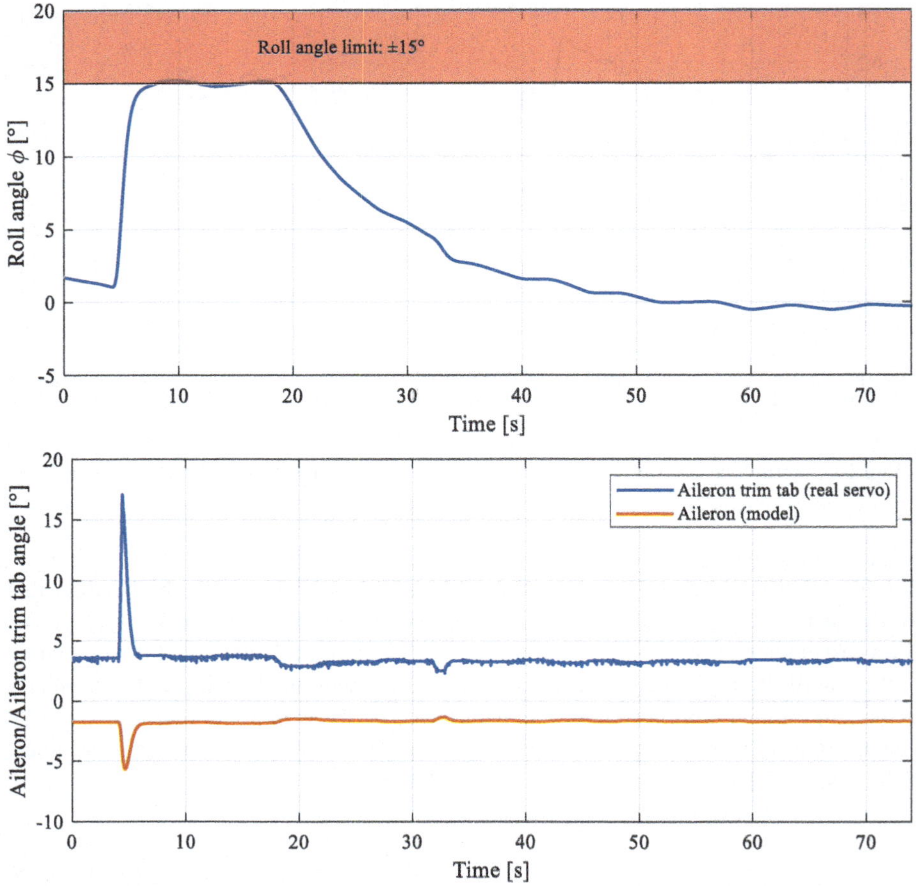

Fig. 12. The aeroplane roll angle and aailerone and ailerone trim tab deflections angles during the heading stabilisation process

- Heading settling accuracy of $\pm 1°$,
- Undershoot within the range of $\pm 2°$,
- Roll angle not exceeding $\pm 15°$.

During the heading stabilisation, like while stabilising the altitude, airspeed change is observed that causes changes in the slipstream effect, and consequently generation of a non-zero sideslip angle. These disturbances are compensated with properly adjusted rudder trimmer operation, as shown in Fig. 13.

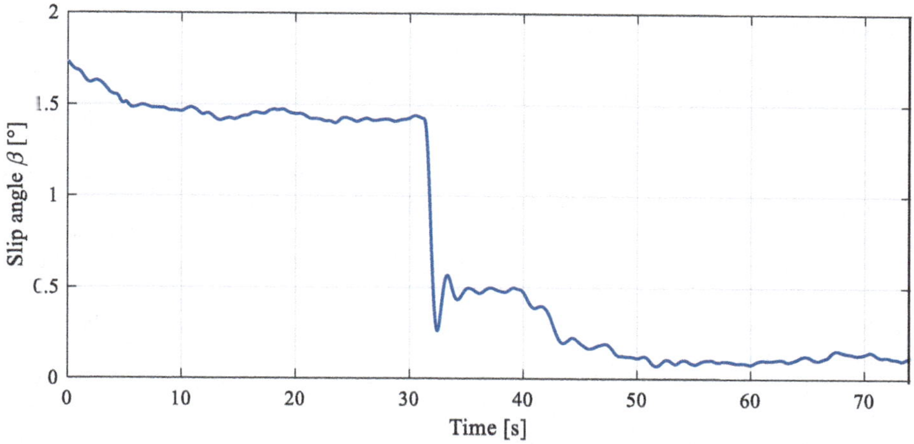

Fig. 13. Aeroplane sideslip angle generated by the slipstream effect in heading stabilisation

4 Concluding Remarks

The possibility to stabilise/control an aircraft with coordinated deflections of the trimming surfaces is a beneficial alternative to those currently used for more complex direct, fly-by-wire and autopilot systems. In particular, the carried-out HIL research showed the following benefits that arise in connection with the presented system:

- Flight stabilisation system using trimming tabs is capable of automatically 3D control of the aeroplane flight even in scenarios typical for AP;
- Simple servomechanism structure compared to those used in AP, allowing for manual trimming of primary flight control surfaces and automatic control of flight;
- Model-based development of the flight control system gives the possibility to replace simulation models with real hardware, e.g. actuators gradually;
- Simplified introduction of control system design changes – changes need to be done only in one place, the model. A new code is generated automatically and can be uploaded into the aeroplane's hardware to the final test;
- A pilot can test the system in the engineering simulator at an earlier stage of the design process;
- Moreover, HIL simulations facilitate the elaboration of the flight test programme and enhance the awareness of expected results.
- Aeroplane flight stabilisation system assures the safety conditions thanks to the limits introduced by the software controlling the flight.

Acknowledgements. Researches described in that paper have been funded under the EU co-financed project number POIR.04.01.02-00-0006/17-00, titled "Innovative system of flight stabilisation with use of trimmers" – ISSLOT.

References

1. Krawczyk, M., Zajdel, A., Szczepański, C.: The airplane trim system – new functionalities. In: 9th EASN International Conference on Innovation in European Aeronautics Research, Athens, Greece, September 2019
2. Krawczyk, M., Zajdel, A.: Trimming surfaces in the automatic plane control process, VII Konferencja Awioniki. J. KONES Powertrain Transp. **24**(1) (2017)
3. Krawczyk, M., Graffstein, J.: A proposition of control augmentation system for dumping the harmful impact of slipstream in turboprop airplanes. Sci. Pap. Rzeszów Univ. Technol.-Mech. **288**, 287–295 (2013)
4. Krawczyk, M.: Conditions for unmanned aircraft reliability determination. Mainten. II Reliabil. **15**(1), 31–36 (2013)
5. Samolej, S., Rogalski, T., Nowak, D.: Flight Control system on the VxWorks real-time operating system platform – selected implementation issues. Sci. Pap. Rzeszów Univ. Technol.-Mech. **288**, 515–524 (2013)
6. European Aviation Safety Agency, CS-23 - Certification Specifications for Normal, Utility, Aerobatic, and Commuter Category Aeroplanes (2012)
7. Krawczyk, M., Rogalski, T.: General overview of experimental control system developed on I-23 manager aircraft. Trans. Inst. Aviat. **4**(206), 93–102 (2010)
8. Blajer, W., Graffstein, J., Krawczyk, M.: Modeling of aircraft prescribed trajectory flight as an inverse simulation problem. In: Modeling Simulation and Control of Nonlinear Engineering Dynamical Systems: State of The Art, Perspectives and Applications, pp. 153–162. Springer (2009)
9. Phillips, W.F.: Mechanics of Flight. Wiley, Hoboken (2004)
10. MIL-F-8785C – Military specification flying qualities of piloted aeroplanes (1980)
11. Levy, D.: Design of a full time wing leveler system using tab driven aileron controls. Hilton Head (1992)
12. Fiszdon, W.: Mechnika Lotu, Państwowe Wydawnictwo Naukowe, Warszawa (1961)
13. Perkins, D.C., Hage, E.R.: Airplane performance stability and control. University of Michigan, Detroit, Wiley (1949)
14. Jaśkowski, P., Wolf, T.: Analiza właściwości aerodynamicznych powierzchni trymujących samolotu demonstratora technologii. Grant no POIR.04.01.02-00-0006/17 "Innowacyjny system stabilizacji lotu za pomocą trymerów", Warszawa (2019)
15. Rogalski, T., Rzucidło, P., Prusik, J.: Unmanned aircraft automatic flight control algorithm in a spin maneuver. Aircr. Eng. Aerosp. Technol. **92**(8), 1215–1224 (2020). https://doi.org/10.1108/AEAT-05-2019-0099

Robotics

Application of Safety Systems Based on Laser Scanners for Installations with Automatically Relocatable Industrial Robots

Piotr Falkowski[1,2(✉)], Michał Smater[1], and Jakub Koper[1]

[1] Łukasiewicz Research Network – Industrial Research Institute of Automation and Measurements PIAP, Warsaw, Poland
piotr.falkowski@piap.lukasiewicz.gov.pl
[2] Institute of Aeronautics and Applied Mechanics, Faculty of Power and Aeronautical Engineering, Warsaw University of Technology, Warsaw, Poland

Abstract. The fourth industrial revolution brought a concept of interchangeability in manufacturing. Thanks to the diversity of sensors applied in the production lines, manipulators may perform their tasks at various workstations. Their relocation may be solved with the application of automated mobile robots. However, this generates a problem of preventing workers in the factories from potential harm caused by machines. Furthermore, traditional safety systems are not effective, as they restrict too much area form human access. Hence, a flexible method of creating safety system based on laser scanners is considered. The following paper is based on a real case study. Even though a used setup is dedicated to training in automotive, the results of simulations and analysis are scalable and applicable to the factories of the future. The paper compares traditional and non-conventional approaches towards creating safety systems, presents capabilities of the ones based on laser scanners and assesses possibilities of their practical application in different sectors of industry.

Keywords: Industry 4.0 · Laser scanners · Mobile industrial robots · Safety systems · Training in automotive

1 Introduction

In the era of Industry 4.0 flexibility became one of the leading trends in the design of industrial robots [1, 2]. Possibility of changing the current task, not only at one workstation but also across several stands or machines, can increase the efficiency of the robotic set-up. Moreover, it enables dynamic changes in the production line, thereby eliminating the robot's downtime. These aspects are highly desirable in industry, and in particular in its automotive sector [3].

While using industrial robots, it is possible to easily change their function. It typically involves their relocation, exchange of the tool and the switch of the program. While the program may be shifted automatically via the PLC controller and the additional fast tool change system may be assembled to the robot's flange, only transferring of the

R. Szewczyk et al. (Eds.): AUTOMATION 2021, AISC 1390, pp. 199–208, 2021.
https://doi.org/10.1007/978-3-030-74893-7_19

manipulator to a new location remains inefficient and problematic. It is most often achieved by the demounting and assembling to another workstation manually or with the use of linear tracks. The first method is not convenient due to the need of lifting the manipulator and its precision setup at another position. Thus, it is also not applicable to heavier robots. On the other hand, tracks enable motion of almost any type of industrial machines but at the same time, their motion is limited to just a single direction linear displacement. Moreover, these devices tend to be expensive.

Another possible solution is to assemble the robots on mobile constructions and dock them to workspaces. This approach is particularly useful for the lighter manipulators. For this, either simple trolleys or advanced autonomous mobile robots may be applied as the mentioned mobile constructions [4, 5].

Mobile industrial robots are generally equipped with enough sensors to move freely within the range of human. If only cobots are being placed on them, safety systems for such setups are not a problem. However, due to the lower payloads and velocities permitted for this type of robotic arms, the application of industrial manipulators may be necessary. Then, human access to the robot's range zone shall be limited. It may be realised with a physical barrier or with sensors plugged into the setup's safety system. Such a solution is consistent with relevant standards only if no human can enter a dangerous zone while the robot works in automatic mode without immediate turn-off of the machine.

Considered solution with industrial robots moving around the factory and docking to different workstations in order to realise their tasks increases a flexibility of production with low costs [6]. However, due to safety standards it requires a potential exclusion of a significant part of area from human access. As it may not be acceptable, the flexible production system may need tailored adaptable safety system, so to avoid undesirable situations.

The following paper describes an idea of implementing laser sensors for a safety system of a setup consisting of numerous workstations and industrial manipulators placed on fully automatic industrial mobile robots. It is divided into six sections covering the case description, comparison of various solutions towards safety and a presentation of one's real implementation. Its main aim is to assess the applicability of adaptable safety systems based on laser scanners for manufactures and plants of the future.

2 Setup Description

The setup used for the tests of safety system consisted of two robotic arms ABB IRB 1200–7/0.7, one assembled on a manual trolley, and the other one mounted onto an autonomous mobile robot MiR 100, and four workstations arranged in a rectangular pattern (see Fig. 1). The practical task of the manipulators is to manipulate cubic objects with the edge length of 60 mm, drawn from the storage, and to transport them on the upper panel of the mobile robot's casing.

The sets of industrial robots on either a trolley or MiR 100 move freely in a limited region. It is marked on the floor and bordered in a map of an autonomous robot. To perform their tasks, the movable devices have to dock to the workstations. Then, they are positioned precisely, prevented from undesired movement relative to the workspace and supplied with electric and pneumatic supply. The chosen mobile robot does not require a constant supply, but it has to be charged every few hours. This is realised automatically at the charging station.

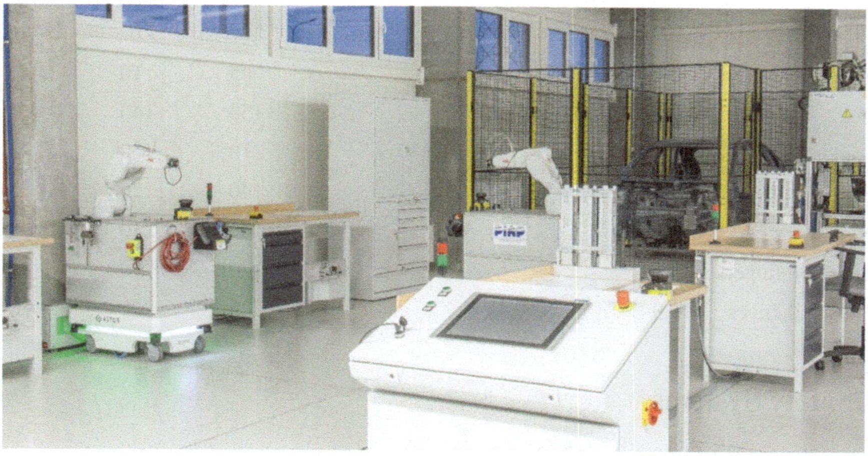

Fig. 1. Setup with implemented safety system

Even though the installation used for tests is a setup dedicated for training in a field of robotics in the automotive sector, it realises similar tasks and generates similar problems to the flexible production lines present in big factories and plants. In a real-life, there are typically more workstations placed further from each other. Often, also a fleet of many mobile robots may be involved. However, these just scale the main problem of creating a flexible safety system but do not affect its essence.

3 Law Regulations

All the robotised installations need to be compliant with international standards of safety. These pieces of legislation regulate the possible accessibility to the machine while working with automatic mode on. In case that an impact of a robot on a particular part of the body may cause an energy transfer or a force load exceeding the maxima presented in corresponding standards (see Table 1), such installation requires additional safety barriers applied [7, 8].

These may consist of various devices and physical barriers described in the further paragraph. However, it is not the only method of increasing safety of workers. Every employee whose job is connected with the machine has to undergo a special training tailored to their responsibilities and risks [8].

Table 1. Maximum values of energy, momentum and force transferable during an impact by a robot not to require additional safety barriers [7]

Segment of body	Maximum permissible value of a parameter transferred during an impact of a robot		
	Energy [J]	Force [F]	Pressure [N/cm^2]
Face	0.11	65	110
Neck	0.84	150	140
Shoulders and back	2.5	210	160
Chest	1.6	140	120
Pelvis	2.6	180	210
Hands and fingers	0.49	140	190
Tights and knees	1.9	220	220

4 Safety Systems

4.1 Cobots and Industrial Manipulators

Collaborative robots called cobots gained their popularity thanks to the possibility to cooperate with humans without additional safety barriers. Their construction guarantees low impacts on a body in case of any collision; thus, they do not cause mechanical danger on the contrary to industrial robotic arms. However, this comes with relatively low payloads and short reaches [9]. While the capabilities are more important than flexibility to cooperate freely with other workers, conventional manipulators are much more likely to be applied. Combining them with linear tracks or autonomous industrial robots makes them interoperable between various workstations. Nevertheless, it also creates a need to safeguard much bigger area what may cause problems with the free motion of operators and suppliers. To avoid such a situation, adaptable safety systems shall be considered only.

4.2 Physical Barriers

Conventional installations are typically surrounded by a fence limiting access to the machinery. Of course, every fence has a gate enabling service or programming in a manual mode. However, to safeguard a user from physical harm, every gate is equipped with additional protection. These may be light curtains activated with an automatic mode of a robot and stopping it in case of intersecting any laser beams or with a physical door with a limit switch and a lock activated with an automatic mode of a robot. Of course, a natural infrastructure of a production hall may be used as a physical barrier – e.g. fence of a robot cell may also consist of walls.

4.3 Sensors and Other Elements

Safety systems may also consist of other elements such as pressure mats, visual systems or optical sensors. Their application enables definition of zones which intrusion affects behaviour of an installation. Typically entering an outer zone by a human only slows own a robot to its safe speed, while entering the inner zone totally stops it.

Moreover, safety systems can be based on even more advanced technologies such as artificial skin for robots or cognitive algorithms, enabling more complex reactions to potential danger [10, 11].

All safety systems also contain safety stop buttons. They are placed within a human arm range (from 5th percentile female population to 95th percentile male) from a control panel, gate and workstations. They may be also placed along the fence in case an access to the closest emergency stop button is not convenient from a whole surrounding.

5 Real Implementation

5.1 Main Safety System

Main safety system consists of safety modules connected in series to two parallel circuits plugged to the safety programmable controller. These are a safety stop on a control panel and other modules placed on workstations or on robots and activated while being docked to one of the workstations (see Fig. 2 and 3). Workstations 1 and 2 have one safety stop placed in their corner each and laser scanners placed in the middle of their long edges. Workstations 3 and 4 do not need any optic sensors as they are placed next to the hall's wall, but they also have safety stop buttons placed on them. Both movable devices with industrial robots are equipped with laser sensors. In case that any of the circuits O1-I1/O2-I2 is interrupted, the whole installation turns into emergency stop mode and brakes immediately.

SICK S32B-3011BA (see Table 2) is a model of the laser scanners used in all the workstations and movable devices with industrial robots. Their field of view was divided into two zones and fitted not to detect any constant elements of installation or any part of industrial robots performing their main tasks (see Fig. 3). A signal generated while interfering a warning field seen by laser is matched with a system signal of a robot slowing it down to the safety speed. On the contrary, entering a protected field interrupts a safety circuit.

Protected fields of the laser scanners are shaped in a way which does not allow a user intersect a range zone of an operating manipulator. It works for both, a set of a scanner on the mobile device and either Workstation 1 or Workstation 2 as well as just for a mobile device docked to the Workstation 3 or 4. The warning fields are kept minimal just to warn unaware operator coming too close to machines without the need of emergency stop.

5.2 Additional Safety Systems for Robots

Every industrial manipulator is provided with its own minimal safety systems. It consists of a mode switch (from manual to automatic), dead man's switches and a safety button placed on its teach pendant. An additional safety button is also placed on a controller. Therefore, any of the robots in automatic mode may be stopped from the teach pendant of another manipulator. This may be used by an online-programming operator who wants to stop a robot working at a distant workspace.

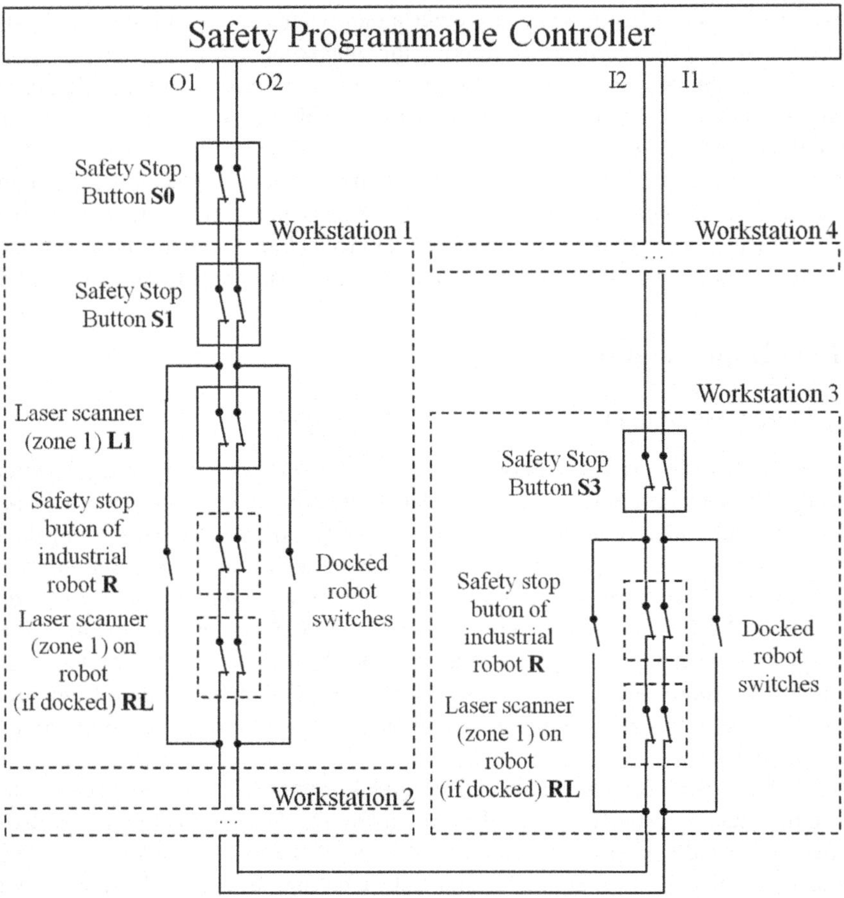

Fig. 2. Scheme of the safety system

Table 2. Parameters of the SICK S32B-3011BA laser scanner

Parameter	Value
Protected field range	3 m
Warning field range	8 m
Distance measuring range	30 m
Scanning angle	270°
Response time	80 ms

5.3 Cell Arrangements

As mentioned above, the main advantage of applying a flexible safety system based on laser scanners is decreasing area with restricted access. Applying movable industrial

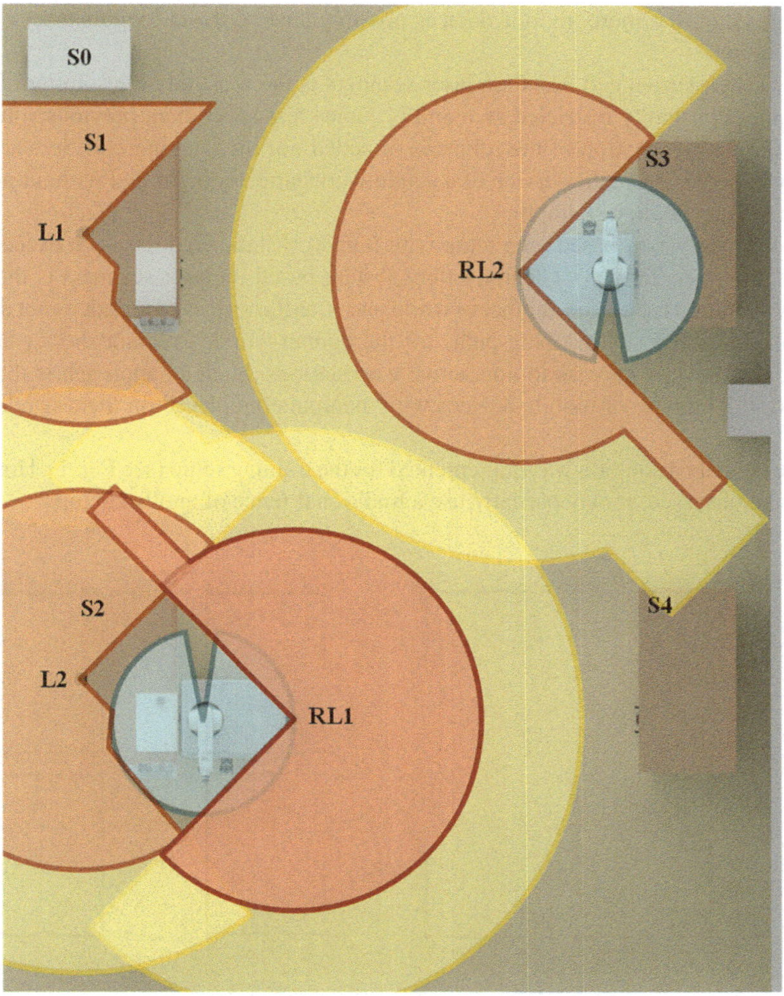

Fig. 3. Placement of the safety stop buttons (S and R), laser scanners (L and RL), ranges of the lasers' zones (warning field – yellow and protected field – red) and a maximum reach of docked industrial manipulators (blue)

robots operating at various workstations gives numerous possibilities of designing of a safety system. For the purpose of this case, a single robotic cell, separate cells for each workstation and virtual cells based on laser zones are considered (see Fig. 4).

Creating a single cell is the simplest method which does not affect the motion of robots between workstations. However, it requires exclusion of the big area and makes all the workstation useless in case of servicing any of them considered (see Fig. 4).

On the other hand, separate cells are more convenient when it comes to human intervention. They require a much smaller area. Nevertheless, they may require longer fence in total what generates additional costs and still block walking corridors in factories (see Fig. 4). Furthermore, mobile devices possibly have to travel long distances in this configuration.

Creating a virtual cell based on laser scanners is the most flexible solution. It does not require constantly restricted area and occupies less space than previous solutions. Moreover, neither motion of the robots is impeded nor the walking corridors are necessarily blocked. Also, the service of a single workstation is permitted without need of turning off the whole installation.

As this analysis is based on a setup with four workstations, it has been exceeded in a simulation (see Table 3). Clearly, safety system based on laser scanners is the most effective solution for a large number of workspaces, as the restricted area does not depend on them. Moreover, the walking-paths for the human workers remain shorter and the operators have direct access to non-active workstations. Such an approach is the most effective if a number of mobile devices with manipulators placed on them is relatively small.

The described solution was implemented for the training setup (see Fig. 1). However, as may be observed, it may partially use a traditional fence of another setup.

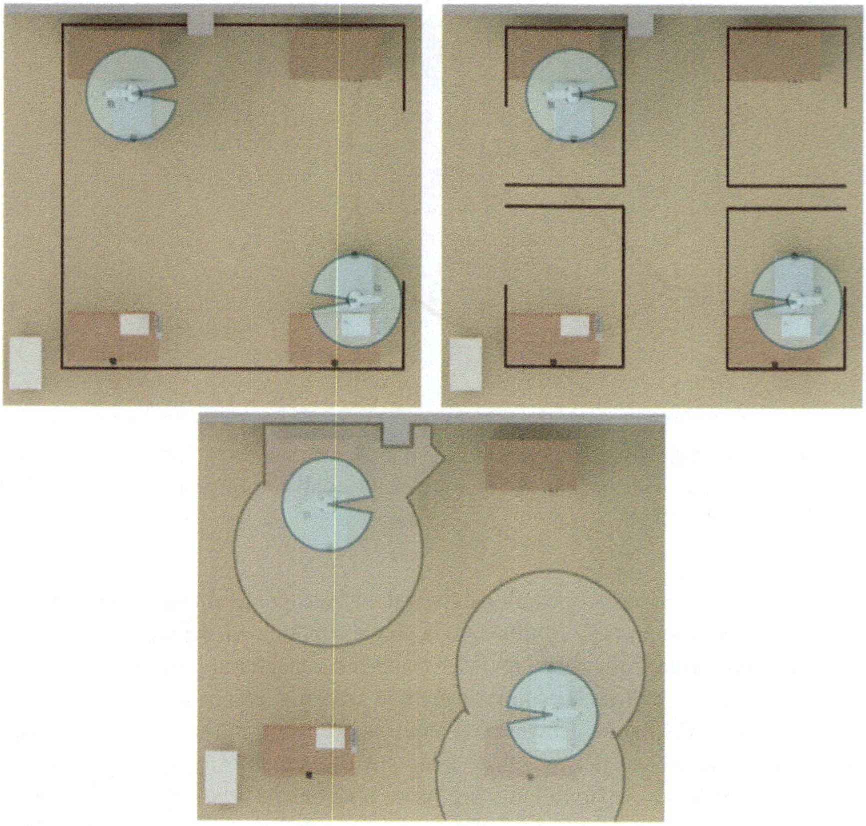

Fig. 4. Possible arrangements of robot cells (from the top-left: one cell, separate cells, flexible virtual cells based on laser scanners)

Table 3. Restricted area dependent on the arrangement of robot cells

Arrangement	Restricted area [m^2]		
	Four workspaces	Six workspaces	Eight workspaces
One cell	39.5	59.2	78.9
Separate cells	24.3	36.5	48.7
Flexible safety system	15.1/18.1 per robot		

6 Summary

As Industry 4.0 requires maximum flexibility of manufacturing, moving industrial robots between workspaces is unavoidable. Apart from the problem of precise docking, also solved during the trials, this generates a need for adjustable safety system. It may be implemented with the set of laser scanners assembled to both, the moving subsystems and the workstations. This method is especially useful for installations with many workspaces spread around a big area. Application of such a system facilitates the motion of a human in the production hall and possible service interventions. While necessary, the operators may also perform their tasks at free workstations and supply or empty storages relevant for the robots' tasks. Moreover, this approach increases the lifetime of mobile industrial robots due to the shorter distances they travel. Not only is it a safe and efficient solution, but this approach also decreases the costs of enlarging production by creating additional workspaces. Such setup configurations can be effective even for the processes with high-precision requirements, while accompanied with a dedicated docking system [12]. For all these reasons and the highest potential of implementing industrial manipulators on autonomous mobile robots in automotive, this idea is the most applicable for this sector. However, it may be also used in pharmaceutics and advanced manufacturing [2, 13].

References

1. Weyer, S., et al.: Towards industry 4.0-standardization as the crucial challenge for highly modular, multi-vendor production systems. IFAC-PapersOnLine **48**(3), 579–584 (2015)
2. Evjemo, L.D., et al.: Trends in smart manufacturing: role of humans and industrial robots in smart factories. Curr. Robot. Rep. **1**, 35–41 (2020)
3. Angerer, S., et al.: Give me a hand – the potential of mobile assistive robots in automotive logistics and assembly applications. In: 2012 IEEE International Conference on Technologies for Practical Robot Applications (TePRA), pp. 111–116. IEEE (2012)
4. Virk, G.S.: Industrial mobile robots: the future. In: Industrial Robot, no. 24, pp. 102–105 (1997)
5. Schneier, M., Bostelman, R.: Literature review of mobile robots for manufacturing. US Department of Commerce. National Institute of Standards and Technology (2015)
6. Schneier, M., et al.: Literature review of mobile robots for manufacturing. US Department of Commerce, National Institute of Standards and Technology (2015)
7. ISO TS 15066 – Robots and robotic devices – Collaborative robots (2016)

8. ISO 10218-2: 2011 – Robots and robotic devices – Safety requirements for industrial robots– Part 2: Robot systems and integration (2011)
9. Niku, S.B.: Introduction to Robotics: Analysis, Control Applications. Wiley, Hoboken (2020)
10. Falkowski, P., et al.: Applicability of artificial robotic skin for industrial manipulators. In: Advanced, Contemporary Control. Springer, Cham (2020)
11. Bryndin, E.: Increase of safety use robots in industry 4.0 by developing sensitivity and professional behavioral skills. Am. J. Mech. Ind. Eng. **5**(1), 6–14 (2020)
12. Falkowski, P., et al.: An Approach towards high-precision docking of the mobile robots for industrial purposes. In: Conference on Automation. Springer, Cham (2020)
13. Wan, J., et al.: Reconfigurable smart factory for drug packing in healthcare industry 4.0. IEEE Trans. Ind. Inf. **15**(1), 507–516 (2018)

Robotic Swarm Shape Control Based on Virtual Viscoelastic Chain

Jakub Wiech(✉) and Zenon Hendzel

Faculty of Applied Mechanics and Robotics, Rzeszow University of Technology, Rzeszów, Poland
j.wiech@prz.edu.pl

Abstract. We present a physicomimetic method for self-organization and swarm shape control of nonholonomic robotic swarm, based on length controlled spring damper chain. The article starts with explanation of two wheel robot dynamics and relation between virtual forces allowing for robot control and swarm formation of a desired convex shape. Two methods of swarm control are analyzed. In the first case the swarm shape is achieved by virtual spring damper chain with analytically defined length of springs in rest. In the second case a numerical method is introduced allowing for forming any closed convex shape converted form picture to point cloud. The paper ends with numeric simulations for performance evaluation of proposed control method.

Keywords: Swarm robotics · Swarm shape control · Artificial potential fields

1 Introduction

1.1 Swarm Control Objective

The main objective of swarm control is to achieve desired swarm behavior by local interaction between swarm robots. A collective behavior emerges from all combined decisions made by robots according to relatively simple rules [2]. Robots take into consideration their relative position to local neighbors and environment (obstacles) that the robots can sense or interact with. Also another requirements are swarm scalability and robustness that have to be ensured in design of swarm control algorithm. Available methods of swarm control that met all the requirements are either bioinspired, physicomimetic or based on various optimization approaches [5, 6, 8, 9, 11, 14]. Other solutions using virtual spring based methods are, for example, Urcola et al. [10] presenting a formation control based on virtual structure composed of virtual spring dampers connecting following robots to leader. Robotic swarm interconnected by virtual spring and damper in parallel configuration was proposed by Chen et al. [3] where formation control of satellites under a gravitational potential field was investigated. Belkacem and Foudil [1] study a swarm aggregation model based on virtual springs dampers equilibrium emerging from robots interconnections. Other work by Shucker and Bennett [7] propose a fully Distributed Robotic Macrosensor which control algorithm was influenced by spring damper mesh.

R. Szewczyk et al. (Eds.): AUTOMATION 2021, AISC 1390, pp. 209–218, 2021.
https://doi.org/10.1007/978-3-030-74893-7_20

Our article presents a swarm control algorithm that is inspired by the behavior of point mass particles interconnected with nonlinear springs and dampers. In this case a swarm mimics the behavior of nonlinear viscoelastic materials which falls into category of physics based method. The controls of nonholonomic robots are defined by the relation between virtual net force acting on the robot and robots linear and angular velocity.

1.2 The Idea of Controlling a Swarm of Robots

The coordinated control of swarm robots comes down to maintaining an inter robots distances, collisions avoidance and optionally, to maintain the desired shape of the swarm during its movement.

The synthesis of the swarm's control revolves around generation and preservation of the swarm's shape located in nonlinear artificial potential fields. By modeling a swarm as a virtual deformable chain changed by forces coming from those fields, the problem of coordinated robot movement can be solved.

The interpretation of nonlinear artificial potential fields can be nonlinear virtual elastic damping connections which connect adjacent robots R_i, and chosen reference point D. The robots form a virtual elastic chain enabling symmetrical arrangement of the robots with a given swarm shape.

2 Swarm Robots Dynamics

The most commonly used robots in swarm research are nonholonomic two wheel robots (see Fig. 1).

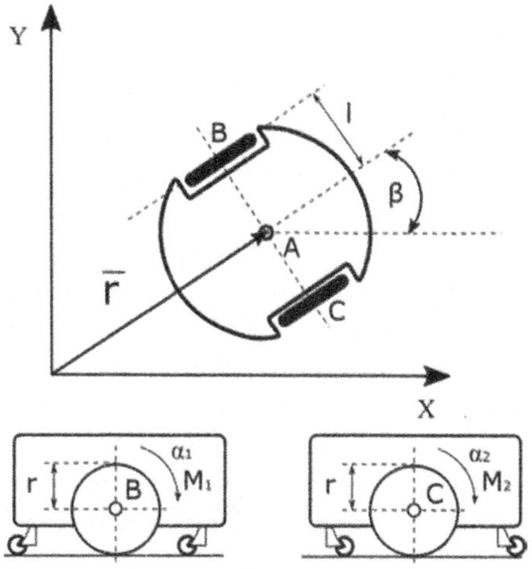

Fig. 1. Two wheel robot schematic

Presented robot has two degrees of freedom [15]. In the word coordinates a posture is defined as $\left[x_A, y_A, \beta\right]^T$, where (x_A, y_A) is the position of the point A, and β is the heading angle of the robot with respect to absolute coordinates (x, y). Taking α_1, α_2 - angles of the self-turn of the driving wheels as independent coordinates, then from velocity vectors of points A, B, C we receive velocities expressed in robot frame coordinates

$$\dot{\alpha}_1 = \frac{v_A}{r} + \dot{\beta}\frac{l}{r}, \dot{\alpha}_2 = \frac{v_A}{r} - \dot{\beta}\frac{l}{r}, \tag{1}$$

By applying Maggi's mathematical formalism, the dynamic equations of motion of the MWR, as shown in Fig. 1 can be expressed as [15, 16]

$$\boldsymbol{M}\ddot{\boldsymbol{\alpha}} + \boldsymbol{C}(\dot{\boldsymbol{\alpha}})\dot{\boldsymbol{\alpha}} + \boldsymbol{F}(\dot{\boldsymbol{\alpha}}) = \boldsymbol{u}, \tag{2}$$

where the matrices $\boldsymbol{M}, \boldsymbol{C}(\dot{\boldsymbol{\alpha}})$ and vectors, $\boldsymbol{F}(\dot{\boldsymbol{\alpha}}), \boldsymbol{u}$ result from the description of the MWR dynamics with Maggi's formalism. The matrices and vectors have the following forms

$$\boldsymbol{M} = \begin{bmatrix} a_1 + a_2 + a_3 & a_1 - a_2 \\ a_1 - a_2 & a_1 + a_2 + a_3 \end{bmatrix}, \boldsymbol{C}(\dot{\boldsymbol{\alpha}}) = \begin{bmatrix} 0 & 2a_4(\dot{\alpha}_2 - \dot{\alpha}_1) \\ -2a_4(\dot{\alpha}_2 - \dot{\alpha}_1) & 0 \end{bmatrix},$$
$$\boldsymbol{F}(\dot{\boldsymbol{\alpha}}) = \begin{bmatrix} a_5 sgn(\dot{\alpha}_1) \\ a_6 sgn(\dot{\alpha}_2) \end{bmatrix}, \boldsymbol{u} = \begin{bmatrix} M_1 \\ M_2 \end{bmatrix}, \tag{3}$$

Vector $\boldsymbol{a} = [a_1, a_2, a_3, a_4, a_5, a_6]^T$ are parameters derived from the geometry, the distribution of mass, the movement resistances of the MWR, and the dynamic properties of the actuator systems.

3 Swarm Control

3.1 Swarm Control Method

The main task of a swarm of robots is to organize themselves into the desired convex shape in an unobstructed environment. The robot's trajectory is defined for the virtual reference point, i.e. the swarm's initial position. Swarm robots are designed to maintain a certain distance between themselves and from a reference point.

The shape of the swarm and its location result from the virtual forces acting on robots. The movement of the i-th robot is the result of the virtual forces of the neighboring robots and the reference point acting on the i-th robot (Fig. 2).

The swarm's robots are connected with each other by springs and dampers that form a chain. We assume that the robots have a mass m and they are connected by springs of stiffness k and dampers with a damping coefficients c. The rest length of the springs is equal to d_0.

Virtual springs and dampers are connected at the centers of mass of each robot, marked as point A_i.

Each virtual connection between the robots as well as between robots and reference point D exerts respectively forces F_{ij} and F_{iD} equal to the sum of the spring forces F_{Sij}, F_{SiD} and the viscous forces of the dampers F_{Tij}, F_{TiD} (forces with lower index T). We assume that the spring stiffness and damper damping coefficients are linear, but their

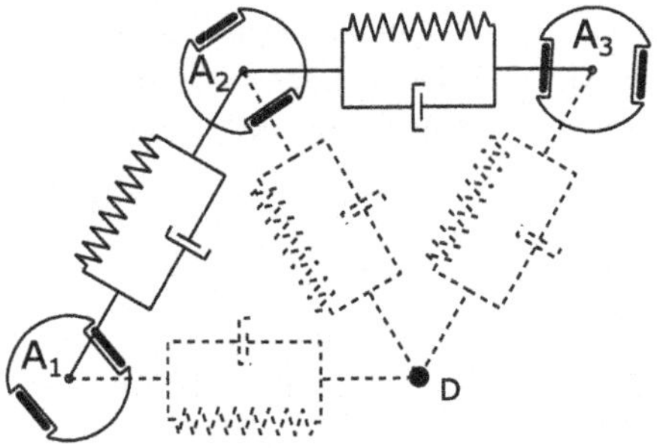

Fig. 2. Spring-Dampers mesh connecting mobile robots

rest length is calculated using a non-linear equation. Their forces are proportional to the displacement and the relative speed between the ends of the springs.

$$F_{ij} = F_{Sij} + F_{Tij} = k_R\left(d_{0R} - d_{ij}\right) + c_R\frac{d}{dt}\left(d_{0R} - d_{ij}\right) = k_R e_{ij} + c_R \dot{e}_{ij}, \tag{4}$$

$$F_{iD} = F_{SiD} + F_{TiD} = k_D(d_{0D} - d_{iD}) + c_D\frac{d}{dt}(d_{0D} - d_{iD}) = k_D e_{iD} + c_D \dot{e}_{iD}, \tag{5}$$

The inter robot and reference point parameters are defined as follows: k_R, k_D stand for spring stiffness coefficients, c_R, c_D are the damping coefficients, d_{ij}, d_{iD} are the current spring lengths and d_{0R}, d_{0D} are the resting spring lengths.

Each robot has a limited sensing range of SR_r, which determines the number of virtual spring damper connections created. In addition, robots are only able to create a virtual link with the two closest robots. Such rules force robots to create a virtual viscoelastic chain. To establish a connection only with the closest neighbors, the sensing range must meet the inequality

$$d_{max} > SR_r > d_{0R}, \tag{6}$$

The total vector force exerted on the robot i is equal to the sum of all forces in the connections between the robot and its visible neighbors N_r.

$$\overline{F_i} = \sum\nolimits_{j \in N_r} \overline{F_{ij}} + \overline{F_{iD}} = \sum\nolimits_{j \in N_r} \left(\overline{F_{Sij}} + \overline{F_{Tij}}\right) + \overline{F_{SiD}} + \overline{F_{TiD}}, \tag{7}$$

From Eq. (1) we conclude that we need the value of the linear velocity v_A and the angular velocity $\dot{\beta}$ to determine the input signals of the robot control with net force F_i as a known value. To determine the values of v_A, $\dot{\beta}$, we first need to find the appropriate forces causing the robot to move forward or backward (force F_{ip}) and rotate around point A_i (force F_{iq}) (see Fig. 3).

The origin of the vectors $\overline{F_{ip}}$, $\overline{F_{iq}}$ is at off-center point E, located in front of the robot at a distance a from point A_i. Such placement of the origin is only to illustrate that the force F_{iq} will cause the robot to rotate around point A_i and force F_{ip} will pull the robot forward. The nature of the used nonholonomic robot makes it impossible for the robot to move sideways. If we know the values of the chain link and reference point forces

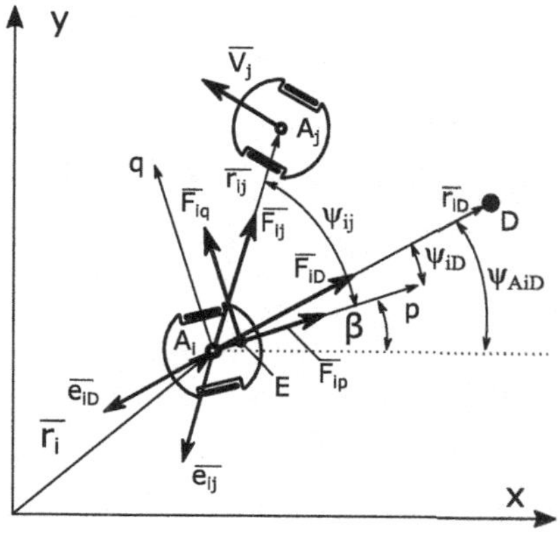

Fig. 3. The i-th robot driving forces

with their direction, we can find the values of $\overline{F_{ip}}$ and $\overline{F_{iq}}$ vector components of the total force vector $\overline{F_i}$ given by

$$\begin{aligned} F_{ip} &= \sum_{j\in N_r} F_{ij} \cos \Psi_{ij} + F_{iD} \cos \Psi_{iD}, \\ F_{iq} &= \sum_{j\in N_r} F_{ij} \sin \Psi_{ij} + F_{iD} \sin \Psi_{iD}, \end{aligned} \tag{8}$$

Using above equations the values of desired linear velocity v_A and the angular velocity $\dot{\beta}$ are determined by

$$\begin{aligned} v_{id} &= \lambda_2 \int (F_{ip} - \lambda_3 v_i)dt = \lambda_2 \int (\sum_{j\in N_r} F_{ij} \cos \Psi_{ij} + F_{iD} \cos \Psi_{iD} - \lambda_3 v_i)dt, \\ \dot{\beta}_{id} &= \lambda_1 F_{iq} = \lambda_1 \sum_{j\in N_r} F_{ij} \sin \Psi_{ij} + F_{iD} \sin \Psi_{iD}, \end{aligned} \tag{9}$$

where, $\lambda_2 > 0, \lambda_2[m/(Ns)], \lambda_1 > 0, \lambda_1\left[\frac{rad}{Ns}\right], \lambda_2, \lambda_1$ are the gain constants, $\lambda_3 > 0, \lambda_3\left[\frac{Ns}{m}\right]$ – is the virtual viscosity constant.

The virtual viscosity $\lambda_3 v_i$ is used to prevent swarm oscillations. As another method of control the behavior of the swarm and its stability we introduce limitations:

Limit of the total value of force $F_{ip} - \lambda_3 v_i$, velocities v_{id} and $\dot{\beta}_{id}$

$$\begin{aligned} F_{max} &> F_{ip} - \lambda_3 v_i > F_{min}, \\ v_{max} &> v_{id} > v_{min}, \\ \dot{\beta}_{max} &> \dot{\beta}_{id} > \dot{\beta}_{min}, \end{aligned} \tag{10}$$

The constant gains, virtual viscosity constant, damping coefficients and stiffness coefficients can be selected by the designer so that the values of v_{id} and $\dot{\beta}_{id}$ are adequate for the performed shape. The sample schematics of the control loop is shown below (Fig. 4).

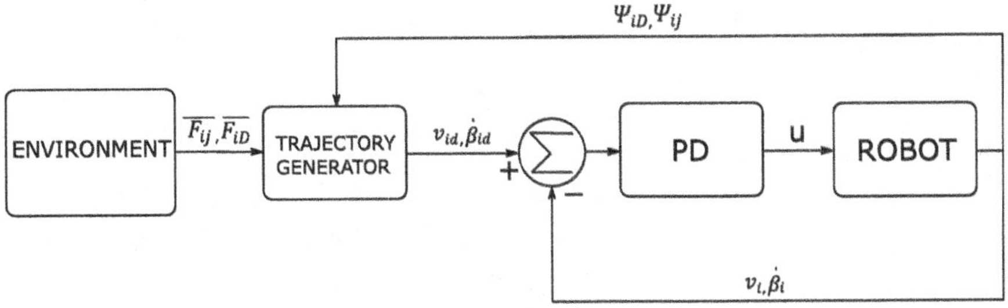

Fig. 4. The i-th robot control loop

3.2 Swarm Shape Control

The method of virtual spring-damper connections enables, self-organization of the swarm, i.e. an ordered grouping of robots in the form of a grid consisting of equilateral triangles or a circle with center in the reference point [4, 12, 13].

The value of the virtual force acting from the reference point D defines the shape of the swarm. In the simplest case, the force value is equal to:

$$F_{iD} = k_D e_{iD} + c_D \dot{e}_{iD}, \tag{11}$$

where: $e_{iD} = d_{0D} - d_{iD}$, is the difference between the desired distance (d_{0D}) and the actual distance (d_{iD}) from the reference point.

If the error value e_{iD} is defined by shape function $f(d_{0D})$ then we can obtain the desired shape of the swarm. In order for the robots to form a given shape and maintain the cohesion of the swarm, there must be two types of virtual connections. Connections between neighboring robots (solid line) and the reference point D (dashed line) (Fig. 2).

3.3 Analytical Approach

The shape of the swarm is defined by specifying the frame of reference and the shape function in polar coordinates. The shape function $f(d_{0D})$ describes the dependence of d_{0D} (the desired distance from the reference point D), at which the robot is to be positioned as a function of the deviation angle Ψ_{AiD} from the reference system.

The simplest example of a shape function is a circle (Fig. 5) resulting from the relationship:

$$d_{0D} = R, R = const, \ for \ \Psi_{AiD} = (-\pi, \pi), \tag{12}$$

Another example is an ellipse (Fig. 6):

$$d_{0D} = \frac{ab}{\sqrt{(a \sin \Psi_{AiD})^2 + (b \cos \Psi_{AiD})^2}}, a, b = const, \tag{13}$$

3.4 Numerical Approach

Another method of defining the shape of a swarm is to transform the outline of a desired figure into a point cloud. By transforming the Cartesian coordinates of points into polar coordinates, we can determine the distances d_{OD} corresponding to the angles Ψ_{AiD}.

Knowing the value of the angle Ψ_{AiD} between the robot and the reference point D, we can find the closest point in the point cloud corresponding to this angle.

Examples of self-organization of two data shapes by the shape function approximated by a point cloud are a rounded rectangle (Fig. 8) and a heart (Fig. 9).

4 Simulations

For simulations bellow the parameters were chosen as follows:

$$a_1 + a_2 + a_3 = 0.0111, \quad a_1 - a_2 = 3.7224e-06,$$

$$a_5 = 1.0995, \quad a_6 = 1.0995,$$

$$c_D = 3.8, c_R = 3.8, \quad k_D = 0.9, k_R = 0.8,$$

$$a = 0.7, b = 1.6, R = 1, \quad d_{OD} = 1.29, d_{OR} = 0.4,$$

$$\lambda_1 = 120, \lambda_2 = 1, \lambda_3 = 1.1, \quad F_{max} = 10, F_{min} = -10$$

$$v_{max} = 0.05, v_{min} = -0.05, \quad \dot{\beta}_{max} = 5, \dot{\beta}_{min} = -5,$$

Simulation discretization step: 0.03 [s].

4.1 Analytical Approach, Circle and Ellipse

In both cases, the robotic swarm was able to form the desired shapes in less than 200 [s]. In Eqs. (12, 13) we can determine the radius of the circle and the semi-major and semi-minor axes of the ellipse. Due to the analytical form of the shape function, the value of virtual force F_{iD} remains continuous throughout the duration of the simulations. The jump discontinuities in desired linear and angular velocities v_A and $\dot{\beta}$ are the result of change of neighbors within the 1st robot's sensing range (Fig. 7).

4.2 Numerical Approach, Rounded Square and Heart

In the numerical approach the orientations of robots are less uniform. The different robots' orientations are the result of constant changes in force values from the reference point. The real value of angle Ψ_{AiD} is rounded to the nearest discretized value of the angle Ψ_{AiD}. This approximation leads to a sudden change of the virtual force F_{iD} and, consequently, to a sudden change in the orientations of the robots (Fig. 9). As in the analytical approach, the virtual force from the neighboring robots oscillates between zero and non zero value, which is the result of creating and breaking a virtual link with the one of the two closest robots (Fig. 10).

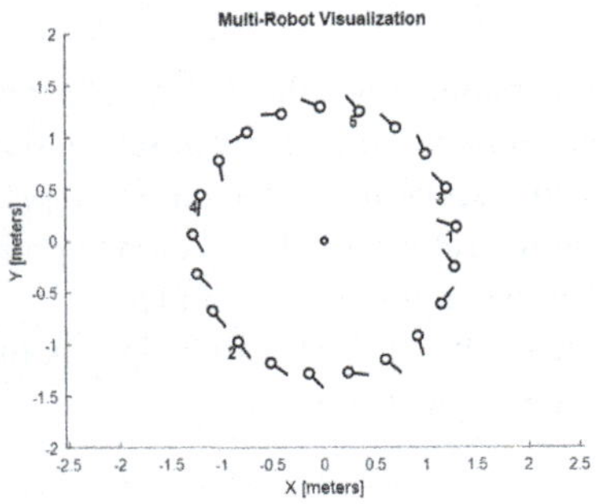

Fig. 5. Circle-shaped swarm

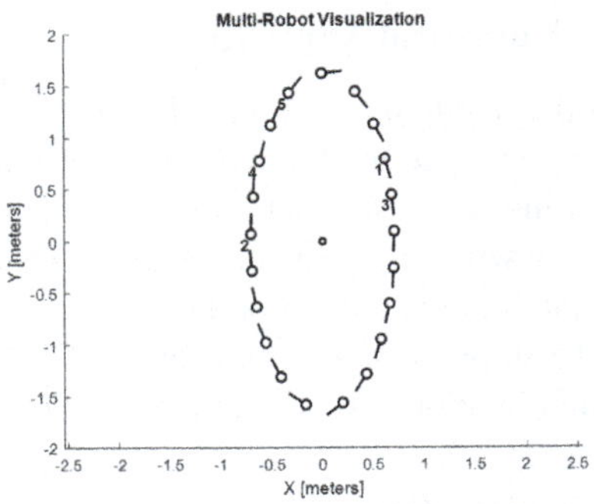

Fig. 6. Ellipse-shaped swarm

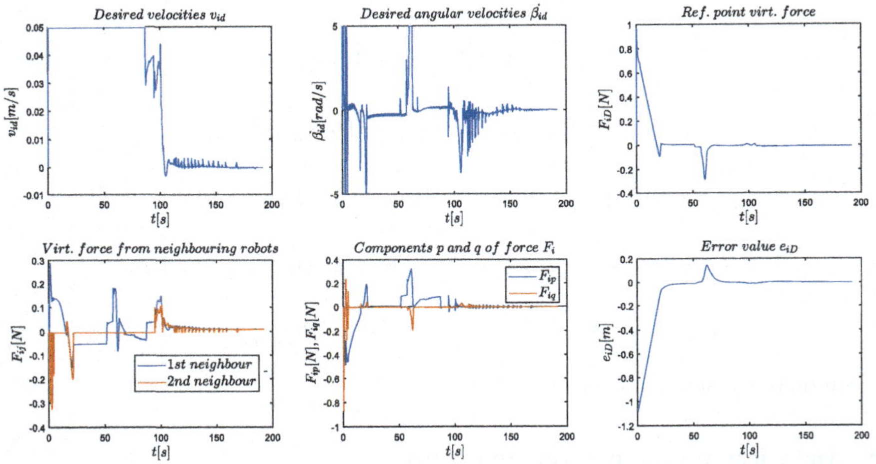

Fig. 7. Graphs of velocity, virtual forces and the difference in distance between robot no. 1 and the reference point in shape of a circle

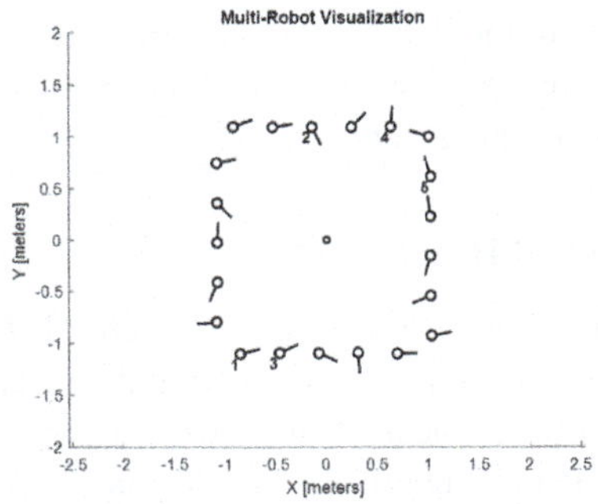

Fig. 8. Rounded square swarm

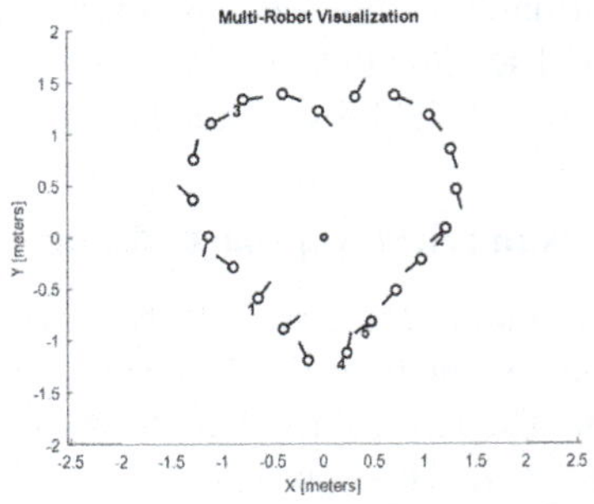

Fig. 9. Heart-shaped swarm

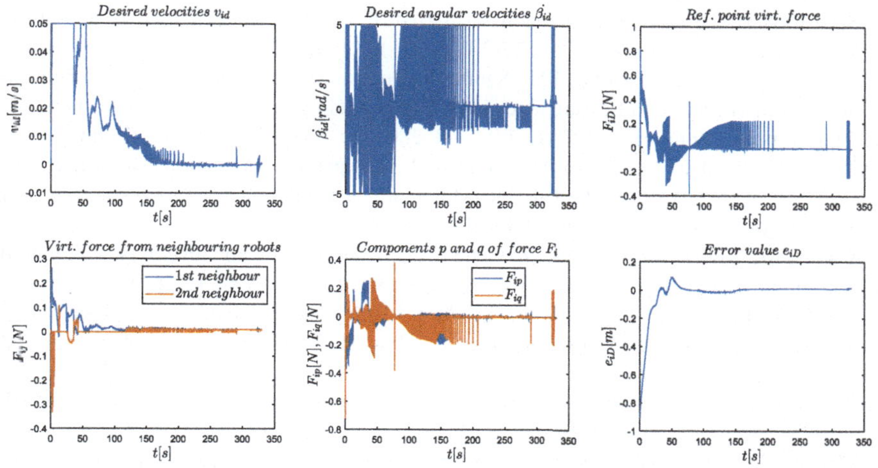

Fig. 10. Graphs of velocity, virtual forces and the difference in distance between robot no. 1 and the reference point in shape of a rounded rectangle

5 Summary

Both the analytical shape function and the function approximated by the point cloud allow for obtaining a specific shape of a swarm. The analytical approach requires defining the shape function in polar coordinates, which is possible only for certain convex figures. The advantages of analytical shape functions are continuous control signals, whereas in the case of numerical approach there are jump discontinuities in the desired linear and angular velocities v_A and $\dot{\beta}$. Nevertheless, the numerical approach is the simplest method for defining a swarm shape. The presented shapes of the swarm are convex and not filled shapes, which requires further research in swarm shape formation.

References

1. Balkacem, K., Foudil, C.: A virtual viscoelastic based aggregation model for self-organization of swarm robots system. In: TAROS 2016: Towards Autonomous Robotic Systems, pp. 202–213. Springer (2016)
2. Brambilla, M., Ferrante, E., Birattari, M., Dorigo, M.: Swarm robotics: a review from the swarm engineering perspective. Swarm Intell. **7**, 1–41 (2013). https://doi.org/10.1007/s11721-012-0075-2
3. Chen, Q., Veres, S.M., Wang, Y., Meng, Y.: Virtual spring, -damper mesh-based formation control for spacecraft swarms in potential fields. J. Guid. Control Dyn. **38**(3), 539–546 (2015)
4. Hendzel, Z., Wiech, J.: Robotic swarm self-organisation control. Acta Mech. Autom. **13**(2), 130–134 (2019)
5. Nouyan, S., Campo, A., Dorigo, M.: Path formation in a robot swarm: self-organized strategies to find your way home. Swarm Intell. **2**(1), 1–23 (2008)
6. Reif, J.H., Wang, J.: Social potential fields: a distributed behavioral control for autonomous robots. Robot. Auton. Syst. **27**(3), 171–194 (1999)
7. Shucker, B., Bennett, J.K.: Virtual spring mesh algorithms for control of distributed robotic macrosensors. University of Colorado Boulder, Technical Report CU-CS-996-05 (2005)

8. Spears, W.M., Spears, D.F., Hamann, J.C., Heil, R.: Distributed, physics-based control of swarms of vehicles. Auton. Robot. **17**(2–3), 137–162 (2004)
9. Trianni, V., Groß, R., Labella, T.H., Şahin, E., Dorigo, M.: Evolving aggregation behaviors in a swarm of robots. In Lecture notes in Artificial Intelligence. Advances in Artificial Life: 7th European Conference—ECAL 2003, vol. 2801, pp. 865–874. Springer, Berlin (2003)
10. Urcola, P., Riazuelo, L., Lazaro, M., Montano, L.: Cooperative navigation using environment compliant robot formations. In: IEEE/RSJ International Conference on Intelligent Robots and Systems, IROS 2008, pp. 2789–2794. IEEE (2008)
11. Waibel, M., Keller, L., Floreano, D.: Genetic team composition and level of selection in the evolution of cooperation. IEEE Trans. Evol. Comput. **13**(3), 648–660 (2009)
12. Wiech, J., Eremeyev, V.A., Giorgio, I.: Virtual spring damper method for nonholonomic robotic swarm self-organization and leader following. Continuum Mech. Thermodyn. **30**, 1091–1102 (2018)
13. Wiech, J., Hendzel, Z.: Overhead vision system for testing swarms and groups of wheeled robots. J. Autom. Mob. Robot. Intell. Syst. **13**(2), 15–22 (2019)
14. Mohan, Y., Ponnambalam, S.G.: An extensive review of research in swarm robotics. In: 2009 World Congress on Nature & Biologically Inspired Computing (NaBIC), Coimbatore, pp. 140–145 (2009)
15. Żylski, W.: Motion planning for wheeled mobile robot using potential field method. J. Theor. Appl. Mech. **42**(3), 695–705 (2004)
16. Żylski, W., Giergiel, J.: Description of a mobile robot by Maggie's equations. J. Theor. Appl. Mech. **43**(3), 511–521 (2005)

Robust Tracking Control of Omni-Mecanum Wheeled Robot

Zenon Hendzel[1] and Maciej Kołodziej[2(✉)]

[1] Faculty of Mechanical Engineering and Aeronautics, Department of Applied Mechanics and Robotics, Rzeszow University of Technology, Rzeszów, Poland
zenhen@prz.edu.pl

[2] Doctoral School of Engineering and Technical Sciences at the Rzeszow University of Technology, Rzeszów, Poland
d515@stud.prz.edu.pl

Abstract. This paper presents an approach to the problem of controlling mechanical objects of unspecified description, considering variable operating conditions. The controlled object is a mobile robot with mecanum wheels. The control algorithm is based on the input-output linearization method and variable structure control approach. The major contribution of this paper lies in application of new approach to control wheeled robot along desired trajectory in the present parametric and non-parametric modelling inaccuracies. To solve the control task, taking into account compensation for nonlinearity and the object variable operating conditions, the Lyapunov stability theory is applied. Basing on numerical analysis, it has been demonstrated that the tested control algorithm is robust to assumed inaccuracies in modelling.

Keywords: Mechatronics · Mobile robot · Mecanum wheels · Robust control

1 Introduction

Classic methods of designing control systems, which include optimal control methods and various algebraic methods, require knowledge of the mathematical model of the object and the parameters of this model. However, determining the exact model of an object with the evaluation of its parameters is often a very difficult task, especially when the object's properties change during the work. For example, the mass moment of inertia of a wheeled robot frame varies depending on the value of the mass to be transported and the way it is distributed. Motion resistances can also be variable. Therefore, the models used in practice only approximate the properties of objects. In the robust control method, based on the knowledge about the object, the accuracy of the model is determined using the predicted range of variability for each parameter. Next, a controller is designed, consisting of a compensating element for the nonlinearity of the object and a stabilizing element, so that the closed system works correctly for each object within the assumed limits of parameter variability. The regulator obtained in this way is called a robust regulator. The advantage of this method of synthesizing control algorithms is

R. Szewczyk et al. (Eds.): AUTOMATION 2021, AISC 1390, pp. 219–229, 2021.
https://doi.org/10.1007/978-3-030-74893-7_21

its simplicity and high operational reliability. In recent years, the study of mobile robots with mecanum-wheeled has gained popularity [1–3]. Advantages such as an ability to reach every point of workspace or turning in any direction with zero turning radius made omnidirectional robots prominent replacement to classical robots in industrial applications such as fork lifters, manipulator driving platforms [4, 5]. Mecanum-wheeled robot is able to move in any direction without even along a curved path keeping frame orientation in set values. Nevertheless, on account of the use of separate motors per each wheel, uncertainties arise, which makes the motion control more demanding, especially when moving over surfaces with very low or very high friction coefficient [6]. Well-known kinematic planning that is commonly used to generate trajectories [7] does not take account into nonlinearity of robots. In recent years complex control systems of mecanum-wheeled robots have already been reported in the literature [8, 9]. For instance, in [10] authors carried out research about adaptive robust second-order sliding mode control [10]. In [11], the design of the adaptive sliding mode controller is shown. In [12] four continuous Sliding-Mode Control strategies are presented, the position and orientation of the platform are assumed to be the only available information. This paper presents an approach to the problem of robust control of a mobile robot with mecanum wheels. The control algorithm is based on the input-output linearization method and variable structure control approach. To solve the control task Lyapunov theory of stability is applied. In Sect. 2, kinematics and dynamics equations for mecanum wheel mobile robot are formulated. Sect. 3 provides the synthesis of robust control of a mobile robot and its description is given. Section 4 contains a simulation. Finally, conclusions are presented in Sect. 5.

2 Kinematics and Dynamics of Wheeled Robot

While describing the parameters of a robot's motion, we are interested in the issues of kinematics and dynamics of such systems [16]. To solve the problem of controlling such objects we solve the reverse task. In the description of the kinematics of a mobile wheeled robot with mecanum wheels [13], kinematics equations should be determined from which it is possible to determine linear motion parameters such as sector, speed, and acceleration of any selected point of the robot or angular parameters of motion. To describe the kinematics of a mobile robot with mecanum wheels, the model shown in Fig. 1 is adopted [13].

In this figure, x, y, and z are the axes of the fixed system. The basic units of this model are frame 5 and the driving units. The driving unit consists of wheels 1, 2, 3, 4, which are mounted on semi-axles and set in motion by a drive module associated with a given wheel. Points A1, A2, A3, A4 lie in their centers of symmetry. These wheels rotate around their own axes which do not change their position relative to the frame. The rollers are placed on the periphery of the wheels, and they are set at an angle of $\alpha = \pi/4$ radians, relative to the axis of the driving wheel. Point S is the center of mass of the frame, point H is the point lying on the axis A1A2 halfway between these points. Point B, like point H, is the point belonging to the frame. Knowing the geometry of the

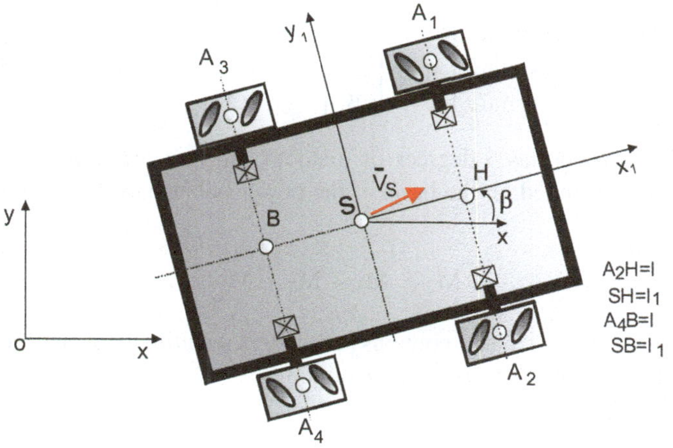

Fig. 1. Wheeled robot model

system and applying the classical methods used in mechanics, we obtain a description of the kinematics of a mobile wheeled robot in the form [13]:

$$\begin{aligned}
\dot{x}_s\cos(\beta-a)+\dot{y}_s\sin(\beta-a)-\dot{\beta}(l\cos a+l_1\sin a)&=\omega_1(R+r)\cos a\\
\dot{x}_s\cos(\beta-a)+\dot{y}_s\sin(\beta-a)-\dot{\beta}(l\cos a+l_1\sin a)&=\omega_2(R+r)\cos a\\
\dot{x}_s\cos(\beta-a)+\dot{y}_s\sin(\beta-a)-\dot{\beta}(l\cos a+l_1\sin a)&=\omega_3(R+r)\cos a\\
\dot{x}_s\cos(\beta-a)+\dot{y}_s\sin(\beta-a)-\dot{\beta}(l\cos a+l_1\sin a)&=\omega_4(R+r)\cos a
\end{aligned} \tag{1}$$

where: β is the angle of temporary rotation of the frame, ω_i are angular velocity vectors of wheels. The obtained equations allow determining the inverse kinematics task which will be the set motion trajectory for the point S of the mobile-wheeled robot in the control system. To determine the dynamic equations of motion [14, 15] of the mobile wheeled robot (Fig. 1), Maggi's equations [16] with an additional disturbance convenient during the synthesis of the control system were used. General form of Maggi's equations is:

$$M(q)\ddot{q}+C(q,\dot{q})\dot{q}+F(\omega)+\tau_d(t)=u \tag{2}$$

Where disturbances vector meets the restriction $\|\tau_d(t)\| < b, b = \text{const} > 0$, matrices M, C and vector F, take form

$$M(q)=\begin{bmatrix} a_1\sin b+a_2\cos(\beta-a) & a_2\sin(\beta-a)-a_1\cos\beta & -a_3\\ a_2\cos(\beta+a)-a_8\sin(\beta-a) & a_2\sin(\beta+a)-a_8\cos(\beta-a) & -a_4\\ a_1\cos\beta+a_2\cos(\beta-a) & a_2\sin(\beta-a)+a_1\sin(\beta-a) & a_3 \end{bmatrix}$$

$$C(q,\dot{q})=\begin{bmatrix} -a_2\sin(\beta-a)\dot{\beta} & a_2\cos(\beta-a)\dot{\beta} & 0\\ -a_2\sin(\beta+a)\dot{\beta} & a_2\cos(\beta+a)\dot{\beta} & 0\\ -a_2\sin(\beta-a)\dot{\beta} & a_2\sin(\beta-a)\dot{\beta} & 0 \end{bmatrix} \tag{3}$$

$$F(\omega) = \begin{bmatrix} a_5 \operatorname{sgn}\omega_1 \\ a_6 \operatorname{sgn}\omega_3 \\ a_7 \operatorname{sgn}\omega_4 \end{bmatrix} \quad q = \begin{bmatrix} x_s \\ y_s \\ \beta \end{bmatrix} \quad u = \begin{bmatrix} M_1 \\ M_3 \\ M_4 \end{bmatrix} \quad \tau_d(t) = \begin{bmatrix} \tau_{d1} \\ \tau_{d3} \\ \tau_{d4} \end{bmatrix}$$

The analyzed robot has 3 degrees of freedom and four drive motors, the missing moment M_2 is determined on the basis of the power balance of the drive systems [14, 20].

$$M_2 = M_3 + M_4 - M_1 \tag{4}$$

The vector $a = [a_1, \ldots, a_8]^T$ contains parameters resulting from the geometry, mass distribution, and the robot's resistance to motion.

3 Synthesis of Robust Control

From a mathematical point of view, robust control systems have a structure consisting of compensation and feedback. In the synthesis of a robust algorithm for controlling the follow-up motion of wheeled mobile robots, we will use the linearization method of the input-output type system and the theory of sliding motion of systems with a variable structure [17–19]. Control systems with a variable structure [17, 18] are a certain class of systems in which the control algorithm changes according to certain rules, depending on the current state of the control object. In the synthesis of a robust tracking system of a wheeled robot, the control law u should be determined such that for the initial conditions $\left[q(0), \dot{q}(0)\right] \in \Omega = [(q, \dot{q}) : q_d(0) = q(0), \dot{q}_d(0) = \dot{q}(0)]$ solution $q\left[t, q(0), \dot{q}(0)\right], \dot{q}\left[t, q(0), \dot{q}(0)\right]$ it coincides with the manifold Ω_d defined as: $\Omega_d = [(q, \dot{q}) : q = q_d, \dot{q} = \dot{q}_d$. The general scheme of the robust control system used is shown in Fig. 2.

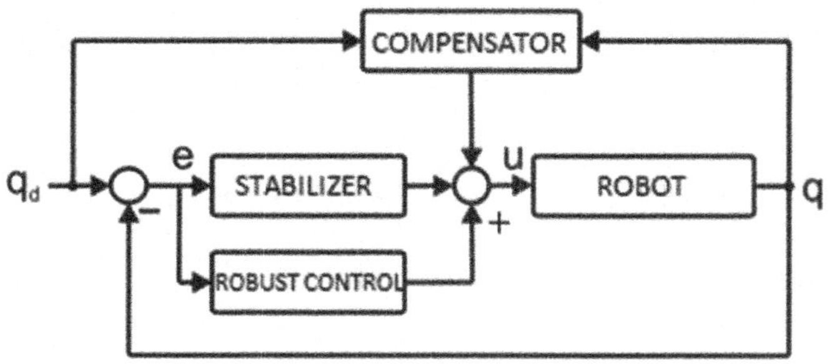

Fig. 2. The structure of the robust control system

The task of the compensator is to compensate for the non-linearity of the object, while the task of the stabilizer is to eliminate the lagging error resulting from incomplete compensation of non-linearities and changing working conditions of the robot. On the other hand, the task of the robust control is to ensure the correct operation of the closed system for each object within the assumed limits of parameter variability.

The adopted mathematical model of the robot (2) is linear due to the parameters a. Using this property, the dynamics equation was written in the form

$$Y(\ddot{q}, \dot{q}, q)a + \tau_d(t) = u \tag{5}$$

where $Y(\ddot{q}, \dot{q}, q)$ is the matrix of known functions, a is the vector of parameters of the mathematical model of the robot. As mentioned earlier, the task of the compensator is to compensate for the nonlinearities of the robot. Perfect compensation requires precise knowledge of the parameter vector a. In practice, this requirement is very difficult to meet. Let us assume that in a given class of mathematical models (5) there are parametric inaccuracies in modeling, which means that there is a vector of parameters $\hat{a}$ and a vector γ with positive elements that satisfy the relationship

$$\left|\hat{a}_i - a_i\right| \leq \gamma_i, i = 1, 2, \ldots 8 \tag{6}$$

Value $\hat{a}_i$ is the evaluation of the parameter a_i, and γ_i is the limitation of changes in the i-th parameter value.

By defining the tracking error, the generalized error, and the auxiliary vector as

$$e = q_d - q \tag{7}$$

$$s = \dot{e} + \Lambda e \tag{8}$$

$$v = \dot{q} - \Lambda e \tag{9}$$

where Λ is a diagonal matrix positively defined. Then Eq. (2) can be written as

$$M(q)\dot{s} = -u + \tau_d(t) - C(q, \dot{q})s + Y_v(q, \dot{q}, v, \dot{v})a \tag{10}$$

where $Y_v(q, \dot{q}, v, \dot{v})a = M(q, a)\dot{v} + C(q, \dot{q}, a)v + F(\dot{q}, a)$.

As a candidate for the Lyapunov function, the following function has been selected

$$V = 0.5s^T M(q)s \tag{11}$$

The derivative of the function V(s) is

$$\dot{V} = s^T M(q)\dot{s} + 0.5s^T \dot{M}(q)s \tag{12}$$

Substituting from Eq. (10) to (12) and assuming that there is no change in frame orientation while the robot is moving $\beta = \text{const}$ we will get

$$\dot{V} = s^T[-u + \tau_d(t) + Y_v(q, \dot{q}, v, \dot{v})a \tag{13}$$

Control signal with a variable structure was selected:

$$u = u_{eq} + k\,\text{sgn}(s) + K_D s \tag{14}$$

Where u_{eq} is equivalent control whose task is to compensate for the nonlinearity of the object, signal ksgn(s) is defined as vector of elements kjsgn(sj), KD is constant, positive-definite design matrix of controller gains, the element $K_d s$ is an equation of PD controller

$$K_d s = K_d \dot{e} + K_d \Lambda e \tag{15}$$

Introducing the Eqs. (14), (15) to Eq. (13), the result was obtained

$$\dot{V} = s^T[-u_{eq} - k\,sgn(s) - K_D s + \tau_d(t) + Y_V(q, \dot{q}, v, \dot{v})a] \tag{16}$$

Bearing in mind the properties of sliding motion of systems with a variable structure, in the task of tracking a given motion trajectory, an equivalent control signal in the form of

$$u_{eq} = Y_V(q, \dot{q}, v, \dot{v})\hat{a} \tag{17}$$

Derivative $\dot{V}(t)$ is now

$$\dot{V} = s^T[Y_V(q, \dot{q}, v, \dot{v})\,\tilde{a} - k\,sgn(s) - K_D s + \tau_d(t)] \tag{18}$$

where $\tilde{a} = a - \hat{a}$. Since the expression in square brackets can be written as

$$Y_V(\dot{q}, v, \dot{v})\,\tilde{a} - k\,sgn(s) + \tau_d(t) = \left\{\sum\nolimits_{i=1}^{N} Y_{Vij}(q, \dot{q}, v, \dot{v})\tilde{a}_i - k_j sgn(s_j) + \tau_d(t)\right\}^T \tag{19}$$

If sliding movement gain factor is selected

$$k_j = \sum\nolimits_{i=1}^{N} \left|Y_{Vij}(q, \dot{q}, v, \dot{v})\right|\gamma_i + b_j + \eta_j \tag{20}$$

where $\left|\tau_{dj}(t)\right| \le b_j$, we gain

$$\dot{V} \le -s^T K_D s - \sum\nolimits_{j=1}^{3} \eta_j \left|s_j\right| \tag{21}$$

Condition (21) guarantees that the trajectory of the system reaches the sliding surface in a finite time and will remain there. So the tracking error will tend to zero.

A function $sgn(s_j)$ is not continuous for $s_j = 0$ and causes a high frequency of executive system switching. Therefore it is approximated with

$$sat\left(\frac{s_j}{\Phi_j}\right) = \begin{cases} sgn(s_j), & \left|s_j\right| > \Phi_j \\ \frac{s_j}{\Phi_j}, & \left|s_j\right| \le \Phi_j \end{cases} \tag{22}$$

where Φ_j is phase plane $S_j = 0$.

4 Simulation

To demonstrate the effectiveness of the proposed controller, simulation has been carried out using MATLAB/Simulink environment. The selected point S of the mobile robot on the trajectory of curved shape involves stages of motion [20]: moving on a rectilinear track with a fixed speed $V_S^* = 0.4\,m/s$, moving on a curved path of radius $R = 0.7$ m, exit from the curve taking into account a transitional period, then moving on the rectilinear track and braking. Initial conditions of motion: $x_S(0) = 1\,m, y_S(0) = 5\,m, \beta(0) = 0$. The angle of the robot frame $\beta(t)$ does not change during motion. In time $t \geq 9$ s, point S of the robot starts a curved trajectory with radius $R = 0.7\,m$. Then, the self-rotation angle of wheel 2 increases, and the self-rotation angle of wheel 1 decreased resulting from the structure of the robot, as shown in Fig. 3a. The change in motion is confirmed by a change in the angular velocity of the wheels, as shown in Fig. 3b, and the course of angular accelerations of wheels, see Fig. 4a. Braking starts in time $t \geq 33$ s. Values of kinematic parameters during motion are shown in Fig. 3. Values of robot dynamic parameters have been adopted as specified in Table 1.

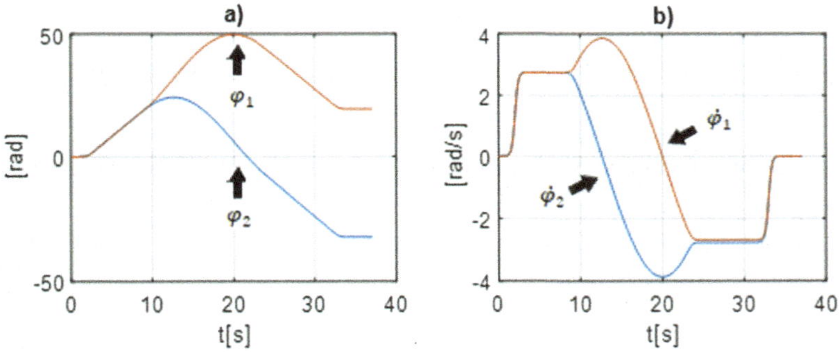

Fig. 3. The parameters set for the movement of the robot: a) values of self-rotation angles of the driving wheels, b) angular velocity of self-rotation of the driving wheels

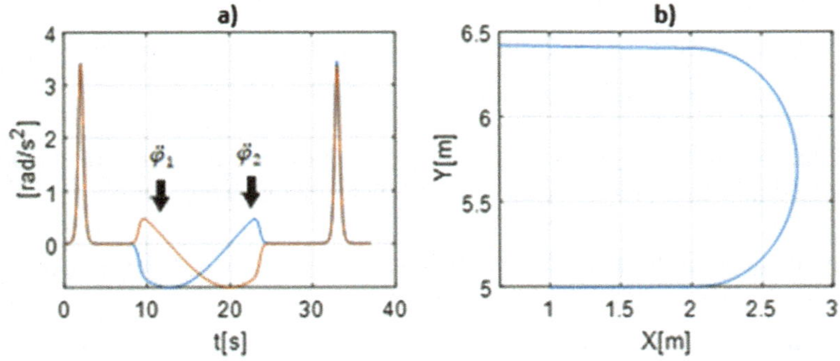

Fig. 4. The parameters set for the movement of the robot: a) angular accelerations, b) trajectory set for the point S

Table 1. Values of nominal parameters a_i of a mobile robot

a_1	a_2	a_3	a_4	a_5	a_6	a_7	a_8
0.044	11.5708	2.4558	2.4545	3.001	3.001	3.001	0.0622

The following robust control system parameters were used in the numerical test: $\Lambda = \mathrm{diag}[2, 2, 2]$, $K_D = 3I$, $\tau_d = 0$, hence $b_1 = b_2 = b_3 = 0$ and furthermore $K_D = \mathrm{diag}\{k_{D1}, k_{D2}, k_{D3}\} = \mathrm{diag}\{3, 3, 3\}$, $\Lambda = \mathrm{diag}\{1, 1, 1\}$, $\eta_1 = \eta_2 = \eta_3 = 0.03$, $\Phi_1 = \Phi_2 = \Phi_3 = 0.1$, $\gamma = 0.2$. In order to verify the robustness of the proposed solution, there also has been assumed that for the time $t \geq 15s$ there is a disturbance $a+\Delta a$, where $\Delta a = [0, 0, 0, 0, 2, 2, 2, 0]^T$, associated with the change of rolling friction of the robot wheels. In this case, the matrix $Y_V(q, \dot{q}, v, \dot{v})$ has the form

$$\begin{bmatrix} \dot{v}_1\sin\beta - \dot{v}_2\cos B & \dot{v}_1\cos(\beta - \alpha) + \dot{v}_1\sin(\beta - \alpha) - \dot{v}_1\sin(\beta - \alpha)\dot{\beta} + \dot{v}_2\cos(\beta - \alpha)\dot{\beta} \\ 0 & \dot{v}_1\cos(\beta + \alpha) + \dot{v}_1\sin(\beta + \alpha) - \dot{v}_1\sin(\beta + \alpha)\dot{\beta} + \dot{v}_2\cos(\beta + \alpha)\dot{\beta} \\ \dot{v}_1\cos\beta - \dot{v}_2\sin B & \dot{v}_1\cos(\beta - \alpha) + \dot{v}_1\sin(\beta - \alpha) - \dot{v}_1\sin(\beta - \alpha)\dot{\beta} + \dot{v}_2\cos(\beta - \alpha)\dot{\beta} \end{bmatrix}$$

$$\begin{bmatrix} -\dot{v}_3 & 0 & \mathrm{sgn}\omega_1 & 0 & 0 & 0 \\ 0 & -\dot{v}_3 & 0 & \mathrm{sgn}\omega_2 & 0 & -\dot{v}_1\sin(\beta - \alpha) - \dot{}_2\cos(b + \alpha\dot{)} \\ \dot{v}_3 & 0 & 0 & 0 & \mathrm{sgn}\omega_3 & 0 \end{bmatrix}$$

The slide control gain coefficients determined according to (20) have the form

$$\begin{aligned} k_1 = & |\dot{v}_1\sin\beta - \dot{v}_2\cos B|\gamma_1 + \left|\dot{v}_1\cos(\beta - a) + \dot{v}_1\sin(\beta - a) - \dot{v}_1\sin(\beta - a)\dot{\beta} + \dot{v}_2\cos(\beta - a)\dot{\beta}\right|\gamma_2 \\ & - |\dot{v}_3|\gamma_3 + |\mathrm{sgn}\omega_1|\gamma_5 + \eta_1 \end{aligned}$$

$$\begin{aligned} k_2 = & \left|\dot{v}_1\cos(\beta + a) + \dot{v}_1\sin(\beta + a) - \dot{v}_1\sin(\beta + a)\dot{\beta} + \dot{v}_2\cos(\beta + a)\dot{\beta}\right|\gamma_2 \\ & - |\dot{v}_3|\gamma_4 + |\mathrm{sgn}\omega_2|\gamma_6 - \left|\dot{v}_1\sin(\beta - a) - \dot{}_2\cos(b + a\dot{)}\right|\gamma_8 + \eta_2 \end{aligned}$$

$$\begin{aligned} k_3 = & |\dot{v}_1\cos b - \dot{v}_2\sin B|\gamma_1 \\ & + \left|\dot{v}_1\cos(\beta - a) + \dot{v}_1\sin(\beta - a) - \dot{v}_1\sin(\beta - a)\dot{\beta} + \dot{v}_2\cos(\beta - a)\dot{\beta}\right| \\ & + |\dot{v}_3|\gamma_3 + |\mathrm{sgn}\omega_3|\gamma_7 + \eta_3 \end{aligned}$$

In order to assess the quality of the adopted control, the following indicators have been adopted, limited to motion evaluation for wheel 2 only (Fig. 5):

- maximum value of the angle of rotation error $e_2 = (\varphi_{2d} - \varphi_2)$[rad], $e_{max(.)} = \left|e_{(.)}\right|$ [rad],
- root-mean-squared error for wheel rotation angles $\varepsilon_2 = \sqrt{\left(\frac{1}{n}\right)\sum_{k=1}^{n} e_{2k}^2}$, where k is the number of subsequent discrete measurements,
- root-mean-squared error for tracking the given angular velocity $\dot{\varepsilon_2} = \sqrt{\left(\frac{1}{n}\right)\sum_{k=1}^{n} \dot{e}_{2k}^2}$ [rad/s], $n = 3701$ – total number of discrete measurements,

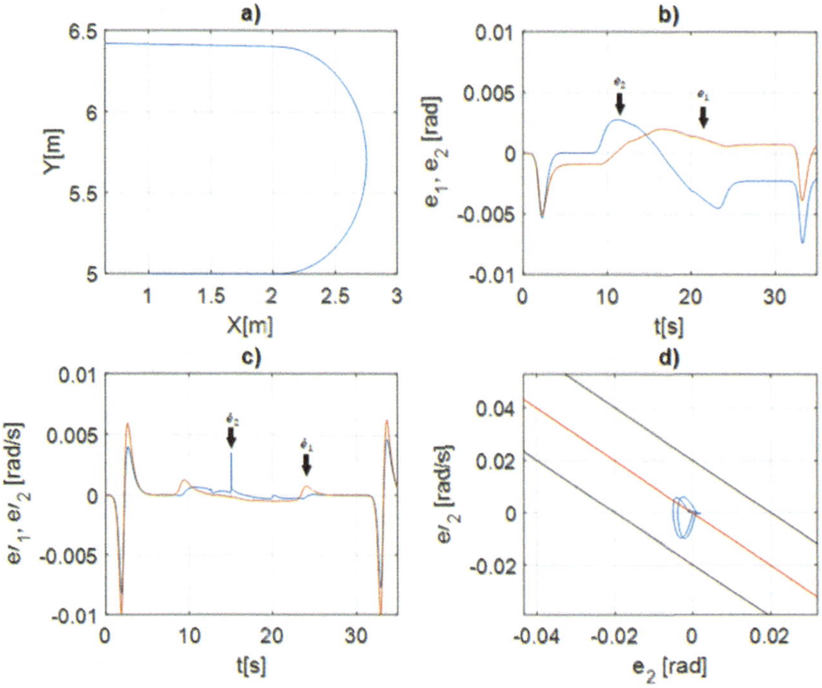

Fig. 5. Simulation results: a) obtained trajectory of point S, b) errors of tracking the rotation angles of wheels 1 and 2, c)errors of tracking the angular velocity of wheels 1 and 2, d) error in the phase plane for generalized error s_2

- generalised root-mean-squared error: $s_2 = \dot{e}_2 + \lambda e_2$, $v_2 = \sqrt{(\frac{1}{n}) \sum_{k=1}^{n} e_{2k}^2}$ [rad/s], $d_k = \sqrt{(x_{Sk} - x_{dSk})^2 + (y_{Sk} - y_{dSk})^2}$ [m],
- maximum distance $d_{max} = \max(d_k)$ [m].

Table 2. Values of the quality indicators for robust control of wheel 1

e_{imax} [rad]		ε_i [rad]	$\dot{e}_{imax}$ [rad/s]	$\dot{\varepsilon}_i$ [rad/s]	v_i [rad/s]	d_{max} [m]
0.00821	0.00724		0.01121	0.01575	0.02369	0.003284

Figure 6 presents patterns of control signals in the presence of variable operating conditions of the object resulting from the assumed resistance to motion in the form of a change in the rolling friction. Equivalent control has the biggest share in the control signal, which compensates for non-linearities of the control object. From the moment a parametric disturbance occurs, the PD and slide control contribution increases, because equivalent control does not fully compensate for non-linearities of the robot.

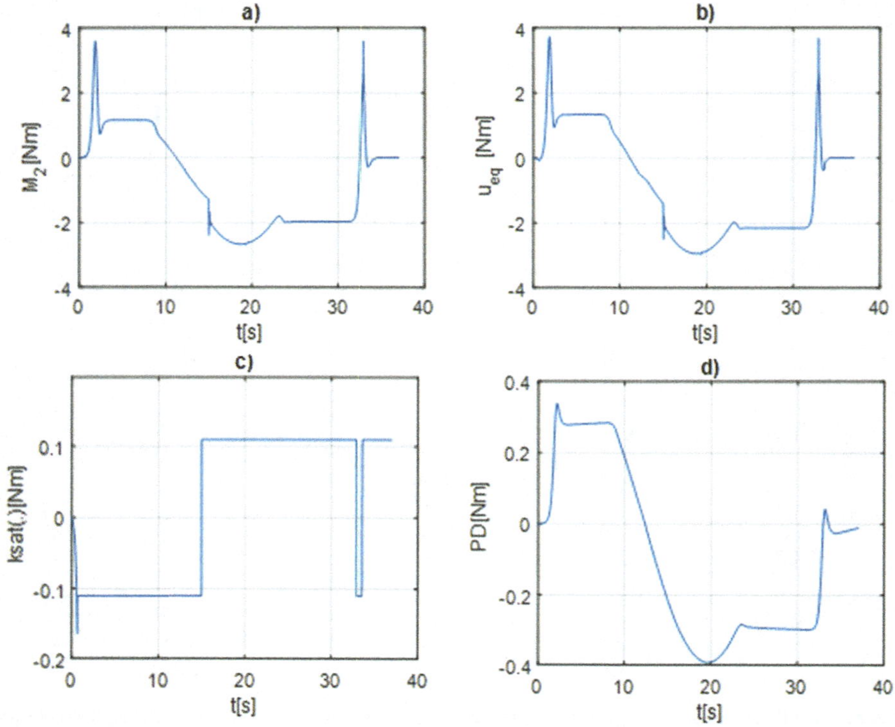

Fig. 6. Simulation results: a) total control, b) equivalent control, c) slide control, d) PD control

By analyzing the obtained indicators of the quality of motion, it can be concluded that the obtained accuracy is satisfactory. The parametric disturbance is particularly visible for wheel 2. The stability of the adopted solution is confirmed by the limitations of signals shown in Fig. 4. according to Lyapunov's interpretation of stability.

5 Conclusions

In this paper, a robust trajectory tracking controller for a Mecanum-wheeled mobile robot has been proposed. Based on the derived 3-DOF dynamics model considering friction uncertainty, a nonlinear stable robust control law is derived using a sliding mode control law via Lyapunov stability theory. Computer simulations are presented to illustrate the control system performance. Real implementation study using the suggested robust control law with microcontroller deserves future consideration.

References

1. Pin, F., Killough, S.: A new family of omnidirectional and holonomic wheeled platforms for mobile robots. IEEE Trans. Robot. Autom. **10**(4), 480–489 (2002)

2. Xie, L., Scheifele, C., Xu, W., Stol, K.A., Heavy-duty omni-directional Mecanum-wheeled robot for autonomous navigation: system development and simulation realization. In: IEEE International Conference on Mechatronics, Nagoya, Japan, pp. 256–261 (2015)
3. Qian, J., Zi, B., Wang, D., Ma, Y., Zhang, D.: The design and development of an omni-directional mobile robot oriented to an intelligent manufacturing system. Sensors **17**(9), 2073 (2017)
4. Adamov, B.I.: Influence of mecanum wheels construction on accuracy of the omnidirectional platform navigation (on example of KUKA youBot robot). In: 2018 25th Saint Petersburg International Conference on Integrated Navigation Systems (ICINS), St. Petersburg, pp. 1–4. IEEE (2018)
5. Hsu, P.-E., Hsu, Y.-L., Lu, J.-M., Jerry, Tsai, J.-H., Chen, Y.-S.: iRW: an intelligent robotic wheelchair integrated with advanced robotic and telehealth solutions. In: 1st Asia Pacific eCare and TeleCare Congress, Hong Kong, China, 16–19 June 2011
6. Li, D., Ye, J.: Adaptive robust control of wheeled mobile robot with uncertainties. In: 2014 IEEE 13th International Workshop on Advanced Motion Control, pp. 518–523 (2014)
7. Li, T.H.S., Chen, C.-Y., Hung, H.-L., Yeh, Y.-C.: A fully fuzzy trajectory tracking control design for surveillance and security robots. In: 2008 IEEE International Conference on Systems, Man and Cybernetics, pp. 1995–2000. Singapore (2008)
8. Lin, L.-C., Shih, H.-Y.: Modeling and adaptive control of an omni-mecanum-wheeled robot. Intell. Control Autom. **4**, 166–179 (2013)
9. Thi, K., Manh, N., Vo, H., Tran, V., Nguyen, D., Bui, A.: Trajectory tracking control for four-wheeled omnidirectional mobile robot using Backstepping technique aggregated with sliding mode control, pp. 131–134 (2019). https://doi.org/10.1109/ICA-SYMP.2019.8646041
10. Alakshendra, V., Chiddarwar, S. S.: A robust adaptive control of mecanum wheel mobile robot: simulation and experimental validation. In: Proceedings of the International Conference on Intelligent Robots and Systems, pp. 5606–5611 (2016)
11. Lu, X., Zhang, X., Zhang, G., Jia, S.: Design of adaptive sliding mode controller for four-mecanum wheel mobile robot. In: 37th Chinese Control Conference (CCC) (2018)
12. Ovalle, L., Ríos, H., Llama, M., Santibáñez, V., Dzul, A.: Omnidirectional mobile robot robust tracking: sliding-mode output-based control approaches. Control Eng. Pract. **85**, 50–58 (2019)
13. Hendzel, Z.: A description of the motion of a mobile robot with mecanum wheels – kinematics. In: Szewczyk, R., Zieliński, C., Kaliczyńska, M. (eds.) Automation 2019. AUTOMATION 2019. Advances in Intelligent Systems and Computing, vol. 920. Springer (2020)
14. Hendzel, Z.: A description of the motion of a mobile robot with mecanum wheels – dynamics. In: Szewczyk, R., Zieliński, C., Kaliczyńska, M. (eds.) Automation 2019. AUTOMATION 2019. Advances in Intelligent Systems and Computing, vol. 920. Springer (2020)
15. Hendzel, Z., Rykała, Ł.: Modelling of dynamics of a wheeled mobile robot with mecanum wheels with the use of Lagrange equations of the second kind. Int. J. Appl. Mech. Eng. **22**(1), 81–99 (2017). https://doi.org/10.1515/ijame-2017-0005
16. Żylski, W.: Kinematics and dynamics of wheeled mobile robots (in Polish). Rzeszow University of Technology Publishing Mouse, Rzeszow (1996)
17. Slotine, J.J., Li, W.: Applied Non-linear Control. Prentice Hall, New Jersey (1991)
18. Utkin, I.U.: Variable system with sliding modes. IEEE Trans. Autom. Control **22**(2), 212–222 (1977)
19. Sira-Ramirez, H.: Differential geometric methods in variable-structure control. Int. J. Control **48**(4), 1359–1390 (1988)
20. Hendzel, Z.: Robust neural networks control of omni-mecanum wheeled robot with Hamilton-Jacobi inequality. J. Theor. Appl. Mech. **56**(4), 1193–1204 (2018). https://doi.org/10.15632/jtam-pl.56.4.1193

Multi-stream Fusion in Image Sets Comparison

Pawel Piwowarski(✉) and Włodzimierz Kasprzak

Warsaw University of Technology, Institute of Control and Computation Engineering, ul. Nowowiejska 15/19, 00-665 Warszawa, Poland
pawel@piwowarski.com.pl, w.kasprzak@elka.pw.edu.pl

Abstract. A unique problem is considered: how to automatically determine whether two images (pictures, drawings, or photos) show the same 3D object pictured in different views. Similarity or equality of the objects seen despite the spotting, size or angle of view is one of the biggest challenges. Thus, a unique proposition of solving this task is proposed. In this complex solution image sets are compared and following question is answered: are these sets similar in the sense that they present an object of the same type? The approach uses CNN-like deep neural networks for object region masking, specific feature extraction and proper type classification, and similarity measures for data matching. Experimental validation is performed on a public car image database.

Keywords: CNN · Deep ranking · Image sets · Image classification · Late fusion · Similarity metric

1 Introduction

The evaluation of "image similarity" is the ability to define how two images - pictured objects - are different from each other. This type of comparison is not challenging if having a static picture of the same object visualized/taken from the same place, differ with few details. However, what if we do not analyze "the same" image but the set of pictures. What if we would like to quantify how similar are image sets, not only a single image? This problem appears in various real-world business scenarios, i.e. fraud detection, visual navigation or product search. It is also useful in visual search or data duplication avoidance.

Direct comparison of entire images (i.e. a front-to-end approach) is a new approach to image analysis. The growing demand for this type of solution is observed and coming from companies which seek effective answers to their business problems. Two significant papers were published in recent years. The first one gives a review of a few methods developed for finding similar images in datasets [3]. The second paper proposes deep learning techniques for this task [11]. All other research work focuses on entire and single image comparison

The research work was supported by Warsaw University of Technology.

R. Szewczyk et al. (Eds.): AUTOMATION 2021, AISC 1390, pp. 230–240, 2021.
https://doi.org/10.1007/978-3-030-74893-7_22

(Fig. 1) or on extracting features from the images and comparing them using some relevant metrics (e.g. the Euclidean distance in vector space). We are not aware about any work dealing with image sets similarity and comparison (Fig. 2). It is essential to set a benchmark for this kind of similarity problem.

In this paper, we focus on the image set comparison problem and propose a new approach. It is still based on the similarity of individual images but extended to a general-purpose approach to image set similarity. The solution requires not only the use of some category-level similarity, but also an Instance-to-Instance metric. Thus the similarity score will be built to combine the semantic content of an image with the image features comparison. A verification of our solution is expected in business contexts. The proposed solution may be deployed for comparison of sets of images showing cars, cats, buildings, etc.

The next two sections present the problem (goal of research) (Sect. 2) and our approach to image set-to-image set comparison (Sect. 3). Experimental validation is reported in Sect. 4. Conclusions are drawn in Sect. 5.

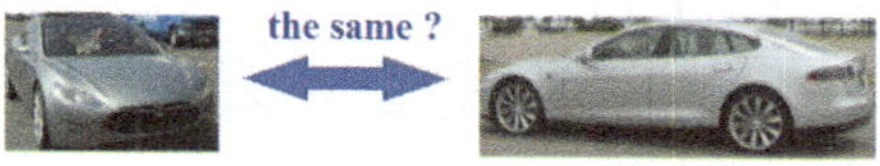

Fig. 1. The 3D object-to-object comparison problem in 2D images

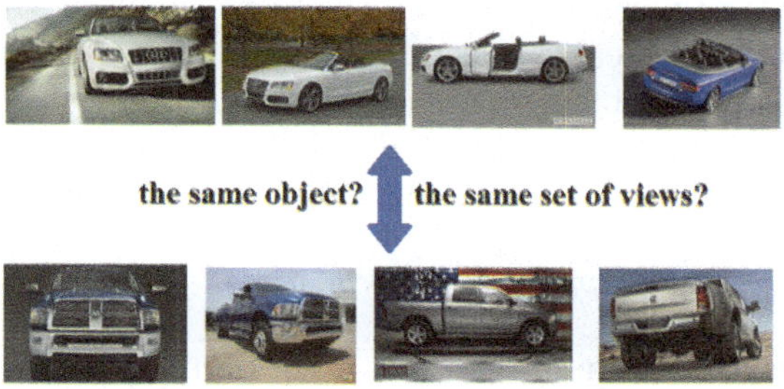

Fig. 2. The "object set - to - object set" comparison problem

2 Problem

The problem considered here is to estimate the similarity of 3D objects contained in two image sets. The core of this process is the similarity of objects visible in every pair of images from both sets. Hence, in this section we introduce the application domain (context) and similarity metrics.

2.1 Context

First, we need to define what kind of 3D objects we would like to compare. We shall distinguish here the "image type" from "object type or category" and from the "context". "Are the images of the same type", means whether the object views are the same, i.e. a "front view", a "side view" are examples of image types. "Are the objects from the same type" is the question related to the single 3D objects classified according to our "context" (application domain). Object "categories" are the generalization of types having the same basic properties. Examples of object types in the category of "cars" can be: "vans", "sedans", "trucks", "lorries". Examples of an application domain (context) are: cars for fraud detection or buildings and other cities infrastructure for visual navigation. The context can refer to single object categories (e.g. cars) or many object categories that can appear in the same image(e.g. "car next to the tree" or "car on the bridge"). In this work we focus on the context of single cars per image.

2.2 Similarity of Image Pairs

Scoring of image pair similarity is defined with quantitative metrics. There exists several metrics typically used in image analysis: Euclidean distance, structural similarity index, distance between image descriptors and many others. In 2015, the "FaceNET" was introduced [9] proposing the idea of triplet loss as learning criterion. The triplet loss minimizes the distance between the anchor and the positive example and maximizes the distance between the anchor and the negative example.

One of the general purposes image of the image similarity models, which are based on the triplet loss idea, is "Deep Ranking" [11]. It is one of the best neural network architectures for image matching (based on published results). Hence, we selected "DeepRanking" as the image similarity solution in our approach. The triplet loss implementation in Deep Ranking is as follows:

$$D(f(p_i), f(p_i^+)) < D(f(p_i), f(p_i^-)), \forall p_i, p_i^+, p_i^- \; ; \; r\left(p_i, p_i^+\right) > r(p_i, p_i^-) \quad (1)$$

where r is the image-pair similarity score. Deep Ranking metric is the Euclidean distance between two feature vectors.

3 Similarity of Image Sets

3.1 Structure

We propose a solution for the estimation of content of image set similarity. Our proposition is generic for a non-fixed count of image pairs, and it will be proper even for single pair. The core of this solution consists of three image processing streams for application-dependent image-pair comparison: the classification of object views, image-pair similarity based on semantic image segmentation and on feature-based similarity metrics (Fig. 3). After accumulating similarity scores

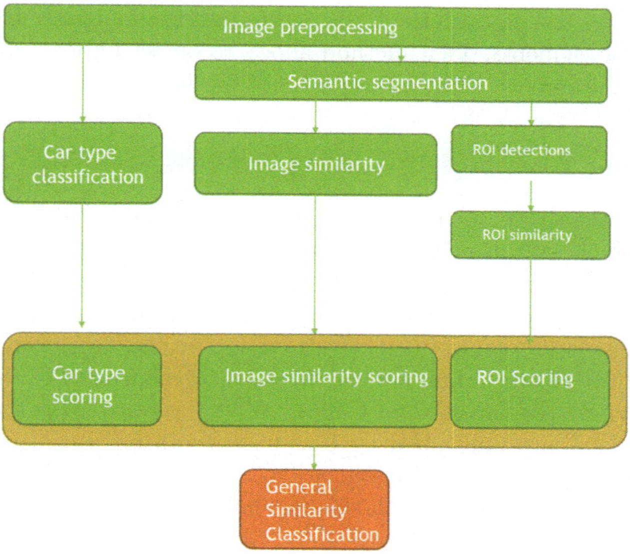

Fig. 3. The structure of our approach to image set comparison

obtained for all image-pairs from two analysed image sets, the individual results are fused and further classified to obtain a final likelihood whether both sets contain the same object.

The approach performs multi-score late fusion, where the following three individual scores are fused:

1. Car type score (class likelihood),
2. Comparison of the entire image similarity (based on Deep Ranking),
3. Comparison of the ROI similarity (based on Deep Ranking).

3.2 Input Data

The input image sets contain 4 images each. The order of the images is important. On the first position, there should be a front picture, on the second - a front/side view, the third is a side view and the last - a rear/side view (see Fig. 2).

3.3 Semantic Segmentation

The purpose of "semantic image segmentation" is to find in the image all possible object instances that belong to object types considered in our context (application domain). Quality of this step is crucial because the extracted and classified image areas will be then used next to calculate similarity scores for object instances.

In this paper, we use the "Mask RCNN" approach [5] as a state-of-the art solution to image segmentation. Mask RCNN is based on the Faster-RCNN net [8] and allows to find both the ROIs and to perform image segmentation and classification of segments. That method allows to detect important car features in the image and remove unnecessary parts of the image.

3.4 Car Type Classifier

The car type identity is an important precondition for image set recognition. Thus an image classification stream is considered, that gives car type likelihood values as its result. It is implemented as a deep neural network based on the Inception V3 architecture [10] that we have trained on the Stanford Car Dataset. We used weights from a pre-trained model, trained on the "Imagenet" dataset [1].

Last layer was re-trained for car classification. It recognizes 41 classes of cars. Eight basic types are: Van, Sedan, Coupe, SUV, Cabriolet, Hatchback, Combi, and Pickup. They are multiplied by 5 camera views for each type. One final class is "Other". During image classification, in the post-classification step final 8-elementary vector is created with the likelihood value for each basic class.

3.5 Similarity of Image Pairs

The second data analysis stream is focused on estimation of image pair similarity. Analyzed pair consists images assigned to the same position in both sets. We are using the "Deep Ranking" model, as presented in previous section. The final result is a 4-elementary vector containing similarity scores for every image pair.

3.6 Detecting and Comparing Image ROIs

The comparison of whole car images is supported by the third data analysis stream which focuses on the detection of specific car parts. Thus, regions of interest (ROI) are selected in every image and compared to images from the second set. In this way some parts of secondary importance are eliminated from comparison. For example, this allows to avoid the homogeneous region of the door and focus on the interesting texture of the rear or front part of the car. In this stream, we use the "Selective Search" algorithm. It explores color, texture, size and shape information to define an image ROI. The final comparison is again based on "Deep Ranking".

3.7 Late Fusion and Final Classification

We are using a late fusion of individual stream results by concatenating them into a vector of all available scores. Results from each stream obtained for four image-pairs are merged together into one vector of length 112. The final classifier is a neural network with fully connected layers. Its input is the fused data vector and it produces the final similarity score of the two image sets (Table 1).

Table 1. Architecture of the final classifier - a deep neural network

Number	Layer (type)	Output	Parameter number
1	Dense (dense)	64	7232
2	Batch_normalization	64	256
3	Dense_1 (dense)	128	8320
4	Batch_normalization_1	128	512
5.	Dense_2 (dense)	196	25284
6	Batch_normalization_2	196	784
7	Dense_3 (dense)	128	25216
8	Batch_normalization_1	128	512
9	Dense_4 (dense)	2	258

Total parameters: 68.374.
Trainable parameters: 67.342. Non-trainable parameters: 1.032

4 Results

4.1 ROI Detection

For training the Mask RCNN model, we have used the COCO Dataset (the 2017 version) [7]. COCO contains 81 categories of objects sufficient for training and testing the entire model. Semantic segmentation also allows to remove unnecessary parts of images and focus only on consistent with the context.

The Mask RCNN implementation is using the ResNET 101 [4] as backbone. For our research we selected the "Stanford Car Dataset" [6] and focused on "Car" and "Truck" types from the COCO Dataset.

Fig. 4. Illustration of car mask extraction for car images from the COCO database

On Fig. 4 the example of car mask detection results from two steps:

1. the rectangular boundary box for car region extraction comes from the annotations given in the COCO dataset,
2. the Mask RCNN is generating the proper envelope of the car image region within the boundary box.

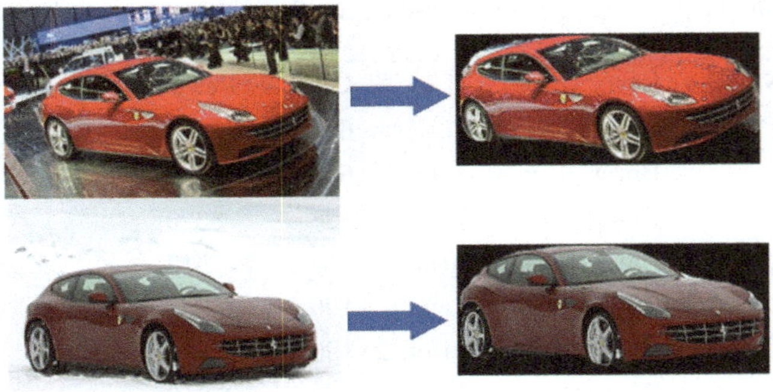

Fig. 5. Illustration of car mask detection in single-car images

We evaluated visually the correctness of produced masks for our car dataset and found that 89.7% of them were fully correct (see further illustration on Fig. 5). This step of image preprocessing has significantly improved the results of our image similarity stream. It is obvious, that the comparison of object's ROIs gives a higher similarity score than the comparison of entire images. For the pair of left images (see Fig. 5), we got a low similarity score of 0.092, while for the masked ROI image pair on the right side, the score was 0.392.

4.2 Car Type-View Classifier

Car type-view classifier was trained on the Stanford Car Dataset. This dataset was manually split into subsets for different car body types and camera views. The summary of the training and validation results is shown on Fig. 6.

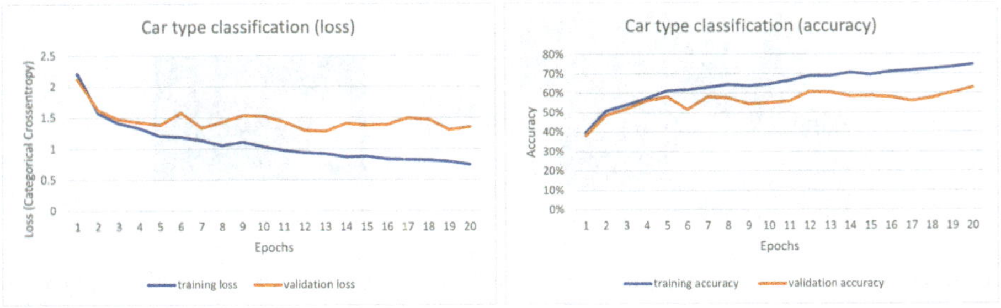

(a) (b)

Fig. 6. Summary of the training and validation results of the car type classifier (41 classes): (a) classification loss, (b) classification accuracy

4.3 Single-Stream Evaluation of Image Set Content

We evaluated this type classifier (41 classes) in the context of 5 picture-sets (holding 4 views mentioned earlier and an additional full rear view) Fig. 7. Based on expert judgement, we set a threshold of 3 images. Hence, if 3 (or more) images from the set are correctly classified, such set is correctly classified.

Fig. 7. Illustration of one 5-image set

Table 2. Summary of car type-view classification results, summarized for basic 8 types, decided for 5-image sets assuming a 3+ individual correctness threshold

Number of photos	Van	Sedan	Coupe	SUV	Cabriolet	Hatchback	Combi	Pickup	Out of
3+	44	49	50	68	58	48	51	66	72

As we can see from Table 2, the success rate varies between types. But even with partly unperfect results, it should be an important help in the final classification of image set similarity. For the total number of 41 classes we got an accuracy of 74.7%. Comparative result of such a classification problem doesn't exist. We know from the literature that a simple classification of the basic car type (8 classes) on this dataset can reach a 93.6% success rate [2].

4.4 Final Two Set-Similarity Classifier

The question whether two sets are similar and all their images show the same object is more difficult to answer, than to classify the object in a single image to defined classes. Adding every next photo pair increases the complexity. On the other hand in the eyed applications this accumulation of images allows to get the expected useful information.

The final classifier was trained on sets of cars images randomly selected from whole datasets (car model was kept). That classifier gives satisfying results for ordered four-view sets. Some examples below show three decision types: proper acceptance of consistent image sets of the same model (Fig. 8), no decision for less consistent sets of the same model (Fig. 9) and rejection of two sets with different models (Fig. 10).

By the moment of publishing this article we reached the success rate of 50.2% by implementation of the multi-stream image pair similarity evaluation approach. It shall be stated/underlined that presented results are very first in this area. We expect to meet a lot of new possibilities to improve these results in future studies (Fig. 11).

Fig. 8. Detecting the same car model in two most consistent image sets

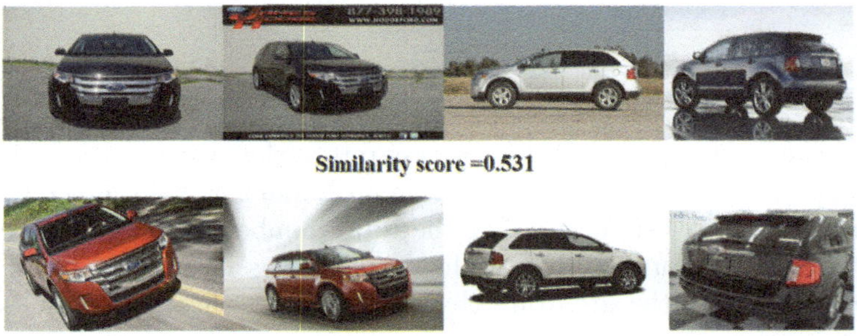

Fig. 9. No-decision for the same car model in two less consistent image sets

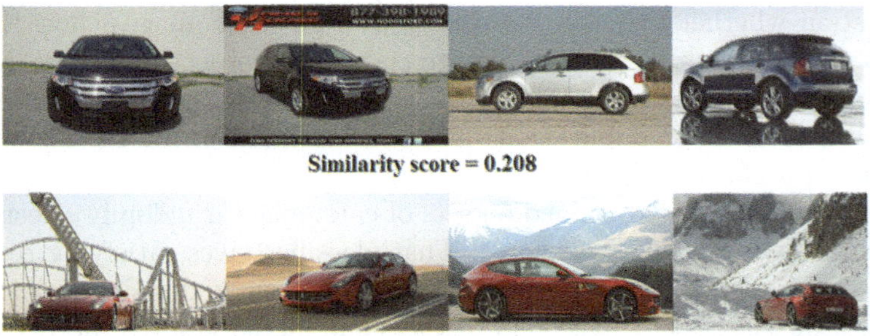

Fig. 10. Rejecting sets with two different car models

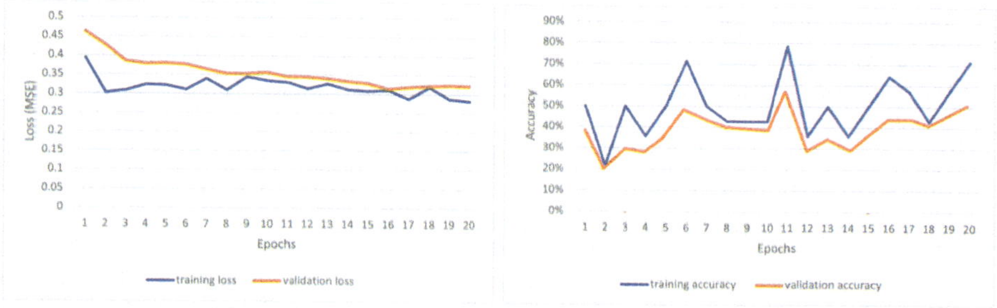

(a) (b)

Fig. 11. Training process of the final classifier: (a) classification loss, (b) classification accuracy

5 Conclusion

In our research work we have not found any paper related to comparison of two image sets. In our opinion we propose the first approach to the problem. This presented solution allow to compare two image sets and answer questions like: are they similar in structure (views and car types)? Whether they show the same car instance? Our approach is sufficient to be implemented in various domains. In further research it shall be developed and improved by application of other additional streams analysing selected image content.

References

1. Chollet, F., et al.: Keras (2015). http://keras.io
2. Dehghan, A., Masood, S.Z., Shu, G., Ortiz, E.G.: View independent vehicle make, model and color recognition using convolutional neural network (2017). https://arxiv.org/abs/1702.01721
3. Gaillard, M., Egyed-Zsigmond, E.: Large scale reverse image search: a method comparison for almost identical image retrieval. In: Proceedings - INFORSID 2017, p. 127 (2017). https://hal.archives-ouvertes.fr/hal-01591756
4. He, K., Zhang, X., Ren, S., Sun, J.: Deep residual learning for image recognition, pp. 770–778 (2016). https://doi.org/10.1109/CVPR.2016.90
5. He, K., Gkioxari, G., Dollár, P., Girshick, R.: Mask R-CNN. In: 2017 IEEE International Conference on Computer Vision (ICCV) (2017)
6. Krause, J., Stark, M., Deng, J., Fei-Fei, L.: 3D object representations for fine-grained categorization. In: 4th International IEEE Workshop on 3D Representation and Recognition (3dRR-13) (2013)
7. Lin, T.Y., Maire, M., Belongie, S., Hays, J., Perona, P., Ramanan, D., Dollár, P., Zitnick, C.L.: Microsoft COCO: Common objects in context. In: Computer Vision ECCV 2014, pp. 740–755 (2014)
8. Ren, S., He, K., Girshick, R., Sun, J.: Faster R-CNN: towards real-time object detection with region proposal networks. IEEE Trans. Pattern Anal. Mach. Intell. **39**(6), 1137–1149 (2017)

9. Schroff, F., Kalenichenko, D., Philbin, J.: FaceNet: a unified embedding for face recognition and clustering. In: 2015 IEEE Conference on Computer Vision and Pattern Recognition (CVPR), June 2015. https://doi.org/10.1109/CVPR.2015.7298682
10. Szegedy, C., Vanhoucke, V., Ioffe, S., Shlens, J., Wojna, Z.: Rethinking the inception architecture for computer vision. In: Proceedings of the IEEE Conference on Computer Vision and Pattern Recognition (CVPR), June 2016
11. Wang, J., Song, Y., Leung, T., Rosenberg, C., Wang, J., Philbin, J., Chen, B., Wu, Y.: Learning fine-grained image similarity with deep ranking. In: Proceedings of the IEEE Conference on Computer Vision and Pattern Recognition (CVPR), June 2014

Evaluation of Hand-Eye Calibration Algorithms in Application to Robotic Flexible Manufacturing Systems

Jakub Chudziński[1,2(✉)], Dominik Belter[1,2], Filip Jarecki[2], Michał Spławski[2], Paweł Drapikowski[1,2], and Piotr Kaczmarek[1,2]

[1] Institute of Control, Robotics and Information Engineering, Poznan University of Technology, Poznan, Poland
jakub.chudzinski@put.poznan.pl
[2] A.S. Adrian Stern, Wilczak 45/47, 61-623 Poznan, Poland
http://robotics.put.poznan.pl
https://rob-tech.pl/en

Abstract. The article is focused on two vital practical aspects related to robotic flexible manufacturing system which can be used for example in small craft factory to support the production of short series: the automatic visual system calibration with respect to the workspace and visual work-piece localization. The test procedure was performed in the system consisted of a collaborative robot UR-10, HD RGB camera and 8-cameras Optitrack motion capture system used as a reference. For calibration three hand-eye algorithms were investigated, where the best accuracy enabling to localize workspace coordinate system was achieved for dual-quaternions approach with the error below 0.4 mm in each direction. The work-piece localization was performed by means of 40×40 mm ArUco tag, the mean localization error was below 0.6mm in every dimension.

Keywords: Robot sensing and perception · Localization · Hand-eye calibration · Flexible manufacturing systems

1 Introduction

Robotic flexible manufacturing systems (RFMS) are a dynamically developing branch of industry 4.0. In the case of a system that can be localized in various places, the arrangement of the workspace changes. Additionally, to enable handling by technical staff with basic competences, the RFMS can be equipped with a vision system that allows for the configuration and manufacturing tasks in a semi-automatic mode. The calibration procedure of such a system, which is the addressed in this work, includes both the calibration of the robot in relation to the working space and the calibration of the vision system in relation to the robot and the working space.

R. Szewczyk et al. (Eds.): AUTOMATION 2021, AISC 1390, pp. 241–250, 2021.
https://doi.org/10.1007/978-3-030-74893-7_23

Standard approaches to intrinsic camera calibration are based on a planar pattern that is observed by the camera from various orientations and identifying the parameters of the camera model [1,2]. The popular methods for identifying the parameters of the camera model are based on the optimization with the Levenberg-Marquardt algorithm, gradient descent [3,6], Particle Swarm Optimization or genetic algorithm [5]. The external calibration of a camera attached to a robot can also be performed with the factor graph optimization. This approach is demonstrated on a quadruped walking robot [4]. A similar approach based on the factor graph optimization is also applied to a service robot, but in this case, the robot calibrates itself and localizes itself in the natural environment simultaneously [5]. Calibration of the external parameters of the robot and camera (hand-eye calibration) consists in determining the transformation between the robot's tip and the camera attached to it (eye-in-hand), or between the robot's base and the camera mounted in a fixed position and observing the scene (eye-to-hand). The accuracy of robotic stations using vision systems depends, among other things, on the quality of the calibration. Additionally, this process applies not only to stationary robots, but also to mobile [4], as well as multi-camera systems [7] or the integration of motion tracking systems (motion capture) such as Optitrack with an external camera [8,9].

The purpose of the eye-in-hand calibration is to determine the X transformation between the robot's tip and the attached camera, whose relative poses are described by the transformation of A and B, respectively, satisfying the equation:

$$AX = XB \tag{1}$$

Having a set of $N \geq 2$ transformations, the system of equations can be solved, however, the task is difficult due to the limitations of the camera, because the visibility of the calibration object (checkerboard, ChArUco board, visual tags, etc.) depends on the viewing angle, resolution and sharpness of the image. This leads to the development of various optimization methods based directly on the Eq. 1, the first by Tsai [10] and newer ones, e.g. dual-quaternions based approach [11] and it's modification proposed by Furrer et al. [12], 4D Procrustes Analysis [13] or indirectly by minimizing the reprojection error [14].

Current work is focused on two vital practical aspects related to robotic flexible manufacturing system which can be used for example in small craft factory to support the production of short series: the evaluation and selection of hand-eye calibration algorithms based on its accuracy, and assessment of visual-tag localization accuracy for localization of the work-pieces in the robot workspace. However, the hand-eye algorithms are well known usually the accuracy is evaluated as an error of a relative movement of the camera [13]. For the sake of the current application, it is crucial to assess the absolute error of the workspace calibration and work-pieces localization.

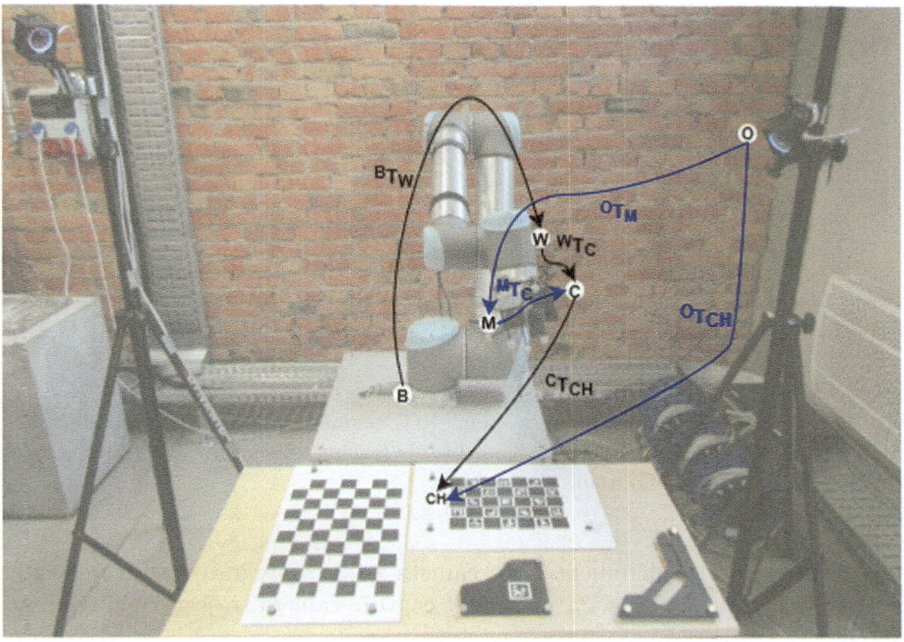

Fig. 1. The kinematic structure of the system and spatial relations between the base of robot B, robot's wrist W, camera C, ChArUco board CH. The Optitrack reference system consists of base frame O and tracked markers M. The markers on the calibration board has been mounted and configured to match the detected frame.

2 Methods

2.1 Experimental Setup

The whole system and its kinematic structure were presented in Fig. 1. The model of robotic flexible manufacturing system (RFMS) consists of the collaborative robot UR-10 (manufacturer Universal Robots), HD RGB camera (ELP 5MP Webcam). The calibration procedure is performed by using 5×7 squares ChArUco board (Fig. 2c). The RFMS should also be capable of localizing with respect to objects (namely work-pieces) in operation mode. It is assumed that the work-pieces will be localized by using ArUco visual tag of size 40×40 mm (Fig. 2b) therefore, tags localization accuracy is also considered in the experiment.

The reference system is motion capture Optitrack system with eight Prime 13W cameras. The system is capable of sub-20 μm accuracy and thus let's develop a reliable testing environment. The ChArUco board and tags are integrated with three spherical, reflective markers by using dedicated frames (Fig. 2b,c), where the transformation between markers coordinate system and tag/board is known. It let to calculate the tag/board position in the motion capture system coordinate system directly. The frames manufacturing error and thus markers localization in tag/board holder frames is below 0.1 mm assuming

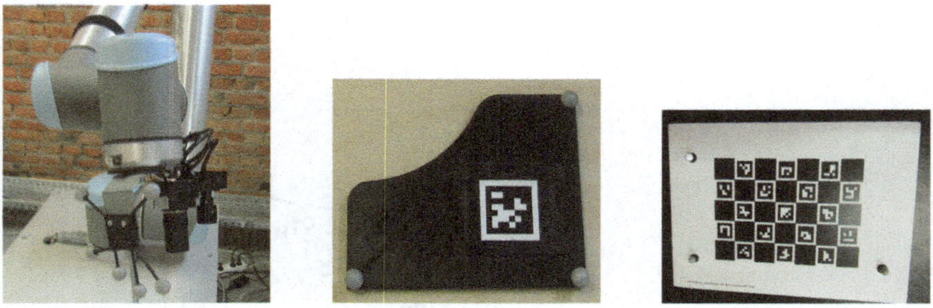

(a) The camera with special rigid body markers set attached.

(b) Holder with replaceable tag and Optitrack markers.

(c) Calibration board integrated with Optitrack markers.

Fig. 2. Individual elements of the system. RGB camera (a) mounted on robot and calibrated previously with ChArUco board (c) is tracking visual tag (b) giving a localization's estimate of the system.

the accuracy of tag alignment. The camera is also integrated with dedicated rigid body marker set (Fig. 2a) which let capture camera trajectory in reference system.

2.2 Experimental Procedure

The procedure consisted of two test. First the accuracy of three hand-eye calibration algorithms was assessed. Each has open-source implementation available for ROS environment and the high accuracy is claimed by authors. The tested algorithms are:

- `visp_hand2eye_calibration` [15], called shortly *visp*;
- `hand_eye_calibration` [12], called shortly *hand*;
- `hand_eye_SO4` [13], called shortly *procrust.*

In the second test, the accuracy of ArUco tag localization was investigated. To detect and estimate robot's position in relate to ArUco tag, `aruco_ros` package was used ([16]). In both tests, the robot with the camera attached followed a trajectory composed of ca. 130 points located on the hemisphere. During stops, images of the observed tag or board and the current position of the robot were collected. In the case of the first test, the images were first used to calibrate the intrinsic parameters of the camera and then to estimate the position of the calibration board. The estimated camera with respect to the board and the recorded robot positions were used as input to the position hand-eye procedure. Mutual calibration allowed to localize all objects the robot's base coordinate system. The accuracy of the calibration algorithms was evaluated by comparing the estimated board localization with reference from the motion capture system.

2.3 Accuracy Evaluation

The system integrates the robot localization based on visual tags with the Optitrack reference system. The kinematic structure and spatial relationships presented in the Fig. 1 let to assess the accuracy of both the hand-eye calibration and visual tags-based localization. The procedure consist on determining the calibration of the robot-camera ${}^{W}T_C$ and the Optitrack-camera ${}^{M}T_C$ transformations, and then the mutual calibration of the Optitrack-robot-camera ${}^{B}T_O$.

$$ {}^{B}T_O = {}^{B}T_W \cdot {}^{W}T_C \cdot {}^{M}T_C^{-1} \cdot {}^{O}T_M^{-1} \quad (2) $$

The reference and estimated transformation defining the position of the calibration board in the robot's base coordinates are calculated as:

$$ {}^{B}T_{CH} = {}^{B}T_O \cdot {}^{O}T_{CH} \quad (3) \qquad {}^{B}\hat{T}_{CH} = {}^{B}T_O \cdot {}^{O}T_M \cdot {}^{M}T_C \cdot {}^{C}T_{CH} \quad (4) $$

The localization error (ΔD) is expressed as the difference of reference and estimated board translation with respect to robot's base:

$$ \Delta D = {}^{B}\hat{T}_{CH} \cdot [0 \quad 0 \quad 0 \quad 1]^T - {}^{B}T_{CH} \cdot [0 \quad 0 \quad 0 \quad 1]^T \quad (5) $$

Further it is presented as euclidean norm and denoted as euclidean distance error or as an error in x, y, z coordinates. The angular error ($\Delta\alpha$) is expressed as an angle between reference and estimated normal vectors of the ChArUco:

$$ \Delta\alpha = arccos[({}^{B}\hat{T}_{CH} \cdot [0 \quad 0 \quad 1 \quad 0]^T)^T \cdot ({}^{B}T_{CH} \cdot [0 \quad 0 \quad 1 \quad 0]^T)] \quad (6) $$

3 Results

3.1 Hand-Eye Algorithms Accuracy Comparison

The accuracy values recorded for three hand-eye algorithms were presented in Fig. 3 as box plots and in Table 1. It can be observed (Fig. 3) that the error variability decreases with the distance from the board (it is true for *visp* and *hand* algorithms). Moreover, the smallest variability and lowest mean error were observed for all algorithms in the distance of 0.81 m. The mean error (Table 1) is the smallest for *hand* and is significantly ($p<0.05$) better than for other algorithms, moreover the result obtained for *procrust* is significantly worse than other algorithms. The average angular error for all methods is 0.4 deg. Regarding the distribution of the localization error with respect to coordinates, it can be observed (Table 1), that the error in x and y coordinates is similar for all algorithms and is ca. 0.2 mm. The substantial differences are observed in z axis where the *hand* error is 0.4 mm and performs significantly better ($p < 0.05$) than other algorithms.

Table 1. Average errors based on ChArUco detection and calculated variability expressed as Standard Error (SE). Algorithm *hand* presents the smallest euclidean error and it has been chosen for the next experiment.

Algorithm	ΔD_x [mm]	ΔD_y[mm]	ΔD_z[mm]	$\overline{\lvert\Delta D\rvert}$[mm]	$\Delta\alpha$[deg]
visp	-0.6 ± 0.1	-0.2 ± 0.1	0.7 ± 0.1	0.9 ± 0.2	0.41 ± 0.01
procrust	-0.4 ± 0.1	-0.2 ± 0.1	1.2 ± 0.1	1.3 ± 0.2	0.40 ± 0.01
hand	-0.3 ± 0.1	-0.2 ± 0.1	0.4 ± 0.1	0.6 ± 0.2	0.43 ± 0.01

Table 2. Euclidean error of the tag position with Standard Error ($\lvert\overline{\Delta D}\rvert$ ± SE [mm]) for samples in range between 0.18 m–0.59 m and two view angles ranges.

Angle range [deg] / Max. dist. [m]	0.24	0.42	0.59
<0-15)	0.7 ± 0.5	3.3 ± 0.5	6.0 ± 0.5
<15-30>		1.2 ± 0.5	2.6 ± 0.5

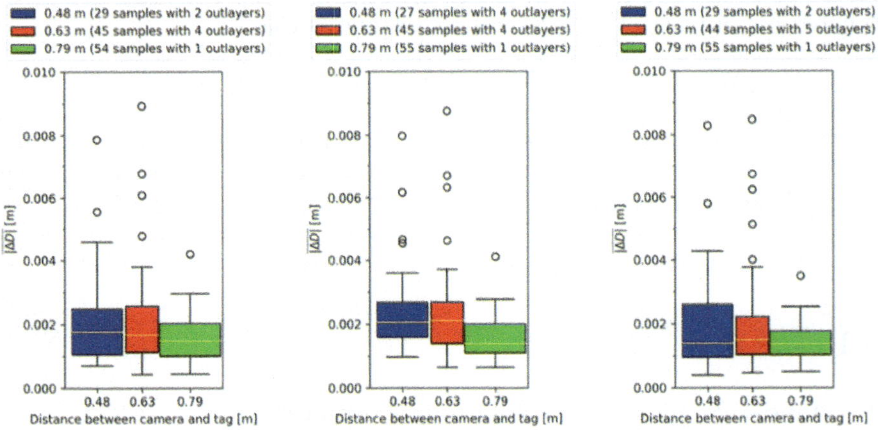

(a) Accuracy for *visp* (b) Accuracy for *procrust* (c) Accuracy for *hand*

Fig. 3. Results of accuracy tests for hand-eye algorithms. The graphs show box plot of the euclidean error depending on the distance of the camera from the board.

3.2 Accuracy of Visual Tags Pose Estimation

The tag was observed in distance from 0.18 to 0.6 m and for view angles varied from 0–30°. All observation points were evenly distributed over the hemisphere. The results were presented in Fig. 4 and in Table 2.

The mean errors of all samples in x and y axes of Fig. 4a are about 0.5 mm. Still, the error in z-direction and variance of estimates grow significantly with observation distance. The error in z-axis can be assumed as the most prominent factor contributing to euclidean distance error (Fig. 4a). It can be observed that

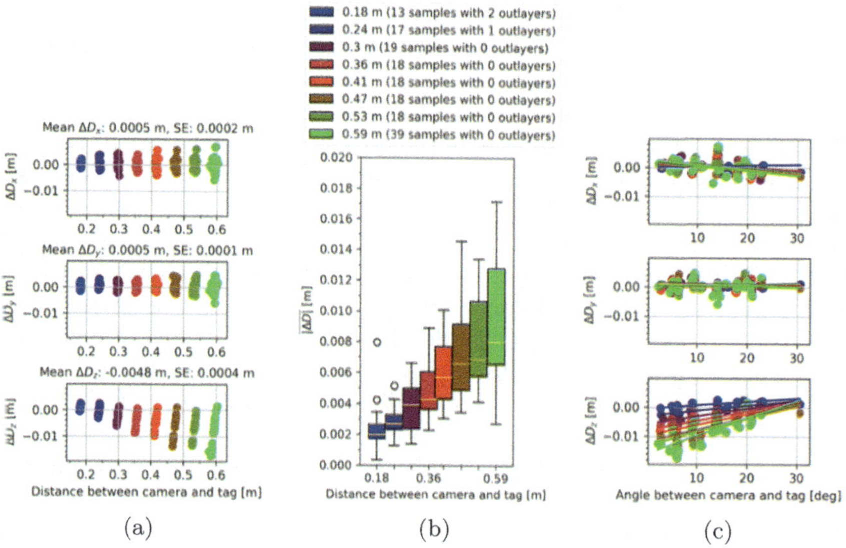

Fig. 4. Translation errors in axes (a,c) and euclidean error (b) for ArUco tag.

in the tested range of view angles (0–15°) the error is linearly dependent on the view angle, while it is the highest for 0deg and the distance of 0.6 m It decreases with the angle increase. The slope of linear function increase with the distance, and varies from (0.1 mm/deg at a distance 0.18 m to 0.2 mm/deg at the distance 0.6 m). This effect does not occur for x and y axes. To determine the factors influencing the accuracy the samples were organized and aggregated in cross-table, where the accuracy was calculated for three different maximum distances and for two distinct view angle ranges (Table 2). Due to a low number of samples acquired in the distance 0.18–0.24 m the error is calculated for the full range of view angles and is 0.7 mm which is significantly smaller (based on one-way ANOVA and Kruskal-Wallis methods) than the estimates obtained for larger ranges of distance ($p < 2e{-}5$). Moreover, it can be observed that when that position is estimated from a distance above 0.24 m the estimates calculated for view angles in the range 15–30° presents significantly higher accuracy than in the range 0–15°.

4 Discussion

The results presented in current work suggest that for the sake of robot and camera calibration with respect to the workspace coordinates system the best among tested hand-eye methods is dual-quaternion algorithm (*hand*). Assuming that the system acquire evenly distributed samples from different view angles, it let to localize the workspace origin with error less than 0.4mm in all directions. Surprisingly the *procrust* method turned to be the worst, even the results presented by authors suggested that it should be one of the most promising approaches [13]. The potential explanation of this issue is that the article comparing *procrust* with other methods was focused only on relative changes of orientation while here the error is calculated as a mean of samples with respect to the absolute, fixed point. Therefore the errors values cannot be directly compared. The second difference is that the used samples contained outliers which might be generated due to changes of scene illumination or shadows generated by the robot arm or partial calibration board clipping. For the system aimed to be used in different places and calibrated for different workspaces, the presence of outliers has to be considered. According to our observations, the *procrust* method is not resistant to outlier samples. It seems that the *hand* utilizing simultaneous minimization of the translation and rotation error combined with the refinement step to find maximum likelihood solution works better in the presence of outliers.

The ArUco tag mean localization error is below 0.6 mm for every axis which is comparable to human operator precision. The error calculated as the norm of the mean localization is 0.7 mm for observation from the distance of the 0.24 m. Regarding the marker size, this result is even better than the results presented by Abbas et al. [17]. The marker seen from the distance of 0.18 m occupies rectangle of size 210×210 px while for distance 0.6 m its size is only 67×67 px. For the sake of the application, the tag localization accuracy should be improved. It can be potentially done by increasing marker size, using several joined markers, or by increasing the camera depth of field, or camera resolution. The camera depth of field is limited by lens properties and expected range of observation. In the experiment, this parameter was adjusted to observe both the board and the tag. The observed higher accuracy for view angles higher than 15° could be explained based on the subpixel approximation. For measuring the distance in z-direction when the view angle is 0° the marker size is determined only by rectangle fit, for small markers, therefore the sub-pixel approximation error may contribute significantly to the distance estimation error. In case of non zero view angle, the perspective deformation is observed, and thus the homography might be better conditioned.

5 Conclusion

The presented work was focused on analyzing the accuracy of the calibration camera and robot with respect to the workspace coordinate system, which is a vital element for reconfigurable, robotic flexible manufacturing systems. It was shown that the highest localization accuracy of the coordinate system represented by ArUco board is below 0.4 mm in every direction and was achieved by the dual-quaternion algorithm (*hand*) [12]. This algorithm was significantly better than two other tested algorithms and revealed higher robustness for the presence of outlier samples. The second vital aspect was the assessment of accuracy for work-pieces localization. It was assumed that the work-piece is localized by using 40×40 mm ArUco tag. For camera enabling to observe the whole scene with depth of field in range 0.16–0.8 m, the localization accuracy was smaller than 0.6 mm in each direction which is comparable with human operator precision. It can be concluded that the proposed approach can be successfully applied in a robotic flexible manufacturing system. It let to accomplish automatic calibration of the camera intrinsic parameters and the mutual calibration of the camera and robot with respect do the external coordinate system and to automatically localize work-pieces with the precision compared to human operator.

Acknowledgments. This work was supported by the European Regional Development Fund (ERDF) Wielkopolska Regional Operational Programme for 2014-2020, Measure 1.2.

References

1. Zhang, Z.: A flexible new technique for camera calibration. IEEE Trans. Pattern Anal. Mach. Intell. **22**(11), 1330–1334 (2000)
2. Heikkila, J., Silven, O.: A four-step camera calibration procedure with implicit image correction. In: Proceedings of IEEE Computer Society Conference on Computer Vision and Pattern Recognition, pp. 1106–1112 (1997)
3. Bouguet, J.-Y.: Camera calibration toolbox for matlab. http://www.vision.caltech.edu/bouguetj/calib_doc/
4. Reinke, A., Camurri, M., Semini, C.: A factor graph approach to multi-camera extrinsic calibration on legged robots. In: 2019 Third IEEE International Conference on Robotic Computing (IRC), pp. 391–394 (2019)
5. Kümmerle, R., Grisetti, G., Stachniss, C., Burgard, W.: Simultaneous parameter calibration, localization, and mapping for robust service robotics. In: Advanced Robotics and Its Social Impacts, pp. 76–79 (2011)
6. Gwak, S., Kim, J., Park, F.C.: Numerical optimization on the Euclidean group with applications to camera calibration. IEEE Trans. Robot. Autom. **19**(1), 65–74 (2003)
7. Kalibr. Calibration tools repository with example of multiple-camera calibration (2020). https://github.com/ethz-asl/kalibr
8. Basalt. Calibration tools repository with example of optitrack-camera calibration (2020). https://gitlab.com/VladyslavUsenko/basalt/blob/master/doc/Calibration.md
9. Soławski, M., Jarecki, F., Chudziński, J., Staszak, R., Kaczmarek, P., Drapikowski, P., Belter, D.: Motion planning of the cooperative robot with visual markers (2019)
10. Tsai, R.Y., Lenz, R.K., et al.: A new technique for fully autonomous and efficient 3D robotics hand/eye calibration. IEEE Trans. Robot. Autom. **5**(3), 345–358 (1989)
11. Daniilidis, K.: Hand-eye calibration using dual quaternions. Int. J. Robot. Res. **18**(3), 286–298 (1999)
12. Furrer, F., Fehr, M., Novkovic, T., Sommer, H., Gilitschenski, I., Siegwart, R.: Evaluation of combined time-offset estimation and hand-eye calibration on robotic datasets. Springer, Cham (2017). ISBN 978-3-319-67361-5
13. Jin, W., Sun, Y., Wang, M., Liu, M.: Hand-eye calibration: 4-D procrustes analysis approach. IEEE Trans. Instrum. Meas. **69**(6), 2966–2981 (2019)
14. Koide, K., Menegatti, E.: General hand-eye calibration based on reprojection error minimization. IEEE Robot. Autom. Lett. **4**(2), 1021–1028 (2019)

15. Lagadic. ROS package that estimates the camera position with respect to its effector using the ViSP library. https://github.com/lagadic/vision_visp/tree/master/visp_hand2eye_calibration
16. Pal robotics. Software package and ROS wrappers of the Aruco Augmented Reality marker detector library. https://github.com/pal-robotics/aruco_ros
17. Abbas, S.M., Aslam, S., Berns, K., Muhammad, A.: Analysis and improvements in apriltag based state estimation. Sensors **19**(24), 5480 (2019)

LiDAR Localization and Mapping for Autonomous Vehicles: Recent Solutions and Trends

Piotr Skrzypczyński(✉)

Institute of Robotics and Machine Intelligence, Poznań University of Technology, ul. Piotrowo 3A, 60-965 Poznań, Poland
piotr.skrzypczynski@put.poznan.pl

Abstract. This paper presents a brief survey of the current achievements in LiDAR SLAM and discusses some recent trends and new ideas in this area. The focus is on LiDAR SLAM applied to autonomous vehicles, which still strugle with real-world complexity. We identify the challenges in efficient environment representation, robust estimation over large state spaces, and real-time handling of the scene dynamics and diversified semantics. Some of these issues are illustrated by preliminary results of our recent research in LiDAR SLAM.

Keywords: SLAM · LiDAR · Autonomous cars · Machine learning

1 Introduction

Limited availability of the Global Navigation Satellite Systems (GNSS) signal in a number of environments (e.g. tunnels, underground parking lots, and so-called urban canyons) frequently makes the autonomous vehicles dependent on their on-board sensors for localization. The vehicle pose estimate and a map of the surroundings can be obtained using Simultaneous Localization and Mapping (SLAM). While a significant part of the SLAM-related research is devoted to visual SLAM [4], technological innovations in the area of 3-D laser scanners (Light Detection And Ranging – LiDAR) resulted in a number of LiDAR-based SLAM solutions. Although cameras are certainly the most affordable and easy to use sensors, active laser sensors have such practical advantages as longer range, larger field of view, and the ability to work under any lighting conditions [35]. Therefore, LiDARs are considered an enabling technology in the growing autonomous vehicles sector [21].

In this paper we review the recent developments in LiDAR SLAM in the context of autonomous vehicles, focusing on the methods used for 3-D scan registration, efficient map representations, and the approaches based on machine learning. We also advocate two new concepts in large-scale LiDAR SLAM: using higher-level geometric features for map representation, and applying the direct formulation to the scan registration problem. We demonstrate some results of our recent research to exemplify these ideas.

R. Szewczyk et al. (Eds.): AUTOMATION 2021, AISC 1390, pp. 251–261, 2021.
https://doi.org/10.1007/978-3-030-74893-7_24

2 Methods for Registration of 3-D Scans

State-of-the-art range-based SLAM and localization solutions either extract some geometric features from the scans [39], or employ raw point clouds in order to register together the consecutive 3-D scans. The most popular approaches are variants of the ICP (Iterative Closest Points) method [31], that are surveyed in [27]. In spite of its wide use and many improvements, the ICP has high computational cost because of the need to find the nearest neighbor according to the applied metrics. This process can be improved using proper data structures (kd-trees for instance) and parallelized on GPU [1], but still prevents the ICP-based approaches from using all the points in the data stream of multi-beam LiDARs. Employing normal vectors [32] improves convergence, but all ICP-like methods suffer from data association problems due to the heuristic nature of the nearest neighbor metrics and the need to re-establish the associations by search once the sensor/scan pose is updated. The Lidar Odometry And Mapping (LOAM) [41] applies different ICP metrics to points classified as planar patches and linear edges (point-to-plane and point-to-line, respectively). IMLS-SLAM [8] uses implicit moving least squares surfaces and matches points selected by a criterion that exploits the normal vectors information to constrain all six degrees of freedom of the sensor. The more recent MC2SLAM [23] selects points from the scans to maintain a certain distance between them, and uses tightly integrated IMU measurements to compensate distortions in the in-motion acquired scans.

Neither of these approaches is a full SLAM system in the sense of computing a maximum-a-posteriori estimate of the sensor trajectory and map. Such a SLAM formulation is now de-facto standard in visual SLAM [4] and treats the estimation problem as a factor graph [18], applying nonlinear optimization to solve it. The visual SLAM systems follow one of two approaches: the indirect approach extracts salient features (usually points) from images and minimizes the reprojection errors between the corresponding features, while the direct approach is feature-less, minimizing the photometric error over the whole image or selected pixels, e.g. those with the salient gradient. The leading representatives of these two approaches, ORB-SLAM (now available in version 3) [5] and DSO [12], respectively, have similar performance with respect to the accuracy on benchmark data, but DSO is more robust in different environments, while ORB-SLAM needs to extract a considerable number of features from each image to work properly. The direct approach was already applied to RGB-D data exploiting dense depth [36], while the more recent system from [9] can use camera images, depth, and normal information at the same time with a generalized formulation of the photoconsistency constraints.

In order to apply the direct approach to LiDAR data one needs an ordered representation of the scanned points with efficient access to individual "pixels". Such a representation is the range image – a cylindrical projection of the points obtained from a rotating LiDAR (Fig. 1A). The popular Ouster and Velodyne LiDARs measure range at selected horizontal and vertical angles, returning also the corresponding intensity (reflectance) values that capture environment information similar to regular images (Fig. 1B). To represent a single scan as an

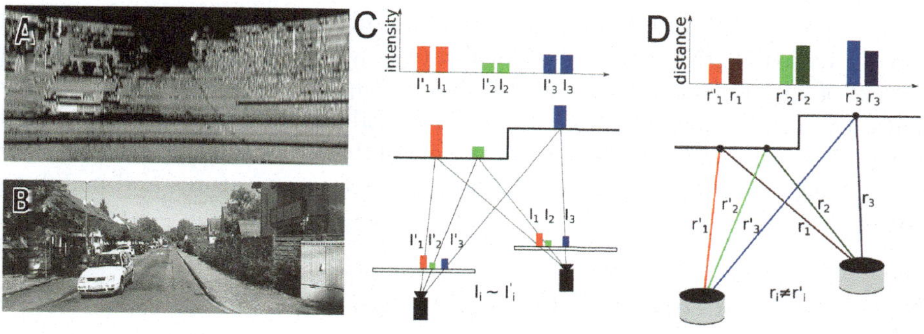

Fig. 1. Direct formulation of the SLAM problem is possible for modern LiDAR sensors that provide images (A) similar to low-resolution grayscale images from regular cameras (B). The direct registration principle for a camera (C) differs from the same principle for a LiDAR (D), because the range values measured by a LiDAR depend on the transformation between the sensor poses

image, each pixel has to be mapped according to the angle increment of its corresponding laser beam using a non-linear projection, as shown in our recent paper [24]. While vision-based direct SLAM solutions differ in many important details, all of them exploit the photo-consistency assumption. The intensity values in the pixels being projections of the same scene point into two camera images should be the same if the cameras are of the same type and parameters, and there are no occlusions or noise (Fig. 1C). This principle can be applied to range measurements (Fig. 1D). In this case not only the pixel positions but also the measured values depend on the transformation between the sensor poses. The idea of estimating the rigid transformation $[\mathbf{R}|\mathbf{t}]^*$ between two scans using the direct method can be written in a general form:

$$[\mathbf{R}|\mathbf{t}]^* = \underset{\mathbf{R},\mathbf{t}}{argmin} \sum_{u,v} \mathbf{e}_{u,v}^T \left(S_{k-1}, S_k, [\mathbf{R}|\mathbf{t}]\right) \mathbf{e}_{u,v} \left(S_{k-1}, S_k, [\mathbf{R}|\mathbf{t}]\right), \quad (1)$$

where $\mathbf{e}_{u,v}$ is the residual error at the u, v scan (range image) coordinates computed between the scan S_k and its counterpart obtained by transforming the previous scan S_{k-1} by the rotation and translation $[\mathbf{R}|\mathbf{t}]$ (k is time index). The point-wise error term $\mathbf{e}_{u,v}$ is the difference between a predicted value of the image pixel $\mathcal{I}(u, v)$ and its measured value: $\mathbf{e}_{u,v} = \mathcal{I}_{S_{k-1}}(u, v) - \mathcal{I}\left(\text{Proj}_{S_k}^{S_{k-1}}(u, v)\right)$, where Proj_A^B projects a (synthetic) image obtained at pose A onto a sensor located at pose B (cf. Fig. 1D). With this formulation an efficient mechanism of local (windowed) optimization for the transformations between consecutive images can be employed to obtain accurate tracking of the sensor [12]. What is important, in this approach both the range data and intensity data can be used, adding also the consistency between the normal vectors computed from the range data. However, to ensure that the consistency assumptions hold for the processed data, efficient methods for detecting occluded and dynamic points in the synthesized images need to be implemented.

The direct LiDAR SLAM is a new idea, a similar approach was so far applied only to mixed LiDAR and vision localization systems [34] in the application context of autonomous vehicles. Multi-sensor setups for vehicles require accurate spatiotemporal calibration [25], hence direct LiDAR SLAM with synthesized images is a promising idea, because of the possibility to use all the available data at real-time without extended calibration, extraction of features and matching heuristics.

3 Map Structure and Optimization

In most of the contemporary LiDAR odometry or SLAM systems the map is not structured and optimized, because its point cloud representation does not permit efficient handling of individual features/objects. On the other hand, in visual SLAM efficient management of the historical data in the map is crucial to the achieved accuracy [5]. Nowadays, the optimization is preferred to the filtering solutions (EKF, Particle Filter) due to the ability to update the linearization points and alleviate the negative effects of incorrect measurements by robust cost functions [38]. While the bundle adjustment (BA) idea is used in the vision community for long time, solving this problem efficiently for real-time SLAM became possible due to the emergence of efficient least-squares solvers, e.g. g^{2}o [17]. These solvers represent the least-squares optimization problem in the form of a factor graph, where vertices represent optimizable parameters, while edges represent the constraints (measurements) imposed on these parameters:

$$\underset{\mathbf{v}_1,\mathbf{v}_2,\ldots,\mathbf{v}_n}{\operatorname{argmin}} E = \sum_i \mathbf{e}_i^T \boldsymbol{\Omega}_i \mathbf{e}_i, \tag{2}$$

where E denotes the minimized cost function, $\mathbf{e}_i$ stands for the i-th measurement (usually residual error between the real measurement and prediction or map component) that joins one or multiple vertices, while each vertex $\mathbf{v}_j$ is the parameter that is optimized. The information matrix $\boldsymbol{\Omega}_i$ weights the uncertainty of i-th measurement. Notice that formula (1) describing direct transformation estimation can be considered a special case of (2).

The SLAM optimization problem has a special structure that occurs when recent poses and observations are jointly optimized [16]. Without loss of generality we assume that each observation joins a pose of the initial observation, current pose, and a selected observation, and thus the Jacobian of that measurement, $\mathbf{J}_i$, has a sparse structure. By combining all of the considered observations, we approximate the Hessian matrix $\mathbf{H}$, achieving the sparse, block-diagonal structure of the optimization problem. With this formulation the Gauss-Newton or Levenberg-Marquardt optimizers compute the update as:

$$\Delta\mathbf{x} = \mathbf{H}^{-1}\mathbf{b}, \tag{3}$$

where $\mathbf{b}$ is a vector stacked from the measurements (residuals) multiplied by the respective Jacobians [16]. The update step can be efficiently determined by separating the sensor pose part $\mathbf{x}_p$ and observation part $\mathbf{x}_f$:

$$\mathbf{H}\Delta\mathbf{x} = \begin{bmatrix} \mathbf{A} & \mathbf{B} \\ \mathbf{B}^T & \mathbf{C} \end{bmatrix} \begin{bmatrix} \Delta\mathbf{x}_p \\ \Delta\mathbf{x}_f \end{bmatrix} = \begin{bmatrix} \mathbf{b}_p \\ \mathbf{b}_f \end{bmatrix} = \mathbf{b}. \tag{4}$$

Due to the special structure, and the use of Schur complement, this problem can be solved as:

$$(\mathbf{A} - \mathbf{B}\mathbf{D}^{-1}\mathbf{B}^T)\Delta\mathbf{x}_p = \mathbf{b}_p - \mathbf{B}\mathbf{D}^{-1}\mathbf{b}_f, \tag{5}$$

which can be computed efficiently, because $\mathbf{D}$ is a block diagonal matrix and its inverse is easy to obtain. The solution for $\Delta\mathbf{x}_f$ is also easy to determine once $\Delta\mathbf{x}_p$ is known (Fig. 2).

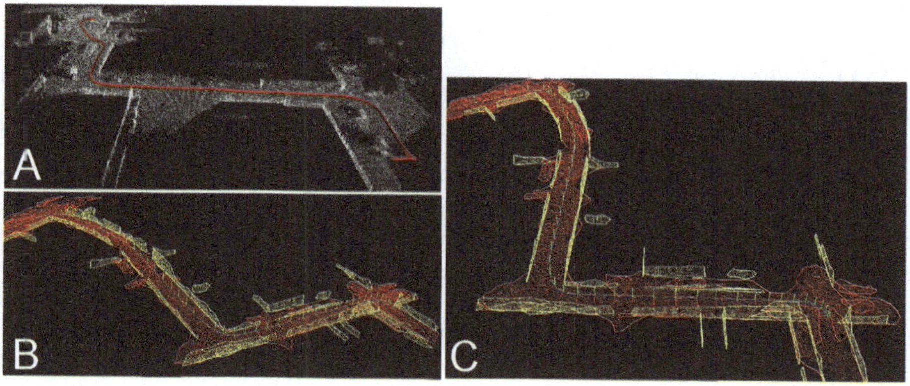

Fig. 2. Environment maps made of raw points (A), and geometric features for fragments of the KITTI Benchmark [14] sequences no. 03 (B) and no. 04 (C)

Implementation of efficient optimization requires a proper map structure. In visual SLAM the map is either organized as a collection of images (direct approach), or made up of sparse feature points within a graph structure [5]. In contrast, maps in LiDAR SLAM often contain raw measurement points. Surfel-based maps are also used with LiDAR SLAM for autonomous vehicles, e.g. [3]. Neither of these approaches makes it possible to efficiently implement global optimization and marginalization of the historical states, what keeps the size of the map bounded.

Few LiDAR localization systems attempt to structure the map beyond raw points. In LOAM the acquired point clouds are segmented into semantically distinct planar patches and edges (line segments), but the distinction is used only locally with different variants of ICP, while LOAM does not create geometric features at the object level. The LeGO-LOAM [33] system makes a step further, explicitly detecting the ground plane, and distinguishing between points found

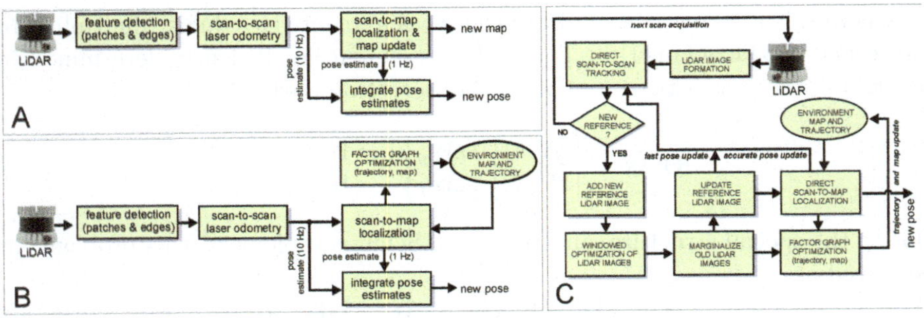

Fig. 3. Three different architectures of LiDAR SLAM: the LOAM pipeline with separated laser odometry and mapping (A), the full SLAM architecture with joint map and trajectory optimization (B), and the direct LiDAR SLAM architecture (C). Loop detection modules not shown for the sake of clarity

on the ground and on other objects. The architecture of LOAM and LeGO-LOAM combines real-time scan-to-scan sensor pose tracking (odometry) and slower, but more accurate scan-to-map localization (Fig. 3A).

A natural approach for LiDAR data in partially structured environments, such as urban scenes typical to autonomous vehicles are higher-level geometric features, such as planar and linear segments. Planar features in EKF-based SLAM with 3-D laser data were used by Weingarten and Siegwart [40]. More recently, Salas-Moreno *et al.* [30] employed planar segments and surfels for dense environment mapping. A map representation with planar segment as features is used by our PlaneLoc global localization system [39]. A similar concept of planar features was used in our recent PlaneLOAM SLAM [7]. This system is intended for autonomous vehicles and inherits some elements of the open-source LOAM, but uses a map of planar segments and line segments that can be globally optimized using a BA-like approach implemented with g^2o (Fig. 3B).

Replacing the LOAM-style ICP data registration with the direct SLAM concept should make it possible to implement windowed local BA [16] for improved odometry and global optimization of the map and trajectory with marginalization. Although such an architecture is more complicated (Fig. 3C) it takes advantage of all the recent techniques that contributed to the high accuracy of visual SLAM.

4 Machine Learning in LiDAR SLAM

The enormous success of deep learning approaches in computer vision, also for LiDAR data processing [10], motivates research on the application of learning methods to visual odometry and SLAM.

However, the application of deep learning to LiDAR SLAM turned out to be much more problematic, mostly because of the lack of a uniform representation of the acquired scans that are sensor-specific, and the fact that standard representations of the range data are not differentiable. State-of-the-art neural models for processing of point clouds ensure invariance to the point-order permutation, like PointNet [26], or employ a voxel representation [42]. While the PointNet approach and its derivatives (e.g. PointPillars [19]) make it possible to process point clouds with convolutional networks, they are often considered too complicated and memory-hungry to be applied in real-time SLAM. Systems conceived specifically for self-driving cars often use LiDAR data representations that assume the geometry of the laser scans (all-around scanning) and motion in 2-D [19,22].

Among the few attempts to solve LiDAR-based localization in an unknown environment (either odometry or SLAM) with a neural network architecture, Velas et al. [37] transformed the pose estimation problem into a classification task, while the recent L^3-Net system [22] combines a lightweight variant of the PointNet (mini-PointNet) that extracts sparse features from point clouds with another neural network that accumulates the matching costs from all the features and outputs the pose. Unfortunately, casting the pose estimation problem as classification or cost accumulation leads to decreased accuracy of the pose estimate, because the result is bounded to discrete values depending on the data structure being used. In contrast, the LO-Net [20] implements a full laser-based odometry pipeline with learnable models. LO-Net offers pose estimation accuracy comparable to state-of-the-art model-based systems (LOAM in particular), but requires supervised learning with ground-truth pose labels, and 3D object semantic labels for pre-training of the consistency prediction masks used in this system. The recent learnable LiDAR odometry method [6] applies unsupervised learning. The unsupervised loss is based on ICP-like formulation, leveraging also the normal vectors consistency to obtain a confidence map that provides weighting factors to the loss function. The learned confidence map shows higher values for smooth surfaces and lower for trees or vehicles. This system was tested successfully on three different data sets with different LiDARs, but was trained separately on each of these data sets. This points to the problem with entirely learnable LiDAR odometry or SLAM: they may experience problems when confronted with a previously unseen environment.

Learned models also turned out to be effective in solving the topological (appearance-based) localization problem for loop closing [13]. Place recognition is also useful for LiDAR-based SLAM, once a proper data representation is established. An efficient representation of LiDAR range data for appearance-based localization was proposed in the SegMap system [11], in the form of point cloud segments, with attached learned descriptors. SegMap identifies previously visited places matching the local map of segments obtained from the current perception to a global map of segments. The segments are incrementally grown from the LiDAR points accumulated in a dynamic voxel grid. A deep neural network is used to generate the descriptors of segments. Its input is a 3-D binary voxel grid describing the segment, while the descriptor is obtained by taking the activations of the last layer of the model. There are also few attempts to use combined range/intensity place recognition with LiDAR data [15].

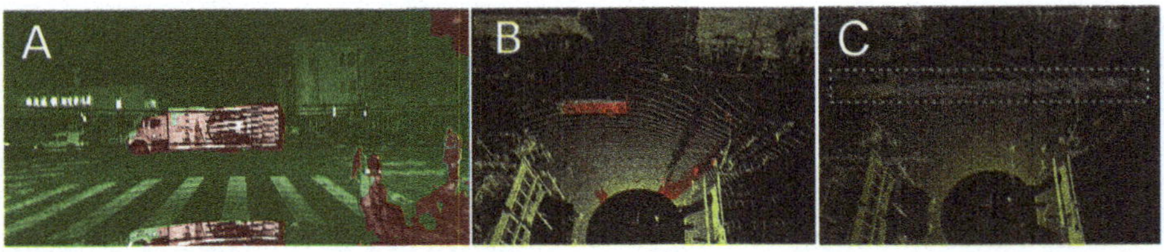

Fig. 4. Results demonstrating the feasibility of using images synthesized from LiDAR data with the deep learning paradigm – motion segmentation on a sequence from Ouster OS1-128: segmented intensity image (A), rejected points (red) visualized in the point cloud (B), and the same scans integrated by PlaneLOAM without removal of non-stationary objects (C). Note that the slowly moving truck left a wake of spurious points in the map (blue rectangle)

Because autonomous vehicles operate in environments with non-stationary objects, often it is required to remove the dynamic clutter or to distinguish between measurements belonging to different classes of objects. Deep learning neural models are currently considered most suitable for semantic segmentation of images and object detection. Recently, we have proposed a method for real-time motion segmentation in LiDAR data without a need to learn from labelled datasets of point clouds, which are laborious in preparation [2]. Our method [24] employs the intensity output of the LiDAR to synthesize camera-like images which are then feed to a neural network for semantics-based motion segmentation. For training we exploit grayscale images from commonly available datasets with class or instance-level point-wise annotations. The approach has been tested with two different neural networks: the Efficient Residual Factorized convNet (ERFNet) [28], a fast supervised deep learning model for semantic segmentation of images, and the Competitive Collaboration [29], a framework for unsupervised learning of optical flow and segmentation of image sequences into the static scene and moving objects. With the Competitive Collaboration we even do not need annotated images, because the network learns the segmentation task from the sequence of grayscale images in unsupervised manner, and then transfers successfully to the LiDAR intensity domain using synthesized images. The obtained segmentation (static and non-static pixels) is projected to the depth data domain and used in PlaneLOAM to mask the 3-D points that should not be used for localization (Fig. 4).

5 Conclusions

This paper briefly surveyed the recent approaches to LiDAR SLAM, particularly in the context of application to autonomous vehicles. We pointed out the open problems in scan registration for modern multi-beam LiDARs, map representation and efficient learning, and proposed new solutions to these problems, namely the direct formulation of scan registration, a map structured in high-level geometric entities, and unsupervised learning for motion segmentation in

LiDAR data. We believe that these proposals define promising directions of future research in SLAM.

References

1. Będkowski, J., Röhling, T., Hoeller, F., Shulz, D., Schneider, F.E.: Benchmark of 6D SLAM (6D simultaneous localization and mapping) algorithms with robotic mobile mapping systems. Found. Comput. Decis. Sci. **42**(3), 275–295 (2017)
2. Behley, J., Garbade, M., Milioto, A., Quenzel, J., Behnke, S., Stachniss, C., Gall, J.: SemanticKITTI: a dataset for semantic scene understanding of LiDAR sequences. In: IEEE/CVF International Conference on Computer Vision, pp. 9296–9306 (2019)
3. Behley, J., Stachniss, C.: Efficient surfel-based SLAM using 3D laser range data in urban environments. In: Robotics: Science and Systems (2018)
4. Cadena, C., Carlone, L., Carrillo, H., Latif, Y., Scaramuzza, D., Neira, J., Reid, I., Leonard, J.: Past, present, and future of simultaneous localization and mapping: toward the robust-perception age. IEEE Trans. Robot. **32**(6), 1309–1332 (2016)
5. Campos, C., Elvira, R., Gómez Rodríguez, J.J., Montiel, J.M.M., Tardós, J.D.: ORB-SLAM3: an accurate open-source library for visual, visual-inertial and multi-map SLAM. arXiv, cs.RO 2007.11898 (2020)
6. Cho, Y., Kim, G., Kim, A.: Unsupervised geometry-aware deep LiDAR odometry. In: IEEE International Conference on Robotics and Automation, pp. 2145–2152 (2020)
7. Ćwian, K., Nowicki, M.R., Nowak, T., Skrzypczyński, P.: Planar features for accurate laser-based 3-D SLAM in urban environments. In: Bartoszewicz, A., et al. (eds.) Advanced, Contemporary Control. AISC, vol. 1196, pp. 941–953. Springer (2020)
8. Deschaud, J.: IMLS-SLAM: Scan-to-model matching based on 3D data. In: IEEE International Conference on Robotics and Automation, pp. 2480–2485 (2018)
9. Della Corte, B., Bogoslavskyi, I., Stachniss, C., Grisetti, G.: A general framework for flexible multi-cue photometric point cloud registration. In: IEEE International Conference on Robotics and Automation, pp. 4969–4976 (2018)
10. Dewan, A., Oliveira, G.L., Burgard, W.: Deep semantic classification for 3D LiDAR data. In: IEEE/RSJ International Conference on Intelligent Robots and Systems, pp. 3544–3549 (2017)
11. Dube, R., Cramariuc, A., Dugas, D., Sommer, H., Dymczyk, M., Nieto, J., Siegwart, R., Cadena, C.: SegMap: segment-based mapping and localization using data-driven descriptors. Int. J. Robot. Res. **39**(2–3), 339–355 (2020)
12. Engel, J., Koltun, V., Cremers, D.: Direct sparse odometry. IEEE Trans. Pattern Anal. Mach. Intell. **40**(3), 611–625 (2018)
13. Facil, J., Olid, D., Montesano, L., Civera, J.: Condition-invariant multi-view place recognition. arXiv, cs.CV 1902.09516 (2019)
14. Geiger, A., Lenz, P., Urtasun, R.: Are we ready for autonomous driving? In: The KITTI Vision Benchmark Suite IEEE/CVF International Conference on Computer Vision and Pattern Recognition, pp. 3354–3361 (2012)
15. Guo, J., Borges, P., Park, C., Gawel, A.: Local descriptor for robust place recognition using LiDAR intensity. IEEE Robot. Autom. Lett. **4**(2), 1470–1477 (2019)
16. Konolige, K.: Sparse sparse bundle adjustment. In: British Machine Vision Conference, pp. 102.1–102.11 (2010)

17. Kümmerle, R., Grisetti, G., Strasdat, H., Konolige, K., Burgard, W.: g2o: a general framework for graph optimization. In: IEEE International Conference on Robotics and Automation, pp. 3607–3613 (2011)
18. Kschischang, F., Frey, B., Loeliger, H.-A.: Factor graphs and the sum-product algorithm. IEEE Trans. Inf. Theory **47**(2), 498–519 (2001)
19. Lang, A., Vora, S., Caesar, H., Zhou, L., Yang, J., Beijbom, O.: PointPillars: fast encoders for object detection from point clouds. arXiv, cs.LG 1812.05784 (2018)
20. Li, Q., Chen, S., Wang, C., Li, X., Wen, C., Cheng, M., Li, J.: LO-net: deep real-time LiDAR odometry. In: IEEE/CVF International Conference on Computer Vision and Pattern Recognition (2019)
21. Li, Y., Ibanez-Guzman, J.: Lidar for autonomous driving: the principles, challenges, and trends for automotive lidar and perception systems. arXiv, cs.RO 2004.08467 (2020)
22. Lu, W., Zhou, Y., Wan, G., Hou ,S., Song, S.: L^3-net: towards learning based LiDAR localization for autonomous driving. In: IEEE/CVF International Conference on Computer Vision and Pattern Recognition, pp. 6382–6391 (2019)
23. Neuhaus, F., Koß, T., Kohnen, R., Paulus, D.: MC2SLAM: real-time inertial lidar odometry using two-scan motion compensation. In: Brox, T., et al. (eds.) Pattern Recognition GCPR 2018. LNCS, vol. 11269, pp. 60–72. Springer (2019)
24. Nowak, T., Ćwian, K., Skrzypczyński, P.: Cross-modal transfer learning for segmentation of non-stationary objects using LiDAR intensity data. In: IEEE International Conference on Robotics and Automation, Workshop on Sensing, Estimating and Understanding the Dynamic World (2020)
25. Nowicki, M.R.: Spatiotemporal calibration of camera and 3D laser scanner. IEEE Robot. Autom. Lett. **5**(4), 6451–6458 (2020)
26. Qi, C., Su, H., Mo, K., Guibas, L.J.: PointNet: deep learning on point sets for 3D classification and segmentation. In: IEEE/CVF International Conference on Computer Vision and Pattern Recognition, pp. 652–660 (2017)
27. Pomerleau, F., Colas, F., Siegwart, R.: A review of point cloud registration algorithms for mobile robotics. Found. Trends Robot. **4**(1), 1–104 (2015)
28. Romera, F., Alvarez, J., Bergasa, L., Arroyo, R.: ERFNet: efficient residual factorized ConvNet for real-time semantic segmentation. IEEE Trans. Intell. Transp. Syst. **19**(1), 263–272 (2018)
29. Ranjan, A., Jampani, V., Balles, L., Kim, K., Sun, D., Wulff, J., Black, M.: Competitive collaboration: joint unsupervised learning of depth, camera motion, optical flow and motion segmentation. In: IEEE/CVF International Conference on Computer Vision and Pattern Recognition, pp. 12232–12241 (2019)
30. Salas-Moreno, R.F., Glocken, B., Kelly, P.H.J., Davison, A.J.: Dense planar SLAM. In: IEEE International Symposium on Mixed and Augmented Reality, Munich, pp. 157–164 (2014)
31. Segal, A., Haehnel, D., Thrun, S.: Generalized-ICP. In: Robotics: Science and Systems (2009)
32. Serafin, J., Grisetti, G.: NICP: Dense normal based point cloud registration. In: IEEE/RSJ International Conference on Intelligent Robots and Systems, pp. 742–749 (2015)
33. Shan, T., Englot, B.: LeGO-LOAM: lightweight and ground-optimized LiDAR odometry and mapping on variable terrain. In: IEEE/RSJ International Conference on Intelligent Robots and Systems, pp. 4758–4765 (2018)
34. Shin, Y., Park, Y.S., Kim, A.: Direct visual SLAM using sparse depth for camera-LiDAR system. In: IEEE International Conference on Robotics and Automation, pp. 5144–5151 (2018)

35. Skrzypczyński P.: Mobile robot localization: where we are and what are the challenges? In: Szewczyk, R., et al. (eds.) Automation 2017. Innovations in Automation, Robotics and Measurement Techniques. AISC, vol. 550, pp. 249–267. Springer (2017)
36. Steinbrücker, F., Sturm, J., Cremers, D.: Real-time visual odometry from dense RGB-D images. In: International Conference on Computer Vision, Workshop on Live Dense Reconstruction with Moving Cameras, pp. 719–722 (2011)
37. Velas, M., Spanel, M., Hradis, M., Herout, A.: CNN for IMU assisted odometry estimation using Velodyne LiDAR. In: International Conference on Autonomous Robotic Systems and Computing, pp. 71–77 (2018)
38. Strasdat, H., Montiel, J., Davison, A.: Real-time mococular SLAM: why filter? In: IEEE International Conference on Robotics and Automation, pp. 2657–2664 (2010)
39. Wietrzykowski, J., Skrzypczyński, P.: PlaneLoc: probabilistic global localization in 3-D using local planar features. Robot. Auton. Syst. **113**(3), 160–173 (2019)
40. Weingarten, J., Siegwart, R.: 3D SLAM using planar segments. In: IEEE/RSJ International Conference on Intelligent Robots and Systems, pp. 3062–3067 (2006)
41. Zhang, J., Singh, S.: Low-drift and real-time LiDAR odometry and mapping. Auton. Robot. **41**(2), 401–416 (2017)
42. Zhou, Y., Tuzel, O.: VoxelNet: end-to-end learning for point cloud based 3D object detection. In: IEEE/CVF International Conference on Computer Vision and Pattern Recognition, pp. 4490–4499 (2018)

Convolutional Neural Network-Based Local Obstacle Avoidance for a Mobile Robot

Milena Molska and Dominik Belter(✉)

Institute of Robotics and Machine Intelligence, Poznan University of Technology, ul. Piotrowo 3A, 60-965 Poznan, Poland
dominik.belter@put.poznan.pl
http://lrm.put.poznan.pl/

Abstract. This paper presents collision avoidance and local motion planning modules for a mobile robot equipped with a depth camera. In this paper, we identify some limitations of the existing neural controller, and then we propose the extensions which improve the behavior of the robot. We show that the knowledge about control history is crucial to efficiently avoid collisions with the obstacles if the robot is equipped with a narrow field of view camera. We propose the architecture which utilizes CNN-based neural modules to plan the local motion of the robot. Finally, we provide the results of the experimental verification on the real robot.

Keywords: Deep learning · Visual perception · Motion planning

1 Introduction

Mobile robots operating in an indoor environment should be capable to detect and avoid obstacles while moving to the goal pose. To this end, depth sensors like laser range finders or RGB-D cameras are used. The traditional approach to motion planning and collision avoidance requires a model of the environment. The robot is localized while moving in the environment and then the measurements are accumulated in the map. The obtained map is used to find a collision-free path to the goal pose. Local planners like Dynamic Window Approach (DWA) [4] use a local model of the environment to define the motion of the robot in the dynamic environment. However, tuning the parameters of the planner requires the knowledge about dynamical properties of the robot and very often is time-consuming.

In this research, we explore the problem of motion planning for a robot equipped with an RGB-D camera (Fig. 1). The robot should use the depth images

This work was supported by the National Centre for Research and Development (NBR) through project LIDER/33/0176/L-8/16/NCBR/2017. We would like to thank Lei Tai, Shaohua Li and Ming Liu for sharing raw data from their experiments.

R. Szewczyk et al. (Eds.): AUTOMATION 2021, AISC 1390, pp. 262–271, 2021.
https://doi.org/10.1007/978-3-030-74893-7_25

Fig. 1. Labbot robot equipped with RGB-D Orbec Astra Pro camera and Slamtec Mapper for 2D localization and mapping system during experiments in the indoor environment

only to find the local path to the goal configuration. In the proposed approach, we avoid tuning the parameters of the planner. Instead, we are looking for a method that can be trained from the demonstration. The robot is controlled by the human operator and avoids obstacles during the demonstration stage. We expect that after the training, the robot will mimic the presented behavior.

In this paper, we extend the solution given in [13]. We identify situations that are difficult for the mobile robot in the indoor environment. Then, we propose the modification to the architecture of the neural network and the training procedure to deal with situations that are problematic for the original solution [13]. We show that the history of commands provided to the input of the neural network allows the robot to deal with the camera's narrow field of view and makes the motion of the robot more natural without explicitly using the map.

2 Related Work

Traditional methods for motion planning of mobile platforms can be represented by A*, RRT* [8], and Informed RRT* [5]. They utilize various data structures like grids, graphs, or trees. However, to plan the motion of the robot a model of the environment is required. Data from the range sensors are used to obtain various environment models like OctoMap [7], mesh model [12], or sufels [15]. However, this approach requires careful selection of the mapping, motion planning algorithm, and tuning various parameters.

Recently, convolutional neural networks became popular in the task of autonomous driving [1]. In [13] the neural network takes depth images and performs classification tasks. The neural network classifies the input image and produces five output commands (one for going forward, two for turning left, and two for turning right). To train the neural network a set of depth images and corresponding control commands are collected when the human operator controls the robot. During inference, the neural network mimics the behavior of the human operator and the robot avoids obstacles. The obtained method can be used to replace the Dynamic Window Approach (DWA) local planner for collision avoidance [4]. Similar behavior can be obtained when the reinforcement learning algorithm is used to train the robot [14]. In this case, the training of

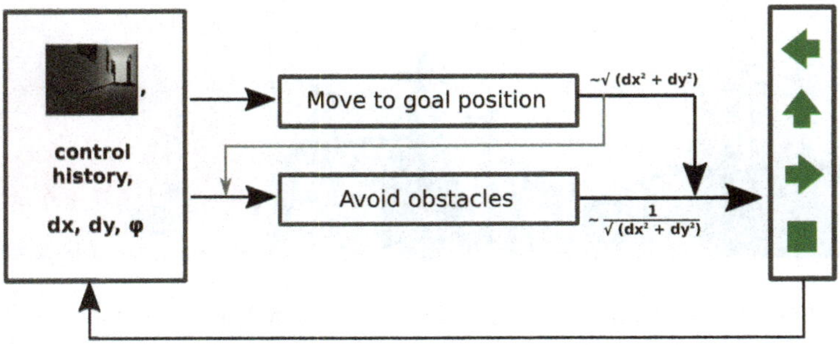

Fig. 2. Subsumption architecture utilized to integrate the control system of the robot. The neural network used for obstacles avoidance is presented in Fig. 5 and described in Sect. 3.1. The block responsible for moving the robot to the global pose is described in Sect. 3.2.

the controller is performed in the simulation and the robot utilizes data from a laser range sensor.

An example collision avoidance method for the off-road robot was presented in the early stage of the development of the convolutional neural networks [9]. The steering angle for the robot is produced by the CNN according to the RGB image provided to the input [9]. Similarly to our approach the reference training samples are collected when the robot is controlled by a human operator. The neural network can be also trained using the reference motion planner [11]. The robot used a laser scanner to build a 2D map and the neural network trained on the grid-based Dijkstra and DWA local planner for collision avoidance and following the path to the goal at the same time. The goal position is provided to the neural network directly through the fully-connected layers. In [3], the neural network provides information about the affordance indicators (angle between the car's heading and the tangent of the road, distance to the preceding cars, distance to the left lanes markings, etc.). Then, the controller uses the affordance indicators to compute the steering commands. The controller can be obtained by training in the simulation using deep reinforcement learning algorithm [10] on the synthetic images or the real street views [6].

3 Navigation with Deep Neural Network

All methods presented in this paper are implemented on our Labbot mobile robot. The robot is presented in Fig. 1. The robot is equipped with Orbec Astra Pro for 3D environment perception. It also uses Slamtec Mapper M1M1 ToF Laser Scanner Kit for 2D localization and mapping. The robot also utilizes the Robot Operating System (ROS) to control the differential drive mobile base and integrate all modules into a common framework.

The architecture of the proposed system is presented in Fig. 2. It consists of two main blocks. The first module is responsible for guiding the robot to the goal position. The second module allows the robot to avoid obstacles. We utilize

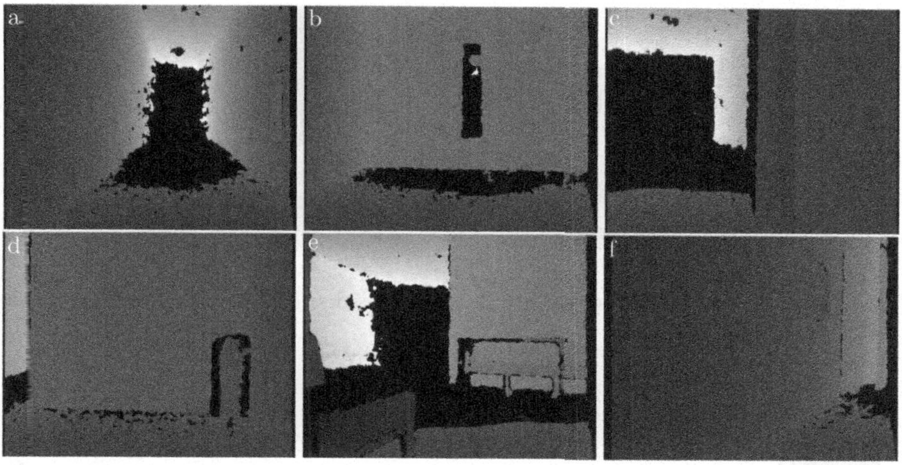

Fig. 3. Example depth images from the training set and the reference output command: go straight (a, b), turn left (c, d), turn right (e, f)

a subsumption architecture [2] to merge the output from both modules but the priority of the modules depends on the distance from the goal pose. To define the history of the commands, we use the outputs produced by both modules. Our experience shows that this approach stabilizes the motion of the robot, and as a result, the robot prefers detours that avoid obstacles. At the same time, the robot moves closer to the goal pose.

3.1 Collision Avoidance

In the first step, we used the neural network architecture and the images from [13] to train the neural network. Then, the neural network was used to control our mobile robot in various indoor environments. We identified the situations when the obtained control system does not work. In some cases, the robot tries to get through the door with a small glass window in the middle. The robot misclassifies this region with the corridor and moves forward. The robot also has problems with corners. When the robot gets closer to the corner it observes the wall on one side and turns to the other side to avoid the obstacle. Then, it observes the second wall and turns in the opposite direction. The procedure repeats until the robot hits the wall. In the following sections, we show how to deal with these situations.

Most of the collisions are caused by the narrow field of view of the camera. The robot does not see the obstacles on the sides. The robot also loses information about the obstacle when the obstacle is no longer in the camera view. As a result, it hits objects which are no longer observed by the camera. Moreover, it is not easy to determine the control command for a single input depth image. Example depth images and corresponding output commands are presented in Fig. 3. The images can be very similar but they produce completely different

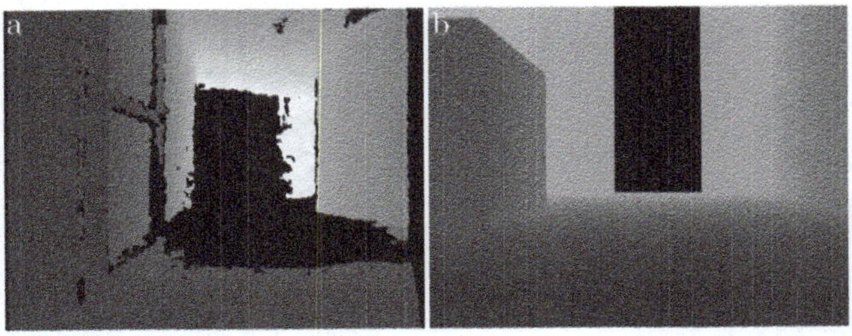

Fig. 4. Difference between the real (a) and synthetic depth image (b) used to train the neural network

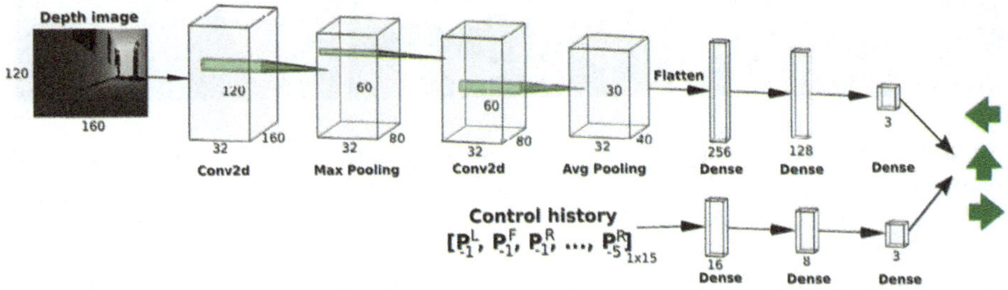

Fig. 5. The proposed architecture of the neural network for collision avoidance using depth image and control history on the input and producing control commands

commands according to the training samples given by the human operator. This comes from the fact that the human operator remembers the previous configuration of the obstacles and is aware of the dimensions of the robot. It also means that the previously observed obstacles are sometimes more important than the current view.

Taking into account our experience with the baseline method, we propose to provide information about the previous motions of the robot to the input of the neural network. Our goal is to keep the neural network simple so instead of providing the set of previous images, we provide the set of previous control commands (a set of all previous output probabilities from the neural network). For the same reason, we provide the control commands directly to the input of the fully connected layers instead of using LSTM neural network architecture. The number of previous commands is set to 5 which means that we utilize additional 15 float values on the input of the neural network. We also claim that some errors are caused by an insufficient number of training examples. The environment used to train the robot in [13] differs from the environment where our robot operates. Thus, we collected additional samples in our building and in the Gazebo simulator to randomize the data in the training set. The example images collected on our robot and in the simulator are presented in Fig. 4.

Finally, we propose a new architecture of the neural network dedicated to avoiding obstacles presented in Fig. 5. The depth image from the RGB-D camera

Table 1. Results of the comparison of three architectures of the neural networks in the collision avoidance task

Architecture	Training		Validation		Test	
	Loss	Acc [%]	Loss	Acc [%]	Loss	Acc [%]
Baseline [13]	0.0224	93.90	0.0833	76.97	0.0844	77.71
Modified	0.0252	95.63	0.0583	91.22	0.0380	94.04
With history	0.0239	96.50	0.0370	93.85	0.0275	95.08

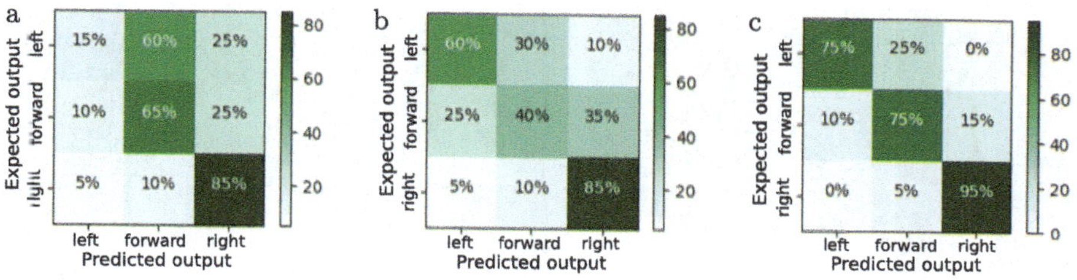

Fig. 6. Comparison between confusion matrices obtained for three various CNN architectures: reference method [13] (a), modified architecture (b) and the proposed architecture (c)

is scaled to the size 120 × 160 px. Then, the convolutional layers of the neural network are used to extract the features from the images. The history of control commands is provided to the separate input of the neural network. The fully connected layers classify the images and history of control commands and provide three commands: move forward, turn left and turn right.

3.2 Motion to the Goal Pose

To generate the motion of the robot to the goal pose we utilize a simple PID controller. However, our goal is to build a fully NN-based controller. Thus, we decided to collect input and output data from the PID controller and train the neural network. The input to the neural network is the position and orientation error. The neural network is based on the fully-connected layers only and returns four motion commands (turn left, turn right, move forward, and stop). The accuracy of the trained neural network on the testing dataset is 99.98%. Note that this neural network mimics the PID controller only and does not work better than the original algorithm.

4 Results

We trained the proposed neural network on the generated dataset consisting of 2257 samples. The results of the training are presented in Table 1. The accuracies for the training and testing datasets are 96.50% and 95.08%, respectively.

Table 2. Obtained recall, precision, and accuracy for the methods verified during the experiments

Architecture	Recall			Precision			Accuracy [%]
	Left	Straight	Right	Left	Straight	Right	
Basic	15.0	65.0	85.0	50.0	48.15	62.96	55.0
Modified	60.0	40.0	85.0	66.67	50.0	65.38	61.67
With history	75.0	75.0	95.0	88.23	71.42	86.36	81.67

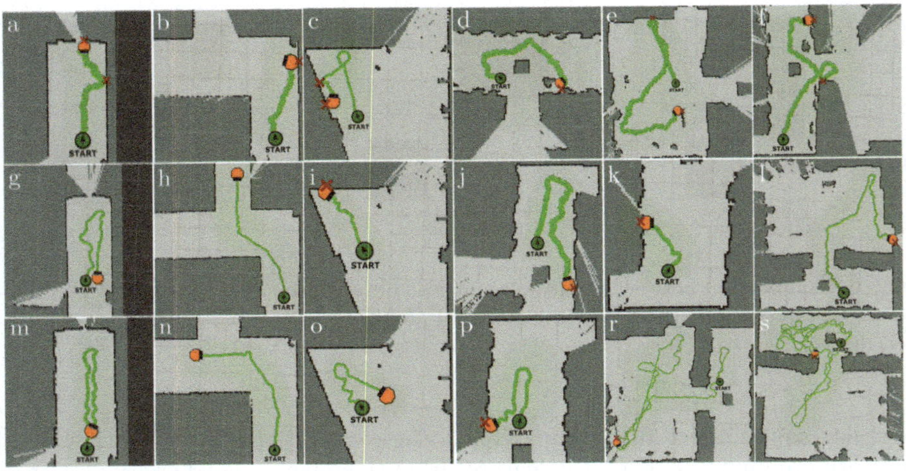

Fig. 7. Results of the collision avoidance module on the real robot for the reference architecture (a–f), modified architecture (g–l), and the proposed architecture (m–s)

Additionally, we train the baseline neural network from [13] and a modified version of the neural network. In the modified version of the neural network, we removed convolution and max-pooling layers. We also reduced the number of outputs from five to three, which is sufficient in our case. The training results presented in Table 1 show that the simplification improves the results on the testing dataset. It means that the neural network can better generalize. Further improvement is possible when the history of commands is added. It confirms our hypothesis that the previous commands allow the neural network to better explain the data and better mimic the behavior of the human operator.

In the next experiment, we compare the same architectures but in this case, we use data generated in the simulation only. For the reference architecture [13] we merge the classes which generate turning left into one class and right turns into the second class so for each method we have three classes on the output. To show the results, we generated the confusion matrix (Fig. 6). In Fig. 6 we can observe that the original and the modified architecture of the neural network misclassifies the turn left and turn right commands. For the modified method the misclassification for the completely different classes is reduced but this behavior

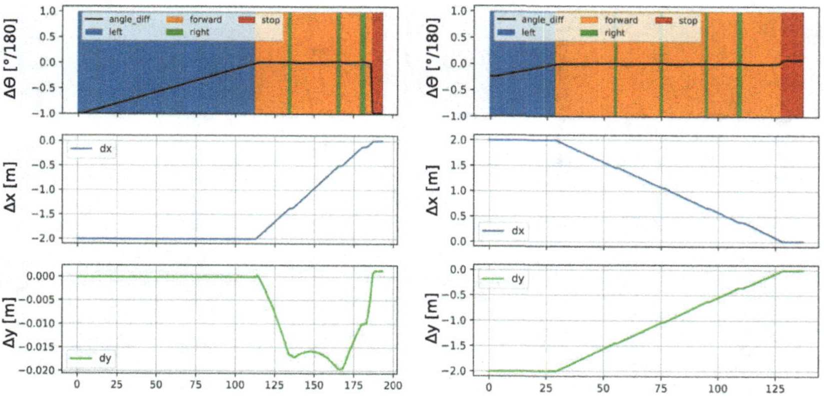

Fig. 8. State of the robot and the controller during moving to the goal position. Each column represents a separate experiment.

is eliminated only for the neural network with a history of commands (Fig. 6c). Recall, precision, and the average accuracy obtained for this experiment are presented in Table 2. Again, the proposed method obtained the best results.

Finally, we compare the results on the real robot equipped with the RGB-D camera and SLAM system[1]. The neural networks work on-line on the robot and control the motion of the robot. The results are presented in Fig. 7. In the top row of Fig. 7 we present the trajectories obtained for the baseline method [13]. During each experiment, the robot hits the obstacles. The modified architecture of the neural network improves the results (middle row in Fig. 7). For the modified architecture of the neural network, the robot moves longer without hitting obstacles. However, it still has problems with the corners. The proposed solution allows avoiding obstacles for a significantly longer time and the trajectory of the robot is more smooth. In Fig. 7 we also show failure cases, when the robot hits the legs of the chairs or the legs of the table (Fig. 7r and Fig. 7s) or a tabletop which is not visible in the camera (Fig. 7p). However, this comes from the fact that the robot was trained to avoid obstacles like walls and tested in a different environment that contains narrow obstacles.

In the next experiments, we verified the proposed architecture which generates motion to the goal pose. The example control commands and pose errors for two experiments are presented in Fig. 8. The neural network and the priority strategy forces the robot to rotate first and then moves the robot to the goal position. If required, the robot avoids obstacles, and the error for the position and orientation increases but in the end, the pose error is reduced to zero.

We also compared various approaches to generate the motion of the robot to the goal pose. In Fig. 9a–c we show the results of the end-to-end neural network which was trained to avoid obstacles and following to the goal pose at the same time. The following to the goal pose is a dominant behavior and the robot very

[1] short video from experiments is available at https://youtu.be/6ssP9L52_WQ.

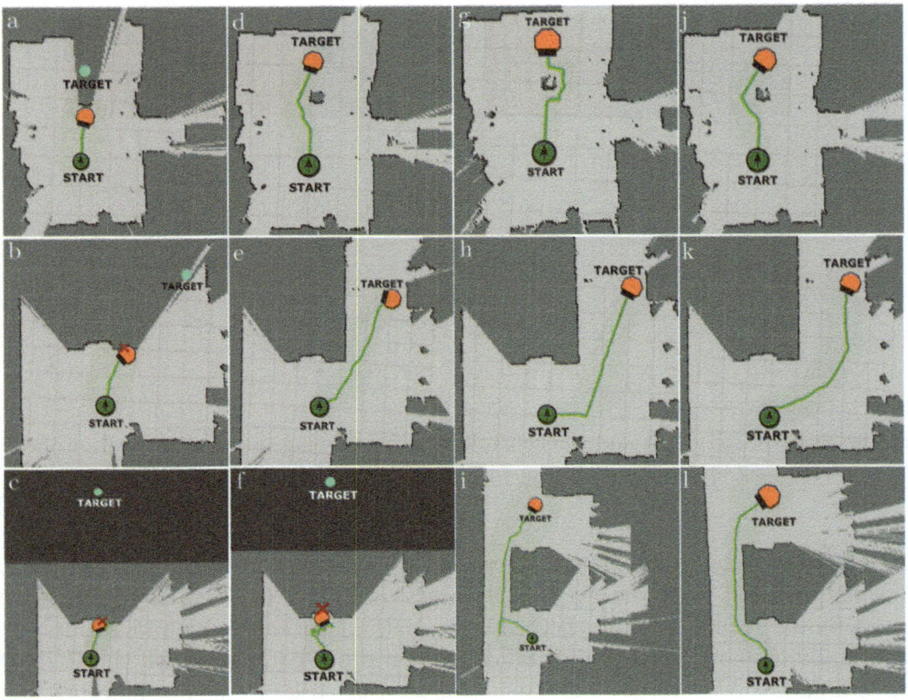

Fig. 9. Results for the experiments for the end-to-end neural network (a, b, c), basic subsumption architecture (d, e, f), subsumption architecture with changing priority (g, h, i), subsumption architecture with the noise (j, k, l)

quickly hits the obstacle. Then, we tested the proposed architecture without priorities (Fig. 9d–f) and with the priorities depending on the distance to the goal pose (Fig. 9g–i). For the last method, we can observe some unnecessary motions. The proposed architecture which also mixes the history from both neural networks provides a smooth path to the goal pose (Fig. 9j–l).

5 Conclusions and Future Work

In this paper, we present a neural network-based controller for mobile robots. We extend the method proposed in [13]. We improve the performance of the collision avoidance controller for the differential-drive mobile robot by including the history of control commands on the input of the neural network. We also extended the training dataset by examples from the real robot and the robot in the simulated environment. We also propose the architecture which allows reaching the goal pose and avoiding obstacles. The properties of the proposed system are verified on the real robot in the various indoor scenarios. In the future, we are going to work on neural-based motion planning with knowledge about the global and local map of the environment.

References

1. Bojarski, M., Testa, D.D., Dworakowski, D., Firner, B., Flepp, B., Goyal, P., Jackel, L.D., Monfort, M., Muller, U., Zhang, J., Zhang, X., Zhao, J., Zieba, K.: End to end learning for self-driving cars. CoRR **abs/1604.07316** (2016). http://arxiv.org/abs/1604.07316
2. Brooks, R.: A robust layered control system for a mobile robot. IEEE J. Robot. Autom. **2**(1), 14–23 (1986)
3. Chen, C., Seff, A., Kornhauser, A., Xiao, J.: DeepDriving: learning affordance for direct perception in autonomous driving. In: 2015 IEEE International Conference on Computer Vision (ICCV), pp. 2722–2730 (2015)
4. Fox, D., Burgard, W., Thrun, S.: The dynamic window approach to collision avoidance. IEEE Robot. Autom. Mag. **4**(1), 23–33 (1997)
5. Gammell, J.D., Srinivasa, S.S., Barfoot, T.D.: Informed RRT*: optimal sampling-based path planning focused via direct sampling of an admissible ellipsoidal heuristic. In: 2014 IEEE/RSJ International Conference on Intelligent Robots and Systems, pp. 2997–3004 (2014)
6. Hermann, K.M., Malinowski, M., Mirowski, P., Banki-Horvath, A., Anderson, K., Hadsell, R.: Learning to follow directions in street view. In: Proceedings of the AAAI Conference on Artificial Intelligence, vol. 34, pp. 11773–11781 (2020)
7. Hornung, A., Wurm, K.M., Bennewitz, M., Stachniss, C., Burgard, W.: OctoMap: an efficient probabilistic 3D mapping framework based on octrees. Auton. Robot. **34**(3), 189–206 (2013)
8. Karaman, S., Frazzoli, E.: Sampling-based algorithms for optimal motion planning. Int. J. Robot. Res. **30**(7), 846–894 (2011)
9. LeCun, Y., Muller, U., Ben, J., Cosatto, E., Flepp, B.: Off-road obstacle avoidance through end-to-end learning. In: Proceedings of the 18th International Conference on Neural Information Processing Systems (NIPS 2005), pp. 739–746 (2005)
10. Mirowski, P., Pascanu, R., Viola, F., Soyer, H., Ballard, A., Banino, A., Denil, M., Goroshin, R., Sifre, L., Kavukcuoglu, K., Kumaran, D., Hadsell, R.: Learning to navigate in complex environments. In: International Conference on Learning Representations (2017)
11. Pfeiffer, M., Schaeuble, M., Nieto, J., Siegwart, R., Cadena, C.: From perception to decision: a data-driven approach to end-to-end motion planning for autonomous ground robots. In: 2017 IEEE International Conference on Robotics and Automation (ICRA), pp. 1527–1533 (2017)
12. Steinbruecker, F., Sturm, J., Cremers, D.: Volumetric 3D mapping in real-time on a CPU. In: Proceedings of the IEEE International Conference on Robotics and Automation, pp. 2021–2028 (2014)
13. Tai, L., Li, S., Liu, M.: A deep-network solution towards model-less obstacle avoidance. In: 2016 IEEE/RSJ International Conference on Intelligent Robots and Systems (IROS), pp. 2759–2764 (2016)
14. Tai, L., Paolo, G., Liu, M.: Virtual-to-real deep reinforcement learning: continuous control of mobile robots for mapless navigation. In: 2017 IEEE/RSJ International Conference on Intelligent Robots and Systems (IROS), pp. 31–36 (2017)
15. Whelan, T., Salas-Moreno, R.F., Glocker, B., Davison, A.J., Leutenegger, S.: ElasticFusion: real-time dense SLAM and light source estimation. Int. J. Robot. Res. **35**(14), 1697–1716 (2016)

Benchmark and Analysis of Path Planning Algorithms of "ROS MoveIt!" for Pick and Place Task in Tomato Harvesting

Filip Jędrzejczyk(✉), Jarosław Bajer, Grzegorz Gawdzik, Jakub Główka, and Agnieszka Sprońska

ŁUKASIEWICZ Research Network – Industrial Research Institute for Automation and Measurements PIAP, Warsaw, Poland
{filip.jedrzejczyk,jaroslaw.bajer,grzegorz.gawdzik, jakub.glowka,agnieszka.spronska}@piap.lukasiewicz.gov.pl

Abstract. This article describes a benchmark of motion planners available in ROS MoveIt!. The benchmark exercise has been done with planners' configured to be optimal for a planning scene imitating conditions in a greenhouse during the real process of picking a tomato with a robotic arm. It provides an overview of the available motion planners, analysis of benchmark results and further validation of the most suitable planners with the use of a real 6 DOF robotic arm. Presented experimental results include time of planning, time of generated plan simplification, time of plan execution and time of processing in a cycle composed of start-goal-start poses movement. Obtained results allow to compare efficiency, repeatability and usefulness of particular planners in a use case of robotic tomato harvesting or in similar pick and place tasks.

Keywords: Manipulator · Robotic arm · Path planning · Pick and place · MoveIt! · OMPL · CHOMP · STOMP

1 Introduction

ROS MoveIt! [12] is a widely used tool for motion planning and control of multiple types of robotic arms and end effectors. It provides interfaces for utilizing planning algorithms from open source trajectory planning libraries of the Open Motion Planning Library (OMPL), Covariant Hamiltonian Optimization for Motion Planning (CHOMP) and Stochastic Trajectory Optimization for Motion Planning (STOMP). State of the art algorithms from these open source libraries can, in specific configuration, be used in real life applications for many pick and place scenarios, what can be seen in numerous solutions already introduced by the robotics community [13].

The specific use case related to the tomato harvesting with a robotic arm has been defined in a course of technical collaboration between the authors

R. Szewczyk et al. (Eds.): AUTOMATION 2021, AISC 1390, pp. 272–284, 2021.
https://doi.org/10.1007/978-3-030-74893-7_26

of the research described in this paper and the company that develops and sells the autonomous tomato harvester for passive greenhouses. To increase the efficiency of the harvesting tasks the robot system required optimization of the path planning algorithms used by the manipulator for picking the tomato and placing it in the box. The required time for execution of one harvesting cycle needed to be decreased as much as possible and at the same time the collision avoidance had to be improved.

In order to address these problems and select the optimal planning algorithm and its configuration, a benchmark exercise of available motion planners from the ROS MoveIt! framework has been performed. Its results allowed to select the most suitable planners for the specific scenario of pick and place task in the given use case and possibly for similar pick and place tasks.

Evaluation of planners' parameters and overview of the times of execution can be helpful in selecting the most suitable planner. Solving the problem of harvesting a tomato can be done in several ways: using inverse kinematic solvers, parallel planning (multi-threaded applications) or by dividing single movement into several stages (the shorter the distance, the faster the planning and execution of the movement - parallel execution of the movement and seeking solution for the next stage). This paper describes the benchmark of the planners in a simple sequential cycle for a robotic manipulator performing a pick and place task as a set of subsequent steps, without any additional instruments.

The use of end effector (gripper) was beyond the scope of this research and therefore it has not been included.

Results of the benchmark analysis are presented in the following chapters.

2 Benchmark Description

2.1 PC Specification

All computation described in the study was done using a single PC with the following parameters:

- CPU: Intel® Core™ i7-3520M CPU @ 2.90 GHz
- Graphics: Intel® Ivybridge Mobile
- Hard drive: 512 GB SSD SATA
- Memory: 8 GB DDR3
- OS: Ubuntu 18.04
- ROS: Melodic

2.2 Evaluation Parameters

The following evaluation metrics were defined:

- **Planing time (s):** time of seeking solution.
- **Simplification time (s):** time of simplification of the solution found.

- **Total planning time (s):** time needed to generate, interpolate, simplify and process the solution.
  total_planning_time = planning_time + interpolation_time + simplification_time + process_time
- **Execution time (s):** time required for the actual movement execution, based on the previously generated plan.
- **Total time (s):** entire time required for moving the robot from the current to a requested pose.
  total_time = total_planing_time + execution_time

2.3 Test Scene

The test scene utilized in this benchmark analysis was chosen based on the experience gathered during the technical collaboration between the authors of the research and the autonomous tomato harvester producer. It reflects the most difficult case that the planning algorithms might encounter while performing the real tomato harvesting task. In this task the robot is about to harvest the tomatoes from the plant located in the corner of the growing area.

The scene, which is a 3D virtual environment used for motion planning, consists of the robot, a platform that the robot is mounted on and a number of obstacles (spacial collision objects) that the robot has to omit. The robot, that is used in the research is a robotic arm with 6 DoF (degrees of freedom). Initial pose, in which the robot begins each cycle is shown in Fig. 1a. Pose of the robot shall be understood as a set of values, describing position in radians for each robotic arm's joint. In the initial pose the robotic arm is positioned over the box for tomatoes, close to the robot basis. The target (the tomato to pick) is located on the opposite site, in the middle of the cubic object, which represents the obstacles all around the tomato (being simplified reflection of the real environment with branches and leaves of tomato bushes). Target pose, that allows to pick the selected tomato using an end effector, is shown in Fig. 1b. The vertical plane simulates another row of tomato bushes or a wall of a greenhouse, whereas the horizontal plane reflects a surface the robot is placed on.

2.4 Approach

All available planners in MoveIt! were benchmarked in order to find the one which is the most suitable in the tomato harvesting use case and other similar pick and place tasks.

The scenario assumed performing one full harvesting cycle that consists of the series of subsequent steps:

- Robot starts at a pose over the box for tomatoes (initial pose).
- Robot plans and executes path from the initial pose to the tomato (target pose, the end effector is parallel to the ground).
- Robot plans and executes path from the tomato (target pose) to the box for tomatoes (initial pose).

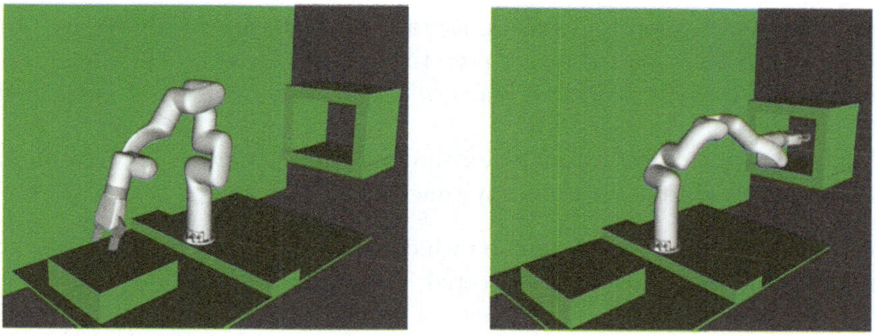

(a) Initial pose. **(b)** Target pose

Fig. 1. Initial and target pose.

As mentioned previously, the step of picking the tomato (with the use of end effector) was not relevant for this research.

Before advancing to the actual benchmark analysis, the optimal setup of each planner for the respective scene and scenario had to be defined.

According to the authors experience planners available in ROS MoveIt! generate initial version of the plan quickly, but usually these are not the most efficient and valid paths. Because of that, after the planning is finished the path is post-processed in order to make it smoother, faster to execute and collision free. With additional cost of time spent on post-processing, most of the planners can be tuned to output smooth, fast and reliable plans. However, a balance should be kept. The amount of time saved on the plan execution cannot be less than the amount of time spent on the optimization of that plan. Otherwise, it is better to switch to another planning algorithm or just to execute a slower but valid and already possessed path. While most of the considered planners are able to finish the planning task after finding a feasible solution, some of them (LazyPRMstar, PRM, PRMstar) fully use the allowed timeframe for optimizing the solution, even if an achievable plan has been found earlier [7]. To eliminate this time consuming behaviour from the benchmark and to filter out the planners which are not worth to be considered for introduction in a real life solution, an upper bound limit for the duration of the plan generation had to be assumed.

It has to be noted that there are also other ways to obtain the most suitable solution: creating intermediate and approach points [2], reversing the plan for the way back, using IK solver [1] or parallel planning, but these were beyond the scope of this benchmark exercise and therefore not considered.

Multiple ways of configuration of every planner were checked. Each one was tested by running in simulation, in order to collect the evaluation parameters and check if they are valid. Each setup was run at least 10 times, which is enough to compare, reason and select the best one. At first, to be optimal, the setup of the planner had to be valid, i.e. meet the following requirements for every run:

- generate a plan from the initial pose to the target pose within 5 s,
- generate a plan from the target pose to the initial pose within 5 s,
- execute both of the generated plans with tolerance of joint positions equal to 0.0001 rad.
- both paths avoid obstacles (there is no physical interaction between any collision object and the robot at any time).

From the list of valid setups the one which achieved the shortest average total time of the harvesting cycle was selected.

The analysis consisted of two stages. The first one was comparison of evaluation parameters of the optimally configured planners, which were collected in 10 simulation runs for the respective scene and scenario. Basing on average results 3 most suitable planners were to be selected for each of 6 categories:

- Total time to tomato
- Total time to the tomato box
- Total time of a full cycle
- Execution time to tomato
- Execution time to the tomato box
- Execution time of a full cycle

The second stage of the benchmark was the validation of the selected planners in a lab setting, using a real 6 DoF robotic arm. Data were collected for 10 runs per each planner. Evaluation of the validation results allowed to select the most suitable planners to be used for a 6 DoF manipulator in the tomato harvesting use case as well as potentially in other similar pick and place use cases.

3 Available Planners

Algorithms available in MoveIt! are not deterministic as it is the case of for example the Search-Based Planning Library (SBPL) [9], but they are suitable to be considered for a use case where the environment contains multiple collision objects around the target.

The OMPL [5] is flexible and applicable to a wide variety of robotic systems. As a consequence to this flexibility, the library does not explicitly represent the geometry of the workspace or the robot operating in it. There is no default collision checker in the OMPL. This allows the library to operate with relatively few assumptions, enabling it to plan for a huge number of systems while remaining compact and portable. A conscious effort is also made to limit the number of external dependencies necessary to compile and execute the code in the OMPL. The majority of the third party code utilized in the library comes from Boost [11], which allows the OMPL to function in most major operating systems. Every OMPL planner has its own set of parameters [3] that help to obtain the optimal path [6]. These parameters can be tuned to make the robotic arm take the shortest path or to avoid obstacles in a certain distance.

Stochastic Trajectory Optimization for Motion Planning (STOMP) [4] generates scattered trajectories in order to explore the space around the initial guess.

The initial guess [8] is a joint space trajectory, which can be created in three different ways: linear interpolation, cubic interpolation and a minimal control cost trajectory. Then it combines them into an updated trajectory after applying the cost function.

Covariant Hamiltonian Optimization for Motion Planning (CHOMP) [14] works similarly to STOMP, it is based on initial trajectory optimization. When initial guess meets a collision object, CHOMP tries to avoid obstacle by discretizing the trajectory into a set of waypoints in space and uses finite differencing to obtain velocities and accelerations on joints [10].

4 Benchmark

4.1 OMPL Planners

Optimal configuration of the planners obtained during the experiment setup process is shown in the Planner/Parameters table.

The following OMPL planners did not find the valid path to the target within 5s timeframe: RRT, RRTstar, TRRT, LBTRRT, LazyPRM, SPARStwo, SPARS. Average simplification time value was similar for all planners and took around 0.1s.

4.2 CHOMP/STOMP Planners

It was not possible to obtain any valid solution for the respective scene and scenario with the use of CHOMP and STOMP.

At the time of performing this research, bugs in MoveIt! STOMP API made it impossible to carry out the requested planning and execution actions. There was a problem with initial pose of planning. The start point deviated from the current robot state so it was impossible to perform the benchmark in a form described in this paper. Therefore, STOMP had to be excluded from the analysis.

According to relevant documentation [10], CHOMP generates paths that are not guaranteed to be collision free. Taking into account the assumed constraints, the results obtained from path generation in CHOMP would not prove adequate to be considered as a recommended solution. Multiple runs exposed, that after creating an initial trajectory CHOMP couldn't find a valid solution during it's optimization phase. Although it is not designed as an independent planner for complex scenes, it might be considered to be used as a tool for smooth trajectories computed by other planning algorithms.

5 Results Analysis

We have selected the top 3 planners for each category. Since some of the planners got to the final 3 in multiple categories, the evaluation gave us in total 6 planners to be validated on the real robot, shown in the Category/Planners table.

Planner	Parameters
SBL	Range = 2.0, longest valid segment fraction = 0.005
EST	longest valid segment fraction = 0.005, Range = 2.0, goal bias = 0.05
LBKPIECE	Longest valid segment fraction = 0.005 range = 2.0, border fraction = 0.9, min valid path fraction = 0.8
BKPIECE	Range = 0.2, border fraction = 0.9, failed expansion score factor = 0.5, min valid path fraction = 0.8 longest valid segment fraction = 0.005
KPIECE	Range = 0.2, goal bias = 0.05, border fraction = 0.9, longest valid segment fraction = 0.005 failed expansion score factor = 0.5, min valid path fraction = 0.8
RRTConnect	Range = 2.0, longest valid segment fraction = 0.005
PRM	Max nearest neighbours = 8, planning time = 2s, longest valid segment fraction = 0.015
PRMstar	Longest valid segment fraction = 0.015, planning time = 2s
FMT	nearest k = 1, cache cc = 1, heuristics = 1, extended fmt = 1, num of samples = 1000, radius multiplier = 1.1, longest valid segment fraction = 0.015
BFMT	Num of samples = 1000, radius multiplier = 1.0, nearest k = 1, balanced = 1, optimality = 1, heuristics = 1, cache cc = 1, extended fmt = 1, longest valid segment fraction = 0.015
PDST	Longest valid segment fraction = 0.015
STRIDE	use projected distance = 0, degree = 16, min degree = 12, max pts per leaf = 6, estimated dimension = 0.0, range = 0.2, goal bias = 0.05, max degree = 18, longest valid segment fraction = 0.015, min valid path fraction = 0.2
BiTRRT	Range = 0.0, temp change factor = 0.1, init temp= 100, frountier threshold = 0.0, cost threshold = inf, longest valid segment fraction = 0.015 frountier node ratio = 0.1
BiEST	Range = 0.0, longest valid segment fraction = 0.015
ProjEST	Range = 2.0, goal bias = 0.05, longest valid segment fraction = 0.015
LazyPRMstar	longest valid segment fraction = 0.015, planning time = 3s

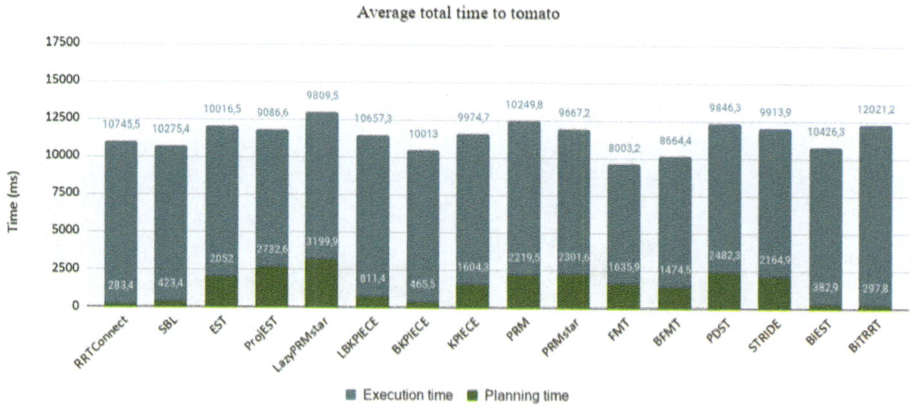

Fig. 2. Average of planning times and execution times to tomato in simulation.

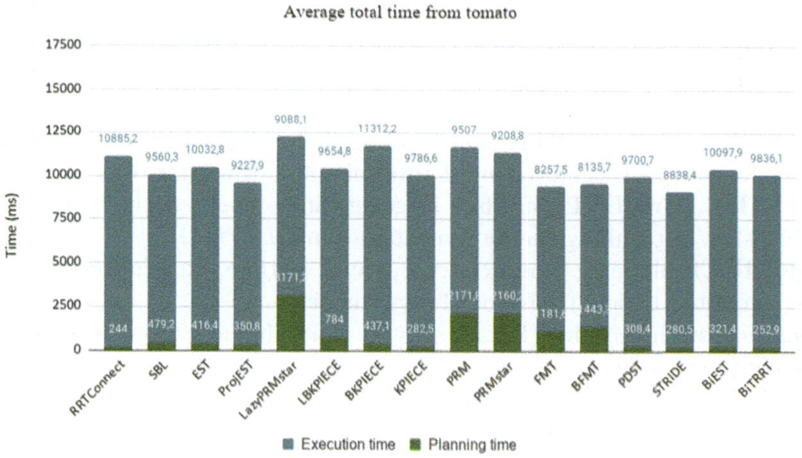

Fig. 3. Average of planning times and execution times from tomato in simulation.

Category	Planners
Average execution time to tomato	FMT, BFMT, ProjEST
Average execution time to the tomato box	BFMT, FMT, STRIDE
Average execution time both directions	FMT, BFMT, ProjEST
Average total time to tomato	FMT, BFMT, BKPIECE
Average total time to the tomato box	STRIDE, FMT, ProjEST
Average total time both directions	FMT, BFMT, SBL

FMT, BFMT planners are good in creating smooth and optimal path for both directions with average total planning around 1.5s. The times required to generate a plan to tomato for these two planners are one of the longest from

Fig. 4. Average of planning times and execution times of the full cycle in simulation.

among the OMPL planners. Nevertheless, the short times of execution allowed to obtain short total planning times (Fig. 4). Therefore, these planners can be considered the fastest in simulation.

SBL has high repeatability, which resulted in the short total time of the full cycle. Moreover, SBL achieved short total planning times for both directions.

STRIDE, ProjEST were fast in generating smooth paths to the tomato box, but slow in planning the movement to tomato. (Fig. 2).

BKPIECE planning time is short, but average time of execution takes longer as the computed trajectory is not always smooth.

Analysis of the results (Fig. 3) showed that planning from the tomato to the tomato box is less time consuming. This is caused by the higher number of complex collisions in the surrounding of the tomato than around the tomato box. It can be concluded that the presence of obstacles at the beginning - close to the pose the planners start planning from - has less impact on the planning time than the number of obstacles at the end - in the surroundings of the target. As it was mentioned previously, there are other ways to perform the harvesting cycle faster. If only the scenario assumes that the poses at the tomato box are always the same it is possible to plan only once (to the tomato), and then to reverse the plan to return to the box. Application of such solution might be attractive because only one planning would be required and the time needed for reversing the plan on modern computers is negligible - much less then the time required for another planning. The obtained results also allow to improve the trick. Generating the plan might be done only for the movement from tomato to the tomato box. Then the plan may be reversed and in that form used for reaching the tomato. This approach might in fact be considered the fastest way of planning.

6 Validation of Results

Performing the harvesting cycle with a real robot has revealed issues that were not present in the simulation. Some planners, e.g. STRIDE, often exceeded the joint speed limits what resulted in stops during the movement execution.

SBL proved to be the most universal planner with short times of planning and movement execution. It achieved the shortest total time of the full cycle (Fig. 7), the shortest total time to the tomato (Fig. 5) and one of the shortest total times from the tomato (Fig. 6).

Execution times of **FMT** and **BFMT** were noticeably shorter than for other planners. However, their longer plan generation times made them slower than SBL.

The results achieved by **STRIDE** and **ProjEST** revealed the difference in task complexity between planning to and from the tomato. In the case of STRIDE the average total time to tomato was longer by 2.5s. Nevertheless, STRIDE turned out to be the fastest in the second part of the cycle, i.e. movement from tomato to the tomato box.

BKPIECE appeared to be an exception that proves the rule. It was the only planner that has achieved similar times for both directions of the movement.

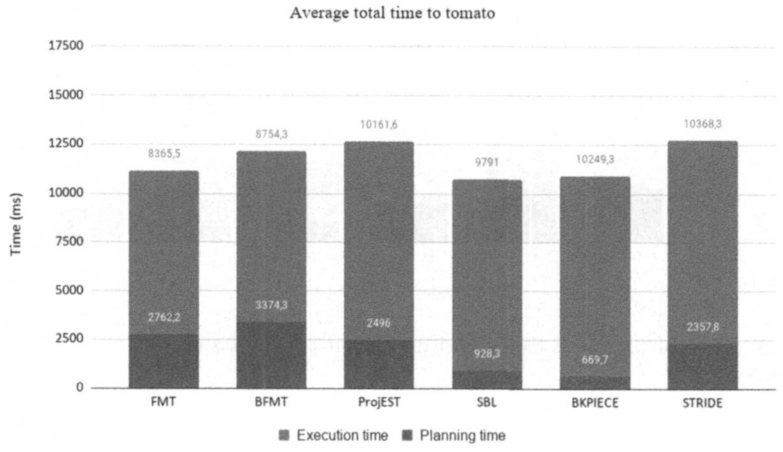

Fig. 5. Average of planning times and execution times to tomato validated on manipulator.

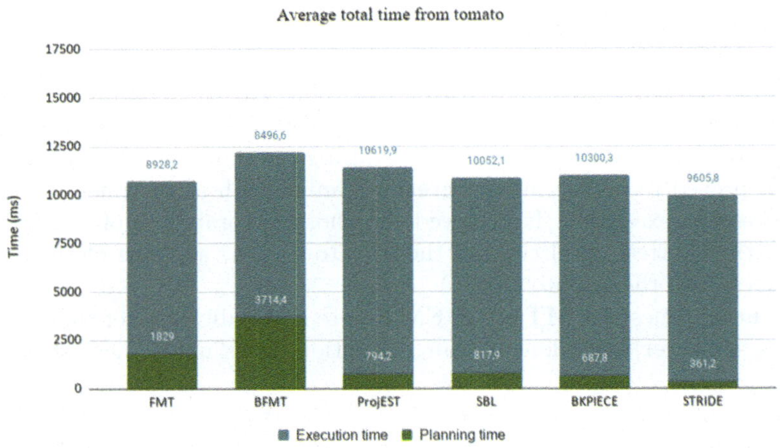

Fig. 6. Average of planning times and execution times from tomato validated on manipulator.

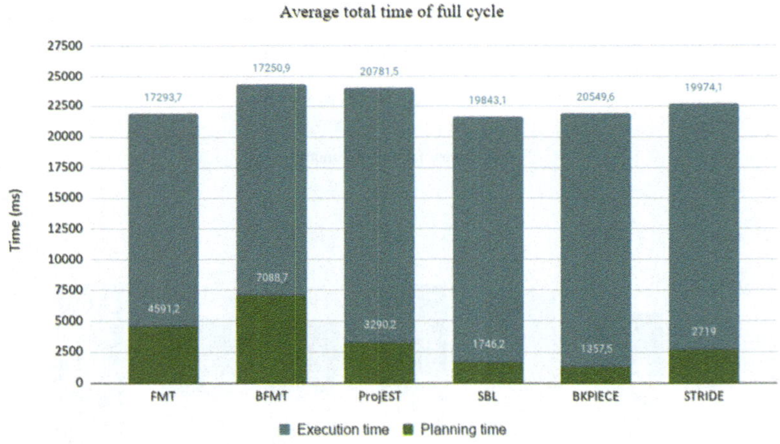

Fig. 7. Average of planning times and execution times of the full cycle validated on manipulator.

7 Summary

The described benchmark of different trajectory planners available in ROS MoveIt! has been performed for an environment that reflects the conditions of the tomato harvesting use case. Constraints imposed by this specific task, that apply to each planner, were adopted and explained. The main purpose of this work was to find the optimal setup of each planner, compare their efficiency and select the ones that could be the most suitable for tomato harvesting scenario or similar pick and place tasks.

The benchmark exercise showed that some planners require more time for preparing the plan, but they are able to output optimal, smooth and fast paths. The stage of searching for optimal setups showed that only multiple runs can prove if the planner is configured optimally and can be used safely - in a lot of cases several promising attempts were followed by an attempt that ignored the obstacles or exceeded robotic arm joint limits what led to an execution fail. For nine out of 25 planners it was not possible to find the setup producing valid solutions and thus they were excluded from being considered for the use case. As the first half of the cycle was planning the path to the tomato, which can be deemed a more demanding task due to the number of obstacles, it took a noticeably longer time. The planners for which it was not possible to find a valid setup still might be considered for use cases with no collision environments or to be used with the reversing trajectory approach.

Results obtained in the validation showed differences in duration of the movement execution between the simulation and the real robot. This may be related to the imperfection of the robot model used in the simulation. Execution of the movement on the real arm has taken slightly longer than in the simulation - the speed and acceleration limits of the robotic arm joints have made the smoothness of the generated path to play a more important role. Due to that, SBL has proven to be faster than FMT and BFMT, which were the fastest planners in the simulation. Therefore, it can be concluded that the SBL planner is the most suitable for the task of tomato harvesting performed sequentially. However, during the collaboration over the development of the tomato harvester it was found by the authors that the fastest way of performing the harvesting cycle was to utilize the reversing trajectory approach. If solely this approach was used, the planner to be selected as the fastest would be STRIDE, which has achieved the shortest total time for the movement from tomato to the tomato box.

Acknowledgements. The project leading to this results has received funding from the European Union's Horizon 2020 research and innovation programme under grant agreement No 779967.

References

1. Beeson, P., Ames, B.: TRAC-IK: an open-source library for improved solving of generic inverse kinematics (2015)
2. Beeson, P., Hart, S., Gee, S.: Cartesian motion planning & task programming with craftsman (2016)
3. Chitta, S., Hershberger, D., Pooley, A., Coleman, D., Gorner, M., Lautman, M., Suarez, F.: https://ros-planning.github.io/moveit_tutorials/doc/ompl_interface/ompl_interface_tutorial.html#ompl-settings. Accessed 1 July 2020
4. Kalakrishnan, M., Chitta, S., Theodorou, E.A., Pastor, P., Schaal, S.: STOMP: stochastic trajectory optimization for motion planning (2011)
5. Kavraki Lab at Rice University: Open motion planning library: A primer (2020)
6. Kavraki Lab at Rice University (2020). https://ompl.kavrakilab.org/optimalplanning.html. Accessed 1 July 2020

7. Kavraki Lab at Rice University (2020). https://ompl.kavrakilab.org/planners.html. Accessed 1 July 2020
8. Magyar, B., Tsiogkas, N., Brito, B., Patel, M., Lane, D., Wang, S.: Guided stochastic optimization for motion planning. Front. Robot. AI **6**, 105 (2019)
9. Open Source Robotics Foundation (2013). http://wiki.ros.org/sbpl. Accessed 1 July 1 2020
10. Ratliff, N., Zucker, M., Bagnell, J.A., Srinivasa, S.: Chomp: gradient optimization techniques for efficient motion planning. In: IEEE International Conference on Robotics and Automation (2009)
11. Schling, B.: The boost C++ libraries (2011)
12. Sucan, I.A., Chitta, S.: https://moveit.ros.org/. Accessed 1 July 2020
13. Sucan, I.A., Chitta, S.: https://moveit.ros.org/robots/. Accessed 1 July 2020
14. Zucker, M., Ratliff, N., Dragan, A.D., Pivtoraiko, M., Klingensmith, M., Dellin, C.M., Bagnell, J.A., Srinivasa, S.S.: CHOMP: covariant hamiltonian optimization for motion planning (2013)

Preliminary Prototype of a 4-Unit Arm-Z Hyper-redundant Modular Manipulator

Ela Zawidzka[1], Wojciech Kiński[2], and Machi Zawidzki[1(✉)]

[1] Institute of Fundamental Technological Research, Polish Academy of Sciences, Warsaw, Poland
zawidzki@MIT.edu

[2] Industrial Research Institute for Automation and Measurements PIAP, Warsaw, Poland

Abstract. Arm-Z is a concept of a robotic manipulator comprised of linearly joined congruent modules with possibility of relative twist. The advantages of Arm-Z are: economization (mass-production) and robustness (modules which failed can be replaced, also if some fail the system can perform certain tasks). Non-intuitive and difficult control are the disadvantages of Arm-Z. It has been introduced over six years ago. The theretofore research focused on the control of the virtual Arm-Z manipulator through the relative twists of its constituent units. This paper documents the fabrication and testing of the preliminary physical prototype of a simple Arm-Z, which has been built using relatively low-tech approach. The modules have been assembled from 3D-printed elements using PET-G filament. Each module is equipped with a transmission mechanism allowing to pre-set the spin of its twist (either positive/right or negative/left). Although the presented system is very simple it demonstrates certain meaningful kinematic actions and sets clear paths for the future research in the area of Extremely Modular Systems.

Keywords: Extremely Modular System · Arm-Z · Hyper-redundant manipulator · Rapid prototyping · Low-tech

1 Introduction

The snake mechanism is a redundant system which, however, makes snakes supremely adapted for various habitats. Analogously, in irregular environments, bio-inspired trunk-like robots in some cases perform better than more conventional wheeled, tracked and legged forms of robotic mobility. Research on snake robots has been conducted for several decades. Snake locomotion has been studied empirically already in 1940s [1]. 50 years later, the first mathematical model has been developed and trunk-like locomotors and manipulators have been proposed in [2]. Trunk-like manipulators have very specific movement characteristics which makes them especially useful in situations where classic robotic arms cannot be used. Such manipulators can transport working heads in very complex,

R. Szewczyk et al. (Eds.): AUTOMATION 2021, AISC 1390, pp. 285–294, 2021.
https://doi.org/10.1007/978-3-030-74893-7_27

narrow, confined, cluttered spaces which are inaccessible by any other means. Reaching range can be of order of tens of meters and actions can be accomplished in hazardous environments. They can perform tasks like welding, cleaning, visual inspection and any other which can be realized by the head mounted on the manipulator tips. It should be stressed that the existing solutions in this field are very limited. Only a few companies offer such type of manipulators, and these are rather experimental (Fig. 1).

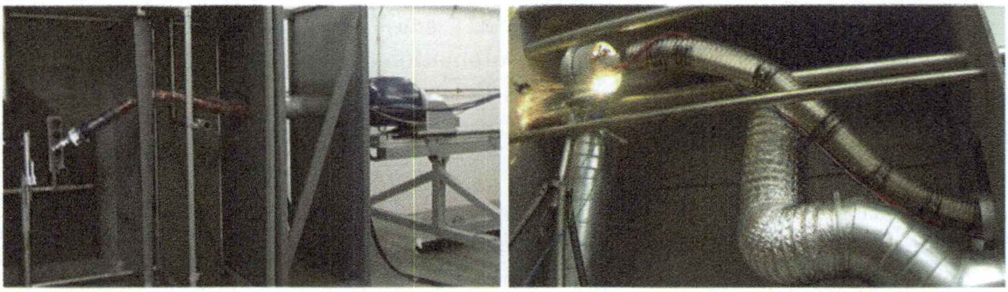

Fig. 1. Two examples of existing trunk-like robotic manipulators. OC Robotics arms during laser in-bore welding (http://www.ocrobotics.com).

Additionally, trunk/snake-like manipulators may have relatively large number of degrees of freedom (DOF) (however, there are exceptions). This contrasts with the classic robotic arms with low number of DOFs which are widely used in the industry. In the case of Arm-Z, the number of DOF is equal to the total number of the connected modules. Besides the possibility of performing complex actions, large number of DOF introduces the possibility of high fault tolerance and robustness. It aligns well with the concept of hyper-redundant manipulators (HRM, [3]). Their advantages come at a price of notoriously difficult control due to their highly non-linear nature. E.g. the inverse kinematics problem for a typical industrial robotic arm can be solved relatively easily [4] and thus the control of such arms is straightforward. On the other hand, the control of a *bionic trunk* requires sophisticated artificial intelligence methods [5–7]. However, in return, one can profit from all the important advantages of trunk-like HRMs. For further details on this class of robotic manipulators see [8].

The concept of extreme modularization is relatively new, and less explored. In general, research on structural optimization of modular structures seems to be relatively sparse. Ref. [9] considers layout optimization of a modular bridge, where the design variables are effectively discrete and encode the choice from a predefined set of local module topologies. Simultaneous optimization of a family of a few vehicle structures at the level of the shared basic beam components is discussed in [10].

2 The Concept of Arm-Z

Complex three-dimensional tubular forms can be built with relatively simple congruent units as presented in [11]. A parametric design system comprised of congruent modules was introduced in [12]. Figure 2 shows examples of mathematical knots constructed with only one type of module.

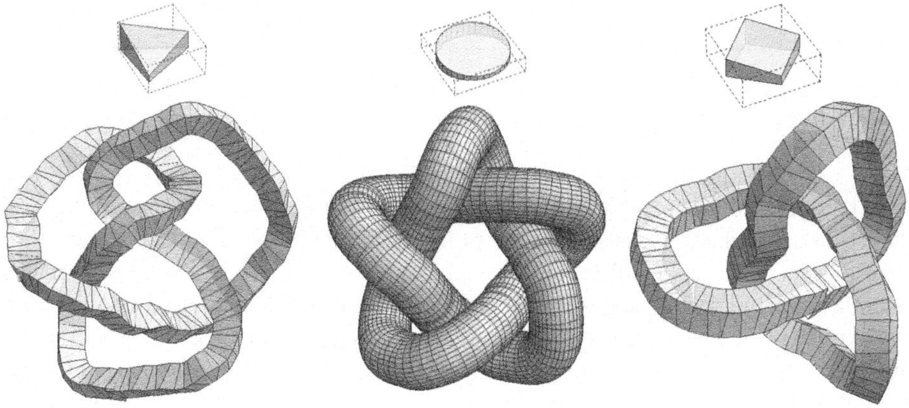

Fig. 2. Three knots assembled with different basic units; from the left: Figure eight (4_1) @ the number of sides: $k = 3$, Cinquefoil (5_1) @ $k = 36$, Trefoil (3_1) @ $k = 4$.

Arm-Z is a conceptual manipulator composed of congruent modules each having one degree of freedom (1-DOF) - a relative twist introduced in 2014 [13]. Simple changes of these twists result in emerging behavior of the entire Arm-Z allowing it to perform complex movements. The relative twists can be continuous or discrete. Arm-Z modules are geometrical objects analog to a sector of circular torus. Each module is defined by parameters: number of sides k, size r, offset d, and angle ζ between bottom (**B**) and top (**T**) surfaces. Figure 3 illustrates these parameters.

The overall shape of Arm-Z is a result of: the geometrical parameters of each module, number of modules and their relative rotation. Figure 4 shows two Arm-Z modules at six consecutive dihedral twists from 0 to π.

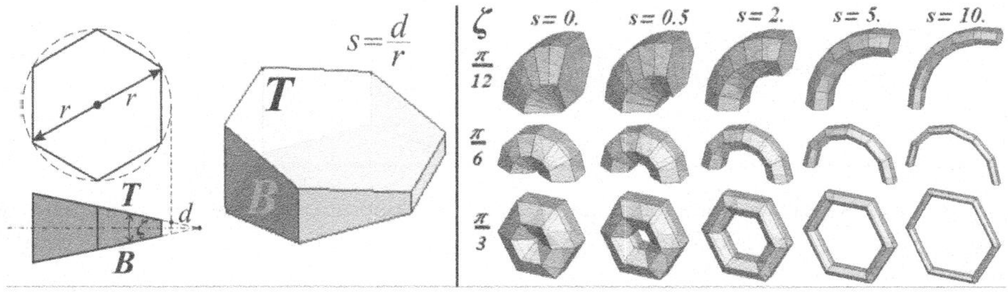

Fig. 3. On the left: visualization of basic hexagonal Arm-Z module defined by parameters: r, d, ζ and $k = 6$ and its axonometric view. On the right: examples of simple assembly of modules with various combinations of their geometrical parameters.

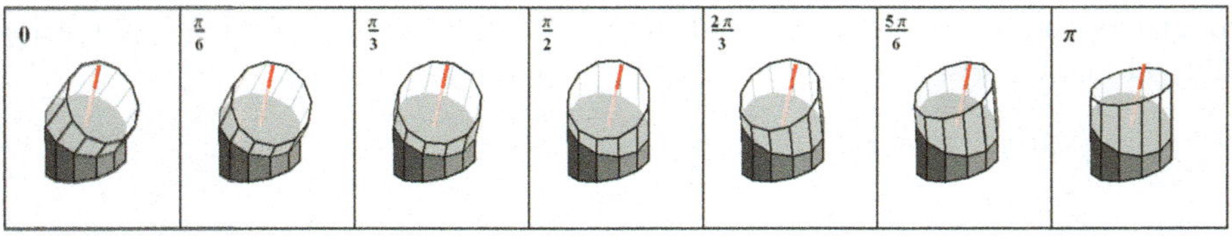

Fig. 4. The relative twist of Arm-Z modules. The first (bottom) module is fixed and indicated in gray. The axis of rotation for the subsequent module is shown in red.

The operational space of a discrete Arm-Z (based on polygonal modules and their dihedral relative twists) forms a point-cloud. Figures 5 and 6 illustrate four examples of an Arm-Z manipulator composed of hexagonal modules with six discrete relative twists and their respective operational spaces. The sizes and "thicknesses" of these point-clouds grow with the number of modules. Moreover, the geometrical parameters of each module influence the properties of the operational space. In some scenarios the task for the manipulator can be localized but high precision would be required (Fig. 5). In other scenarios the task can be less localized at the expense of the tip precision (Fig. 6).

For continuous Arm-Z the operational space forms a complex 3D surface which has voids. This means that in general, the tip of Arm-Z can not reach all points in 3D space within its range, but can reach specific points with certain accuracy (error).

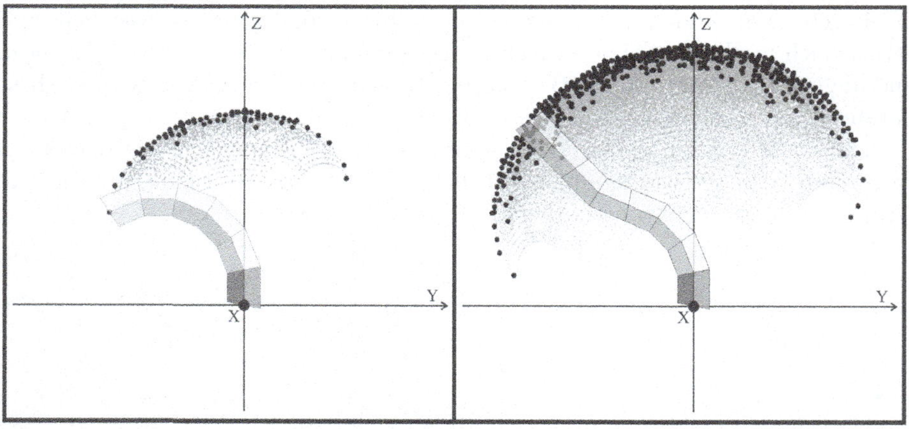

Fig. 5. A cross-section along Y axis through the operational space of a hexagonal Arm-Z with $\zeta = \frac{\pi}{8}$ and six dihedral twists. Manipulators are superimposed in a certain configurations for illustration. On the left: 5-unit manipulator, on the right: 7-unit manipulator. Black and gray dots indicate: the points on the section through the point-cloud, and the side-view of the point-cloud respectively. The fixed Arm-Z module is shown in darker gray.

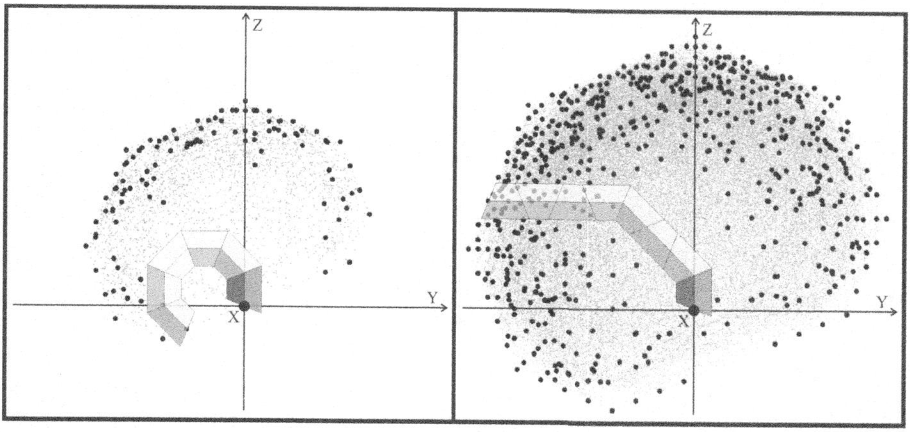

Fig. 6. A cross-section along Y axis through the operational space of a hexagonal Arm-Z with $\zeta = \frac{\pi}{4}$ and six dihedral twists. On the left: 5-unit manipulator, on the right: 7-unit manipulator.

3 Physical Demonstrator

This section presents the making of preliminary physical prototype of a simplified Arm-Z. It has been built using relatively low-tech approach. The framework has been intentionally reduced to a meaningful minimum so the prototype could be completed with simple and inexpensive means.

3.1 (Quasi) Control

In principle, the control of Arm-Z is implemented by relative rotations of individual modules. It is a computationally challenging and highly non-linear problem. For an example of application of Particle Swarm Optimization for controlling Arm-Z to perform desired actions such as bending in a narrow slot, placing the tip of a 12-unit Arm-Z manipulator into a desired position, assuring that environmental constraints are not violated during the transition see [14].

In this project, however, the control of the manipulator is reduced to minimum. During one experiment each module performs a rotation of π about its main axis. The initial position for each module are limited to discrete values of either 0 or π with corresponding notation: 0 and 1, respectively. E.g. a sequence of 0's means that consecutive modules are positioned at angle 0 relative to each other which results in a torus sector shape of Arm-Z. When all the units execute the same relative rotation during the experiment, the term *control* seems not yet relevant. Figure 7 shows a 4-unit Arm-Z in all possible initial configuration with all possible transitions to the after-half-turn configurations. One of the simplest ways of controlling Arm-Z is to set each module to turn either left or right during one experiment.

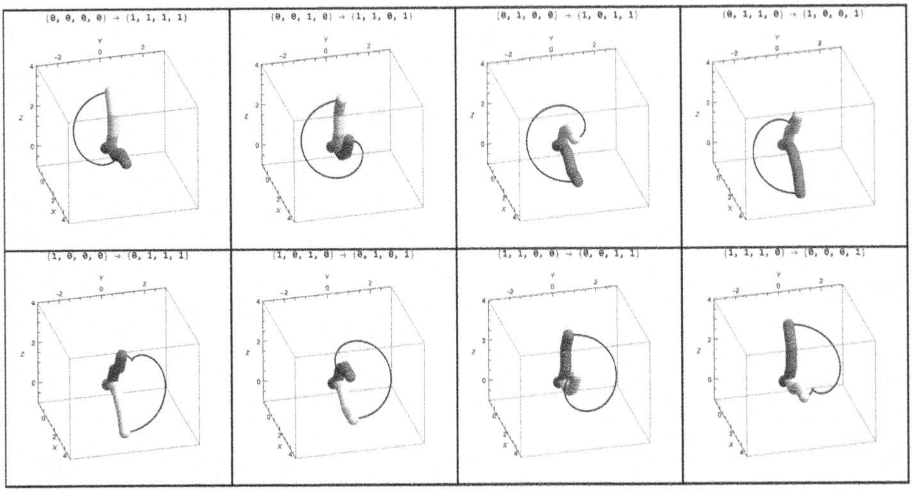

Fig. 7. When all units perform the same relative twist, there is only way of transitioning from the initial to final configurations. The trajectory of the Arm-Z tip is shown as a black line. The initial and final configurations of Arm-Z are shown in light and dark gray, respectively. For each case the transition from initial to final configuration is given on the top.

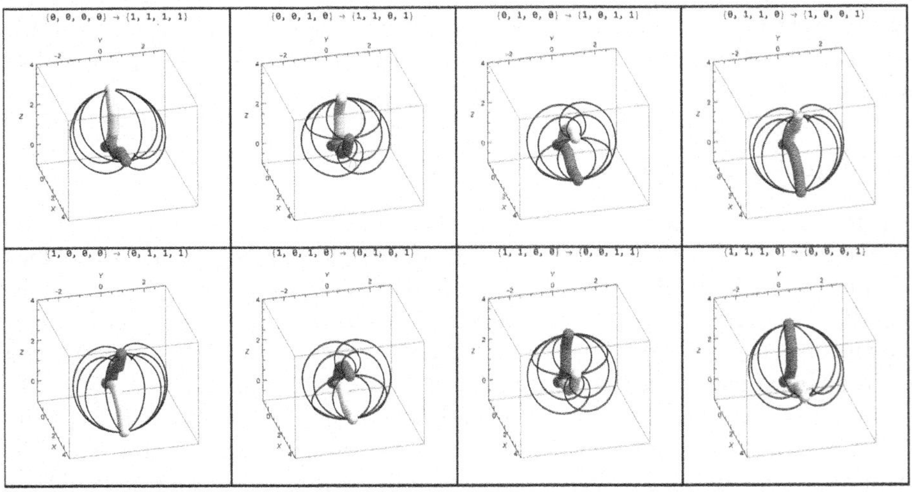

Fig. 8. When all units can turn either left or right, the number of possible trajectories of the Arm-Z tip increases substantially. Color convention as in Fig. 7.

Figure 8 shows the same 4-unit Arm-Z in all possible initial configuration with all possible transitions to the after-half-turn configurations using all possible left/right setups.

3.2 Construction of a Module

The module has been designed in SolidWorks™, the elements of the module have been prepared for printing in Simplify™ and 3D-printed with PET-G (polyethylene terephthalate glycol-modified) on a Dragon3D FDM (fused deposition modeling) industrial machine. Printing of one set took approximately 18 h with 0.8mm nozzle and 0.4096 mm layer height. The internal parts such as gears and levers have been printed with higher precision (0.4 mm nozzle with 0.22 layer) using PLA (polylactic acid). The shafts were made of drawn and ground steel rods and were embedded in rolling barrings in order to reduce friction. The weight of one unit is approximately 700 g. Figure 9 shows the CAD model with visible gears and fabricated physical module.

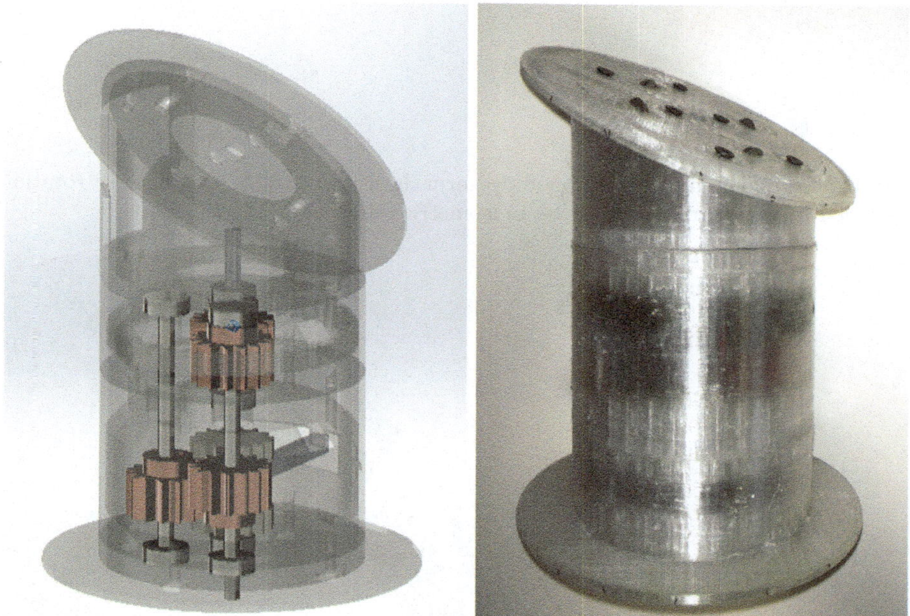

Fig. 9. CAD model and a photograph of assembled physical module of Arm-Z.

3.3 Transmission

Arm-Z manipulator is comprised of linearly joined congruent modules. The first module in the sequence is fastened to a solid base. For simplicity, the drive is applied directly to the first module and transferred to subsequent units by internal gears as shown in Fig. 10. Each module is equipped with a set of cylindrical and bevel gears with straight teeth with involute profile (for connecting the modules).

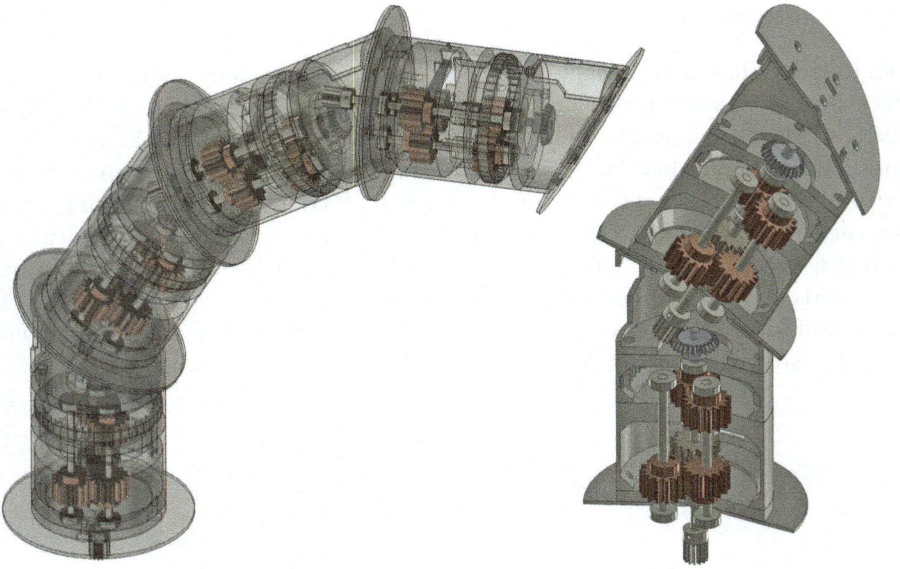

Fig. 10. Chain of four congruent modules form the 4-unit Arm-Z. The drive for rotation is carried by the gear system shown in isometric cutaway on the right

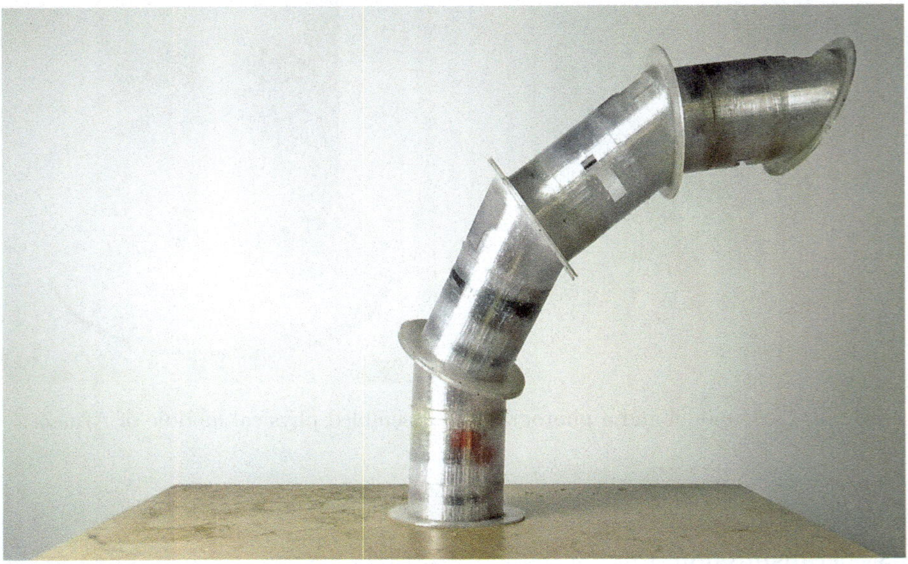

Fig. 11. A photograph of the experimental setup with completed 4-unit operational simplified Arm-Z prototype.

It is possible to change the direction of rotation of a module by selecting one of two gears using a lever: along with or opposite to the direction of rotation of the previous module (drive). The drive shafts are mounted on rolling bearings.

The ratio between the drive shaft and the output shaft is 1:1. The connection between modules is realized by a bearing block. Additionally, each module is reinforced by compressing with pins (threaded rods) passing through its entire height. At the connection between modules gears with bevel teeth are used.

4 Conclusions

In Arm-Z the assembly of congruent units allow for creation of free-form shapes. Therefore it belongs the family of concepts called *Extremely Modular Systems* (EMS for short) introduced in [15]. The two main advantages of all EMSs are: economization (due to mass-production), and robustness (the modules which failed can be easily replaced, also if some fail, the system may still perform some desired tasks).

This paper presents the first approach to making a somewhat operational prototype of Arm-Z. Since the construction of modules is very simple and far from optimal, the weight of the modules is approximately 700 g. Moreover, the gears have been 3D-printed which results in relatively low precision and hardness resulting in high friction and relatively low efficiency. These were the main two factors which limited the size of this prototype to modest four units (Fig. 11).

Nevertheless, this manipulator can be pre-set to perform 2^4 different trajectories of its tip. This project gave invaluable insight to the nature and thus problems of this *Extremely Modular System* and defined the next steps in our research on Arm-Z:

1. Structural optimization of the module and mechanical optimization of the gear system. This will allow to increase the number of units and even to install a head of certain weight on the tip of the manipulator.
2. Optimization of the fabrication process specifically for fused deposition modeling (FDM) [16] which is a well established technology in inexpensive rapid fabrication.
3. Development of more advanced gear switching mechanism allowing for: neutral gear, varying of the angular velocities of the modules, etc.
4. Development of a system to control the rotation of individual modules based on Arduino, step-motors and WiFi technology.

Acknowledgements. This research is a part of the project titled *Arm-Z: an extremely modular hyperredundant low-cost manipulator – development of control methods and efficiency analysis* and funded by *OPUS 17* research grant No. 2019/33/B/ST8/02791 supported by the National Science Center, Poland.

References

1. Gray, J.: The mechanism of locomotion in snakes. J. Exp. Biol. **23**(2), 101–120 (1946)
2. Hirose, S.: Biologically Inspired Robots: Snake-Like Locomotors and Manipulators. Oxford University Press, Oxford (1993)

3. Ning, K., Wöragötter, F.: A novel concept for building a hyper-redundant chain robot. IEEE Trans. Rob. **25**(6), 1237–1248 (2009)
4. Murray, R.M., Li, Z., Sastry, S.S., Sastry, S.S.: A Mathematical Introduction to Robotic Manipulation. CRC Press, Boca Raton (1994)
5. Rolf, M., Steil, J.J.: Efficient exploratory learning of inverse kinematics on a bionic elephant trunk. IEEE Trans. Neural Netw. Learn. Syst. **25**(6), 1147–1160 (2014)
6. Melingui, A., et al.: Qualitative approach for forward kinematic modeling of a compact bionic handling assistant trunk. IFAC Proc. Volumes **47**(3), 9353–9358 (2014)
7. Falkenhahn, V., Hildebrandt, A., Neumann, R., Sawodny, O.: Dynamic control of the bionic handling assistant. IEEE/ASME Trans. Mechatron. **22**(1), 6–17 (2017)
8. Chirikjian, G.S., Burdick, J.W.: A hyper-redundant manipulator. IEEE Robot. Autom. Mag. **1**(4), 22–29 (1994)
9. Tugilimana, A., et al.: Conceptual design of modular bridges including layout optimization and component reusability. J. Bridge Eng. **22**(11), 04017094 (2017)
10. Torstenfelt, B., Klarbring, A.: Structural optimization of modular product families with application to car space frame structures. Struct. Multidiscip. Optim. **32**(2), 133–140 (2006)
11. Fuhs, W., Stachel, H.: Circular pipe-connections. Comput. Graph. **12**(1), 53–57 (1988)
12. Zawidzki, M., Nishinari, K.: Modular pipe-z system for three-dimensional knots. J. Geom. Graph. **17**(1), 81–87 (2013)
13. Zawidzki, M., Nagakura, T.: Arm-Z: a modular virtual manipulative. In: Schröcker, H.-P. (ed.) Proceedings of the 16th International Conference on Geometry and Graphics, pp. 75–80 (2014)
14. Zawidzki, M., Szklarski, J.: Transformations of Arm-Z modular manipulator with particle swarm optimization. Adv. Eng. Softw. **126**, 147–160 (2018)
15. Zawidzki, M.: Discrete Optimization in Architecture: Extremely Modular Systems. Springer, Heidelberg (2017)
16. Kiński, W., Pietkiewicz, P.: Influence of the printing nozzle diameter on tensile strength of produced 3D models in FDM technology. Agric. Eng. **24**(3), 31–38 (2020)

Measuring Techniques and Systems

Measurement of Qualitative Characteristics of Different Types of Wood Waste in the Forestries Zhytomyr Polissya

Igor Korobiichuk[1](✉), Iryna Davydova[2], Valentyn Korobiichuk[2], Volodymyr Shlapak[2], and Andrii Panasiuk[2]

[1] Łukasiewicz Research Network – Industrial Research Institute for Automation and Measurements PIAP, Warsaw, Poland
igor.korobiichuk@piap.lukasiewicz.gov.pl
[2] Zhytomyr Polytechnic State University, Zhytomyr, Ukraine

Abstract. The paper analyzes the prospects of using forestry waste as a raw material for the production of fuel resources (pellets). For each type of wood raw material the main quality parameters were measured: humidity, ash content, volatile substances content, The lower calorific value of the studied samples was calculated on the basis to the experimentally obtained data. The measurement results were compared with the requirements of European standards used in Ukraine. The studies have shown that sawdust from the dry lumber preparation shop without pre-treatment and sawdust from sawmilling and woodworking after pre-drying can be used to make pellets that meet the Swedish quality standard SS 18 71 20.

Keywords: Quality standard · Wood waste · Sawdust · Pellets · Humidity · Ash content · Volatile substances yield · Lower calorific value

1 Introduction

At the present stage of society's development, waste management, along with other environmental issues, is at the forefront of environmental security and sustainable development of the country [1]. Solution of these issues is linked to the need to settle a range of environmental, economic and social challenges and requires ongoing systematic efforts by government, academics and the public. According to the definition given in the Law of Ukraine "On Waste" [2], waste is any substance, material and object that is generated by human activity and has no further use at the place of formation or discovery". Among the main types of waste generated at woodworking enterprises we can distinguish chunks, slabs, rails, offcuts, scraps, chippings, screenings of technological chips, bark residues, sawdust and others.

In the woodworking industry, waste and losses of wood raw materials are generated at all stages of the production process. Forestry waste accounts for 21% of the total timber mass and about 14% of the stock of felling-area resources [3]. In recent years, the amount of logging waste in Ukraine has increased significantly due to climate change and the active development of forest pests [4].

R. Szewczyk et al. (Eds.): AUTOMATION 2021, AISC 1390, pp. 297–308, 2021.
https://doi.org/10.1007/978-3-030-74893-7_28

Unutilized wood can have a negative impact on the environment, although it does not contain any harmful substances. This occurs when large volumes of unprocessed wood waste located in large areas begin to decompose and release methane into the air. Dumps of wood lead to atmospheric dust pollution in the presence of wind. The process of wood decomposition can release greenhouse gases. Rotting, in its turn, contributes to the appearance of insects which can become a source of diseases.

Wood waste should be used to generate energy, since the thermal output of wood biomass fuel is completely in line with the traditional fuel (for example, one ton of coal can be replaced with 4.2 m^3 of such waste) [5, 6]. One way to utilize wood biomass is its direct combustion for getting energy [7], or processing it into fuel materials: pellets, briquettes. The advantages of using biomass as a fuel are the following: it is renewable; has low ash content and small emissions amount; it helps to maintain the balance of carbon dioxide in the atmosphere.

The use of wood waste makes it possible to solve a number of problems, in particular: to improve the sanitary condition of forests, to provide the population with inexpensive energy, to reduce emissions of harmful substances into the environment. The problems of rational use of wood waste were studied and covered in the works of the following scientists: S.V. Gaida, G.G. Geletukha [8], S.A. Gensiruk, P.R. Putzenteilo [9], Ya.V. Koval, A.P. Petrov, I.M. Sinyakevych, Yu.Yu. Tunytsia, E.I. Korotayev, M.E. Matveyev.

The works of some authors analyze the problems of classification of losses and waste of wood. These include the works of S.S. Dykyy, V.S. Yasinsky, V.I. Patyakin. These works provide recommendations for the elimination of losses and complex use of wood raw materials.

Against the background of rising world prices for standard energy, biofuels are a "life jacket" for the domestic energy sector, especially given the potential of Ukraine in this area. In recent years, such domestic and foreign scientists have been solving this problem: O.M. Varchenko, who developed an economic mechanism for regulating the biofuel market in the world's leading countries; G.M. Kaletnik, who in particular considered the state regulation of this industry; Ye. Lebedev, V. Dubrovin, M. Melnychuk, A. Kuznetsova [10].

The introduction of biomass energy technologies is an effective means of reducing the consumption of fossil fuels, which will give real energy and economic independence to both Ukraine as a whole, and its individual regions. Based on the analysis of European experience [11], we can conclude that increasing the use of wood waste allows to raise the growth of marketable products without increasing the volume of forestry production.

Pellet production cannot be considered completely environmentally friendly. During their production, a significant amount of dust is released into the air, which can pose a threat to the health of workers [12]. Biomass has a high reactivity, which is characterized by the yield of volatile substances – 70–86% [13]. The content of non-combustible substances and high humidity leads to a decrease in calorific value and heat productivity of the fuel. With a working ash content of wood biomass of 4% and a change in humidity from 10% to 60%, the heat productivity decreases sharply from 1940 °C to 1182 °C [14].

The level of environmental pollution during the combustion of fuel depends on the content of nitrogen and sulfur in it. According to some data, wood contains less than 1.5% of nitrogen and less than 0.1% of sulfur. The content of calcium, phosphorus, potassium and magnesium, apparently, does not reach such quantities that could negatively affect the energy production process; the resulting ash can be used as fertilizer [15]. Naturally, energy production will be more efficient and more environmentally friendly with less moisture, ash and volatile substances in raw materials, which is why these parameters were studied in our research. Numerous researchers have studied these parameters of wood and pellets made from it [16–20]. However, none of the studies can be considered universal and used for logging companies around the world, as different climatic zones are characterized by their own species diversity of forest ecosystems. Therefore, it is advisable to study the suitability of wood raw materials for the production of pellets for each region separately.

2 Research Methodology

The main purpose of this work is to study qualitative characteristics of wood waste and to determine prospects of its use as bioenergy resource.

Qualitative characteristics of wood waste depend on such parameters as size, moisture, ash content and content of contaminants (soil, stones, etc.). These parameters are very important in terms of the use of wood waste as fuel. In order to exchange information and accumulate knowledge about different types of biomass, the International Energy Agency created a database that currently contains information about 1560 different types of biological materials and ash samples [21].Unfortunately, raw material standards for the production of fuel pellets are not currently approved in Ukraine, thus EU standards are predominantly used in production. Currently in Ukraine there are no state standards for fuel biomass. Some producers of other types of fuel biomass develop and register technical conditions (TC) for their products (fuel chips, wood pellets and briquettes, sunflower husks, rapeseed straw and cereal straw, etc.). Enterprises producing briquettes and pellets for export are guided by the standards of importing countries, in particular such as DIN51 731 and DINplus (Germany) [22], O-Norm M 7135 (Austria) [23], SS 18 71 20 (Sweden) [24], ENplus (pan-European) [25]. That is why the results of our research were compared with the requirements of European standards, the main indicators of which are shown in Table 1.

SE "Emilchine Forestry" is a typical representative of the forestries in Zhytomyr region. Waste from this enterprise was used for the research. The types and volumes of wood waste generated at the enterprise were studied. As a result of the study, there were selected and prepared for further laboratory analysis nine most common types of waste: namely sawdust after sawmilling and woodworking, consisting mainly of pine waste with the addition of a small fraction of other species waste; sawdust after dry materials preparation selected by species types (alder, pine, oak and pear tree; forest residues (needles and pine bark) and pellets produced by forestry with the typical composition of forest species for Zhytomyr Polissya (90–95% of Scots pine and 5–10% of hardwood species: oak, hornbeam, alder, birch, etc.).

The following qualitative parameters of wood raw material were determined during the study: the moisture of air-drywood waste and its analytical samples; the ash content of wood waste; the yield of volatiles and the calorific value of wood waste.

Table 1. Qualitative characteristics of fuelwood pellets [26]

Parameters	DIN 51 731	O-Norm M 7135	DIN plus	SS 18 71 20
Diameter (d), mm	4–10	4–10	–	<25
Length (l), mm	<50	<5 × d	<5 × d	<5 × d
Density, kg/dm^3	>1.0–1.4	>1.12	>1.12	–
Moisture (relative), %	<12	<10	<10	<10
Bulk weight, kg/m^3	650	650	650	>500
Dust, %	–	<2.3%	<2.3%	–
Ash content, %	<1.5	<0.5	<0.5	<1.5
Caloricity (lower), MJ/kg	17.5–19.5	>18	>18	>16.9

Moisture. To determine the moisture content of wood waste, approximately 1.0–2.0 g of the analytical sample was collected into a pre-weighed crucible and put on a laboratory scale and weighed again (Fig. 1). The drying cabinet (Fig. 2) was heated to 110 °C and material was placed there for 30 min. After drying, the crucibles were closed with lids and allowed to cool on a metal support for 2 min. After that, material was put in a desiccator for cooling to a room temperature and then weighed. Simultaneously, control drying for 30 min was performed until the difference between these two weightings was not more than 0.01 g.

Fig. 1. Samples of wood waste prepared for drying

Fig. 2. Drying cabinet 2III-0-01

Ash Content. Before determining the ash content, an empty crucible was calcined in a muffle furnace (Fig. 4) at 550 °C for 60 min. After cooling, the crucibles were weighed. Approximately 1.0–2.0 g of sample was put and evenly distributed in crucible and weighed again. The crucible was placed in a muffle furnace at room temperature. The material was kept there for 60 min (Fig. 3) with the temperature in the furnace gradually rising to (550 ± 15) °C. After cooling, the crucible with ash residue was weighed.

Volatile Substances. To determine the content of volatile substances in wood samples, a sample of raw material in a crucible was heated in a muffle furnace at a predetermined rate to a temperature of (250 ± 10) °C and it was kept at this temperature for exactly 60 min.

Fig. 3. Burning raw material sample in a muffle furnace

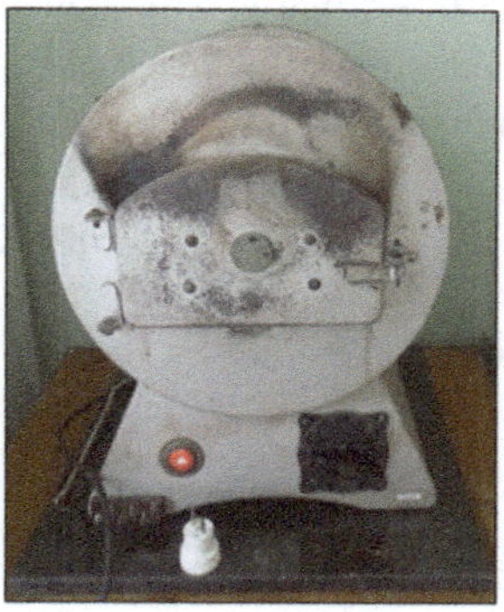

Fig. 4. Muffle furnace

Lower Calorific Value. The calorific value of wood waste was determined by calculation taking into account the type of wood, the determined moisture and the ash content of studied sample. Approximate lower calorific value of wood chips and waste was determined by the formula (1):

$$Q = 18900 - 214W^a - 189A^a, \quad (1)$$

where: Q – the lower calorific value, kJ/kg; W^a – moisture, %; A^a – ash, %.

All indicators of quality of wood fuel raw materials were determined in 10-fold repetition. Samples preparation and measurements were conducted on the basis of specialized hydrogeological laboratory at Zhytomyr polytechnic state university.

Data were processed with application programs STATGRAPHICS, Microsoft EXCEL.

3 The Results of Research

Moisture. The main factor affecting the lower calorific value of wood, including wood chips, is its humidity, which can vary within rather wide ranges (relative humidity from 15% to 60%).

Measurement and calculation data for air-dry wood waste moisture and analytical sample moisture are included in Table 2.

The lowest percentage of moisture was detected in pellets (9.35%). This is due to the fact that the pellets are finished fuel product and raw material for this product has passed certain stages of drying and pressing (Fig. 5).

Among unprocessed wood waste, the lowest percentage of moisture was detected in sawdust samples from workshop for dry materials preparation (alder – 10.22%, pine

– 12.53%, oak – 13.97% and pear tree – 9.68%). Therefore, the sawdust of this workshop is sufficiently dry for pellet production and does not require additional drying. Certain variations of its content are due to the peculiarities of the tree species.

The highest moisture content was found in samples of pine, bark and sawdust after sawmilling and woodworking at SE "Emilchine Forestry". These samples were taken in the forest that is why they are not dry enough. This type of waste can be used for the production of pellets, but it necessarily requires pre-drying, which affects the increase in their cost to some extent.

Table 2. Air-dry wood waste moisture and analytical sample moisture

№	Wood waste type	Crucible mass without material, g	Crucible mass with material for drying, g	Crucible mass with material after drying, g	Crucible mass with material after complete desiccation, g	Moisture of analytical sample (W^a), %	Air-dry wood waste moisture (W), %
1	Alder sawdust	25.880	29.500	29.180	29.130	8.839779	10.22099
2	Pine sawdust	26.410	29.680	29.330	29.270	10.70336	12.53823
3	Oak sawdust	24.610	27.400	27.040	27.010	12.90323	13.97849
4	Pear tree sawdust	26.400	29.345	29.090	29.060	8.658744	9.677419
5	Sawdust from sawmilling	25.770	31.300	29.850	29.200	26.22061	37.97468
6	Wood sawdust from woodworking	26.710	32.520	31.360	30.900	19.96558	27.88296
7	Pellets	28.380	38.540	37.900	37.590	6.299213	9.350394
8	Pine needles	28.330	33.470	32.620	32.440	16.53696	20.03891
9	Pine bark	27.590	30.860	30.340	30.260	15.90214	18.34862

Ash Content. Ash content was determined in all selected samples. Ash is a mixture of minerals that remain after the combustion of all fuel substances of the fuel and after the completion of all transformations that occur with mineral impurities at high temperatures and full access of air. The ash content for different types of biological materials varies widely and is about 0.5% for pure wood. Ash degrades fuel quality. The higher the ash content, the more expensive the transportation, and the higher the cost of slag removal from the furnace. Ash affects the calorific value of the fuel; the higher the ash content, the more difficult it is to achieve complete combustion of fuel. The ash content in the fuel affects the choice of technology for biomass to energy conversion, the choice of equipment, the method of ash removal, as well as it affects the emission of particulate matter in flue gases, the choice and size of the treatment system, the size of ash dumps.

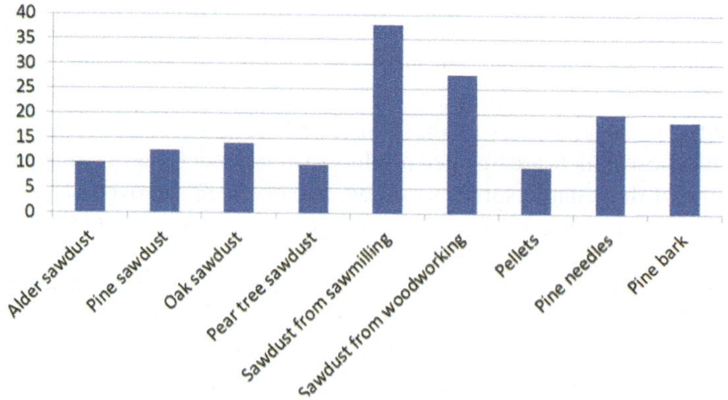

Fig. 5. Air-dry wood waste moisture, %

Analyzing the obtained data (Fig. 6), it can be stated that bark and pine needles have the highest ash content. Their content in raw materials for the production of fuel degrades their pellets quality. Such samples cannot be used for the production of fuel pellets as their ash content exceeds 1.5%.

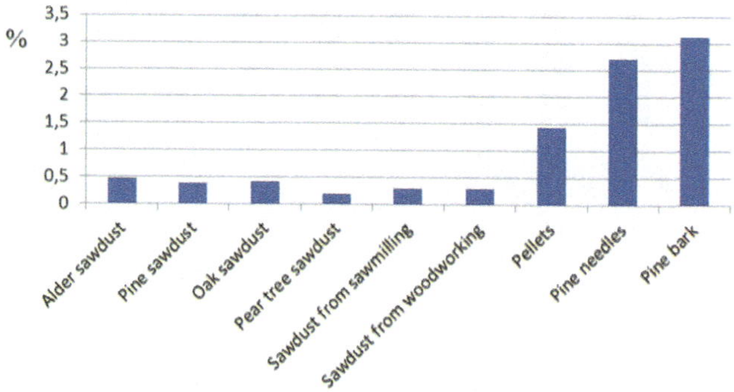

Fig. 6. Ash content of wood waste

All sawdust samples meet the standards and can be used for pellet production due to possible complete combustion of fuel. In some samples, ash content is slightly higher than in pine samples, but it is within normal limits. Thus, adding such material to pellets will not degrade the quality.

Volatile Substances. Volatile substances are gaseous and vapor products released during heating. The yield index of volatile substances characterizes the quality and structural features of the fuel. The content of volatile substances is taken into account when determining their complex and rational use, and also it characterizes the number of thermally unstable atomic groups. The results of calculating the yield of volatile substances of the analytical sample are given in Fig. 7.

Based on the calculations, it can be concluded that sawdust of all samples has a high percentage of volatile substances yield, in particular, samples of sawdust of different sizes that have resinous substances. Meanwhile, finished fuel pellets, pine needles and bark do not show such percentage. The larger the sawdust size, the smaller the percentage of volatile matter yield.

The insignificant release of volatile substances from pellet samples can be explained by the fact that these pellets are finished product that has undergone drying and heat treatment, so some of the volatile substances have already been removed. This necessitates the installation of air purification equipment at the production of pellets and briquettes.

Lower Calorific Value. The calorific value of a fuel, or specific heat of combustion, characterizes the amount of heat released by the complete combustion of a certain volume of fuel. The higher the calorific value of the fuel, the more valuable it is since less fuel is required to produce the same amount of heat. This indicator is most often measured in J/kg (J/m^3; J/l). The calorific value is influenced by the chemical composition and the ratio of combustible and non-combustible elements in the fuel. During the research, the lower calorific value was determined by calculation, taking into account such parameters as humidity and ash content of wood waste.

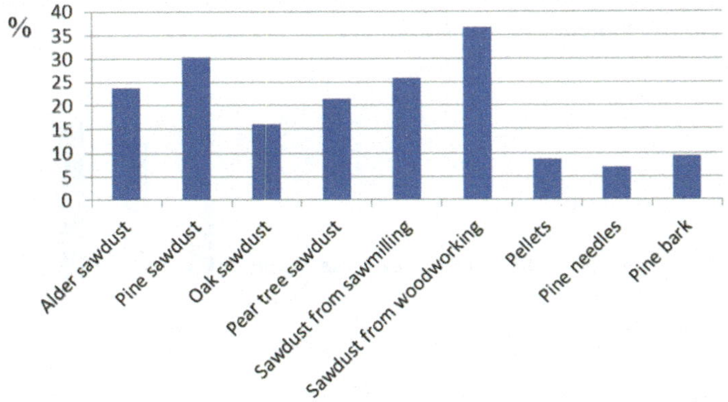

Fig. 7. The yield of volatile substances of the analytical sample

The results of calculation of the lower calorific value for studied wood waste samples are shown in Fig. 8. Finished pellets have the highest calorific value (17279.04 kJ/kg). Among samples of raw material used for the production of pellets the highest calorific value was measured in sawdust of alder, pine, oak and pear tree. The diagram shows that the lowest calorific value was measured in sawdust after sawmilling (13230.3 kJ/kg) and woodworking (14569.21 kJ/kg), as it has the highest percentage of moisture in analytical sample. Therefore, these samples do not meet the standard and require pre-drying before processing.

The Content of Radionuclides. Ukraine has introduced mandatory radiological control of wood and timber products, since much of the forests of Zhytomyr Polissya are still contaminated after the Chernobyl accident [27]. Therefore, each manufacturer is obliged by law to confirm the safety of manufactured products. There is usually no need for additional radiological control in the production of fuels from forest wood waste, as it is

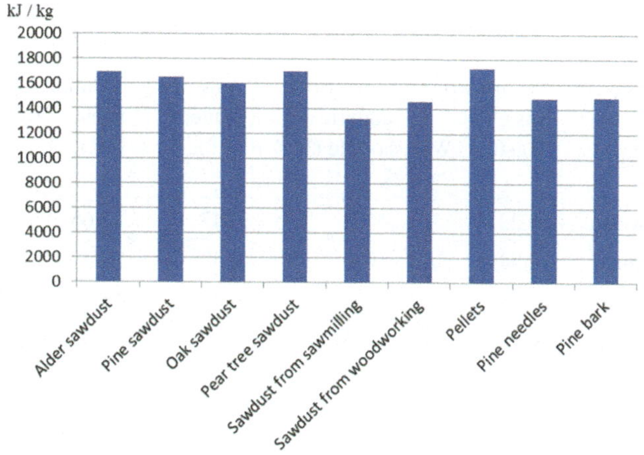

Fig. 8. Lower calorific value

mandatory that wood and timber pass it, and waste is part of forest products and contains the same amount of radionuclides as marketable wood. However, due to the fact that the category of waste usually includes most of the bark, needles and leaves, the root part of the tree and its branches, there can be some variability in the content of radionuclides. Therefore, all selected samples were analyzed in the radioecology laboratory at the university.

Measurement of ^{137}Cs specific activity in samples was made with the complete scintillation spectrometry system GDM-20 (Gammadata Instrument AB, Sweden), which is based on a 3″ × 3″ NaI (Tl) detector on a 14-pin PM-tube. All analyzed samples fully comply with the standards of radiological control and can be used for the manufacture of finished products, including fuels (briquettes, pellets).

4 Conclusions

After analyzing the quality of wood waste on individual indicators, it is necessary to draw generalized conclusions about the possibility of using these wastes as raw materials for the production of pellets. Summary of the research results are shown in Table 3. We do not consider it efficient to conduct a comparative analysis with the results of research by other authors, as their works analyze the relevant parameters for other tree species, which excludes the possibility of comparing them with each other. To determine the compliance of these wood waste with the standards, we used the normative values of European standards, which are most often applied in Ukraine and are used in the export of pellets abroad (Table 3).

Summarizing and analyzing the results of the study, we concluded that the most efficient and appropriate raw material for the production of pellets is sawdust from the dry lumber shop, as the lower calorific value of this waste is the highest among the studied samples. However, even for this category of wood waste, its energy value does not meet any of the analyzed European standards. The lower calorific value directly depends on

Table 3. Summary of research results

№	Wood waste type	Air-dry wood waste moisture (W), %	Moisture of analytical sample (W^a), %	Ash content. (A^a), %	Volatile substances (V^a), %	Lower calorific value (Q), kJ/kg
1	Alder sawdust	10.22	8.84	0.47	23.65	16919.14
2	Pine sawdust	12.54	10.70	0.39	30.47	16535.07
3	Oak sawdust	13.98	12.90	0.43	15.97	16057.77
4	Pear tree sawdust	9.68	8.66	0.20	21.42	17009.60
5	Sawdust from sawmilling	37.97	26.22	0.31	25.80	13230.27
6	Wood sawdust from woodworking	27.88	19.97	0.31	36.53	14569.21
7	Pellets	9.35	6.30	1.44	8.73	17279.04
8	Pine needles	20.04	16.54	2.71	6.95	14848.04
9	Pine bark	18.35	15.90	3.13	9.33	14906.32

the moisture content and ash content of fuels. Humidity standards are applied to finished pellets, which undergo thermal processing during their production and therefore contain less moisture than the raw material. The initial moisture content of sawdust from the dry lumber preparation shop is 9.7–14.0%. These values allow the use of these materials for the manufacture of pellets without prior preparation. As a result of pellet production, wood materials undergo the stages of drying and hot pressing, resulting in their moisture content decrease. Taking into account the technological parameters of production, the type of waste and the initial humidity, the humidity of the finished product will be 9.2–9.4% depending on the wood species. The recalculation of the lower calorific value for moisture after the production of pellets showed that the finished fuel pellets made from waste from the dry lumber shop can in all respects meet the Swedish quality standard SS 18 71 20. The use of labeling according to other standards is unacceptable because their energy value is not so high.

Sawdust from woodworking and sawmills has a fairly high percentage of moisture and cannot be used for the production of pellets without special preparation. After pre-drying and processing, pellets made from these raw materials can meet the quality standard SS 18 71 20. However, during the economic justification of the project of processing this category of waste into pellets, it is necessary to take into account the additional costs of energy and labor resources, which will have an impact on the cost of finished products. As this category of waste is predominant in any forestry enterprise, it is not advisable to install several pellet production lines, especially given their similar final indicators. It will be more rational to mix wood waste: sawdust from woodworking and sawmilling shops after pre-drying and sawdust from dry lumber preparation shop without pre-drying.

The use of needles and bark in the process of making pellets is undesirable, because these samples have a high percentage of ash and moisture, which does not meet the standards and negatively affects the quality of the pellets. Bark can be used to make soil mixtures or mulch or left after grinding in forests to accelerate the restoration of fertility of forest soils.

The results of the research are confirmed by the qualitative indicators of finished pellets made by the neighboring forestry from wood waste with similar species composition. The ash content of the finished pellets is somewhat surprising, as in none of the samples of pure wood this figure exceeded 0.5%, and in the pellets the content of non-combustible components is 1.44%. Probably during the production of pellets, a certain part of the bark and leaves was used, or admixtures of other non-wood materials were added, which facilitate the production of pellets but are not provided by the technological process.

Thus, the introduction of fuel pellets complexes on the basis of forestries is a rational way to solve the problem of using wood waste. Necessary condition for technological processes of waste treatment is their ecological and economic efficiency. The calculation of the processing line payback, taking into account capital and current costs, is 1–1.5 years. Wide range of enterprises of the timber processing complex can provide growth of commodity production not less than by 10–15%.

References

1. Waste management at a woodworking enterprise. https://derevoobrobnyk.com/povodzhennya-z-vidxodami-na-derevoobrobnomu-pidpriyemstvi/
2. On waste, the Law of Ukraine. N 187/98-BP, 05 March 1998. https://zakon3.rada.gov.ua/laws/show/187/98-80. [in Ukrainian]
3. Stecyuk, N.: Some aspects of the analysis of raw material losses and waste use in the woodworking industry. Soc.-Ekon. Probl. Derzhava **2**(7), 81–88 (2012). [in Ukrainian]
4. Shomko, D.V., Shomko, V.V., Davydova, I.V.: Study of the state of forest resources of zhytomyr region. Visnik Zhitomirskogo derzhavnogo Tehnol. Univ. Zhytomyr: ZSTU **1**(83), 299–303 (2019)
5. Dyakova, O.M., Mihalyuk, N.S.: Problems of recycling production and consumption waste in an industrialized area. In: 3rd International Congress on Waste Management VejstTek-2003, Materials of the Congress, Moskva, 3–6 June 2003, 588 p. (2003). [in Russian]
6. Telmo, C., Lousada, J.: Heating values of wood pellets from different species. Biomass Bioenergy **35**(7), 2634–2639 (2011). https://doi.org/10.1016/j.biombioe.2011.02.043
7. Svintuh, M.B.: Use of wood waste for energy, Economic problems of sustainable development, materials of the International scientific-practical conference dedicated to memory prof. Balackogo O.F. SumDU, Sumy **1**, 241–243 (2013). [in Ukrainian]
8. Geletuha, G.G., Zhelezna, T.A., Zhovmir, M.M., et al.: Assessment of biomass energy potential in Ukraine. Part 1. Agricultural waste and wood biomass. Promyslova Teplotehn. **32**(5), 58–65 (2010). [in Ukrainian]
9. Pucentejlo, P.R., Svintuh, M.B.: Ecological and economic aspects of wood waste use. Innov. Ekon. **2**, 135–139 (2013). [in Ukrainian]
10. Zagorna, T.O.: Economic diagnostics. K.: Centr uchbovoyi literatury, 400 p. (2007). [in Ukrainian]

11. Vis, M.: Harmonization of biomass resource assessments. Vol. I: Best practices and methods handbook. Report on WP5 of the EC FP7 Project «Biomass Energy Europe». BTC Biomass Technology Group, B.V., The Netherlands (2010). https://www.eubee.com
12. Hagström, K.: Occupational exposure during production of wood pellets in Sweden. Doctoral thesis, comprehensive summary, 75 p. Örebro University, Örebro (2008)
13. van Loo, S., Koppejan, J. (eds.): The Handbook of Biomass Combustion and Co-firing
14. Golovkov, S.I., Koperin, I.F., Najdenkov V.I.: Energy use of wood waste. Lesn. prom-st: Moskva, 224 p. (1987). [in Russian]
15. Blyum, Ya.B., Geletuha, G.G., Grigoryuk, I.P., et al.: Biological resources and technology of biofuel production, monograph, 408 p. Agrar Media Grup, Kyiv (2010)
16. Chandrasekaran, S.R., Hopke, P.K., Rector, L., Allen, G., Lin, L.: Chemical composition of wood chips and wood pellets. Energy Fuels **26**(8), 4932–4937 (2012). https://doi.org/10.1021/ef300884k
17. Öhmana, M., Bomana, C., Hedmanb, H., Nordina, A., Boströma, D.: Slagging tendencies of wood pellet ash during combustion in residential pellet burners. Biomass Bioenergy **27**(6), 585–596 (2004). https://doi.org/10.1016/j.biombioe.2003.08.016
18. Brackley, A.M., Parrent, D.J.: Production of wood pellets from Alaska-grown white spruce and hemlock. General Technical report. U.S. Department of Agriculture, Forest Service, Portland (2011). https://doi.org/10.2737/PNW-GTR-845
19. Acda, M., Devera, E.: Physico-chemical properties of wood pellets from forest residues. J. Trop. For. Sci **26**(4), 589–595 (2014)
20. Acda, M.: Physico-chemical properties of wood pellets from coppice of short rotation tropical hardwoods. Fuel **160**, 531–533 (2015). https://doi.org/10.1016/j.fuel.2015.08.018
21. Mizhnarodne Energetychne Agentstvo z pytan spalyuvannya biomasy ta sumisnogo spalyuvannya. IEA Bioenergy Task 32, internet sajt. https://www.ieabcc.nl/
22. DIN CEN/TS 14961-2005. https://phyllis.nl/Browse/Standard/CEN-TS-14961
23. ÖNORM M7135. https://www.pelletshome.com/pellets-oenorm
24. Pellet markets in Sweden. www.northernperiphery.eu/files/archive/Downloads/Project_Publications/10/Study_Reports/Pellet_markets_Sweden_2009.pdf
25. ENplus. https://enplus-pellets.eu/en-in/
26. DSTU CEN/TS 15149-1:2009. Tehnichni umovy. Palyvo granulovane
27. Krasnov, V.P., Kurbet, T.V., Korbut, M.B., Bojko, O.L.: Distribution of ^{13}7Cs in the forest ecosystems of the Polissya of Ukraine. Agroekologichnyj zhurnal **1**, 82–87 (2016)

Evaluation of Motion Characteristics Using Absolute Sensors

Marius Šumanas, Darius Urbonis, Algirdas Petronis, Simonas Stankaitis, Tomas Januškevičius, Igor Iljin, Andrius Dzedzickis(✉), and Vytautas Bučinskas

Vilnius Gediminas Technical University, Vilnius J. Basanaviciaus str. 28, Vilnius, Lithuania
{marius.sumanas,vytautas.bucinskas}@vgtu.lt,
andrius.dzedzickis@vilniustech.lt

Abstract. Definition of movement trajectories using absolute sensors is a challenging problem in various application fields due to the appearing additive errors. For the trajectory definition, most often used Inertial measurement units (IMU) consisting of accelerometers, gyroscopes, and magnetometers require the implementation of sensor fusion algorithms and complex data processing. In this research, we investigated the possibility to use cheap 3-axis accelerometer for movement trajectory definition and reproduction. Six degrees of freedom industrial articulated robot was used as a source of ideal trajectories. Obtained results show that using 3-axis accelerometer and relatively simple data processing; raw data mean value offset correction and double integration it is possible to define 2D and 3D trajectories in respect of reference point. In our case, the maximum reproduction error did not exceed 21 mm for, 314 mm length 2D trajectory, and correspondingly 7 mm for the 3D trajectory of the same length. Also, we defined that reproduction errors are proportional to the length of trajectory in the separately taken axis.

Keywords: Absolute sensors · Accelerometers · Signal processing · Position detection · Movement trajectory

1 Introduction

With the development of robotics and digitization, more and more tasks arise when it comes necessary to determine the various characteristics of a moving object, for example: real-time position in respect of reference point; movement speed; movement type and trajectory. Technical challenges requiring solutions to mentioned problems occur in many areas [1]: starting from aviation, where it is necessary to define position and speed of flying objects with respect to the ground, and ending with manufacturing and robotics [2], where it is required to track the position of the tool [3] or control the manipulators according to human hand movements [4, 5]. Determination of the motion trajectory also is very relevant in biomedicine [6] and various virtual reality systems [7], defining human movements.

R. Szewczyk et al. (Eds.): AUTOMATION 2021, AISC 1390, pp. 309–318, 2021.
https://doi.org/10.1007/978-3-030-74893-7_29

Various technical solutions could be implemented to determine movement trajectory, but the most often used technical solution is Inertial measurement units (IMU) [8]. Usually, IMU consists of a combined number of accelerometers, gyroscopes, and magnetometer installed in the majority of newer units. Using IMU, data from accelerometer allows determining the location of the sensor using double integration, taking into account initial position, static and dynamic working conditions [9]. Practical experience shows that in many applications to obtain reliable data is required more than one accelerometer, or instead of few accelerometers, magnetometer, accelerometer and gyroscope data through sensor fusion algorithms could be used, as it is presented in [9]. Where acceleration data, through double integration, was used to determine object position and orientation.

Accelerometer based devices have undergone many trials to determine their accuracy and precision. They are effective in a broad sense, but individual studies and devices require mean and variance determinations and adjustments to gain the most accurate results [10]. It is a known problem of double data integration, and it is assumed that accelerometer data has to be integrated only for a short time, overwise significant positioning error could appear [9].

There are several solutions to solve double integration displacement error, usually, for this purpose, complementary, Kalman or Mahony filters are used [10–13]. In terms of accuracy, Kalman is superior to the complementary filter. However, the complementary filter is more straightforward in terms of implementation difficulty, but, due to accuracy differences, if possible, the Kalman filter should be preferred [10]. The disadvantage of Kalman filter is relatively high computational power requirements – Kalman filter demands substantially more powerful microcontroller. In comparison to the Kalman filter, Mahony filter performs relatively similarly, in terms of accuracy, the difficulty of implementation and computational power [13].

The purpose of this research is to test the viability of MEMS-based acceleration sensors to be used for absolute motion characteristic evaluation. Precise evaluation of motion allows trajectory recreation from recorded kinematic quantities (acceleration).

2 Experimental Research

2.1 Equipment

The experimental setup (Fig. 1) includes a microcontroller, relay module, accelerometer, industrial articulated robot and robot controller. IMU, MPU9250 (TDK Corporation, Japan) with 9 degrees of freedom was used as an accelerometer, meaning that only 3 out of 9 degrees of freedom were used, the other readings from were ignored since our purpose was to evaluate possibilities of the accelerometer. ESP32 microcontroller (Espressif Systems, China) was used for the communication with the accelerometer module, sensor data acquisition and communication with the robot's controller.

Industrial articulated robot "Motoman SSF2000" (Yaskawa, Japan, repeatability 0.08 mm) was used to generate trajectories, which in our case are considered ideal, similarly as in [3]. The robot was controlled by an "NX100" controller to perform the programmed trajectories.

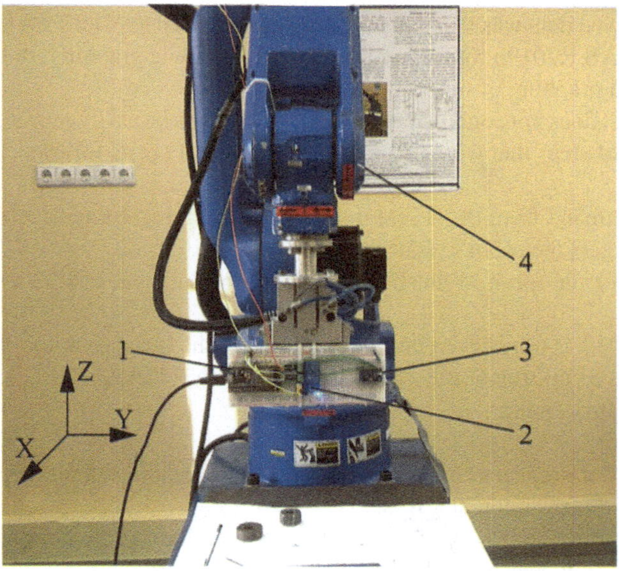

Fig. 1. Experimental setup: 1) ESP 32 microcontroller; 2) relay module; 3) accelerometer; 4) articulated industrial robot.

Accelerometer and microcontroller were installed on the standard development board attached to the robot gripper, as shown in Fig. 1. Such an approach allowed us to adjust the coordinate system of the attached sensor with the main robot coordinate system, keeping the main axes parallelly.

2.2 Methodology

The experiment involved a series of tests during which the robot's end-effector moved through a programmed path. The robot's linear movement speed was kept constant at 150 mm/s throughout all tests. An accelerometer attached to the robot's end-effector was measuring acceleration values in 3 different axes (X, Y, Z) with a sampling rate of 1 kHz. The inbuilt digital low pass filter bandwidth was set to 460 Hz. MPU9250 comes with user-programmable interrupt feature, which was set up to allow the raw sensor data to propagate to the interrupt pin. This means that whenever new acceleration data was available, a 50 μs duration impulse was generated on the interrupt pin of the MPU9250. This feature is used to ensure that no data was lost or recorded by the microcontroller more than once. The monitoring of the interrupt pin state and recording of data was performed by ESP32 microcontroller. I^2C communication protocol with a speed of 400 kHz was used to communicate between the MPU9250 and ESP32. Large SRAM (static random-access memory) capacity of the ESP32 allowed us to store the incoming acceleration values for further use, immediately after receiving a new data set. This feature was used to achieve a higher recording rate on the microcontroller as serial communication between the ESP32 and PC is relatively slow. The recorded acceleration values were sent from ESP32 to PC via USB port and displayed on a serial monitor. Received data was manually copied and imported to Microsoft Excel 2010 and later to MATLAB R2019b for further processing, including graphing, raw data filtering, calculation and graphing of trajectories.

A single test does not contain many steps and thus can be easily repeated many times. The sequence of steps that were taken in the experiment are as follows:

1. A start command from PC is sent to ESP32 to begin recording acceleration values.
2. ESP32 triggers the robot to start moving.
3. PC waits for the robot to finish its movement and for the ESP32 to record all the data.
4. Recorded data is transferred into PC and send to the serial monitor.
5. The displayed information is copied and pasted to Excel spreadsheet or MATLAB for further processing.

Steps 1–5 were repeated many times to generate numerous amounts of data and to avoid accidental measurement errors.

MATLAB script consisting of 4 parts: offsetting, double integration, graphing and accuracy evaluation, was used to generate trajectory graphs.

Typically, the raw acceleration values for individual axes are not centred around zero, i.e., the arithmetic mean is not zero, and even undesired acceleration offsets as slight as 0.01 m/s^2 causes noticeable drifts in position estimation. Those offsets caused by gravity, temperature fluctuations and sensor manufacturing imperfections, were obtained by calculating the arithmetic mean of the raw acceleration values. This procedure was done individually for all three axes. The corrected acceleration values are were obtained by subtracting the offsets from the raw acceleration values:

$$a_{corrected\_i} = a_{raw\_i} - a_{offset} \tag{1}$$

where: $a_{corrected\_i}$ – is the corrected acceleration data set, a_{raw_i} – is the raw acceleration value data set and a_{offset} is the calculated acceleration offset value which could be calculated as:

$$a_{offset} = \overline{a}_{raw} = \frac{\sum_{i=1}^{N} a_{raw\_i}}{N} \tag{2}$$

where: $\overline{a}_{raw}$ – is the arithmetic mean of the raw acceleration value data set, N – the number of discrete points of the signal.

The sensor position in the space in respect of the initial reference point is obtained by iterating through the data points and solving the kinematic equations:

$$vf = v_0 + at \tag{3}$$

$$x_f = x_0 + v_0 t + \frac{1}{2} a t^2 \tag{4}$$

where v_f – is the final velocity, v_0 – is the initial velocity, a – is acceleration, t – is time step, x_f – is the final position and x_0 – is the initial position.

3 Results of the Experiment

Tests are performed to evaluate the possibility to reproduce trajectories of various shapes, using accelerometer from IMU MPU9250. The tests are performed using S-shaped flat (2D) and S-shaped three-dimensional trajectories arranged in different planes. The results obtained by scanning the trajectories in the Z-Y plane are shown in Figs. 2, 3 and 4. The test was repeated 9 times to avoid the impact of random errors, however, in the graphical form we are presenting only the results of the first three, as the errors found in other cases have the same tendencies, and error values were fluctuating about the same final value.

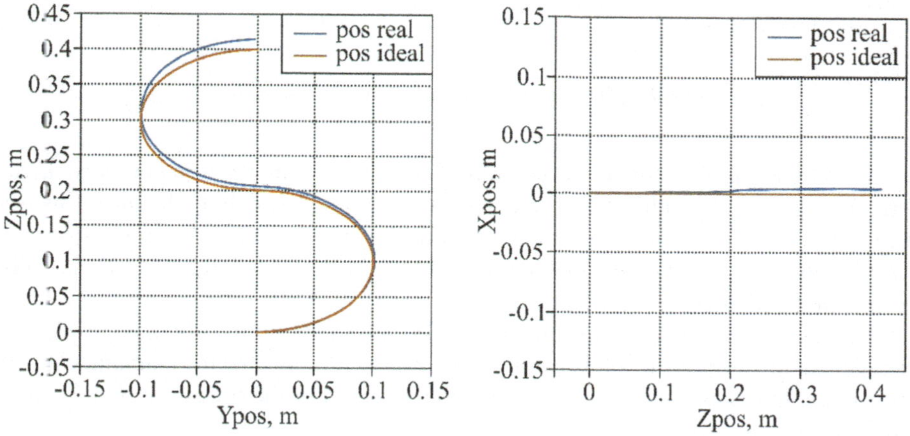

Fig. 2. 2D trajectory determination, 1st test results

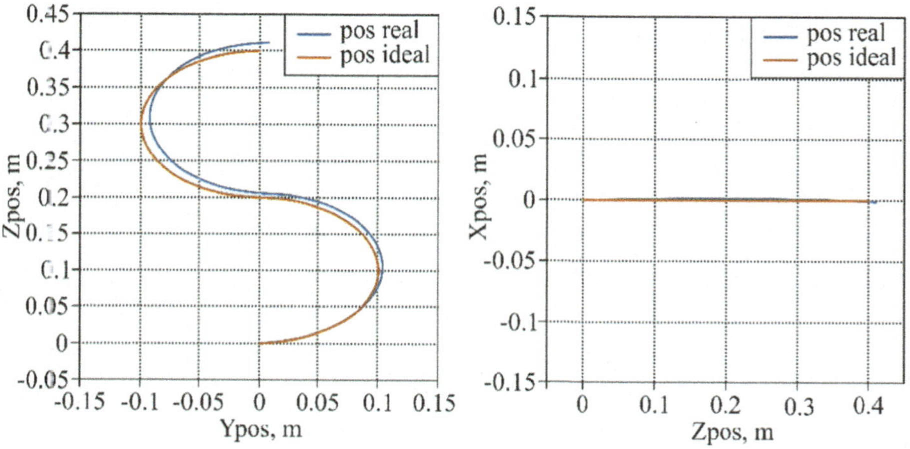

Fig. 3. 2D trajectory determination, 2nd test results

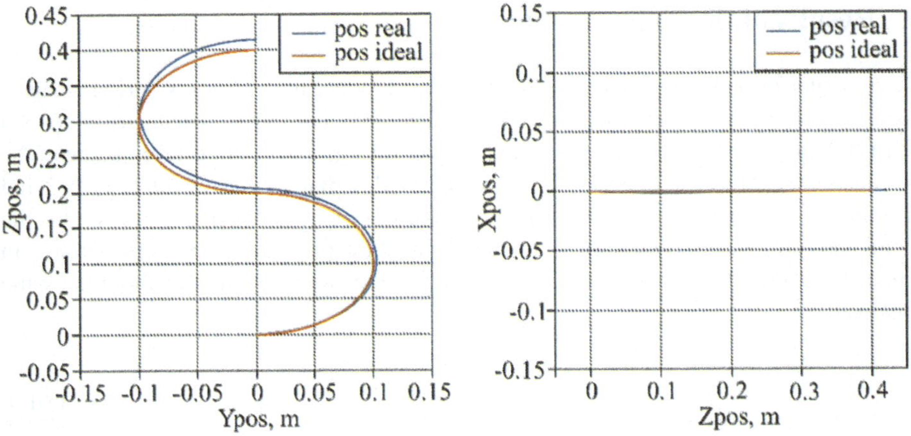

Fig. 4. 2D trajectory determination, 3rd test results

Measured trajectory contains accumulated error during all measuring process; therefore, from provided graphs, it is seen that error value increases along the line of movement. Deviation from known movement trajectory depends on the shape of trajectory; error in vertical plane Z-Y is more significant than in the horizontal plane X-Z, while movement configuration is different.

The results of all 9 tests are summarized in Table 1. From the obtained results we can see that in all cases the total error does not exceed 21 mm, and the magnitude of the error in the individual axes is proportional to the length of the trajectory projection in the measurement axis.

Table 1. 2D trajectory errors.

Experiment No.	Errors in axis			
	X, mm	Y, mm	Z, mm	Total, mm
1	5.00	3.00	17.00	17.97
2	2.00	6.00	19.00	20.02
3	1.00	0.00	20.00	20.02
4	3.00	2.00	15.00	15.43
5	0.00	4.00	16.00	16.49
6	2.00	1.00	10.00	10.25
7	5.00	4.00	17.00	18.17
8	4.00	6.00	14.00	15.75
9	5.00	2.00	20.00	20.71

The results obtained by reproducing the 3D trajectories are shown in Figs. 5, 6, 7 and 8. The experiment was repeated 9 times, and similarly, as in the first case, we are presenting graphs of the first four trials, which represent main tendencies of errors nature and distribution.

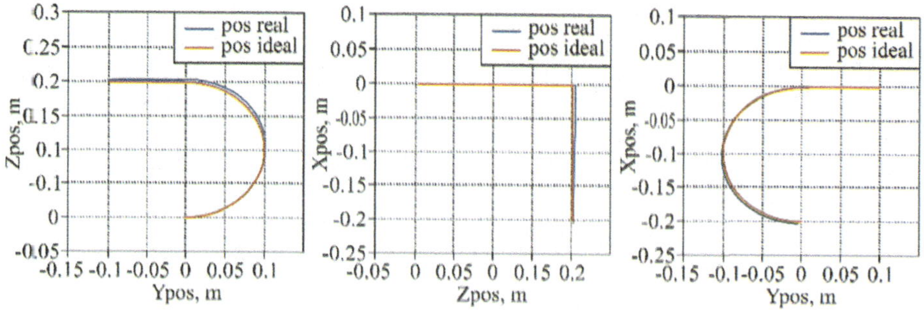

Fig. 5. 3D trajectory setting, 1st test results

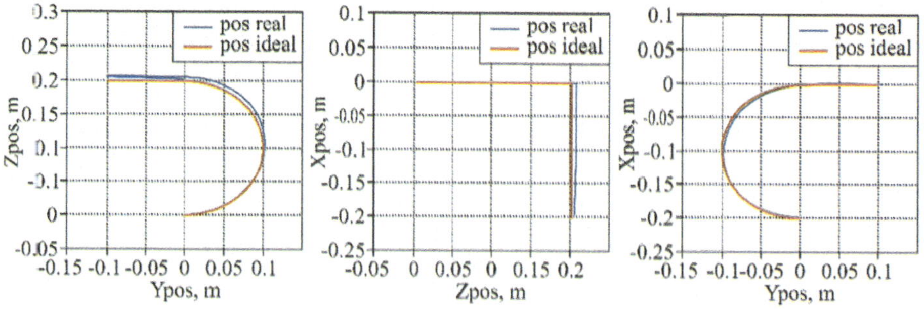

Fig. 6. 3D trajectory setting, 2nd test results

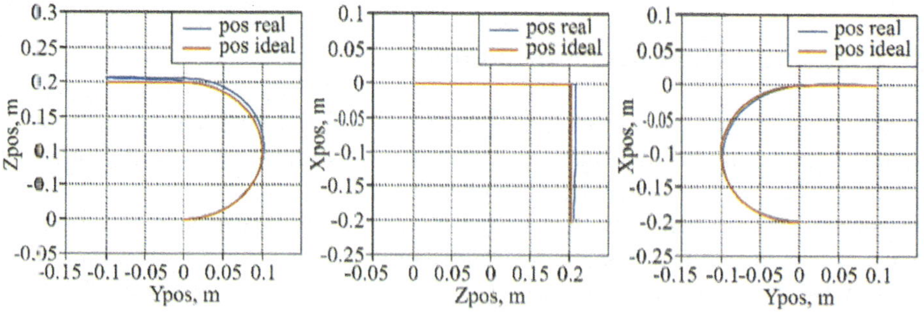

Fig. 7. 3D trajectory setting, 3rd test results

A summary of the test results is given in Table 2. From the results obtained, we can see that the total error did not exceed 7 mm in all cases. The results of this test also confirmed the previous conclusion that the trajectory reproduction error in the separately taken plane is proportional to the length of the projection of the trajectory into that plane.

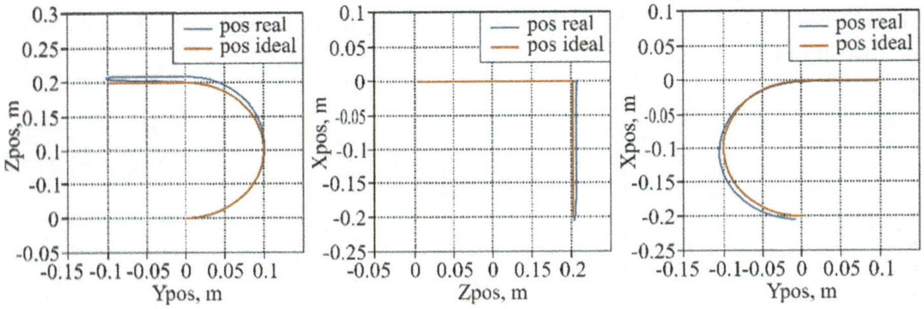

Fig. 8. 3D trajectory setting, 4th test results

Table 2. 3D trajectory errors.

Experiment No.	Errors in planes			
	XY, mm	YZ, mm	XZ, mm	Total, mm
1	4.00	5.00	2.00	6.71
2	1.00	3.00	2.00	3.74
3	2.00	5.00	4.00	6.71
4	1.00	2.00	3.00	3.74
5	5.00	3.00	1.00	5.92
6	2.00	4.00	5.00	6.71
7	2.00	1.00	4.00	4.58
8	3.00	4.00	3.00	5.83
9	5.00	1.00	4.00	6.48

4 Conclusions and Recommendations

Performed research provided perspective findings useful in the areas related to the definition of movement characteristics. The obtained results allow us to draw the following conclusions and recommendations for further research.

Results of performed research prove that it is possible to implement relatively cheap 3-axis accelerometers to define movement trajectory in respect of relative reference point in the 3D space. In our case, all tested 2D and 3D trajectories were successfully reproduced using MPU9250 accelerometer and ESP32 microcontroller for data acquisition. The maximum trajectory reproduction error did not exceed 21 mm for 2D, 314 mm length trajectory, and correspondingly 7 mm for the 3D trajectory of the same length.

Our proposed method of using an industrial articulated robot as a tool to generate ideal trajectories for the research and calibration of IMU parameters has proved its functionality. The articulated robot allows easily create and realize trajectories of various difficulty. Such a solution allows simulating human movements and at the same time can provide much higher speeds and could be used to calibrate wearable and other IMU sensors intended for human-machine related applications.

The analysis of trajectory reproducing errors showed that the errors in the separate planes were proportional to the length of the initial trajectory. Such dependency suggests that in order to increase reproduction accuracy and expand the practical applications of cheap IMU sensors, it is necessary to focus further research on the development of error elimination algorithms or search for solutions based on the implementation of additional reference points.

Acknowledgement. This project has received funding from the European Social Fund (project No 09.3.3.-LMT-K-712–22-0344) under a grant agreement with the Research Council of Lithuania (LMTLT).

References

1. Ahmad, N., Ghazilla, A.R., Khairi, N.M., Kasi, V. : Reviews on various inertial measurement unit (IMU) sensor applications. https://doi.org/10.12720/ijsps.1.2.256-262
2. Botero-Valencia, J., Marquez-Viloria, D., Castano-Londono, L., Morantes-Guzmán, L.: A low-cost platform based on a robotic arm for parameters estimation of inertial measurement units. Meas. J. Int. Meas. Confed. **110**, 257–262 (2017). https://doi.org/10.1016/j.measurement.2017.07.002
3. Hedberg, E., Norén, J., Norrlöf, M., Gunnarsson, S.: Industrial robot tool position estimation using inertial measurements in a complementary filter and an EKF. IFAC-PapersOnLine **50**(1), 12748–12752 (2017). https://doi.org/10.1016/j.ifacol.2017.08.1828
4. Kobayashi, F., Hasegawa, K., Nakamoto, H., Kojima, F.: Motion capture with inertial measurement units for hand/arm robot teleoperation. Int. J. Appl. Electromagn. Mech. **45**(1–4), 931–937 (2014). https://doi.org/10.3233/JAE-141927
5. Ivanov, A.V., Zhilenkov, A.A.: The use of IMU MEMS-sensors for designing of motion capture system for control of robotic objects. In: Proceedings of the 2018 IEEE Conference of Russian Young Researchers in Electrical and Electronic Engineering, ElConRus 2018, vol. 2018-January, pp. 890–893, March 2018. https://doi.org/10.1109/EIConRus.2018.8317231
6. Appelboom, G., et al.: Smart wearable body sensors for patient self-assessment and monitoring. Archives of Public Health **72**(1), 28 (2014). https://doi.org/10.1186/2049-3258-72-28
7. Hachaj, T., Piekarczyk, M.: Evaluation of pattern recognition methods for head gesture-based interface of a virtual reality helmet equipped with a single IMU sensor. Sensors **19**(24), 5408 (2019). https://doi.org/10.3390/s19245408
8. Wagner, J.F.: About motion measurement in sports based on gyroscopes and accelerometers—an engineering point of view. Gyrosc. Navig **9**(1), 3–31 (2018). https://doi.org/10.1134/S2075108718010091
9. Kowalczuk, Z., Merta, T.: Modelling an accelerometer for robot position estimation. In: 2014 19th International Conference on Methods and Models in Automation and Robotics, MMAR 2014, pp. 909–914, November 2014. https://doi.org/10.1109/MMAR.2014.6957478
10. Islam, T., Islam, M., Hossam-E-Haider, M.: Comparison of complementary and Kalman filter based data fusion for attitude heading reference system. ARTICLES YOU MAY BE INTERESTED IN comparison of complementary and Kalman filter based data fusion for attitude heading reference system, vol. 1919, p. 20002 (2017). https://doi.org/10.1063/1.5018520

11. Wilson, S., et al.: Formulation of a new gradient descent MARG orientation algorithm: case study on robot teleoperation. Mech. Syst. Signal Process. **130**, 183–200 (2019). https://doi.org/10.1016/j.ymssp.2019.04.064
12. Narkhede, P., Joseph Raj, A.N., Kumar, V., Karar, V., Poddar, S.: Least square estimation-based adaptive complimentary filter for attitude estimation. Trans. Inst. Meas. Control **41**(1), 235–245 (2019). https://doi.org/10.1177/0142331218755234
13. Al-Fahoum, A.S., Abadir, M.S.: Design of a modified Madgwick filter for quaternion-based orientation estimation using AHRS. Int. J. Comput. Electr. Eng. **10**(3), 174–186 (2018). https://doi.org/10.17706/ijcee.2018.10.3.174-186

Influence of Signal Interference on Determining Direction of Arrival by Using the Indirect Phase Determination Method

Bogdan Kreczmer(✉)

Department of Cybernetics and Robotics, Electronics Faculty,
Wrocław University of Science and Technology, Wrocław, Poland
bogdan.kreczmer@pwr.edu.pl

Abstract. The paper describes the influence of the signal interference phenomenon on determining the azimuth angle of the signal arrival. This influence is especially important when a method based on measuring the phase shift is used. Such a method is applied in the sonar exploited in the presented research. To demonstrate the influence of the interference effect, a series of preliminary experiments are described and reported.

Keywords: Ultrasonic sonar · Direction of arrival · Interference

1 Introduction

Determination of echo arrival direction is the key feature which makes ultrasonic sonars reliable and useful for object detection. This feature means that they can be also effectively used for mobile robot navigation to solve the SLAM problem [10]. There are well-known methods of signal processing that can be used for determining the direction of arrival (DOA) of a signal. One of the most known method is the multiple signal classification (MUSIC) [8] which can be used in the case of simultaneous emission of several signal sources. The estimation of signal parameters via rotational invariance techniques (ESPRIT) [6,7] is another approach to the same problem. Another well-known method is beamforming which exploits array processing techniques. The very effective implementation of this approach is presented in [9]. To receive a signal, a sparse array of analog MEMS microphones was used. This type of sonar was so effective that it was possible to implement a navigation system for a mobile robot [11] and the system was able to solve the SLAM problem. The very important advantage of the beamforming method is that most of the computations can be implemented in hardware. Such a solution based on one-bit MEMS microphones and the architecture developed around an FPGA-SoC is presented in [3]. Further integration and size minimization are presented in [12]. This time a rectangular uniform array of microphones was used instead of a sparse array.

R. Szewczyk et al. (Eds.): AUTOMATION 2021, AISC 1390, pp. 319–328, 2021.
https://doi.org/10.1007/978-3-030-74893-7_30

The important disadvantage of the presented approaches is that they are relatively computationally expensive. In [5], it was presented the approach which is based on measuring signal phase shift. The main advantage is simplicity in software and hardware implementation. The price of its simplicity is that it cannot distinguish between sources simultaneously emitting a signal. This is due to its sensitivity to signal interference. But on the other hand, when the emitted wave packet is not long and the received echoes only partially overlap, the interference effect can provide useful information. It is supposed that bats and dolphins perceive the shape of objects from the interference spectrum generated when multiple reflections from different parts of objects overlap and interfere [1]. To examine such a possibility, preliminary experiments have been performed using a sonar for which the method of indirect phase measurement [5] has been applied. The results of these experiments are presented and discussed in this paper.

2 Determining DOA for the Wave Package

When the received signal is narrowband, some simplification can be made. They enable the use of methods of determining the DOA of the received signal by measuring its phase shift detected by different receivers. The case of phase shift measurement is not trivial when inter-distance between receivers is greater than half the wavelength of the signal. Fortunately, it is not complex either. Such a method is presented in [4] and in more detail together with the theoretical analysis in [5]. The discussed approach is restricted to the 2-D case, it means that the sonar using this method determines distance to an object and the azimuth angle of DOA. The elevation angle is not computed. Despite of this restriction, the method has the important advantage that is its simplicity. However, there is also another important advantage. The azimuth angle can be determined not only for the front of the received wave packet. It can be also found for its other parts. To explain the idea, the example of the sonar containing four receivers will be used (see Fig. 1a). Figure 1b shows the example of signals received by all receivers. When the signal in the receiver channels exceeds the detection threshold, the times of exceeding the zero level by the signal are recorded in the

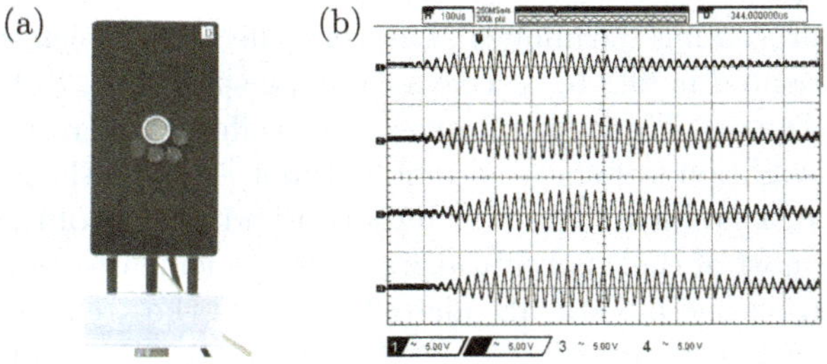

Fig. 1. The sonar with four receivers b) the example of signal received by each of receivers

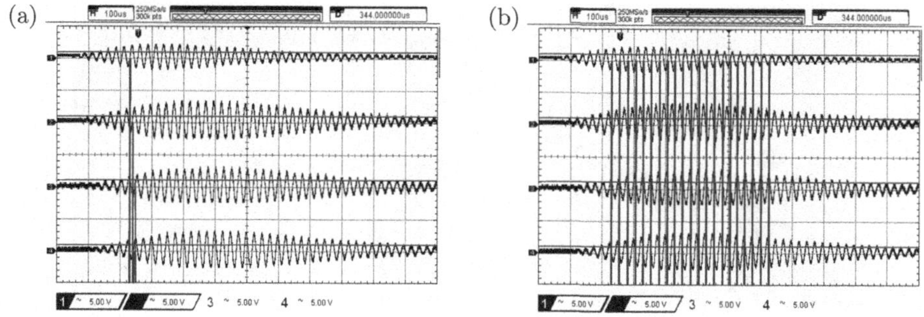

Fig. 2. Measurements of times of signal zero crossings for a) a single detection b) subsequent detection for the entire wave packet, for clarity, moments of zero-crossing detection for the first receiver was marked only

next moment in each of these channels. (see Fig. 2a). They are used to determine the azimuth angle of DOA. The same procedure can be performed for subsequent zero crossing by the signal in this wave packet (see Fig. 2b). When the signal is reflected by a wall or by a post, for all subsequent measurements, the same azimuth angle should be obtained. A similar result should be observed when two objects are almost at the same distance.

This changes when the source of the reflected signals is an object with a more complex surface. In this case, it can be expected that a kind of object profile can be obtained. In general, the variation in azimuth angles of the direction of signal arrival should be observed when two or more echoes partially overlap. Due to changes in the amplitude of the signal in the wave packet (Fig. 1b), the values of the determined azimuth angles will not be stable in the region where the echoes overlap. Conversely, when an echo part is not overlapped by a signal of another echo, it can be expected that subsequent values of the computed azimuth angle should be the same with respect to the error of angle determination. This is the clue that can be used to determine the actual direction to a single object that is the source of the echo. To check these hypothesis, series of preliminary experiments were performed. Their are described and reported in the next section.

3 Experiments

The sonar used in the experiments is shown in Fig. 1a. This device is equipped with a single transmitter BPU-1640T0AH12 and four MA40S4R receivers. All ultrasonic transducers are controlled by the module [2], which exploits the microcontroller MC9S12A64.

The simplest case which exposes interference effect is an arrangement of two aluminum posts. This case was the subject of the described experiment. The arrangement of the experiment stand is presented in Fig. 3. It shows the positions of the posts in relation to the sonar module. The posts were placed

Fig. 3. Experimental stand, measurements were performed for two aluminum posts placed in front of the sonar module

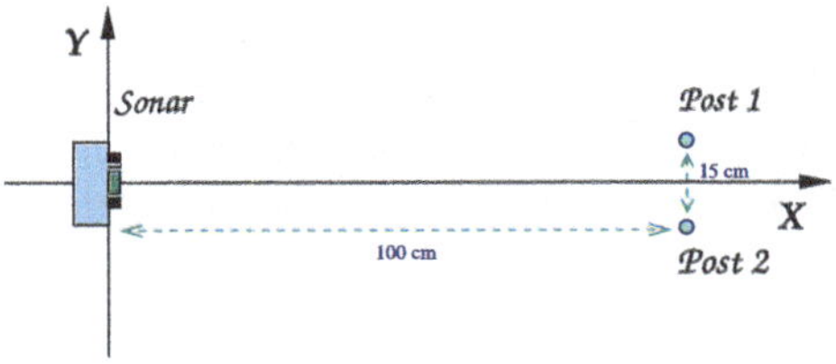

Fig. 4. Scheme of the experimental stand

in front of the sonar module at a distance of 100 cm (see Fig. 4). The distance between posts along the axis OY was constant and equal to 15 cm. In order to compare results of determining the azimuth angles for the signal which is not corrupted by interference and that one which is distorted by this effect, first series of measurements were performed for the single post no. 1. The data obtained from a single measurement of the received echo were used to determine the azimuth angles for 20 successive periods of the registered signal. Then, the same procedure was applied to the single post no. 2. In each series, 100 measurements were done. This made it possible root square deviations to be computed for all determined azimuth angles.

Taking into account that the posts were placed at a distance of 100 cm along the axis OX and 7.5 cm along the axis OY (see Fig. 4), the expected directions to the post no. 1 and the post no. 2 should be $4.29°$ and $-4.29°$, respectively. This gives the angular distance between posts equal to $8.58°$. Moreover, it is expected that the determined azimuth angles will be the same for successive parts of the echo signal.

The results of these measurements for the post no. 1 are presented in Fig. 5, and for the post no. 2 in Fig. 6. In general, the results are very close to the expected values. The azimuth angles determined for the first part of echo for the post no. 1 and 2 are $3.97 \pm 0.11°$ and $-4.55 \pm 0.14°$, respectively. Which gives

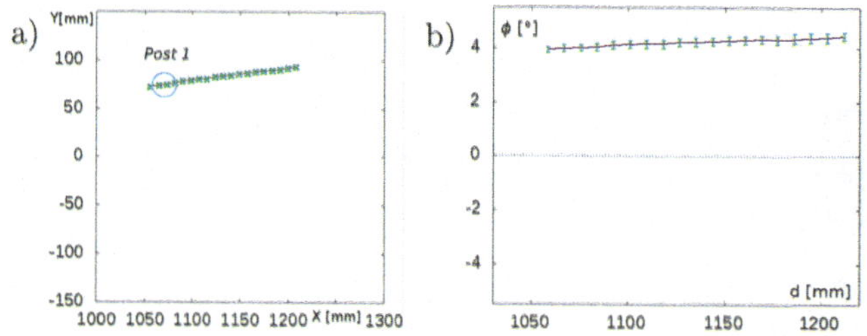

Fig. 5. The results of measurement interpretation when it was only the post no. 1 a) position of points determined for successive zero-crossings of the received echo, b) changes of the determined azimuth angle in relation to the distance

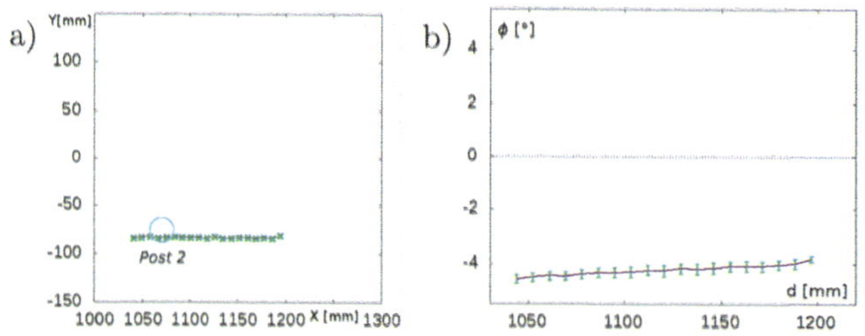

Fig. 6. The results of measurement interpretation when it was only the post no. 2 a) position of points determined for successive zero-crossings of the received echo, b) changes of the determined azimuth angle in relation to the distance

the angular distance of $8.52 \pm 0.25°$. In the range of error, this distance agrees very well with the distance which was theoretically computed. The difference is only $0.06°$. Values of azimuth angles are also very close the theoretical ones. They seem to be shifted about $0.3°$. This accuracy for this experiment is acceptable.

By observing the values of the determined azimuth angle for successive parts of the echoes for the posts no. 1 and 2, some discrepancy can be noticed. As mentioned before, it is expected that azimuth angle values should be the same or nearly the same for all parts of echo signal. Figure 5b and 6b show that there is a slight change. In the both cases, the values of the determined azimuth angles slightly increase. This is probably due to small differences in the receiver channels. Nevertheless, these changes are minor. Therefore, they do not greatly infuence the results of the further experiments presented in this section.

In the first part of the experiment, the position of the post no. 1 was constant. The post no. 2 for each new series of measurements was successively shifted by 1 cm along the OX axis (see Fig. 4). In this way, the measurements were performed for the distances of the post no. 2 equal to 100 cm, 101 cm, and so on up to 110 cm. Then the second part of the measurements was performed. This time, the position of the post no. 2 was constant and equal to 100 cm. Analogously, as it was done previously for the post no. 2, the post no. 1 was shifted 1 cm for each new series of measurements (see Fig. 7). Thus, the successive

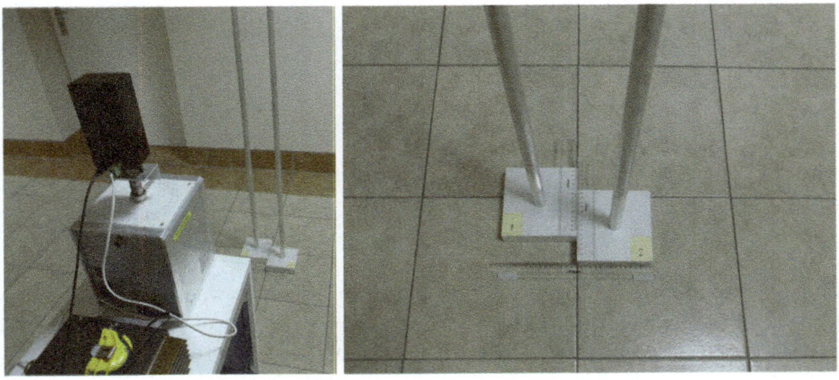

Fig. 7. The measurements were performed for the aluminum posts that were successively shifted relative to each other by a step of 1 cm

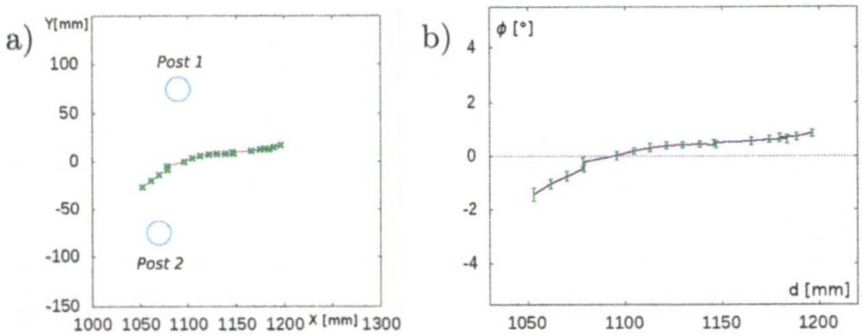

Fig. 8. Results of measurement interpretation for the post no. 1 shifted 20 mm behind the post no. 2 a) position of points determined for successive zero-crossings of the received echo, b) changes of determined azimuth angle in relation to distance

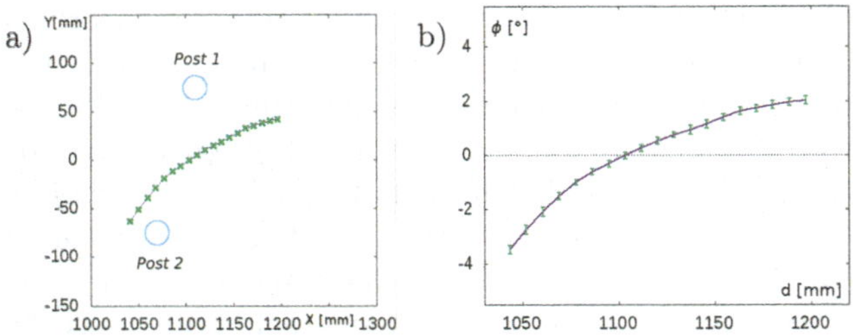

Fig. 9. The results of the measurement interpretation for the post no. 1 shifted by 40 mm behind the post no. 2 a) point positions determined for successive zero-crossings of the received echo, b) changes of the determined azimuth angles in relation to distance

series of measurements were performed when the post no. 2 was at a distance of 101 cm, 102 cm, and so on up to 110 cm.

The results of the first part of the experiment for shifts of 2 cm, 4 cm, 6 cm and 8 cm are presented in Fig. 8, 9, 10 and 11. They show that the interference effect introduces errors of azimuth angle determination. Nevertheless, the

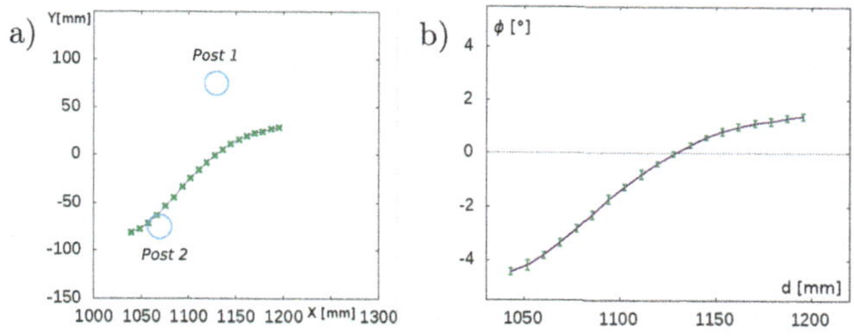

Fig. 10. The results of the measurement interpretation for the post no. 1 shifted by 60 mm behind the post no. 2 a) point positions determined for successive zero-crossings of the received echo, b) changes of the determined azimuth angles in relation to distance

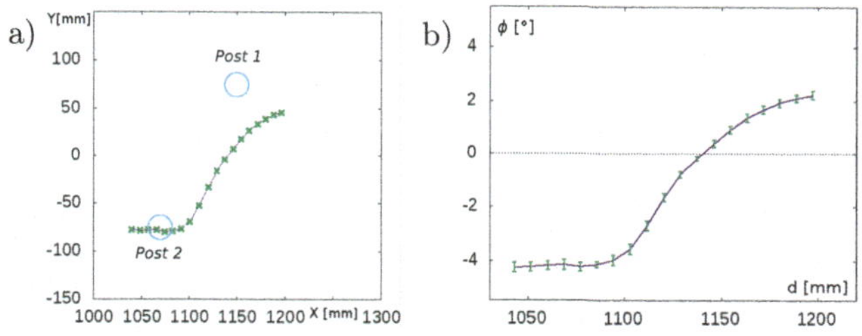

Fig. 11. The results of the measurement interpretation for the post no. 1 shifted by 80 mm behind the post no. 2 a) point positions determined for successive zero-crossings of the received echo, b) changes of the determined azimuth angles in relation to distance

obtained results do not differ very much from the correct ones. It is true when the objects are almost at the same distance from the sonar and they are not far from each other. The points marked in the figures and connected by lines were determined by computing distances and azimuth angles for each selected elementary part of the echo signal. For each echo, 20 successive elementary parts were only taken into account. The figures show the curves created by connecting the points depending on the arrangement of objects. Moreover, their shape is related to the relative positions of the posts. In this sense, it seems that this kind of information can be used to deduce the arrangement of objects.

In the second part of the experiment, the roles of the posts were exchanged and this time the post no. 2 was shifted while the position of the post no. 1 was being constant. The obtained results are presented in Fig. 12, 13, 14 and 15. It was expected that these cases would be symmetrical to the previous ones. The results presented in the figures seems to confirm this expectation. The presented results show that the determination of the azimuth angle for all elementary parts of the echo can create a kind of object alignment profile. Unfortunately, it does not seem to be a unique map between the profile generated by the interfering echoes and the arrangement of the objects. Therefore, such a profile can be seen as an indication rather than certain information.

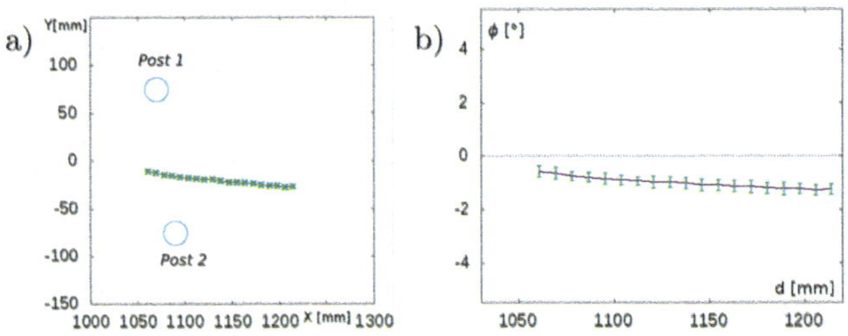

Fig. 12. The results of the measurement interpretation for the post no. 2 shifted by 20 mm behind the post no. 1 a) point positions determined for successive zero-crossings of the received echo, b) changes of the determined azimuth angles in relation to distance

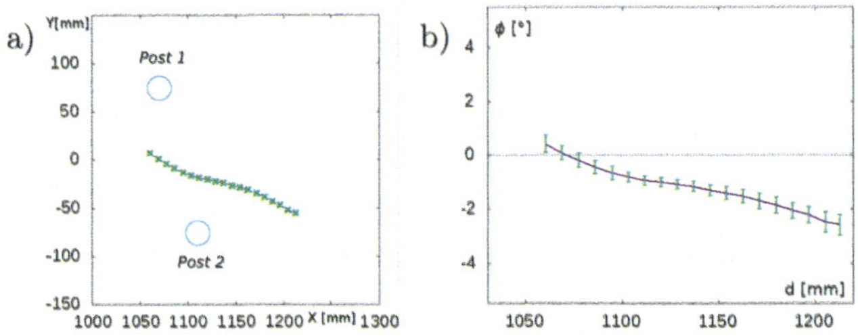

Fig. 13. The results of measurement interpretation for the post no. 2 shifted by 40 mm behind the post no. 1 a) position of points determined for successive zero-crossings of the received echo, b) changes of the determined azimuth angles in relation to distance

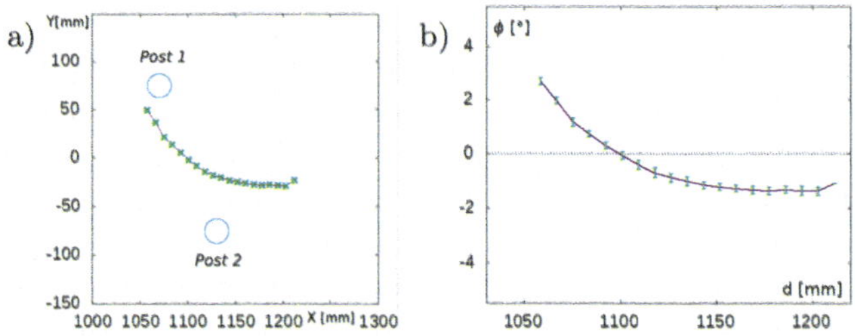

Fig. 14. The results of measurement interpretation for the post no. 2 shifted by 60 mm behind the post no. 1 a) position of points determined for successive zero-crossings of the received echo, b) changes of the determined azimuth angle in relation to distance

In Fig. 16 the summary of obtained results is shown. The positions of the posts are marked with small crosses. To make the diagrams easier to read, the results of single measurements were not marked. Only the lines connecting them are shown. Next to each line is the post shift value that corresponds to this line. Despite the fact that the considered case of shifting pole 1 and then pole 2 is symmetrical, the obtained results do not strictly have this property. The reason is the lack of symmetry in the arrangement of the receivers. Nevertheless, the visible differences do not significantly change the obtained measurement *profiles*.

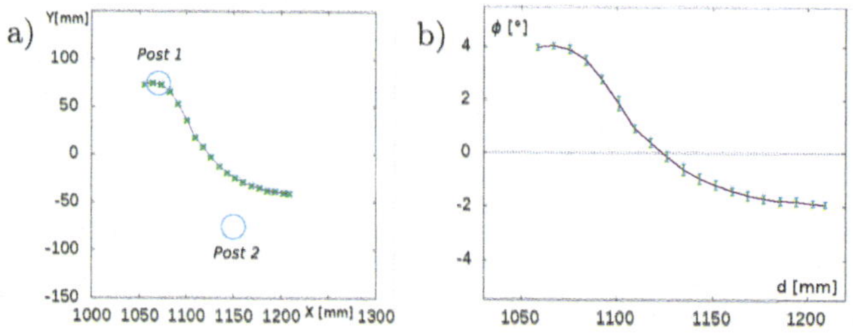

Fig. 15. The results of measurement interpretation for the post no. 2 shifted by 80 mm behind the post no. 1 a) position of points determined for successive zero-crossings of the received echo, b) changes of the determined the azimuth angle in relation to distance

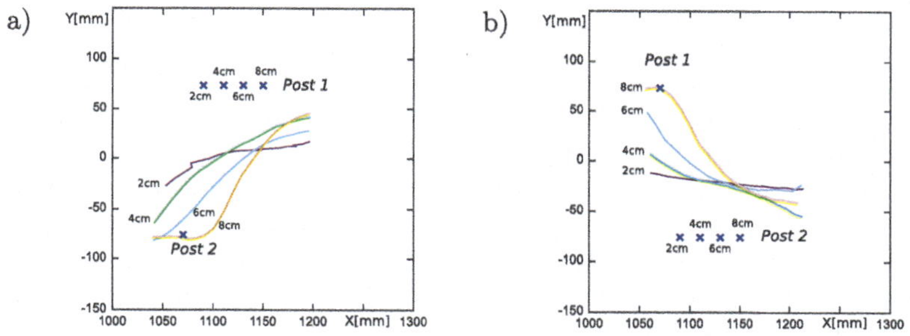

Fig. 16. Summary of the obtained results a) for changes in the position of the post no. 1, b) for changes in the position of the post no. 2

4 Conclusion

When a method based on phase shift measurement is used to determine DOA, interference of overlapping echoes can significantly distort the results. The presented results of experiments seems to show that this type of distortion is smooth and can be useful in determining the area where objects are close together This conclusion seems too optimistic. Nevertheless, a series of stable equal azimuth angles with respect to calculation errors can be taken as a major indication that some of the received signal is from a single object. It therefore increases reliability of such values. The only exception is when two or more objects are almost the same distance. Then it can be expected that almost the same azimuth angle value will be observed even for the entire wave packet. Fortunately, this situation can be identified with additional sonar. Because for the second sonar, these objects will not be the same distance. Regardless, it is a good idea to create a system consisting of several sonars as some redundancy is needed. In the case of signal interference phenomena, this becomes essential.

References

1. Au, W.W.L., Simmons, J.A.: Echolocation in dolphins and bats. Phys. Today **60**(9), 40 (2007)
2. Kabała, M., Wnuk, M.: Moduł z mikrokontrolerem MC9S12A64. Technical report SPR nr 11/2005, Inst. Inform. Autom. Robot. Politechniki Wrocławskiej, December 2005
3. Kerstens, R., Laurijssen, D., Steckel, J.: Low-cost one-bit mems microphone arrays for in-air acoustic imaging using FPGA's. In: 2017 IEEE SENSORS, pp. 1–3, October 2017
4. Kreczmer, B.: Azimuth angle determination for the arrival direction for an ultrasonic echo signal. J. Autom. Mob. Robot. Intell. Syst. **11**(02), 31–41 (2017)
5. Kreczmer, B.: Estimation of the azimuth angle of the arrival direction for an ultrasonic signal by using indirect determination of the phase shift. Archi. Acoust. **44**(3), 585–601 (2019)
6. Roy, R., Kailath, T.: ESPRIT-estimation of signal parameters via rotational invariance techniques. IEEE Trans. Acoust. Speech Signal Process. **37**(7), 984–995 (1989)
7. Roy, R., Paulraj, A., Kailath, T.: Direction-of-arrival estimation by subspace rotation methods - ESPRIT. In: IEEE International Conference on Acoustics, Speech, and Signal Processing, ICASSP 1986, vol. 11, pp. 2495–2498, April 1986
8. Schmidt, R.: Multiple emitter location and signal parameter estimation. IEEE Trans. Antennas Propag. **34**(3), 276–280 (1986)
9. Steckel, J., Boen, A., Peremans, H.: Broadband 3-D sonar system using a sparse array for indoor navigation. IEEE Trans. Robot. **29**(1), 161–171 (2013)
10. Steckel, J., Peremans, H.: Spatial sampling strategy for a 3D sonar sensor supporting BatSLAM. In: 2015 IEEE/RSJ International Conference on Intelligent Robots and Systems (IROS), pp. 723–728, September 2015
11. Steckel, J., Peremans, H.: BatSLAM: simultaneous localization and mapping using biomimetic sonar. PLOS One **8**(1), 1–11 (2013)
12. Verellen, T., Kerstens, R., Steckel, J.: High-resolution ultrasound sensing for robotics using dense microphone arrays. IEEE Access **8**, 190083–190093 (2020)

Volumetric Estimation of Non-structured Heterogeneous Waste

Adam Wołoszczuk(✉), Mateusz Bielecki, Szymon Buś, Mateusz Maciaś, Jakub Główka, and Michał Bryła

ŁUKASIEWICZ Research Network – Industrial Research Institute for Automation and Measurements PIAP, Warsaw, Poland
{adam.woloszczuk,mateusz.bielecki,szymon.bus,mateusz.macias,jakub.glowka, michal.bryla}@piap.lukasiewicz.gov.pl

Abstract. The aim of this paper is to introduce a novel method for volumetric estimation of non-structured heterogeneous waste stored in the trash container of smart waste management application. First, AI-aided method of trash volume estimation based on millimeter wave radar data is presented. Subsequently, the conditions and configuration of experimental setup are described, with research methodology. Further on, insight into experiment outcome in terms of radar sensors and AI-based detector performance benchmarked against legacy sensors is discussed. The final part of the paper summarizes the conclusions of experiment and recommendations for expanding the research.

Keywords: Volumetric estimation · Radar · Neural networks · Artificial intelligence · Waste management · Smart applications

1 Introduction

In this paper we introduce and discuss the volumetric estimation method based on millimeter-wave radar sensors benchmarked against ultrasonic and optical distance measuring methods. The aim of the experiments was to test the hypothesis that data gathered from millimeter wave radar sensor processed by AI-based detector can allow efficient estimation of available volume in trash container. The research case pertains to issues raised by smart waste management professionals who we cooperated with on the course of RobotUnion EU H2020 project. The feedback from professionals indicated the major problems with ultrasonic and optical ToF (Time of Flight) sensors used so far in their applications (further referred to as 'legacy sensors'). The ultrasound sensors performed sub-optimally when the bin was filled with granular objects featuring complex geometry or absorbent materials. Also, a proper adjustment of sensors due to unreliable detection cone was flagged as a problem. The optical sensors displayed difficulties with detection of transparent objects – erroneous readings occurred when plastic, glass or foil were placed in trash bin. General conclusions formulated by cooperating professionals stated that ultrasonic sensors tended to detect false

R. Szewczyk et al. (Eds.): AUTOMATION 2021, AISC 1390, pp. 329–341, 2021.
https://doi.org/10.1007/978-3-030-74893-7_31

full states of container, while optical sensors often omitted full states. These issues encouraged us to search for other means of quantifying the stored waste. The solution consisting of radar and neural network processing was proposed as an alternative and potentially more reliable technique for experimentation.

Even though the experimental case was conducted for waste volume estimation scenario, the approach presented in this paper may be used in the spectrum of smart applications related with storage and logistics.

2 Problem Statement

Ultrasonic and optical sensing are among methods used for solid waste characterisation [1]. Both of these technologies are appropriate for distance measurement, however the indications can be affected either by geometric features of measured object (as in case of ultrasonic sensors) or material properties such as transparency or light absorption (in case of optical sensors) [2]. In case of ultrasonic sensor a beam-width should also be carefully considered to achieve a balance between sensors field of view and measurement resolution [3].

Market survey in the field of waste level detection applications revealed several sonar-based solutions such as MaxBotix and Sensoneo [4,5]. Less common are the commercially available complex ToF solutions for high accuracy 3D topology such as those provided by Nordsense [6].

Based on cooperating professionals experience and our observation, the waste, even within single category, varies in size, shape, density, opacity and spatial arrangement making it significantly difficult to detect and perceive by single type of distance measuring sensor. Also, from practical perspective the most important is the space occupied by waste itself, not including the space in between pieces of waste.

By proposing a radar based approach we aim to avoid the problems related with legacy sensors. Contrary to legacy sensors, that rely on surface reflection measurement, the material-penetrating capabilities of radar sensors allowed us to experiment with reflected signal measurement in order to estimate the volume of waste in the container. State of the art survey revealed some promising research oriented on volume estimation using ground penetrating radar with antenna frequencies of 450 MHz, 900 MHz and 1200 MHz [7].

The millimeter wave radar technology, that corresponds to frequencies of 30GHz and higher [8], is relatively new but it is expected to develop rapidly due to forthcoming phase out of 24 GHz UWB (ultra wide band) radars by year 2022 [9]. Integrated millimeter wave sensors (radar-on-chip sensors) tend to be very compact due to tight antenna array architecture [10], which makes them suitable for robotics [11], UI (user interfaces), motion pattern detection [12] and gesture recognition [13].

Solutions using millimeter radar technology include commercially available products such as Vayyar sensors [14], Acconeer sensors [15], or Texas Instruments IWR6843 [16] as well as in-development projects such as Soli [17], Socionext SC1220AT2 [18] or RadarCat [19] system developed by St Andrews HCI research group. Some of commercial millimeter wave radar solutions are streamlined toward automotive industry with integrated functionalities such as adaptive cruise control, autonomous emergency braking etc. such as TEF8102 [20] radar transceiver by NXP.

There are also some findings in state of the art on neural network based approach to volume estimation related to quantitative food assessment [21].

In order to improve the precision of estimation we decided to combine capabilities of radars with predictions of neural network and implemented an AI-based detector trained with data collected by sensor module.

3 Experiment Configuration

Two main functional blocks of the experimental set-up were defined: sensor module for data collection and AI-based detector for volume estimation performed in post-processing. Data was gathered using custom-designed measurement station hosting sensor module and peripherals. We incorporated a standard trash bin used in smart waste management applications by our cooperating professionals.

3.1 Sensor Module

The purpose of the sensor module was to gather data needed for comparison between radar sensors and legacy sensors, as well as to provide reference from depth-camera. The composition of experimental sensor module was fixed after the analysis of available and potentially useful sensors. In terms of radar selection, the targeted solutions were narrowed to available Acconeer A111 radar-on-chip module and Texas Instruments IWR6843 radar sensor. To simplify the integration of the sensor module both of the radars were used on their developement boards - SparkFun Pulsed Radar Breakout for A111 and IWR6843ISK-ODS with MMWAVEICBOOST interface board for TI IWR6843.

Due to its size and simplicity A111 is more promising in low-range narrow-space applications. However, TI IWR6843 module features more antennae and provides more output formats (including range profile, static and Doppler heatmaps), while A111 returns only range profile data. That is why we decided to integrate both radars in sensor module.

For vision-based 3D reference the Intel RealSense D435i depth camera was selected due to compact size and good software support.

The control software for the module was created to manage measurement cycles and prevent sensors from cross-interference.

Legacy sensors were the ones used by cooperating professionals in trash level detection applications - ultrasound and optical time-of-flight distance sensors. In this case both are of the MicroDetectors brand - UK1D/G1-0ESY Cylindrical Ultrasonic sensor [22] and M18 ToF Sensor: RED LEDModel TFKRD/G1-0EIO [23].

Block diagram of module is presented in Fig. 1a. Intel NUC [24] and Raspberry Pi 4B (RPi) [25] are connected to LAN via Ethernet. Intel Realsense D435i depth camera [26] and Texas Instruments IWR6843 radar module [16] were connected to NUC via USB. A111 radar [27] was connected to RPi via SPI. PCB hat (additional expansion module plugged on top of the RPi) with ADS1115 analog-to-digital (ADC) converter [28] converted output voltages from ultrasonic range sensor UK1D/G1-0ESY and M18 time-of-flight (ToF) sensor TFKRD/G1-0EIO to numeric values which were read by the RPi via I^2C. Real-life photo of the sensors mounted in the module is in Fig. 1c.

Raspbian 10 Buster ran on RPi and Ubuntu 18.04 Bionic ran on NUC. Robot Operating System (ROS) Melodic [29] was used for managing measurements. There was a separate ROS node for each sensor and a separate ROS topic for each type of data from a sensor. For example, there were five types of data from TI IWR6843 radar (detected objects, range profile, noise profile, static heatmap and doppler heatmap) and five corresponding ROS topics.

The dedicated test station, shown on Fig. 1b, was designed and built as a mounting platform for subject sensors. The station allows adjustment of XYZ position of sensor module in relation to waste container. While our A111 radar sensor is integrated on Sparkfun PCB, we needed to create the adapter that would be compatible with Sparkfun PCB and Acconeer hyperbolic lens. Our custom holder design is based on dedicated adapter from Acconeer but modified so that it features 4 slots allowing to fix the lens on predefined chip-lens distance and easy lock that allows to fit and fix the holder on the PCB without the need of using any tools.

3.2 AI-Based Detector

The neural network based detector was created in order to perform actual estimation of volume based on the data gathered by module. Neural networks are modern computational techniques that can resolve many stochastic problems. Based on known and reliable practices [30] it was decided to create neural networks that take data from radars and compute an estimated volume. Training process was realized in two alternative approaches, with different data used as a ground truth (details in Sect. 3.3 section).

The main idea behind the neural network design was to maintain low computational complexity so it could be implemented in computers with low computing power - especially computers without GPU modules. Block diagram of the designed neural network architecture is shown in the Fig. 2. Input data from radars are processed using convolution neural network layers – two-dimensional convolutional layers for data from TI IWR6843 radar and one-dimensional layers for data from A111 radar. The same neural network architecture was used in both approaches.

3.3 Methodology

The hypothesis behind the experiment states that millimeter-wave radar sensors supported with AI-based detector can be effectively used for volume assessment

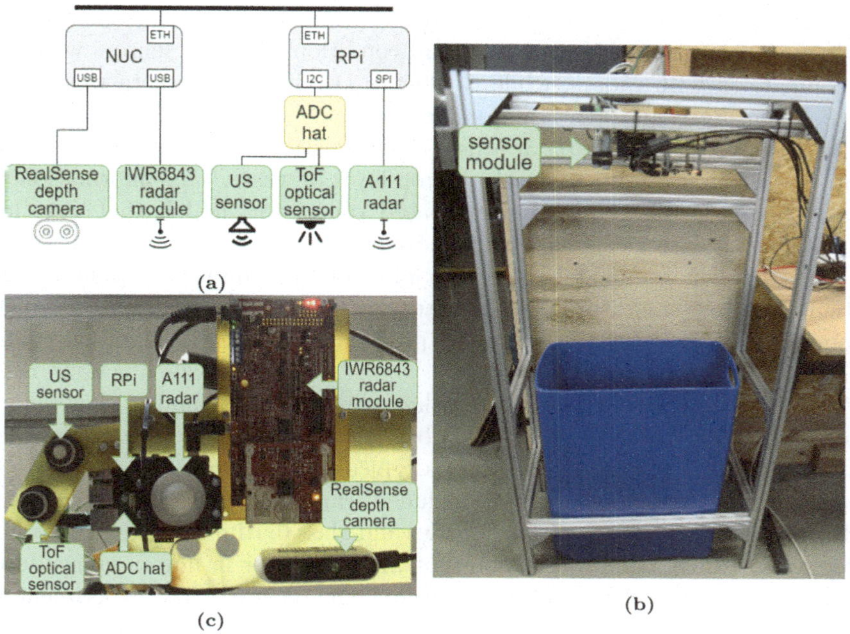

Fig. 1. (a) Block diagram of sensor module, (b) test station, (c) sensors in sensor module.

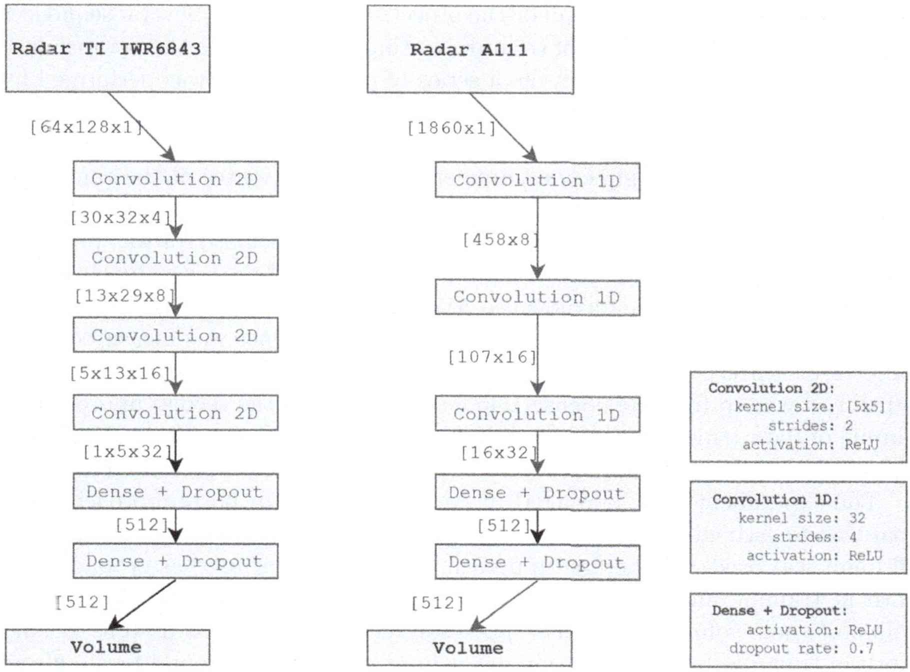

Fig. 2. Architecture of AI-based detector neural networks

of trash stored in the container. The hypothesis was tested on dedicated measurement station in stable environmental conditions. Legacy sensors were used for comparison with radars, while RealSense D435i depth camera was used to obtain visual RGB and depth-map reference.

As described is Sect. 2, radar sensors can penetrate material and deliver the information about objects beneath the top layer of waste. This information was used to build reliable reflected signal-based volumetric models that were later compared with data gathered from legacy sensors and depth camera.

The stable environmental conditions addressed the ambient parameters such as temperature, humidity and light intensity. The experiment was conducted indoors with electric light. The station position in relation to light sources was constant during the whole process.

The data collection process was conducted using sensor module. For the purpose of the experiment several types of objects were selected as waste to be stored in the container. Each type represents a typical sample from a group of objects considered as dry waste. The types included: cardboard boxes of differing sizes ($2.3\,\mathrm{dm}^3$ to $5.6\,\mathrm{dm}^3$), PET $1.5\,\mathrm{dm}^3$ bottles and glass $0.5\,\mathrm{dm}^3$ bottles. The volume of bin was $75\,\mathrm{dm}^3$. The objects were put into container in non organized order and manner while data from all sensors were being recorded. The experiment was conducted in recurring measurement cycles forming a sequence. A sequence covers the series of discrete states of container corresponding with various amounts of waste stored. Starting a sequence from the empty container as reference, a new object was placed in the container at the beginning of each measurement cycle, until the container was full or the objects were depleted. Several sequences were recorded for the purpose of training, testing and validating the performance of AI-based detector. In each cycle, a series of measurements was performed by each sensor in the sensor module, gathering data corresponding to the state of the container. For training and testing the bin was filled 13 times and sensor module collected 248 sample measurements of discrete states of the container. The data from the first 11 fillings of the bin was intended as training data and the remaining 2 fillings as test data. Furthermore validation data was prepared and bin was filled 5 times with additional types of garbage - e.g. lunch boxes, aluminum cans, that were unknown to AI-based detector. Additional 95 samples for validation process were collected. True-volume reference was measured manually (for each type of garbage) in the first approach and calculated based on depth raster map from RealSense D435i sensor data in the second approach. A sample of data (single cycle) containing image and depth map from RealSense, static heatmap from IWR6843 and range profile from A111 is presented in Fig. 3.

The experiment was conducted in two approaches. In both approaches the goal was to estimate net trash volume (sum of manually measured volumes of all items was used). In the first approach, net volume was used as ground truth both in training and in validation. In the second approach, estimate of gross volume (whole volume under the upper surface of objects) was used as ground truth in training, but in validation net volume was used as ground truth. Gross volume estimation was based on RealSense D435i depth camera measurements.

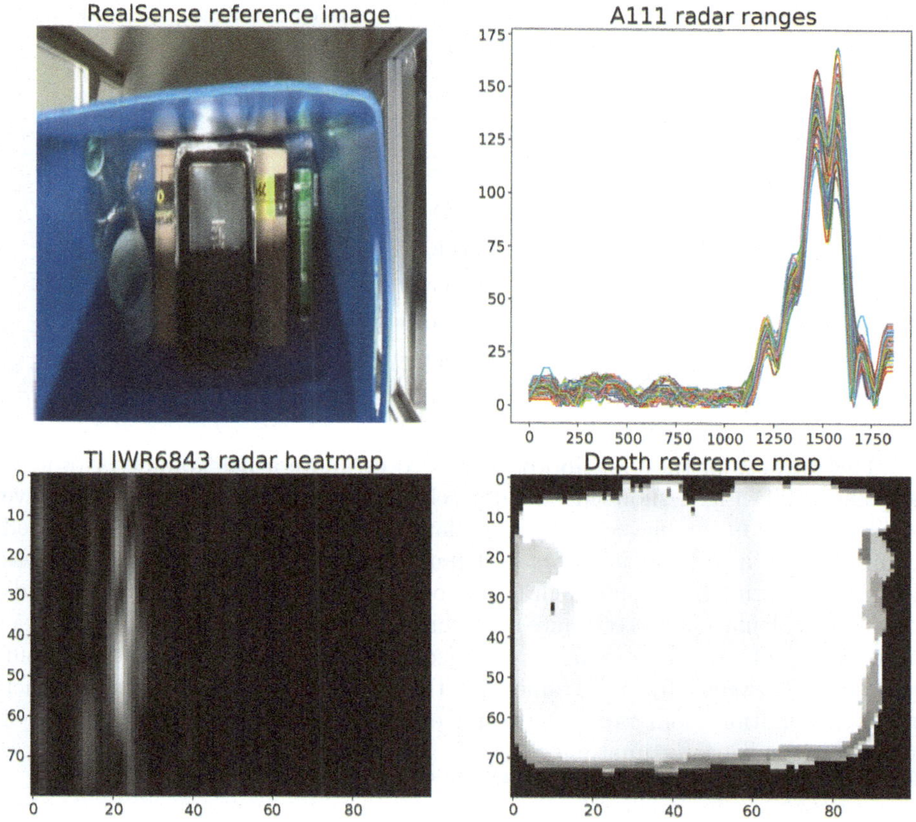

Fig. 3. Sample data: image from RealSense D435i (top left), range profile from A111 radar (top right), static heatmap from TI IWR6843 radar module (bottom left) and depth map from Realsense (bottom right).

4 Outcome

All training processes were conducted using 7000 epochs, with mean squared error as a loss function and mean absolute error function as a metric. In the first approach, 11.5 dm^3 and 15.7 dm^3 mean absolute volume errors (MAE) on validation data were obtained for TI IWR6843 and A111 radar data, respectively. It corresponds to 8.0 and 11.5 mean absolute percentage errors (MAPE). In the second approach, 13.2 dm^3 and 20.5 dm^3 mean absolute volume errors on validation data were obtained for TI IWR6843 and A111 radar data, respectively. It corresponds to 10.0 and 15.0 mean absolute percentage errors. Errors summary is presented in Table 1.

Table 1. Mean absolute error of calculated and estimated cumulative volumes to the true-measured values for different methods for validation data set

Method		MAE $[dm^3]$
Reference volume - RealSense		15.2
Reference volume - M18 ToF		20.2
Reference volume - Ultrasonic UK1D		14.9
The first approach	Estimated volume - TI IWR6843	**11.5**
	Estimated volume - A111	15.7
The second approach	Estimated volume - TI IWR6843	**13.0**
	Estimated volume - A111	18.9

The purpose of the first approach was to estimate a net volume of objects sequence. The Fig. 4 shows net volume changes, corresponding to consecutive objects placed in the bin, for validation data. Reference volumes were calculated using data from different sensors and put together with true (manually measured) and estimated (using neural networks) net volumes. Reference methods of volume calculation overestimate the volumes in most cases. These sensors measure distance to the upper surface of objects and cannot get information about free space between objects. Trained for this task neural networks can extract more information from radar data and estimate approximate net volumes of objects without overestimations.

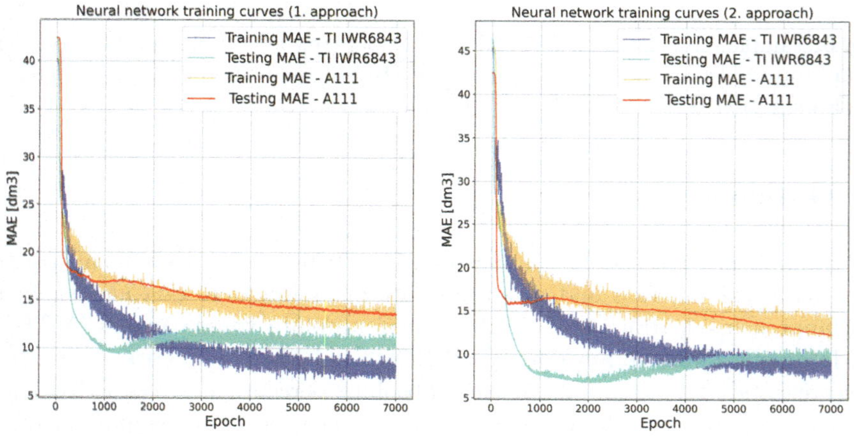

Fig. 4. Neural networks training metrics - mean absolute errors for the first and second approach.

The Fig. 5 presents the comparison between volume estimated by both radar sensors and reference measurements taken by legacy sensors and RealSense D435i

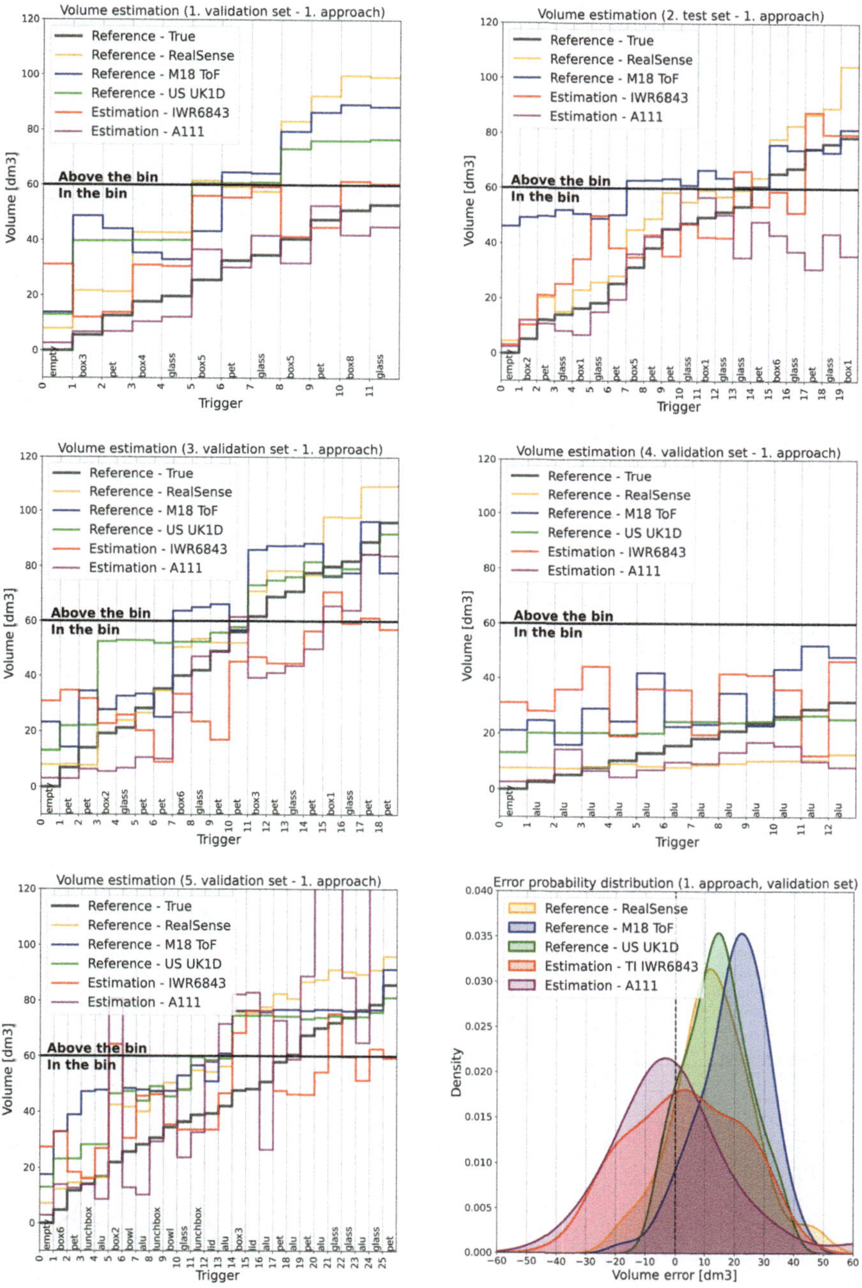

Fig. 5. Volume estimations vs. reference measurements, and volume error density in approach 1

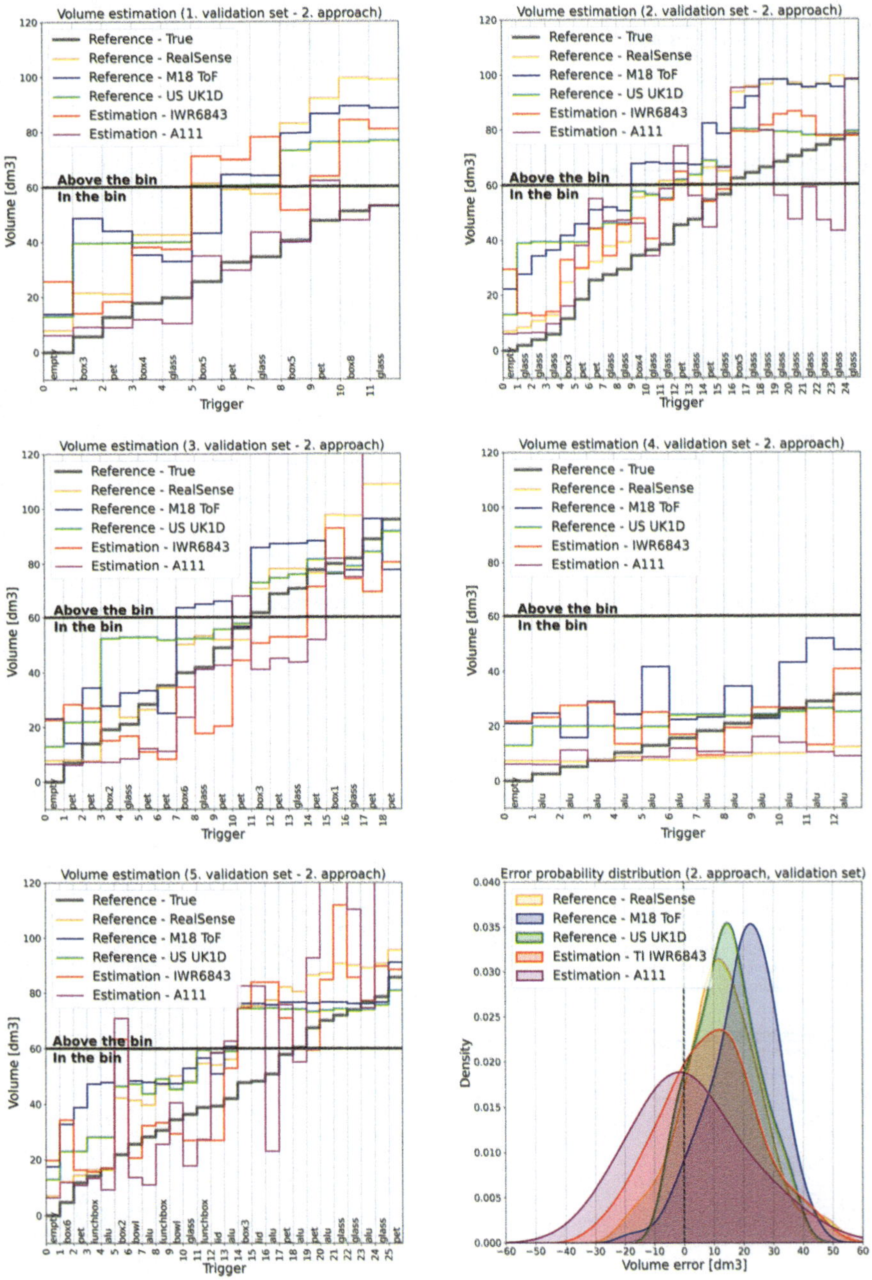

Fig. 6. Volume estimations vs. reference measurements, and volume error density in approach 2

depth camera. The distribution of volume error density shows that neural networks can estimate volume without overestimating values. Approach with using data from TI IWR6843 radar and a convolution neural network stands out as a good alternative to other metrics methods used for comparison and reference - especially to a volume estimated based on RealSense D435i data. In contrary, approach with using A111 radar data is not accurate enough. The difference in radar performance results from the fact that TI IWR6843 radar features multiple antennae and the capability to provide complex output such as radar heatmap, while A111 features single antenna and returns only a range profile. The neural network architecture must be properly designed to get high accuracy of an estimated volume. We used simple architectures to emulate practical aspects of using this approach in microcontrollers and without GPU modules. We can certainly expect that the improvement of the networks and the selection of training parameters will improve the accuracy of the estimation. Dataset preparing process is long and cost-consuming so we tried to find another method not relying on manual volume measurement for each type of garbage separately.

The second method we proposed using other sensor as reference. The idea of the second approach is to use gross volumes calculated from RealSense D435i depth maps, as a true-references for neural network training processes. Training processes were repeated and validation results were stacked again as shown on Fig. 6. The second approach provided worse results than the first approach. We try to fit the volumes from RealSense D435i data, with some errors, so density distribution of estimated volumes error moved too close to RealSense D435i error density. Nevertheless, the second method is still more reliable than legacy sensors methods and data preparation for training process is much faster than manual measurements used in first approach.

5 Conclusions

This paper presents possibility of using millimeter-wave radar sensors and neural networks for estimation of heterogeneous (mixed) waste volume.

Presented methodology provides a basis for implementation in the end-user setup. Conducted experiments gathered 343 trash bin volume measurements represented with data sets obtained from different sensors. Measurements were grouped into 18 recorded sequences. Obtained results show advantages of using radar data and neural network in comparison with legacy sensors for waste volume estimation.

The practical use of recording of training data using depth camera and radar at the same time enables effective and efficient way of obtaining sufficient accuracy of future measurements with use of only one low-cost radar sensor and data processing.

Future research based on lessons learned should be directed towards practical implementation of the obtained results. First step is to conduct more recordings of data sets in relevant conditions using the methodology introduced in this paper. The sensor module and chosen detectors should be deployed in a trash bin that is representing a end-user setup. Another point that is arising from the conducted trials is the possibility to experiment with fusion of data from radar and some legacy sensors in order to provide a sensor module able to measure volume in a wider range of conditions and in more varied configurations.

Acknowledgments. The project leading to the presented results has received funding from the European Union's Horizon 2020 research and innovation programme under grant agreement No 779967.

Data Availability. The data that support the finding of this study and the source code necessary for the reproduction of the results are available at https://github.com/piappl/volumetric.

References

1. Vrancken, C., Longhurst, P.J., Wagland, S.T.: Critical review of real-time methods for solid waste characterisation: informing material recovery and fuel production. Waste Manage. **61**, 40–57 (2017)
2. Mohammad, T.: Using ultrasonic and infrared sensors for distance measurement. World Acad. Sci. Eng. Technol. **51**, 293–299 (2009)
3. Lee, C.Y., Choi, H.G., Park, J.S., Park, K.Y., Lee, S.R.: Collision avoidance by the fusion of different beam-width ultrasonic sensors. In: 2007 IEEE Sensors, pp. 985–988. IEEE, Atlanta (2007)
4. Maxbotix: Easy to use ultrasonic sensors for professional and maker engineers. https://www.maxbotix.com. Accessed 29 Nov 2020
5. Sensoneo: Smart waste management. https://sensoneo.com/. Accessed 29 Nov 2020
6. Nordsense: Smart containers. https://nordsense.com/ns-navigator/ns-pod/. Accessed 09 Oct 2020
7. Simms, J.E., McKay, S.K., McComas, R.W., Fischenich, J.C.: In situ root volume estimation using ground penetrating radar. J. Environ. Eng. Geophys. **22**(3), 209–221 (2017)
8. Wang, X., Kong, L., Kong, F., Qiu, F., Xia, M., Arnon, S., Chen, G.: Millimeter wave communication: a comprehensive survey. IEEE Commun. Surv. Tutor. **20**(3), 1616–1653 (2018)
9. Ramasubramanian, K., Ramaiah, K.: Moving from legacy 24 GHz to state-of-the-art 77 GHz radar. ATZelektronik Worldwide **13**(3), 46–49 (2018)
10. Ghosh, S., Sen, D.: An inclusive survey on array antenna design for millimeter-wave communications. IEEE Access **7**, 83137–83161 (2019)
11. Barrett, D., Alvarez, A.: mmWave radar sensors in robotics applications (2017). https://www.ti.com/lit/wp/spry311a/spry311a.pdf. Accessed 01 Dec 2020
12. Zhang, R., Cao, S.: Real-time human motion behavior detection via CNN using mmWave radar. IEEE Sens. Lett. **3**(2), 1–4 (2019)
13. Smith, K.A., Csech, C., Murdoch, D., Shaker, G.: Gesture recognition using mm-wave sensor for human-car interface. IEEE Sens. Lett. **2**(2), 1–4 (2018)
14. Vayyar: Vayyar sensor website. https://vayyar.com/. Accessed 29 Nov 2020
15. Acconeer: A1 radar sensor. https://www.acconeer.com/products. Accessed 29 Nov 2020

16. Texas Instruments: mmWaveICBoost and Antenna Module User's Guide, April 2020. https://www.ti.com/lit/ug/swru546c/swru546c.pdf. Accessed 26 Nov 2020
17. ATAP, G.: Soli project website. https://atap.google.com/soli/. Accessed 29 Nov 2020
18. Socionext: 60GHz 3D detection model SC1220AT2 website. http://www.socionext.com/en/products/assp/radar-sensor/SC1220/. Accessed 29 Nov 2020
19. St Andrews HCI Research Group: Radar and object recognition. https://sachi.cs.st-andrews.ac.uk/research/interaction/radarcat-exploits-googles-soli-radar-sensor-for-object-and-material-recognition/. Accessed 29 Nov 2020
20. NXP: TEF810X fully-integrated 77 GHz radar transceiver. https://www.nxp.com/products/rf/radar-transceivers/tef810x-fully-integrated-77-ghz-radar-transceiver:TEF810X. Accessed 29 Nov 2020
21. Gao, J., Tan, W., Ma, L., Wang, Y., Tang, W.: MUSEFood: multi-sensor-based food volume estimation on smartphones. In: 2019 IEEE SmartWorld, Ubiquitous Intelligence & Computing, Advanced & Trusted Computing, Scalable Computing & Communications, Cloud & Big Data Computing, Internet of People and Smart City Innovation (SmartWorld/SCALCOM/UIC/ATC/CBDCom/IOP/SCI), pp. 899–906. IEEE (2019)
22. Micro Detectors: UK1 and UKR1 Series Product Catalogue (2020). https://www.microdetectors.com/FileDocumenti/Articoli/Cat-Serie/ENG/UK1-UKR1.pdf. Accessed 26 Nov 2020
23. Micro Detectors: Short Form Catalogue, January 2020. https://www.microdetectors.com/Media/Cat-breve/PDF/cat-breve-ENG.pdf. Accessed 26 Nov 2020
24. Intel: Intel® NUC Kit NUC7i7BNH Product Specifications. https://ark.intel.com/content/www/us/en/ark/products/95065/intel-nuc-kit-nuc7i7bnh.html. Accessed 26 Nov 2020
25. Raspberry Pi Foundation: Raspberry Pi 4 Model B Specifications. https://www.raspberrypi.org/products/raspberry-pi-4-model-b/specifications. Accessed 26 Nov 2020
26. Intel: Intel RealSense Product Family D400 Series, June 2020. https://www.intelrealsense.com/wp-content/uploads/2020/06/Intel-RealSense-D400-Series-Datasheet-June-2020.pdf. Accessed 26 Nov 2020
27. Sparkfun: SparkFun Pulsed Radar Breakout - A111 - SEN-15577 - SparkFun Electronics. https://www.sparkfun.com/products/15577. Accessed 26 Nov 2020
28. Seed Studio: 4-Channel 16-Bit ADC for Raspberry Pi (ADS1115). https://www.seeedstudio.com/4-Channel-16-Bit-ADC-for-Raspberry-Pi-ADS1115.html. Accessed 26 Nov 2020
29. Open Source Robotics Foundation: Melodic - ROS Wiki. http://wiki.ros.org/melodic
30. Goodfellow, I., Bengio, Y., Courville, A.: Deep Learning. MIT Press (2016). http://www.deeplearningbook.org

Uncertainty Bands of the Regression Line for Data with Type A and Type B Uncertainties of Dependent Variable *Y*

Zygmunt Lech Warsza[1] and Jacek Puchalski[2](✉)

[1] Łukasiewicz Research Network – Industrial Research Institute for Automation and Measurements PIAP, Al. Jerozolimskie 202, 02-486 Warsaw, Poland
zlw1936@gmail.com

[2] Central Office of Measures, Elektoralna 2, 00-137 Warsaw, Poland
jacek.puchalski@gum.gov.pl

Abstract. This work concerns on the estimation of the accuracy of function determined by the linear regression method for the description of noncorrelated measured data of Y. Recommendations of the international Guide to the Expression of Uncertainty in Measurement (GUM) are used. The impact of Type B measurement uncertainties is included, which is omitted in the statistical literature on the accuracy of regression method. The introduction presents the essence of the uncertainty calculations used in GUM. The case of random changes of variable Y and the criteria used in linear regression are discussed in detail. For known values of X variable, the parameters and uncertainty bands of regression line are determined for measurements of uncorrelated values of Y with Type A and Type B uncertainties. Considerations are illustrated with four numerical examples of the measurement points with the same coordinates, but different absolute and relative uncertainties.

Keywords: Linear regression · Covariance matrix · Uncertainty of types A and B · Regression line · Uncertainty bands

1 Introduction

The linear regression method is used to determine a linear or linearized function, fitted to the coordinates of points according to a specific criterion [2–4]. The purpose of this work is a method of assessing the accuracy of the parameters of a regression line and its uncertainty bands determined from uncorrelated measurement data. It is based on the recommendations of the latest version of the Guide to the Expression of Uncertainty in Measurement, known under acronym GUM [1], along with Supplements. This guide is now commonly used in Measurement Science[1] and is obligatory in the national service

[1] This common term for metrology and the basics of measurement technique have been proposed in Acta IMEKO in the 1970-s by Ludvik Finkelstein, professor at City University in London, born in Lviv.

R. Szewczyk et al. (Eds.): AUTOMATION 2021, AISC 1390, pp. 342–363, 2021.
https://doi.org/10.1007/978-3-030-74893-7_32

of measures, in all industry measurements, and in other utility research and science. But in the basic literature on regression and even the last one, e.g. [2–4, 8, 17] the influence of type B uncertainty was not noticed and analyzed. This uncertainty results from the Maximum Permissible Error of measuring instruments and systems (MPE) with the symbol $E_{\max}$ [1].

In the introduction of this work, the principle of measurement accuracy assessment based on uncertainties according to the GUM convention is discussed. Then the types of measurement tasks in which linear regression is used and the determination of simple regression parameters and its uncertainty bands for several examples of uncorrelated measurement data with different variants of its uncertainty, known and also unknown, but of the same value for all points, is given.

Let us briefly present the rules for assessing the accuracy of measurements according to the GUM guide [1]. The result of each measurement observation includes the deviation from the unknown real value and its estimate, e.g. the mean value of their set as the sample. Deviations with known and an unknown constant or variable value in a determined manner during the measurements are systematic errors, and deviations that change randomly – are random errors. Systematic errors with known values are removed by corrections, and those with unknown values are randomized with the assumption that the probability of simultaneous occurrence of their maximum values is very small. Then their common influence are described with a random variable, i.e. are treated as random errors. This approach was adopted in the GUM guide since 1993[2].

As an estimator of the measurement result, it is assumed, as for the Gaussian distribution, the estimator of the mean value of the measurement observations in the sample. The assessment of the estimator's accuracy is based on the combined uncertainty u as standard deviation of its measurement data. It is the geometric sum of the type A and B uncertainties, as u_A and u_B, components, i.e.:

$$u = \sqrt{u_A^2 + u_B^2}$$

Uncertainties of type A and B are related to describe the error sets from a different time scale (short time of experiment and long one for the period of instruments' calibration and measurement results validation) and it is assumed that they are not correlated with each other. From the combined standard uncertainty $\underline{u}$, the expanded uncertainty $U = k_P \cdot u$ is determined. It defines a symmetrical interval of width $\pm U$, within which the estimator of the measurement result will be located with the probability P. For normally distributed observation values (Gaussian function) and $P = 0.95$ or 0.99, coverage factor $k_P = (1.96$ and $2.58)$. For other distributions, the Monte Carlo method according to GUM-S1 [1] is used. The u_A component, that is the type A uncertainty, is determined statistically as the standard deviation of the mean value of the n-times repeated measurement results of the quantity investigated in a given experiment. For a small number of $n < 30$ of these measurements, the Student distribution is used, and for a larger number – the normal distribution.

Usually in a measurement experiment, after correction of known systematic errors for the result estimator, its remaining value is not known. It is estimated statistically by

[2] Similar approach was applied in the early 1950-s, in the doctoral dissertation by Stanislaw Trzetrzewiński, later professor of the Gdansk University of Technology, i. e. over 40 years before the publication of the first version of the GUM guide.

the resultant uncertainty of type B from all influences on the measured values and their signals in the tested object, on the parameters of the elements of signal transmission and processing in the measurement system and on the indications devices. The external influences and internal parameter changes occurs within the specified permissible limits of their changes and working conditions of the measured object throughout the rated period of validity of the instrument calibration. When estimating the type B uncertainty of measurements, the knowledge about the predicted ranges, probability distributions, shares of individual interactions and their mutual correlation as well as the influence of changes in internal parameters of the measurement system is used [1, 5–7].

Type B uncertainties of sensors, transducers, meters and other measuring equipment are estimated for the rated values of their parameters on the basis of the maximum permissible errors E_{max} [%]. These errors have different values for instruments dedicated for the laboratory, industrial, field and military applications. They also change with the value of the measured quantity, e.g. uncertainties of measuring current transformers (transformers) depend non-linearly on the ratio and on load. For digital instruments, the permissible absolute error usually has a small initial value and increases linearly with the measured quantity x. These errors are uniformly distributed with a width of $\pm E$. Their standard deviation $u_B = u_{B0} + (x - x_0)E_{max}/\sqrt{3}$ is then its type B uncertainty. If the predicted errors from various influences are smaller than the error E, then its smaller span is assumed accordingly [6, 7]. The correct estimation of type B uncertainty is very important in measurement applications and requires a metrological knowledge and practice. It was not given in statistical publications on the linear regression.

When using the regression in single and variate measurements, the assessment of the accuracy depends on measurement errors and deviations resulting from simplifications of the mathematical model adopted for the tested object and the disturbances occurring in it, which extend the uncertainty band of the regression line. Their influences are jointly described by the residua variable.

It is also necessary to properly plan the experiment, including the selection of the location of tested points and repetition of measurements, estimation the volume of information obtained for each measurement point and the entire range of measurements, and determining the duration of the measurements. In addition, there are also logistical constraints, such as the date and place of tests, the availability of measuring equipment and the test object, including material samples, and the need to automate measurements and data processing. Usually, the number of tests and the costs of their implementation should be also minimized. Such tasks are individual in nature and are not discussed here. Some improvements to the GUM guide have recently been proposed, e.g. median accuracy assessment, Bayes approach [8–10] and robust methods [12]. One must always consider the uncertainty of u_B, correlation and autocorrelation [6, 13–15].

2 Types of Linear Regression and Their Criteria

In general case, the linear regression method is used to determine the relationship $Y = F(X)$ between two multivariate variables Y and X. Based on measurements of the values of both variables at multiple points, or only for measurements of one of them, when the values of the other are known, it determines parameters of the multivariate function $F(\cdot)$ best suited to the measurement results according to a specific criterion. In the linear regression, Y depends linearly on the coefficients of the equations derived from the measurement data.

In this work, the linear regression method is considered only for different cases of one-dimensional variables $Y = [y_1, \ldots y_n]$ and $X = [x_1, \ldots x_n]$. The equation of a straight- line $y = ax + b$, called here the regression line, is sought fitted, according to the criterion for the minimum squared distance from n points with the coordinates (x_i, y_i). The uncertainties of parameters a, b and uncertainty bands of linear regression line within a specified range and with the required probability P are estimated.

Depend on the type and purpose of measurements, the different types of linear regression are used. Here are some basic examples:

a. The values of the x_i coordinates of the variable X for $i = 1, \ldots n$ of the examined points are precisely determined, i.e. with negligibly small uncertainties. The measured values of y_i is treated as random and statistically independent or correlated variables both, when making measurements, and during the validity period of their results. Using the regression method, a straight line is sought, the equation of which will best fit the measurement data according to criterion (11). If the uncertainties $u(y_i)$ of the measurements of the ordinates y and of the examined points are known and not equal, the regression method is applied using the least squares criterion with weights of acronym **WLS**. Weighted Least Squares [5]. Standard and expanded uncertainty bands for simple regression should consider the influence of uncertainty u_B.

b. If the uncertainties $u(y_i)$ of measured points are not known, they are assumed to be the same and the simplest variant of linear regression is applied with the criterion: minimum sum of squared distances of measured points from the regression line in the direction of 0y with weights $w_i = 1$, named by acronym **OLS** - Ordinary Least Squares.

c. The y_i values of each point with a known abscissa x_i undergo a scattering. By measuring them repeatedly, you get a sample with the values $y_{i1}, \ldots y_{in_i}$. For the known probability distribution, or assuming the normal distribution, the local mean value $\overline{y}$ of the sample is calculated as an estimate y_i and its standard deviation $u_A(y_i)$ - as the uncertainty of type A. The effects of autocorrelation on their uncertainties are determined from the local covariance matrices of each point, or the effective number of n_{eff} measurements and is obtained according to [5] or [7–10]. The next step is as for case a.

d. Random dispersions of measurement results and uncertainties occur for both coordinates (x_i, y_i). Then the regression line can be determined for the minimum sum of the normalized squares of the distance of points from the regression line in the 0Y direction or in another direction, e.g. perpendicular to this line [7]. The values of the variable Y (and/or X) of points can be correlated and the propagation of covariance matrix is used. This criterion has the abbreviation **GLS** (Generalized Least Squares).

3 The Regression Line and Its Bands with Type B Uncertainty

In monographs [2–5] and in the latest publications on estimating the accuracy of a simple is determined from measurement data using the linear regression method, e.g. [8, 17], no analysis of the influence of type B uncertainty was found. uncertainty of simple regression for uncorrelated values of y_i obtained from measurements of n examined points ordinates. Figure 1 shows this points and the designated regression line for them

$$y = ax + b \tag{1}$$

This regression line is determined from values of coordinates (x_i, y_i) of the points Q_i, which are at distances $\Delta y = y_i - y_{pi}$ in the direction 0Y from the points $P(x_i, y_{pi})$ on the straight line. The parameters a and b in the Eq. (1) of this line and the course of the boundaries of its uncertainty band are obtained according to the **WLS** criteria. It is the minimum sum of the squares of the distance $(y_i - y_{pi})^2$ for n measured n points Q_i normalized to the square of standard uncertainty $u(y_i) \equiv u_i$ for each point. $y = ax + b$ in model with uncertainties for measurement only the abscissa Y as random variable. This criterion is as follows:

$$SK = \sum_{i=1}^{n} \frac{(y_i - ax_i - b)^2}{u_i^2} \to \min \tag{2}$$

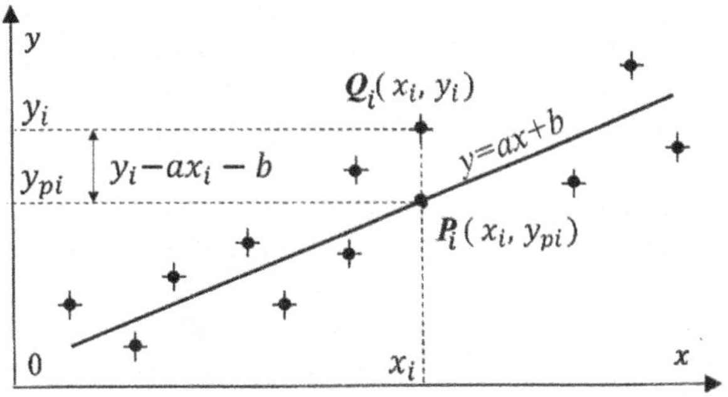

Fig. 1. Illustration of the distance between measured point Q_i and point P_i on the regression straight-line

Formulas that meet this condition become easier after entering auxiliary parameters:

$$\begin{aligned} &S = \textstyle\sum_{i=1}^{n} u_i^{-2}, \quad S_x = \sum_{i=1}^{n} x_i u_i^{-2}, \quad S_{xx} = \sum_{i=1}^{n} x_i^2 u_i^{-2}, \\ &S_{xy} = \textstyle\sum_{i=1}^{n} x_i y_i u_i^{-2}, \quad S_y = \sum_{i=1}^{n} y_i u_i^{-2}, \quad S_{yy} = \sum_{i=1}^{n} y_i^2 u_i^{-2} \end{aligned} \tag{3a–f}$$

After applying formulas (3a–f), Eq. (2) transforms into the form:

$$SK = a^2 S_{xx} + 2abS_x - 2aS_{xy} + b^2 S - 2bS_y + S_{yy} \tag{4}$$

The minimum of (4) occurs for $\frac{\partial SK}{\partial a} = 0$, $\frac{\partial SK}{\partial b} = 0$. Hence the system of equations and its solutions

$$\begin{cases} aS_{xx} + bS_x = S_{xy} \\ aS_x + bS = S_y \end{cases} \tag{5a,b}$$

$$\text{where:}\quad a = \frac{SS_{xy} - S_x S_y}{SS_{xx} - (S_x)^2} \equiv \frac{\Delta_a}{\Delta}\quad b = \frac{S_y S_{xx} - S_x S_{xy}}{SS_{xx} - (S_x)^2} \equiv \frac{\Delta_b}{\Delta} \tag{6a,b}$$

and $\Delta_a = \sum_{i=1}^n u_i^{-2} \cdot \sum_{i=1}^n x_i y_i u_i^{-2} - \sum_{i=1}^n x_i u_i^{-2} \cdot \sum_{i=1}^n y_i u_i^{-2}$,

$$\Delta_b = \sum_{i=1}^n y_i u_i^{-2} \cdot \sum_{i=1}^n x_i^2 u_i^{-2} - \sum_{i=1}^n x_i u_i^{-2} \cdot \sum_{i=1}^n x_i y_i u_i^{-2},$$

$$\Delta = \sum_{i=1}^n u_i^{-2} \sum_{i=1}^n x_i^2 u_i^{-2} - \left(\sum_{i=1}^n x_i u_i^{-2}\right)^2 = \tfrac{1}{2} \sum_{i=1}^n \sum_{j=1}^n (x_i - x_j)^2 u_i^{-2} u_j^{-2}, \tag{7a–c}$$

If the uncertainties $u_i \equiv u$ are of the same value, then they will be shortened in the formulas (6a,b) even if their values are not known and the obtained parameters a and b are the same for the different values of uncertainties u (see chapters 4 and 5).

From (4), (6a,b) and (7a–e) the function of SK_{MJN} results, which is:

$$SK_{\text{MIN}} = -aS_{xy} - bS_y + S_{yy} \tag{8}$$

The uncertainty bands of any regression line extend on both sides of it. They are determined from the uncertainty u_a for the slope a and the uncertainty u_b of the intercept b. It is also necessary to consider the correlation of parameters a and b. The partial derivatives of a and b for i = 1,..., n are described by the formulas:

$$\frac{\partial a}{\partial y_i} \equiv a_i = \frac{1}{u_i^2}\frac{Sx_i - S_x}{\Delta},\quad \frac{\partial b}{\partial y_i} \equiv b_i = \frac{1}{u_i^2}\frac{S_{xx} - S_x x_i}{\Delta} \tag{9a,b}$$

The sensitivity matrix $\boldsymbol{AB}$ and the covariance matrix $\boldsymbol{U_Y}$ for measured n values of y_i points Q_i are:

$$\boldsymbol{AB} = \begin{bmatrix} a_1 \ldots a_n \\ b_1 \ldots b_n \end{bmatrix},\quad \boldsymbol{U_Y} = \begin{bmatrix} u_1^2 & \ldots & 0 \\ \vdots & \ddots & \vdots \\ 0 & \ldots & u_n^2 \end{bmatrix} \tag{10a,b}$$

In Chapter 32 of this book, cases of the Y variable with statistically independent deviations Δy_i of measured points are only considered. On the main diagonal of the covariance matrix $\boldsymbol{U_Y}$ there is a variance u_i^2. For y_i points with uncorrelated coordinates, non-diagonal elements describing the correlation are equal to zero.

The covariance matrix $\boldsymbol{U_{ab}}$ for a and b results from the variance propagation

$$\boldsymbol{U_{ab}} = \boldsymbol{ABU_Y AB^T} = \begin{bmatrix} u_a^2 & \rho_{ab} u_a u_b \\ \rho_{ab} u_a u_b & u_b^2 \end{bmatrix} \tag{11}$$

where: u_a, u_b, ρ_{ab} – uncertainties and correlation coefficient of a and b.

After solving (11), for the uncorrelated values of the variables y_i we get:

$$u_a^2 = \sum_{i=1}^n a_i^2 u_i^2 = \frac{S}{\Delta} = \frac{\sum_{i=1}^n u_i^{-2}}{\Delta},\quad u_b^2 = \sum_{i=1}^n b_i^2 u_i^2 = \frac{S_{xx}}{\Delta} = \frac{\sum_{i=1}^n x_i^2 u_i^{-2}}{\Delta} \tag{12a,b}$$

$$\rho_{ab}u_a u_b = \sum_{i=1}^{n} a_i b_i u_i^2 = \frac{-S_x}{\Delta} = \frac{-\sum_{i=1}^{n} x_i u_i^{-2}}{\Delta} \quad (13)$$

The correlation coefficient for parameters a and b results from (13)

$$\rho_{ab} = -\frac{S_x}{\sqrt{S \cdot S_{xx}}} = -\frac{\sum_{i=1}^{n} x_i u_i^{-2}}{\sqrt{\sum_{i=1}^{n} u_i^{-2} \cdot \sum_{i=1}^{n} x_i^2 u_i^{-2}}} \quad (14)$$

For $x_i > 0$ (where $i = 1, \ldots, n$), coefficient $\rho_{ab} < 0$.

The standard uncertainty u_{ab} of the Y variable for the regression line $y = ax + b$ is determined as a function of x from the equation of uncertainty propagation considering the correlation of a and b:

$$u_{ab}^2 = \left[\frac{\partial y}{\partial a}, \frac{\partial y}{\partial b}\right] \begin{bmatrix} u_a^2 & \rho_{ab}u_a u_b \\ \rho_{ab}u_a u_b & u_b^2 \end{bmatrix} \left[\frac{\partial y}{\partial a}, \frac{\partial y}{\partial b}\right]^T \quad (15)$$

After inserting the derivatives $\frac{\partial y}{\partial a} = x$, $\frac{\partial y}{\partial b} = 1$ and performing the operations is:

$$u_{ab}^2 = x^2 u_a^2 + 2x\rho_{ab}u_a u_b + u_b^2 \quad (16)$$

Expressing the variance u_{ab}^2 through auxiliary parameters from (3a, b) and (7c) we get

$$u_{ab}^2 = \frac{Sx^2 - 2S_x x + S_{xx}}{SS_{xx} - (S_x)^2} = \frac{1}{S} + \frac{S}{\Delta}\left(x - \frac{S_x}{S}\right)^2 \quad (17)$$

For the variance u_{ab}^2 it is the second-order parabola with the axis of symmetry: $x_{min} = \frac{S_x}{S} = \frac{\sum_{i=1}^{n} x_i u_i^{-2}}{\sum_{i=1}^{n} u_i^{-2}}$ with minimum $u_{ab}^2|_{min} = \frac{1}{S} = \frac{1}{\sum_{i=1}^{n} u_i^{-2}}$ and $\frac{S}{\Delta} = \frac{\sum_{i=1}^{n} u_i^{-2}}{\Delta}$ at x^2.

However, the uncertainty u_{ab} is described by the hyperbola of the equation:

$$Su_{ab}^2 - \frac{S^2}{\Delta}\left(x - \frac{S_x}{S}\right)^2 = 1 \quad (18)$$

After substituting the expressions of S and S_x from (13a, b) to (27), a direct dependence of the variance u_{ab}^2 on the parameters of the measured points is obtained

$$u_{ab}^2 = \sum_{i=1}^{n} \frac{(x - x_i)^2 u_i^{-2}}{\Delta} \quad (19)$$

Formulas (12a,b)–(19) can be used for all cases of u_A and u uncertainties. From Eqs. (17) and (19) the following conclusions can be drawn:

- with an increase in the number of measured points Q_i and any of their uncertainty u_{n+i}, the value of the S parameter increases and the A-type uncertainty band narrows, as $u_{ab}|_{min} = 1/\sqrt{S}$ decreases;
- reducing the uncertainties u_i of the measured points also increases the value of S;

- the S/Δ coefficient for the term x^2 of the parabola (18) decreases with the decrease in the uncertainty of measurement points and the increase in their number, so its arms approach to the regression line.

The regression method seeks a straight-line $y = ax + b$ as a mathematical model adjusted to the values of the n coordinates of the measured points $Q_i(x_i, y_i)$. In this paper, the cases where only their ordinates y_i are measured with uncertainties of A type denoted as u_{Ai} are analyzed, and the x_i abscissa are known. Parameters a, b are determined from formulas (6a,b), and their type A uncertainties – from (2a, b), (12, 13). The measure of the accuracy of the points of this line is the area lying on both sides of the line called the band or the corridor [5] of expanded uncertainty with a height of $2U_y$ in the 0y axis direction. Its limits are described by two functions of x as $y_U = ax + b \pm U_y$. They are determined from the combined uncertainty band $u_y(y)$, which is the geometric sum of the type A uncertainty band obtained from the measurements of n points and marked as U_{Aab}, and the band of heuristically estimated type B uncertainty:

$$u(y) = \sqrt{u_{Aaby}^2 + u_{By}^2} \tag{20}$$

For the number of measurement points $n > 30$, the Gaussian distribution is used and $u_{Aab} = u_{ab}$ according to the formula (18) or (19), and for a smaller number of n – the Student's distribution. In the proposal for the new GUM Guide 2 based on the Bayesian approach [8–10, 13 pp. 199], an approximate formula is given:

$$u_{Aab}^2 = \frac{n-1}{n-3} u_{ab}^2 \tag{20a}$$

Uncertainty bands can be presented both as a function of the measured variable Y and converted into a function of the variable X or vice versa. It is converted using the straight-line equation x = $(y - b)/a$ determined from the measurements. And the uncertainty u_B, as already mentioned, is estimated from the value of the maximum permissible error E_{max} of the applied meter. From the $u(y)$ value, the expanded uncertainty $U_P(y)$ is determined with the required confidence level, e.g. with the coverage factor $k_{0.95}$. Only in the special case, when the measurement points are measured with different measures, or under different conditions and at different times (e.g. climate tests), then from multiple measurements for each P_i point, both the y_{pi} estimate is determined as the mean value of $y_{pij}(x_i)$ and its standard deviation as the uncertainty u_{Api}. The combined uncertainty u_{yi} of each point (x_i, y_i) has been calculated with its uncertainty u_{Bi} of type B from the formula:

$$u_{yi} = \sqrt{u_{Ai}^2 + u_{Bi}^2} \tag{20b}$$

Basic types of measurement experiments described by the straight-line pattern are:

1. Measurements of all points are made with a meter so accurate that its type B uncertainty is negligibly small and then the expanded uncertainty is

$$U_y(y) = U_{yA}(y) = k_{0.95}u_{ab}(x) = t_{0.95,n-2}\, u_{ab}(y), \tag{21}$$

where $(n - 2)$ is the effective number of degrees of freedom for n measured points.

2. Typically, the uncertainty of type B is not negligible and when all measurements of y are made with the same meter or measurement system, it increases linearly according to the dependence $u_{By}(y) = u_{B0} + k_B \cdot y$ resulting from the maximum permissible error $E_{\max}$ of this measure. Combined uncertainties $u_y(y)$ of points on regression line are determined from the geometric sum of type A uncertainty $u_{Aab}(y)$ and the type B uncertainty $u_{By}(y)$ for a given x. The expanded uncertainty band U_y is a function of x or y

$$U_y(ax + b) = k_{0.95}\sqrt{u_{Aab}^2(ax + b) + u_{By}^2(ax + b)} \tag{21a}$$

$$U_y(y) = k_{0.95}\sqrt{u_{Aab}^2(y) + u_{By}^2(y)} \tag{21b}$$

The coverage factor $k_{0,95}$ for the convolution of different distributions with uncertainties of type A and type B, e.g. the Student distribution and Gaussian or rectangular, is determined by the Monte Carlo method. Let us assume that coefficient $k_{0.95} \leq t_{0.95,n-2}$ because the coefficient k_P of the rectangular PDF is below 2.
3. In rare cases, the y_i coordinates and some or all of the measurement points are measured with different measures and/or under different influence conditions. Then the sequence of numerical operations in determining the expanded uncertainty is

$$u_i = \sqrt{u_{Ai}^2 + u_{Bi}^2}\ i = (1, \ldots, n); \quad u_{ab} = u_{ab}(u_i(ax + b));$$
$$U_y(ax + b) = t_{0.95,n-2}u_{ab}(ax + b) \tag{21c}$$

That is, the combined uncertainties $u_i(y_i)$ of each point $Q_i(y_i,x_i)$ are counted separately as the geometric sum of $u_{Ai}(y_i)$ and $u_{Bi}(y_i)$, from which the uncertainty $u_{ab}(x)$ and the U_y band of regression line are determined.

The results according to (21b) and (21c) may differ slightly even for all the same uncertainties u_{Ai} of measured points. In both cases the straight-line equation is determined in the same way and with the same values of $y_i(x_i)$ and $u_{Ai}(y_i)$. Only with a special requirement, e.g. that the measured characteristic is averaged over the entire long period of the experiment, combined uncertainties $u_i(y)$ may be used to determine the regression equation, i.e. with the participation of uncertainty u_{Bi} of the measured points. The final result is in the form of a straight-line equation with the uncertainty band U_y (e.g. Fig. 2):

$$y = ax + b \pm U_y(y) \tag{22}$$

4 Examples of Numerical Simulations

Getting to know the specificity of linear regression considering the uncertainties of B-type measurements will be facilitated by several simulated numerical examples taken from other authors' work [16]. They are made for the same set of measured points, but with four different sets of uncertainties of type A for ordinates y_i. The same linear relationship was also assumed for the uncertainty of type B in measurements of the variable y. Influence of uncertainty u_{iA} of the measured points Q_i and uncertainty u_{iB}

of the measurement instrument on parameters and course of the uncertainty band of the characteristic $y = ax + b$ determined by the regression method, will be illustrated with examples.

Table 1a shows the coordinates (x_i, y_i) of ten points Q_i of the same given values x_i and measured values of y_i with different uncertainties u_{Ai}. These are data of simulated four measurement experiments I–IV. It was assumed that in four experiments I–IV for ten measuring points each with coordinates x_i (column 1) there are the same values of y_i estimators (column 2), but with different u_{Ai} uncertainties (columns (I–III) and $u_{Ai} = 0$ in experiment IV). It was assumed that the measurements of y_i of each point were carried out repeatedly, and obtained results were normally distributed, and their mean values and standard deviations are estimates of y_i values and $u_{Ai}(y_i)$ uncertainties. An acceptable simplification was adopted, that in each of the experiments. the measurement is made with the same digital meter and that the same sets of $y_i(x_i)$ coordinates will be obtained. In real measurements they may differ slightly. Uncertainties of experiment III are double values of u_{Ai} from II. In experiments I–IV for ten measuring points with coordinates x_i (column 1) there are the same values of y_i estimators (column 2), but with different u_{Ai} uncertainties (columns (I–III) and $u_{Ai} = 0$ for experiment IV.

It was also assumed that the measurements of y_i of each point were carried out repeatedly, and the obtained results were normally distributed, and their mean values and standard deviations are estimates of y_i values and $u_{Ai}(y_i)$ uncertainties. An acceptable simplification was adopted, that in each of the experiments, the measurements are made with the same digital measure and that the same sets of $y_i(x_i)$ coordinates will be obtained. In real measurements they may differ slightly. Uncertainties of III are double values of u_{Ai} from experiment II, and in IV – uncertainties $u_{Ai} = 0$.

It was also assumed that the uncertainty of B-type measurements by digital meters is the standard deviation of uniform distribution with twice the width of maximum permissible error $E_{\max}$ and increases linearly with the measured value of $y - y_0$. Such dependence of type B uncertainty can also be assumed as the resultant for many measurement systems. It was assumed that for the four data sets from Table 1a the uncertainty dependence u_B is the same and described by the equation: $u_{By}(y) = 0.02(1 + y)$.

The parameters *a, b* of straight lines described by the formulas (17a, b) were determined by the linear regression method. They are the same for data pairs I and IV as well as II and III from Table 1a. The uncertainties u_A of the regression lines are determined for data with the Student's distribution with the coverage factor $k_{0.95} = 2.31$ (28), and $u_B(y)$ – from the maximum permissible error E_{max}. The equations of regression lines and their expanded uncertainties are given in Table 2.

The course of the regression line, common for data I and IV, with the surrounding uncertainty bands $U_{Aab}, U_{0,95}$ and the estimator band $\hat{u}_A^2 = 1.084$ from the formula (40) for data I is shown in Fig. 2a. The uncertainty bandwidths of this regression line as a function of *x* and of *y* (lower scale) are given in Fig. 2b.

Figure 3 compares the uncertainty curves for I and IV data and the type B uncertainty.

The data in Tables 1, 2 and Figs. 2 and 3 show several detailed conclusions:

- type B uncertainty as function of *x* sums up geometrically with determined from the measurements the function of the type A uncertainty $u_{Aab}(y)$ of the regression line;

Table 1. Parameters of regression lines $y = ax + b$ and their expanded uncertainty bands $\underline{U}_{0.95}$ for points with the same x_i and y_i coordinates. The ordinates y_i were measured in 4 experiments with different type A uncertainties u_{Ayi} and type B uncertainties $u_{Byi} = 0.02\ (1 + y_i)$.

Points and their uncertainties					Parameters a, b and points of regression line and their bands with uncertainties u_{Aab}, $U_{0.95}$											
x_i	y_i	$u_B(y) = 0.02(1+y)$			$u_B(x) = 0.02(1+ax+b)$											
		$u_{Ai}(y_i)$			I $a = 1.049$ $b = -0.067$			II $a = 1.038$ $b = -0.389$			III $a = 1.038$ $b = -0.389$			IV $u_A \to 0.$ $a = 1.049$, $b = -0.067$		
		I	II	III	$y(x_i)$	u_{Aab}	$U_{0.95}$	$y(x_i)$	u_{Aab}	$U_{0.95}$	$y(x_i)$	u_{Aab}	$U_{0.95}$	$u_{Bi}(y_i)$	u_{Aab}	$U_{0.95}$
1	2.00		0.25	0.50	0.982	0.176	0.417	2.077	0.128	0.305	0.648	0.256	0.595	0.15		0.091
2	1.00		0.20	0.40	2.030	0.150	0.372	2.868	0.108	0.278	1.686	0.215	0.512	0.10		0.140
3	4.00		0.25	0.50	3.079	0.126	0.346	3.659	0.089	0.268	2.723	0.179	0.447	0.25		0.188
4	3.00		0.15	0.30	4.127	0.107	0.342	4.451	0.075	0.279	3.761	0.150	0.409	0.20		0.236
5	6.00	0.30	0.25	0.50	5.176	0.096	0.361	5.242	0.066	0.308	4.798	0.133	0.407	0.35	0	0.285
6	5.00		0.20	0.40	6.224	0.096	0.400	6.033	0.066	0.350	5.836	0.133	0.439	0.30		0.333
7	8.00		0.25	0.50	7.273	0.107	0.454	6.824	0.075	0.402	6.873	0.150	0.501	0.45		0.382
8	7.00		0.15	0.30	8.321	0.126	0.519	7.616	0.089	0.460	7.911	0.179	0.582	0.40		0.430
9	10.00		0.25	0.50	9.370	0.150	0.590	8.407	0.107	0.521	8.948	0.215	0.675	0.55		0.478
10	11.00		0.25	0.50	10.418	0.176	0.665	9.198	0.128	0.586	9.986	0.255	0.777	0.60		0.527

Table 2. Straight-lines and bands of type uncertainties u_{Aab} and extended uncertainties $U_{0,95}$

No.	Regression lines	Variances of their uncertainties	Uncertainties u_a, u_b and ρ_{ab}
I	$y = 1.048x - 0.067$	$U^2_{Aab}(x) =$ $0.01\left(0.11x^2 - 1.2x + 4.2\right)$	$u_a = 0.033$ $u_b = 0.205$ $\rho_{ab} = -0.89$
		$U^2_{0.95}(x) =$ $0.01\left(0.82x^2 - 5.99x + 22.6\right)$	
II	$y = 1.038x - 0.389$	$U^2_{Aab}(x) =$ $0.01\left(0.06x^2 - 0.66x + 2.24\right)$	$u_a = 0.024$ $u_b = 0.150$ $\rho_{ab} = -0.95$
		$U^2_{0.95}(x) =$ $0.01\left(0.23x^2 - 3.23x + 12.01\right)$	
III	$y = 1.038x - 0.389$	$U^2_{Aab}(x) =$ $0.01\left(0.24x^2 - 2.63x + 8.95\right)$	$u_a = 0.049$ $u_b = 0.30$ $\rho_{ab} = -0.90$
		$U^2_{0.95}(x) =$ $0.01\left(1.5x^2 - 13.77x + 47.82\right)$	
IV	$y = 1.048x - 0.067$	$U^2_{Aab}(x) = 0$	$u_a = 0$ $u_b = 0$ $\rho_{ab} = -0.89$
		$U^2_{0.95}(x) =$ $0.01\left(0.23x^2 - 0.42x + 0.19\right)$	

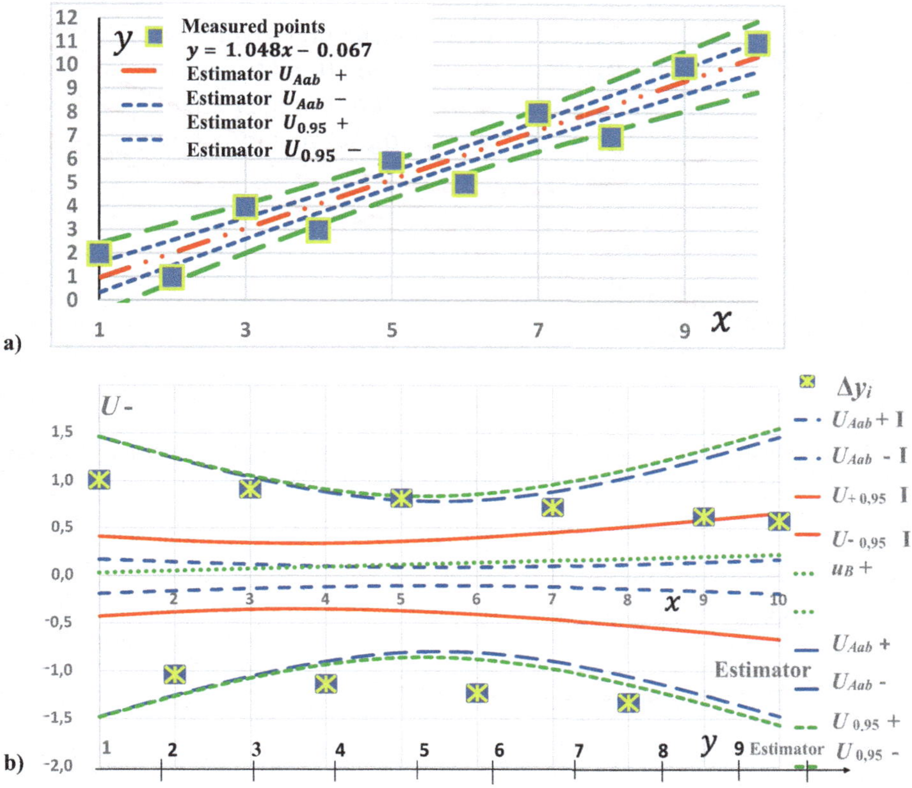

Fig. 2. **a)** Regression line with uncertainties bands ± u_{Aab} (x) and extended uncertainties ± $U_{0,95d}$ (x) for data I of Table 1 in notices under Fig 2 and Fig 3 and estimator $\hat{u}_A = \sqrt{\hat{u}_A^2} =$ *1,084* of Eq. (30); **b)** Uncertainty bands of regression line as function of x or y (bottom scale), its uncertainty ± $u_B(x)$ and Δy of measured points

- for a specific set of coordinates of measured points with the same value of type A uncertainties $u_{Ai} = u_A$, known or unknown, and using the estimator according to formula (30), the same straight-line equation is obtained by the linear regression method, e.g. for data I, IV. The width of its type A uncertainty bands has a minimum in the middle part of the range and increases in both directions from it.
- for a proportional change of any type A uncertainty of all points (e.g. version III data is a duplication of II data), the regression line equation does not change;
- type B uncertainties for measurements with the same instrument are taken as standard deviations of a uniform distribution with a width of 2 maximum permissible errors $\boldsymbol{E}_{\%}$ (MPE), increasing linearly with the value of the measured quantity y, i.e.: $e_{\max}(y) = 0.01E_{max\%}(y - y_0)$. Therefore, the width of the expanded uncertainty band $U_{0.95}$ of the regression line also increases with y and x.

Probability distributions for deviations of the y coordinate from its values on the regression line change with bandwidth of expanded uncertainty (Fig. 4 and Table 3).

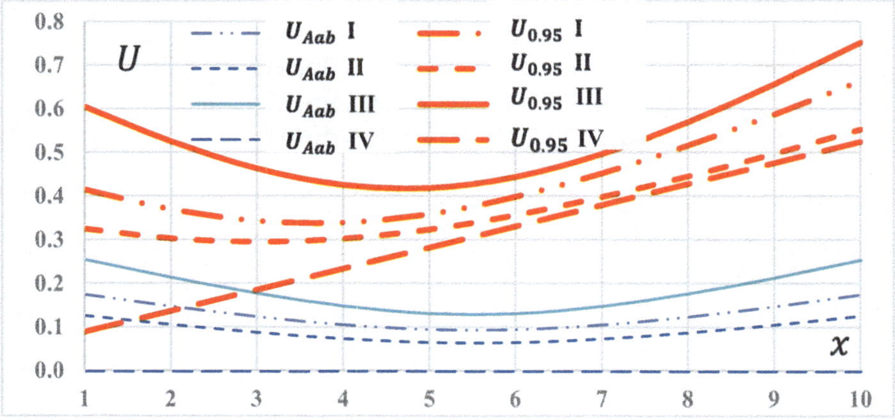

Fig. 3. Uncertainties $U_{Aab\,0.95}$ and $U_{0.95}$ of two regression straight-lines for data I, IV and II, III from Table 1a and their uncertainty type B from (21a), all as functions of x

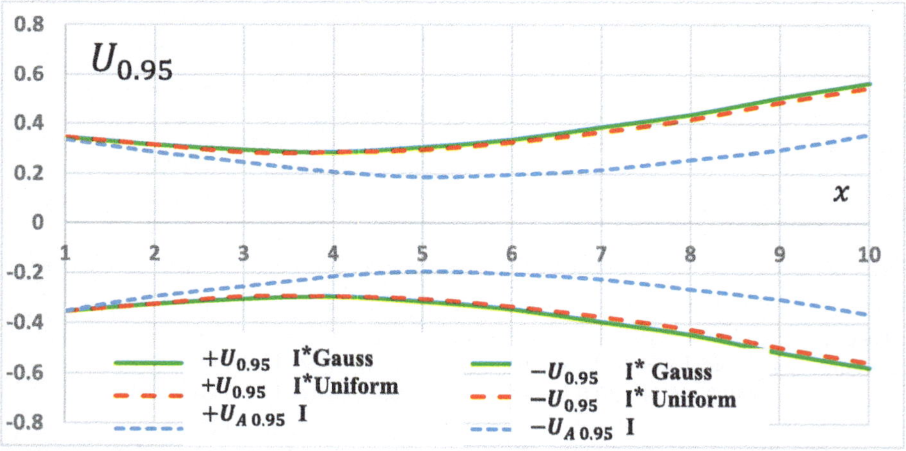

Fig. 4. Uncertainty bands $\pm U_{0.95}$ for deviations Δy from the regression straight-line I, made as the convolution of Gaussian pdf of uncertainty uA and Gaussian or uniform pdf of two different u_B

In Table 3 are shown distributions and histograms obtained by the MonteCarlo method for the data I from Table 1. They depend on what distribution of deviations with type B uncertainty is intertwined with the geometrically summed Gaussian distribution of deviations of type A from the regression line. The distribution of type B uncertainty deviations results from the permissible error E_{max} of the device. It is taken heuristically as uniform under constant influencing conditions, or as normal (Gaussian) for variable conditions with many randomized influences. For the convolution of two Gaussian distributions, the coefficient $k_{0.95} = 2$.

Table 3. Pdf of y from points (x, y) on the regression straight line I formed at gaussian distribution convolution for type A uncertainty with Gaussian or uniform of type B uncertainty

Gaussian distributions for deviations of u_{Aab} uncertainty	**Distribution of deviations in the uncertainty band $U_{0.95}$ for x on the regression line**		
	Convolution of two Gaussian PDF with uncertainties u_{Aab} and $u_B(y) = 0.02\ (1+y)$	Convolution of Gaussian and uniform PDF with uncertainties u_{Aab} and $u_B(y) = 0.02\ (1+y)$	Convolution of Gaussian and uniform PDF with uncertainties u_{Aab} and $u_B(y) = 0{,}04\ (1+y)$
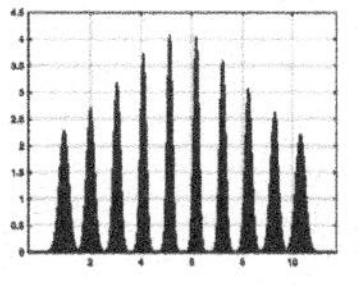	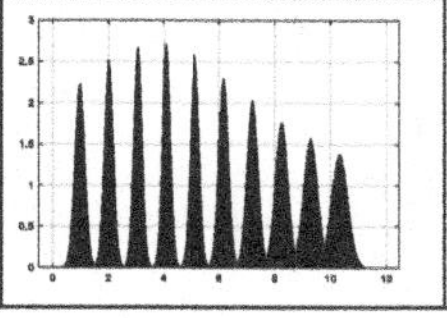	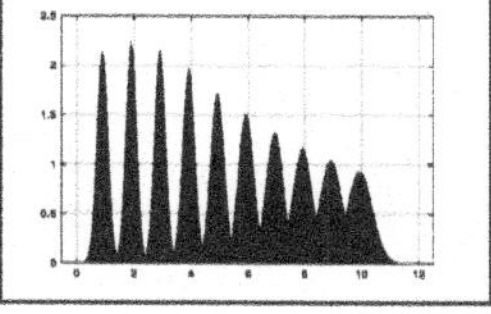	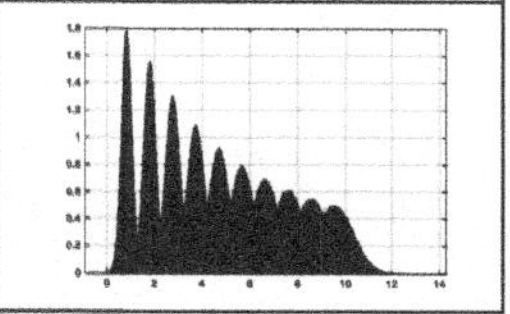

5 The Regression Line for Measurements with the Same Uncertainty

In practice, the regression method is most often used to determine a straight line from the measurements of the variable Y with the same uncertainties u_{yi}, when the values of the variable X are known. We will discuss this case in detail.

5.1 Regression Line Parameter Formulas for Points of Equal Uncertainties

If the linear regression method adjusts the parameters of a straight line to the coordinates of measurement points with the same known or even unknown uncertainties, then criterion (11) is used as the minimum sum of the squared distances of the points from the regression line in the 0Y direction. The straight line and its uncertainty are determined with the use of standard functions, e.g. LINEST, given in EXCEL sheets and workbooks. These functions do not consider type B uncertainties, so their basic formulas need to be completed. The same uncertainties $u_i = \text{const} \equiv u$ of y_i coordinates are a special case of regression with weights. This uncertainty is shortened in the formulas (16a, b) and is obtained

$$a = \frac{n\sum_{i=1}^{n} x_i y_i - \sum_{i=1}^{n} x_i \cdot \sum_{i=1}^{n} y_i}{n\sum_{i=1}^{n} x_i^2 - \left(\sum_{i=1}^{n} x_i\right)^2}, \qquad b = \frac{\sum_{i=1}^{n} y_i \cdot \sum_{i=1}^{n} x_i^2 - \sum_{i=1}^{n} x_i \cdot \sum_{i=1}^{n} x_i y_i}{n\sum_{i=1}^{n} x_i^2 - \left(\sum_{i=1}^{n} x_i\right)^2} \tag{23a,b}$$

These formulas uniquely define the regression line $y = ax + b$ and will simplify even more when the mean values of both variables X and Y are used in their description, marked as follows:

$$\overline{x} = \frac{1}{n}\sum_{i=1}^{n} x_i \quad \text{and} \quad \overline{y} = \frac{1}{n}\sum_{i=1}^{n} y_i \tag{24a,b}$$

Pearson's correlation coefficient ρ_{xy} for these variables is:

$$\rho_{xy} = \frac{\sum_{i=1}^{n}(y_i - \overline{y})(x_i - \overline{x})}{\sqrt{\sum_{i=1}^{n}(x_i - \overline{x})^2 \sum_{i=1}^{n}(y_i - \overline{y})^2}} \tag{25}$$

And the auxiliary parameters from formulas (14)a–f for $u_i = u$ will be as follows:

$$S = \tfrac{n}{u^2}, \quad S_x = \textstyle\sum_{i=1}^{n} x_i = \tfrac{n\overline{x}}{u^2}, \quad S_{xx} = \tfrac{1}{u^2}\textstyle\sum_{i=1}^{n} x_i^2 = \tfrac{1}{u^2}(\textstyle\sum_{i=1}^{n}(x_i - \overline{x})^2 + n\overline{x}^2$$
$$S_y = \tfrac{1}{u^2}\textstyle\sum_{i=1}^{n} y_i = \tfrac{n\overline{y}}{u^2}, \quad S_{xy} = \tfrac{1}{u^2}\textstyle\sum_{i=1}^{n} x_i y_i = \tfrac{1}{u^2}\left(\textstyle\sum_{i=1}^{n}(y_i - \overline{y})(x_i - \overline{x}) + n\overline{y}\overline{x}\right),$$
$$S_{yy} = \tfrac{1}{u^2}\textstyle\sum_{i=1}^{n} y_i^2 = \tfrac{1}{u^2}(\textstyle\sum_{i=1}^{n}(y_i - \overline{y})^2 + n\overline{y}^2)$$

(26a–f)

After using (26a–f) and the proportionality of expressions:

$$\sigma_x = c\sqrt{\sum_{i=1}^{n}(x_i - \overline{x})^2}, \quad \sigma_y = c\sqrt{\sum_{i=1}^{n}(y_i - \overline{y})^2},$$

estimating the standard deviations of the variables X and Y and the proportionality constant c (e.g. c = $1/\sqrt{n}$), the following parameters a, b of the regression line are obtained:

$$a = \frac{\sum_{i=1}^{n}(y_i - \overline{y})(x_i - \overline{x})}{\sum_{i=1}^{n}(x_i - \overline{x})^2} = \frac{\sigma_y \rho_{xy}}{\sigma_x} \quad \text{and} \quad b = \overline{y} - a\overline{x} \tag{27a,b}$$

The equation of the regression straight line for the parameter notations used here is:

$$y - \overline{y} = \rho_{xy}\frac{\sigma_y}{\sigma_x}(x - \overline{x}) \tag{28}$$

Mean values $(\overline{x}, \overline{y})$ mark a point on the regression line. From (14a, b), (15), (23a) and (26a–c) the formulas for the uncertainty of parameters a, b and their correlation coefficient follow:

$$u_a^2 = \frac{u^2}{\sum_{i=1}^{n}(x_i - \overline{x})^2}, \quad u_b^2 = \frac{u^2}{n} + \frac{u^2\overline{x}^2}{\sum_{i=1}^{n}(x_i - \overline{x})^2},$$
$$\rho_{ab} = -\left(1 + \frac{\sigma_x^2}{\overline{x}^2}\right)^{-\frac{1}{2}} \quad \text{where}: \quad \sigma_x^2 = \frac{\sum_{i=1}^{n}(x - \overline{x})^2}{n} \tag{29a–d}$$

The coefficient ρ_{ab} is negative and depends only on the variable x_i. It is the closer to –1 the more $\sigma_x^2 \ll \overline{x}^2$, i.e. for the Q_i points located close to the mean value of $\overline{x}$. On the parameter plane ax, b uncertainties u_a, u_b for $u_{ab} = const$ form an ellipse. (It will be discussed in the next paper). Thus, for the same absolute uncertainty u of all measurement points, the type A uncertainty of points on regression line is described by formula

$$u_A^2(x) = u^2\left(\frac{1}{n} + \frac{(x - \overline{x})^2}{\sum_{i=1}^{n}(x_i - \overline{x})^2}\right) \tag{30}$$

From (22) and (29) result curves of the limits of the combined uncertainty band $u_A(x)$ for the same absolute uncertainty u of measured points. They are parabolic and symmetrical about the mean value $\bar{x}$. For the point $x = \bar{x}$ the middle of the range X, the combined uncertainty has the minimum value $u_{cmin}^2 = \frac{u^2}{n}$. From this value, the uncertainty band extends in both directions to the ends of the X range. For a proportionally increasing uncertainty of type B, the band will be asymmetric with respect to the middle of the range of the variable X and widening towards its end. Combined and expanded uncertainty are determined from formulas (20) and (21) depending on the adopted model. For the same combined uncertainties, i.e. the values of invariant uncertainties of both type A and type B, the uncertainty band described by (21a, 21b) is usually wider than that resulting from (21c), because the expression $-\frac{1}{n} - \frac{(x-\bar{x})^2}{\sum_{i=1}^{n}(x_i-\bar{x})^2}$ is positive, and at the extreme points of the regression line, it has values close to zero.

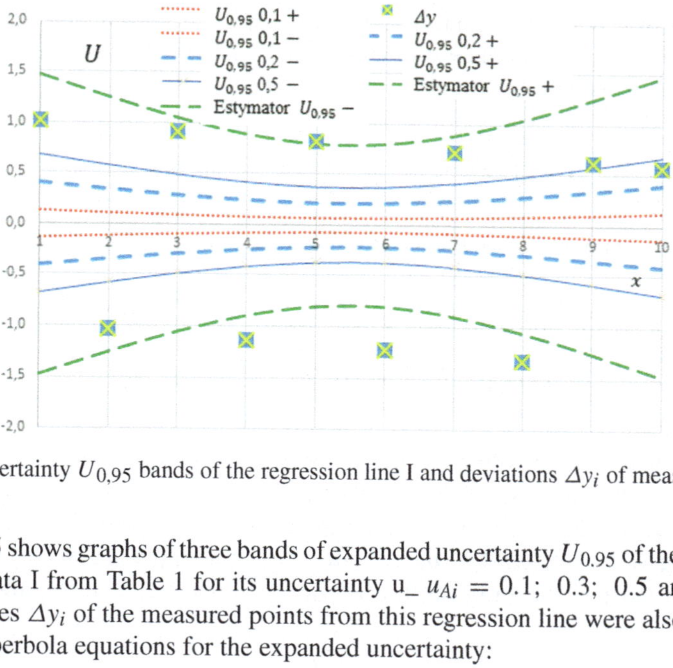

Fig. 5. Uncertainty $U_{0,95}$ bands of the regression line I and deviations Δy_i of measured points

Figure 5 shows graphs of three bands of expanded uncertainty $U_{0.95}$ of the regression line with data I from Table 1 for its uncertainty u_ $u_{Ai} = 0.1;\ 0.3;\ 0.5$ and $u_B = 0$. The distances Δy_i of the measured points from this regression line were also marked.

The hyperbola equations for the expanded uncertainty:

$$U = 2.31\sqrt{0.1 + 0.012(x - 5.5)^2}u_{Ai}$$

The uncertainty bands $U_{0.95}$ shown in Fig. 5 expand proportionally to the uncertainty u_i of measured points. The maximum values are at the ends of the bands. The distances Δy_i of measured points from the straight line (ordinates of points in the figure) exceed the limits of the widest uncertainty band for $u = 0.5$ in both directions.

5.2 Regression Line Using the Variance Estimator $\widehat{u^2}$

If the uncertainties of the measured points are the same but unknown, then the confidence interval for the uncertainties $u(y)$ is assumed to be derived from the mean sum of the squared distances of the points from the regression line for $n-2$ degrees of freedom (uncertainty of model). The unbiased estimator $\widehat{u^2}$ of the variance $E(u^2)$ is

$$\widehat{u^2} = E\left(u^2\right) = \frac{\sum_{i=1}^{n}(y_i - a \cdot x_i - b)^2}{n-2} \tag{31}$$

Based on the formulas (10), (23a,b), (24), (15d–f) and (27) the estimator takes forms:

$$\widehat{u^2} = u^2\frac{-aS_{xy} - bS_y + S_{yy}}{n-2} = \left(1-\rho_{xy}^2\right)\left(\frac{\sigma_y}{\sigma_x}\right)^2\frac{\sum_{i=1}^{n}(x_i-\overline{x})^2}{n-2} = \frac{n}{n-2}\left(1-\rho_{xy}^2\right)\sigma_y^2 \tag{31a}$$

From (31); (31a) it follows that the estimator $\widehat{u^2}$ is the greater the smaller the correlation between the x and y values of the line and the greater is the standard deviation σ_y of the scatter of y_i points Q_i with respect to the mean $\overline{y}$. By replacing the uncertainties u in the formulas (29a,b) with their estimators $\hat{u}$ the following formulas are obtained for the parameters a and b of the simple regression:

$$\hat{u}_a^2 = \left(\frac{\sigma_y}{\sigma_x}\right)^2\frac{1-\rho_{xy}^2}{n-2} = a^2\frac{\rho_{xy}^{-2}-1}{n-2}, \quad \hat{u}_b^2 = u_b^2\frac{\sum_{i=1}^{n}x_i^2}{n} \tag{32a,b}$$

Uncertainty $\hat{u}_A(y_i)$ is determined from following formulas of the variance estimator:

$$\hat{u}_A^2(y_i) = a^2\frac{\rho_{xy}^{-2}-1}{n-2}\left(\frac{\sum_{i=1}^{n}(x_i-\overline{x})^2}{n} + (x_i-\overline{x})^2\right)$$
$$\text{or } \hat{u}_A^2(y_i) = \left(1-\rho_{xy}^2\right)\left(\frac{\sigma_y}{\sigma_x}\right)^2\left(\sigma_x^2 + (x_i-\overline{x})^2\right) \tag{33a,b}$$

From (33a,b, 33a) it follows that the uncertainty of type A of the estimator $\widehat{u^2}$ decreases with the increase of the correlation coefficient ρ_{xy} and increases with the increase in the absolute value of the slope and the regression line. For fully correlated values of the variables X and Y, $\rho_{xy} = 1, a = \sigma_y/\sigma_x$ and the uncertainty $\hat{u}_A$ tends to zero, and for independent variables, i.e. for $\rho_{xy} = 0$ it reaches the maximum:

$$\hat{u}_{Amax}^2(y_i) = \left(\frac{\sigma_y}{\sigma_x}\right)^2\left(\sigma_x^2 + (x_i-\overline{x})^2\right) \tag{34}$$

Combined and extended uncertainties of the regression line and its uncertainty band are determined from the formulas (30)–(32). The estimate $\hat{u} = 1.084 > 0,5$ represents the average distance of the measured points in the 0Y direction from the regression line. Figure 5 shows that its application gives an uncertainty band so large that it covers the location of most of the measured points. Considering the uncertainty band of type B by geometric summation with the band $\hat{u}_A$ extends the bands of complex uncertainty $\hat{u}$ and extended to the degree depending on the course $u_B(x)$.

6 Regression Line for Points with the Same Relative Uncertainty of y_i

The same relative uncertainty $\delta_i \equiv \delta = const$ of type A of the abscissa y_i of measured points Q_i appears in the formulas (4a–f) for auxiliary parameters as a constant value before the sum symbols, i.e.:

$$S = \delta^{-2}\sum_{i=1}^{n} y_i^{-2},\quad S_x = \delta^{-2}\sum_{i=1}^{n} x_i y_i^{-2},\quad S_{xx} = \delta^{-2}\sum_{i=1}^{n} x_i^2 y_i^{-2},$$
$$S_{xy} = \delta^{-2}\sum_{i=1}^{n} x_i y_i^{-1},\quad S_y = \delta^{-2}\sum_{i=1}^{n} y_i^{-1},\quad S_{yy} = n\delta^{-2} \tag{35a–f}$$

Values of the regression line parameters do not depend on the relative uncertainty δ and are as follows:

$$a = \frac{\sum_{i=1}^{n} y_i^{-2}\sum_{i=1}^{n} x_i y_i^{-1} - \sum_{i=1}^{n} x_i y_i^{-2}\sum_{i=1}^{n} y_i^{-1}}{\sum_{i=1}^{n} x_i^2 y_i^{-2}\sum_{i=1}^{n} y_i^{-2} - \left(\sum_{i=1}^{n} x_i y_i^{-2}\right)^2},$$
$$b = \frac{\sum_{i=1}^{n} y_i^{-1}\sum_{i=1}^{n} x_i^2 y_i^{-2} - \sum_{i=1}^{n} x_i y_i^{-2}\sum_{i=1}^{n} x_i y_i^{-1}}{\sum_{i=1}^{n} x_i^2 y_i^{-2}\sum_{i=1}^{n} y_i^{-2} - \left(\sum_{i=1}^{n} x_i y_i^{-2}\right)^2} \tag{36a,b}$$

And the variance $u_c^2(y(x))$ from the formula (29a) will be transformed into the form

$$u_c^2(y(x)) = u_{ab}^2(y(x)) = \delta^2 \frac{\sum_{i=1}^{n}(x - x_i)^2 y_i^{-2}}{\frac{1}{2}\sum_{i=1}^{n}\sum_{j=1}^{n}(x_i - x_j)^2 y_i^{-2} y_j^{-2}} \tag{37}$$

Figure 6 shows three uncertainty bands of the regression line $y = 0.978x - 0.463$ with parameters a, b from the formulas (36a,b) together with of relative uncertainty with the same values $\delta_{yAab}(y)$ for points from the entire x range. The graphs of these uncertainty bands of u_c resulting from the formula (37). The regression line with constant relative uncertainties $\delta_{yi} = \delta = const$ for the measurements of the abscissa of the Q_i points has different parameters than the straight line with constant absolute uncertainties $u_i = u = const$ for the data from Table 1a marked by I. It is more inclined towards the $0x$ axis and its uncertainty bands are significantly narrower. On Fig. 6 you can see a significant increase in the width of the $U_{0.95}$ uncertainty bands with x towards the end of the measurement range, compared to $U_{0.95}$ for $u_i = u = const$ from formula (28). It results from the formula (37). The uncertainty band for constant values of relative uncertainties $\delta_u = 5\%$ in the entire range of the variable x is narrower than the band in Fig. 5 for the absolute uncertainty $u = 0.5$, i.e. 5% of the range. In both cases with constant values of relative uncertainties and absolute uncertainties of the measurement points, the type B uncertainty should also be considered by adding it geometrically to the type A uncertainty for a regression line.

7 Comparison of Two Ways of Considering Type B Uncertainty

Two ways of obtaining the uncertainty of type B for the estimation of the uncertainty band of regression line are compared on the Figs. 7 and 8. The calculations are for the data set I of Table 1a with the same uncertainty of measured points $u_i = u$.

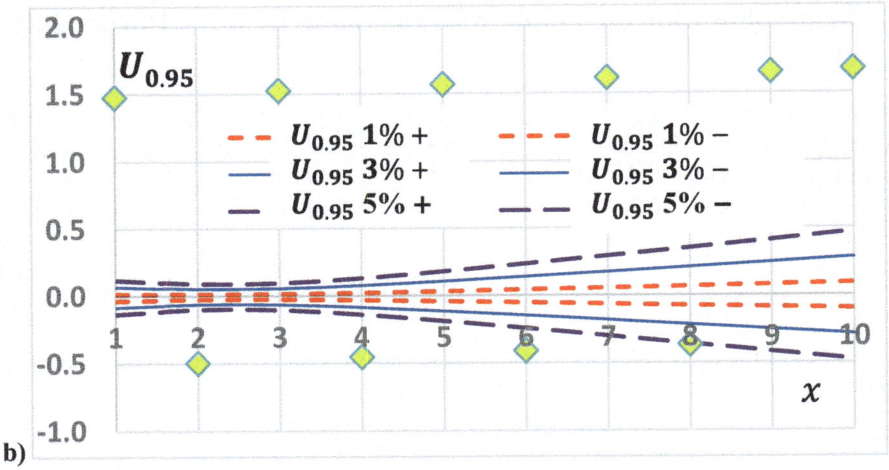

Fig. 6. Uncertainty bands from (31)) of regression line $y = 0{,}978x - 0{,}463$ from (36a,b) with relative uncertainty: $\delta_u = 1\%$, 3% and 5%. for the uncertainty u from (29) as function of x, and distances Δy_i of measurements points from the regression line

The first way, already discussed method with the geometric summation of the type A uncertainty band determined from the measurements, and the heuristically estimated type B uncertainty for the used instruments and experimental conditions as in (21a,b).

The second way is a special method, marked by V. It consists in calculating the line equation and its uncertainty bands from the formulas (21c), i.e. directly from the coordinates y_i and the complex uncertainty u_{Ai} of individual measured points as a geometric sum of u_{Ai} and u_{Bi}. These are the rare cases where each point is measured with a different instrumentation or/and under different conditions.

Results of both methods for the expanded absolute uncertainty are given in Fig. 7, and for the relative uncertainty – in Fig. 8.

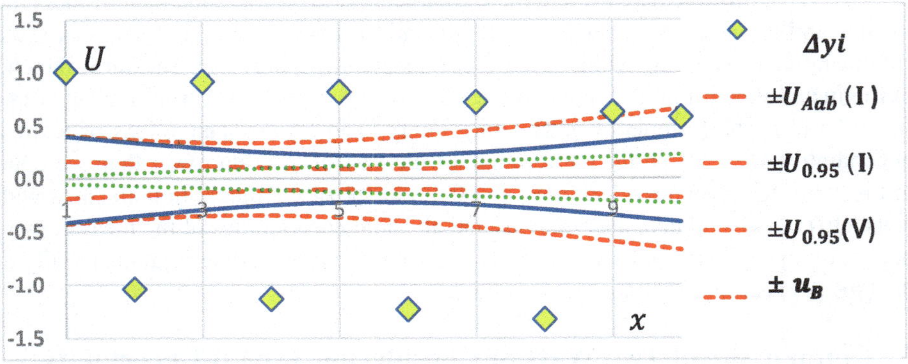

Fig. 7. Comparison of two ways of considering absolute uncertainties type B for determining the expanded uncertainty band for $u_{Ay} = 0.3$ and $u_{By} = 0.02(1 + y)$

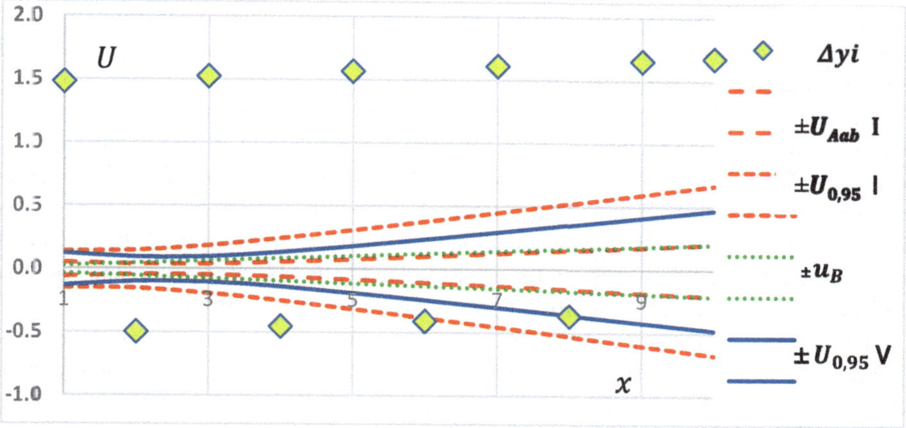

Fig. 8. Two methods for calculate the expanded uncertainty bands $U_{0,95}$ of regression line for points with constant relative uncertainties type A and B of y: $\delta_{Ay} = 5\%$, and $\delta_{By} = 2\%$

8 Summary

The aim of many measurement experiments is to determine the function $Y = F(X)$ between vector random variables, the explained Y and the explanatory X, and to evaluate its accuracy. For one-parameter variables Y and X, using the linear regression method, a straight line, also known as the regression line, is determined, with the equation $y = ax + b$ and set of its uncertainty bands.

The parameters of the regression line equation and of its uncertainty bands, in the general case, depend on the values of the coordinates x_i, y_i of the measured points and their absolute uncertainties $u_{yi}(x_i, y_i) = \sqrt{u_{Ai}^2 + u_{Bi}^2}$.

In this study, the cases where x values of the measured points are known precisely were considered. To assess its accuracy, developed is a method based on the recommendations of the GUM Guide, commonly used in measurements [1]. The accuracy of the variable y of the regression line describes then the band of the expanded uncertainty with the probability P and its limits $\pm\, U_P(y) = \pm\, k_P u(y)$ determined from the combined uncertainty $u(y)$ of the regression line. The uncertainty band $u(y)$ is obtained as the geometric sum of the $u_{yA} = f_A(x)$ characteristic of the type A uncertainty, determined from the measurements, and the $u_{yB}(y) = f_B(x)$ of the type B uncertainty characteristic estimated heuristically from the value of the maximum permissible errors E_{max} of meters and of other elements in given conditions of the measurement system.

The influence of type B uncertainty on the accuracy of the regression line has not been considered in the literature so far. The value of the coverage factor k_P is known only when the combined uncertainty u is determined for data with Gaussian distribution and some other basic distributions. With the distribution obtained from the convolution of deviations with type A and type B uncertainty estimated from a uniform distribution with a width of two maximum permissible errors $2E_{max}$ of the meter, the k_P coefficient is determined numerically using the MonteCarlo method as in Supplement S1of GUM.

If the measurement system and the test object are simultaneously influenced by many independently variable factors, the uncertainty $u_B(y)$ is estimated as the standard deviation of the Gaussian distribution.

For points of the coordinates y_i and measured with the same uncertainty $u_i(y_i) \equiv u$, a regression line is obtained with an equation independent of u. The width $2u_{Aab}(y)$ of its type A uncertainty band is proportional to u. It is similar for the measurements of points with the same relative uncertainty δ_u, but you get a regression line with a different equation. Individual determination of the B-type uncertainty of the measured points should be used only in special cases, e.g. in field climatic or environmental quality studies for measurements with different instruments and under different conditions.

The text presents a series of numerical comparisons of the parameters of the regression line and its bands for the same point coordinates and different absolute and relative uncertainties, type A and B. They enable to learn the specificity of the application of the linear regression method in Measurement Science.

The influence of correlation and autocorrelation of coordinates y of measured points on the equation and uncertainty bands of the regression line describing the measurements is discussed in the next our paper [18] in this book.

References

1. GUM JCGM100:2008, Evaluation of measurement data – Guide to the expression of uncertainty in measurement.+ GUM-S1:JCGM101:2008; Supplement 1 – Propagation of distributions using a Monte Carlo method; +GUM-S2: JCGM102:2011 Supplement 2 – Extension to any number of output quantities. BIPM Paris
2. Draper, R.D., Smith H.: Applied Regression Analysis, 3rd edn. Wiley, New York (1966) (1973)
3. Piotrowski, J.: Theory of Physical and Technical Measurement PWN-Elsevier (1992)
4. Weisberg Sanford Applied Linear Regression, 3rd edn. University of Minnesota, Wiley (2005)
5. Zięba, A.: Analysis of Experimental Data in Science and Technology PWN 2013 (in Polish, extended version to be published in 2021 by Cambridge Scholars Publishing, Newcastle GB) (2013)
6. Dorozhovets, M., Warsza, Z.L.: Uncertainty type A evaluation of autocorrelated measurement observations. Biuletyn Military Tech. Academy Warsaw WAT **LVII**(2), 143–152 (2008)
7. Dorozhovets, M.: Uncertainty of Orthogonal Linear Regression. Pomiary Automatyka Kontrola **53**(9bis 2007), 31–34 (2007). (in Polish)
8. Elster, C., Toman, B.: Bayesian uncertainty analysis for a regression model versus application of GUM Supplement 1 to the least-squares estimate. Metrologia **48**, 233–240 (2011)
9. Bich, W., Cox, M., Michotte, C.: Towards a new GUM-an update. Metrologia **53**, 149–159 (2016)
10. White, D.R.: In pursuit of a fit-for-purpose uncertainty guide. Metrologia **53**, 107–124 (2016)
11. Warsza, Z.L.: Evaluation of the type A uncertainty in measurements with autocorrelated observations. J. Phys. Conf. Ser. **459**, 012035 (2013). 2013 Joint IMEKO TC1+TC7+TC13 Symposium: Measurement Across Physical and Behavioral Sciences, Genova, Italy, 4–6 September
12. Volodarsky, E.T., Warsza, Z.L.: Application of two robust methods in inter-laboratory comparison…. In: Pavese, F., et al. (eds.) Advanced Mathematical and Computational Tools in Metrology and Testing X (Proceedings of AMCTM X in St. Petersburg), Series on Advances in Mathematics for Applied Sciences, vol. 86, pp. 385–391. World Scientific, New Jersey, London, Singapore (2015)

13 Warsza, Z.L.: Methods of extending the analysis of measurement uncertainty. Monograph, PIAP (2016).(in Polish with abstracts of chapters in English)
14. Warsza, Z.L., Puchalski, J.: Estimation of uncertainties in indirect parameter measurements of correlated quantities. In: Proceedings of 12th International Conference “Measurement 2019”, 27–29 May 2019, Smolenice Castle, Center of Slovak Academy of Sciences, Slovakia, pp. 51–57 (2019)
15. Warsza, Z.L., Puchalski, J.: Estimation of uncertainties in indirect multivariable measurements Part 1. Case of correlated quantities. In: Szewczyk, R., et al. (eds.) Proceedings of Automation 2020, Towards Industry of the Future. AISIC Series, vol. 1140, pp. 309–325. Springer (2020)
16. Warsza, Z.L., Puchalski, J., Uncertainty of measurement in linear regression method. Part 1. Straight-line and its uncertainty bands for uncorrelated measurement data. Pomiary Automatyka Robotyka 24(3), 75–86 (2020). https://doi.org/10.14313/PAR_237/79. (in Polish)
17. Ben-Moshe, D.: Identification of Linear Regressions with Errors in all Variables. Econometric Theory, 06 2020 (from Ph.D. thesis in the Hebrew University of Jerusalem, Israel)
18. Warsza, Z.L., Puchalski, J.: Uncertainty bands of the regression line for autocorrelated data of dependent variable Y. (see Chapter 33 in this book)

Uncertainty Bands of the Regression Line for Autocorrelated Data of Dependent Variable *Y*

Zygmunt Lech Warsza[1] and Jacek Puchalski[2(✉)]

[1] Łukasiewicz Research Network – Industrial Research Institute for Automation and Measurements PIAP, Al. Jerozolimskie 202, 02-486 Warsaw, Poland
zlw@op.pl

[2] Central Office of Measures, Elektoralna 2, 00-137 Warsaw, Poland
jacek.puchalski@gum.gov.pl

Abstract. The formalism is proposed for assessing the accuracy of simple regression takes into account both the correlation of the *Y* variable data and the impact of type B uncertainty in routine measurements.This is the continuation of authors' work considering the data with type B uncertainty and uncorrelated random errors. The essence, criteria and dependencies of the regression method were examined. Simulated examples of determining uncertainty bands of the regression line fitted to measured points with different cases of correlated values of dependent ***Y*** variable are considered. The recommendations of GUM Guide [1] was referred to and the type B not discussed yet in the literature was considered. The proposed formalism is illustrated by examples the precisely know abscissa and ordinates with different correlation, and absolute and relative uncertainties type A and type B.

Keywords: Measurement · Autocorrelation · Uncertainty type A and type B · Regression line · Uncertainty bands

1 Introduction

The linear regression method is the basic mathematical tool used to determine a straight-line, or other linearized function, to match a given measurement data according to a specific criterion [2–5]. The accuracy of measurement results in the national metrology service, science, industry and many other fields is now assessed according to uncertainty term recommended in the Guide with acronym GUM [1], and also described in a number of books and other literature, including [5, 6, 14]. In the literature on the linear regression, the accuracy of straight-line parameters is statistically assessed only from the measured data and their uncertainties of the type A in the experiment. The impact of type uncertainty B for results of the measurement experiment and their use in period and permissible conditions of validation data obtained by regression method is not considered [2–4, 8–13]. Correct heuristic assessment of the type B uncertainty as the expected effect of different interactions, unknown in value during the experiment,

R. Szewczyk et al. (Eds.): AUTOMATION 2021, AISC 1390, pp. 364–386, 2021.
https://doi.org/10.1007/978-3-030-74893-7_33

usually requires fairly detailed metrological knowledge and practice in many measuring fields. Even in recent research works on a regression method based on the Bayes [7–9], proposed for the new version of the Guide GUM 2 [13], impact of the uncertainty type B assessment on the accuracy of a regression line was not considered.

In the work [18], analyzed is the determination of the equation of regression line and its uncertainty band for the non-correlated ordinates Y of the measured points. In estimation the accuracy, the type B is also included. The uncertainty bands of the regression line for the autocorrelated coordinates of variable Y are analyzed below.

Effects of known systematic errors are analyzed using the formulas resulting from the total differential and are eliminated by corrections. The accuracy of the measurements is according Guide GUM [1] described by the expanded uncertainty as the range within which the measured value lies with a given probability. This interval is determined on the basis of combined uncertainty $u = \sqrt{u_A^2 + u_B^2}$ as a standard deviation of the distribution of measurement observations values from the estimated value of the measurand. Geometrical components of u are named as uncertainties of types A and B. The uncertainty of type A describes the scatter of repeated measurement results in the experiment. The type B uncertainty estimated the mean measure of the influence of systematic errors of unknown value, constant or slow-varying, which may occur in a given experiment. These errors are randomized, and with using the existing knowledge, their mean and extended values are heuristically estimated.

The type B uncertainty of measuring instruments is usually assessed on the basis of their maximum permissible error (MPE) with the symbol E. For example, for digital meters, the uncertainty u_B is taken as the standard deviation $E/\sqrt{3}$ of a uniform distribution with a span, equal to $2E$. The uncertainty u_B of these instruments increases linearly with the value of the measured quantity. Information on estimating the uncertainties of multi-parameter measurements is provided in Supplement 2 of the GUM and upgrading in few authors‘ works, e.g. in [16, 17].

Also important for the accuracy of the regression line are deviations resulting from the inadequacy of the mathematical model and interferences existing in the test object. They are treated uniformly as uncertainties.

If several values of the same quantity or of different quantities are measured in the same object and in the same time, a correlation may occur between the random sets of their observations, which affects the uncertainty of the results calculated from such measurements. An example is the uncertainty of the sum and difference of two measured values with uncertainties u_1, u_2 and the correlation coefficient ρ, i.e.:

$$u_{\sum} = \sqrt{u_1^2 + 2\rho u_1 u_2 + u_2^2}. \quad (1a)$$

$$u_D = \sqrt{u_1^2 - 2\rho u_1 u_2 + u_2^2}. \quad (1b)$$

The correlation coefficient ρ for the combined uncertainty u of a sum of correlated components is [15]:

$$\rho = \frac{\rho_A u_{1A} u_{2A} + \rho_B u_{1B} u_{2B}}{u_1 u_2} \quad (1c)$$

Similar relations are for relative uncertainties of the product and quotient of two values.

Value of the correlation coefficient ρ_A between two sets of errors described with uncertainties u_A is calculated by classic formula [1–6] from values of observations synchronously performed, and it can take values in the entire range (−1, 1). The coefficient ρ_B is determined similarly to u_B, to a large extent heuristically. Based on the knowledge obtained. The distributions of uncertainties of type A and B are related to a different time scale and it should be assumed that are not correlated with each other.

If two different values are measured with the same or a similar device, and in the same permissible conditions, then when determining the uncertainty u_B for their sum, it can be assumed that in formula (1a) the correlation coefficient $\rho_B = 1$, and for the difference in (1b) $\rho_B = -1$. Then uncertainty type B of the sum and of the difference are

$$u_{B\sum} = u_{B1} + u_{B2} \text{ and } u_{BD} = |u_{B1} - u_{B2}|.$$

For measurements of two values with different instruments or measurement systems and/or under different conditions, the sets of their indications should be treated as uncorrelated, i.e. $\rho_B = 0$, and the uncertainties type B of sum and difference will be equal

$$u_{B\sum} = u_{BD} = \sqrt{u_{B1}^2 + u_{B2}^2}.$$

Estimation the correlation coefficient ρ_B with a value other than 0 and 1 is difficult and requires specially provided long-term statistical studies.

2 Basic Linear Regression Method for Correlated Data of *Y* Variable

2.1 The Criterion in the Regression Method for the Autocorrelated *Y* Variable

In the linear regression method, for a depended random variable ***Y***, the equation of a straight line $y = ax + b$ and its uncertainty band U is determined from n measurements of y_i (where $i = 1,\ldots, n$), made with uncertainties $u(y_i)$, for precisely defined values x_i (for $i = 1,\ldots, n$) of the independent variable ***X***. To describe the estimated uncertainties for the correlated values y_i of variable ***Y***, the covariance matrix is used

$$\boldsymbol{U}_Y = \begin{bmatrix} u^2(y_1) & \cdots & \rho_{y1n}u(y_1)u(y_n) \\ \vdots & \ddots & \vdots \\ \rho_{y1n}u(y_1)u(y_n) & \cdots & u^2(y_n) \end{bmatrix}, \quad (2)$$

where: ρ_{yij}– correlation coefficient of sets of observations of pair y_i, y_j of ***Y*** variable, $u(y_i), u(y_j)$ – their uncertainties, $i, j = 1, \ldots n$.

In creating a regression line for real data, the correlation of the y-ordinates of adjacent points is usually much stronger than that of remote points. The elements with the highest values of the correlation coefficients appear around the main diagonal of the U_Y covariance matrix, and remaining elements can be treated as equal to zero.

When creating the least squares criterion for the multivariate Y variable, the multivariate Gaussian distribution with the probability density is used:

$$f(\boldsymbol{Y}) = \frac{1}{(2\pi)^{\frac{n}{2}} \cdot \sqrt{\det(\boldsymbol{U_Y})}} \cdot \exp\left(-\frac{1}{2}[\boldsymbol{Y_p} - \boldsymbol{Y}]^{\mathbf{T}} \boldsymbol{U}_{\boldsymbol{Y}}^{-1} [\boldsymbol{Y_p} - \boldsymbol{Y}]\right) \quad (3)$$

where:

$\boldsymbol{Y} = [y_1, \ldots y_i, \ldots y_n]^{\mathbf{T}}$ – n-dimensional vector of ordinates y_i of measured points,
$\boldsymbol{Y_p}$ – vector y_{ip} of their parallel projections to regression line in the direction 0y,
$\boldsymbol{U}_{\boldsymbol{Y}}^{-1}$ – inverse of the covariance matrix.

A solution of the linear equation is being sought:

$$\boldsymbol{Y_p} = a\boldsymbol{X_p} + b\mathbf{1} = a\boldsymbol{X_p} + \boldsymbol{b} \quad (4)$$

where: $\boldsymbol{X_p}$ – cut-off vector x_i of measured points, $\mathbf{1} = [1, \ldots, 1]^T$ – unit vector size of n.

The target function P relates to the probability density of a multivariate variable of If only the variable Y is measured, then the function $P = f(Y) \longrightarrow$ max. This will occur for a minimum of the criteria function $SK(a, b)$

$$SK(a, b) = -ln(f(Y)) \rightarrow \min \quad (5)$$

It is assumed that the sum of the normalized squares of distance of the measured points with coordinates (x_i, y_i) for $i = 1, \ldots, n$, from the specified points (x_i, y_{pi}) on a regression line (Fig. 1) will reaches a minimum of zero of the sum of squares of sub-derivatives after the parameters a and b of this line.

In the general case, i.e. when the values of both coordinates are measured, there may be an autocorrelation between values of the y_i and of the x_i coordinates as well correlation between y_i and x_i of measuring points, The determination of a linear regression method requires knowledge of two covariance matrices, $\boldsymbol{U_Y}$ and $\boldsymbol{U_X}$ for both variables $\boldsymbol{Y}$ and $\boldsymbol{X}$. This will be considered in detail in our next paper.

2.2 Regression Line Parameters and Uncertainties for the Autocorrelated Variable *Y*

The equation resulting from criterion (4) is simplified when the x_i values of the measured points do not have a random dispersion, i.e. they are defined precisely, i.e. their uncertainty $u(x_i) = 0$. Then $\boldsymbol{X_p} = \boldsymbol{X}$ and Eq. (4) as the sum of the squared uncertainties normalized to the inverse of the covariance matrix $\boldsymbol{U}_{\boldsymbol{Y}}^{-1}$ will take the form:

$$SK = \boldsymbol{\Delta Y}^{\mathbf{T}} \boldsymbol{U}_{\boldsymbol{Y}}^{-1} \boldsymbol{\Delta Y} = [\boldsymbol{Y_p} - \boldsymbol{Y}]^{T} \boldsymbol{U}_{\boldsymbol{Y}}^{-1} [\boldsymbol{Y_p} - \boldsymbol{Y}] \quad (6)$$

Dependencies of points P_i coordinates lying on the regression line $y = ax + + b$ with abscissas $x_{pi} = x_i$ and ordinates y_{pi}, are described by the vector equation $\boldsymbol{Y_p} = a\boldsymbol{X_p} + \boldsymbol{b}$. Distance of the measured point $\boldsymbol{Q}_i$ from the point P_i is equal to the coordinate differences between this two points (Fig. 1).

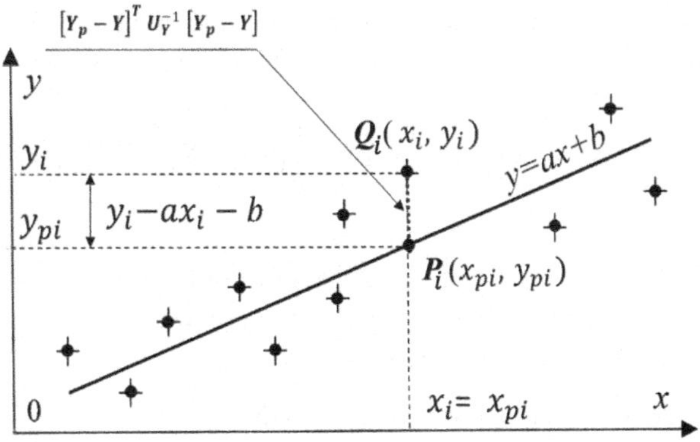

Fig. 1. Regression line for points with measured y_i values of variable Y and known values x_i

In the regression method with parallel projections of measured Q_i points in the 0y direction on the designated straight-line, the point Q_i is therefore at the distance of $y_{pi} - y_i = ax_i + b - y_i$ from the point $\boldsymbol{P_i}$. The dependencies of the regression line will have simpler forms after entering the auxiliary parameters

$$S = 1^T U_Y^{-1} 1 = \sum_{i=1}^{n} \sum_{j=1}^{n} [u_y^{-1}]_{ij},\ S_x = \boldsymbol{X}^T \boldsymbol{U}_Y^{-1} 1 = 1^T \boldsymbol{U}_Y^{-1} \boldsymbol{X},\quad S_{xx} = \boldsymbol{X}^T \boldsymbol{U}_Y^{-1} \boldsymbol{X},$$
$$S_{xy} = \boldsymbol{X}^T \boldsymbol{U}_Y^{-1} \boldsymbol{Y} = \boldsymbol{Y}^T \cdot \boldsymbol{U}_Y^{-1},\quad S_y = \boldsymbol{Y}^T \boldsymbol{U}_Y^{-1} 1 = 1^T \boldsymbol{U}_Y^{-1} \boldsymbol{Y},\quad S_{yy} = \boldsymbol{Y}^T \boldsymbol{U}_Y^{-1} \boldsymbol{Y} \tag{7a-f}$$

Then for $\boldsymbol{X_p} = \boldsymbol{X}$ and $\boldsymbol{Y_p} = a\boldsymbol{X_p} + \boldsymbol{b}$ Eq. (6) takes the form

$$SK = [a\boldsymbol{X} + \boldsymbol{b} - \boldsymbol{Y}]^{\mathrm{T}} \boldsymbol{U}_Y^{-1} [a\boldsymbol{X} + \boldsymbol{b} - \boldsymbol{Y}] = a^2 S_{xx} + 2abS_x - 2aS_{xy} + b^2 S - 2bS_y + S_{yy} \tag{8}$$

The condition for minimizing the expression (8) as the sum of the squares of the distance of n points Q_i from the regression line in the direction of the 0y axis (Fig. 1) results from the solution of the system of two equations for its partial derivatives with respect to the parameters a and$b$ of the regression line, i.e. $\frac{\partial SK}{\partial a} = 0,\ \frac{\partial SK}{\partial b} = 0$

$$\begin{cases} aS_{xx} + bS_x = S_{xy} \\ aS_x + bS = S_y \end{cases} \tag{9a,b}$$

Solving the system of Eqs. (9a,b) we get:

$$a = \frac{SS_{xy} - S_x S_y}{SS_{xx} - (S_x)^2} \equiv \frac{\Delta_a}{\Delta},\qquad b = \frac{S_y S_{xx} - S_x S_{xy}}{SS_{xx} - (S_x)^2} \equiv \frac{\Delta_b}{\Delta} \tag{10a,b}$$

where: $\Delta_a = SS_{xy} - S_x S_y,\ \Delta_b = S_y S_{xx} - S_x S_{xy}$ and $\Delta = SS_{xx} - (S_x)^2$.
The SK_{MIN} criterion from (5) will be met by the parameters a i b, where

$$SK_{\mathrm{MIN}} = -aS_{xy} - bS_y + S_{yy} \tag{11}$$

The uncertainty band $U(y)$ for ordinates y of regression points lies symmetrically on both sides of it. They will be determinate considering the uncertainties u_a of slope a and uncertainty u_b of intercept b. Parameters a, b from formulas (10a,b) are correlated. Partial derivatives $\frac{\partial S_{xy}}{\partial y_i}$, $\frac{\partial S_y}{\partial y_i}$ of the regression straight-line parameters are determined by vectors $\boldsymbol{c}$ and $\boldsymbol{d}$:

$$\boldsymbol{c} = \frac{\partial S_{xy}}{\partial \boldsymbol{Y}} = \boldsymbol{U}_Y^{-1}\boldsymbol{X} = \boldsymbol{X}^T\boldsymbol{U}_Y^{-1}, \boldsymbol{d} = \frac{\partial S_y}{\partial \boldsymbol{Y}} = \boldsymbol{U}_Y^{-1}\boldsymbol{1} = \boldsymbol{1}^T\boldsymbol{U}_Y^{-1} \tag{12a,b}$$

Sensitivity matrix for parameters a, b has the form $\boldsymbol{AB}$= $[\boldsymbol{A}, \boldsymbol{B}]$ of elements defined by:

$$\boldsymbol{A} = \frac{\partial a}{\partial \boldsymbol{Y}} = \frac{S\boldsymbol{c} - S_x\boldsymbol{d}}{\Delta}, \qquad \boldsymbol{B} = \frac{\partial b}{\partial \boldsymbol{Y}} = \frac{S_{xx}\boldsymbol{d} - S_x\boldsymbol{c}}{\Delta} \tag{13a,b}$$

The covariance matrix of parameters a and b follows from the matrix equation for propagation of variance, i.e.:

$$\boldsymbol{U_{ab}} = \boldsymbol{AB}\ \boldsymbol{U}_Y\ \boldsymbol{AB}^T = \begin{bmatrix} \boldsymbol{AU}_Y\boldsymbol{A}^T & \boldsymbol{AU}_Y\boldsymbol{B}^T \\ \boldsymbol{AU}_Y\boldsymbol{B}^T & \boldsymbol{BU}_Y\boldsymbol{B}^T \end{bmatrix} = \begin{bmatrix} u_a^2 & \rho_{ab}u_a u_b \\ \rho_{ab}u_a u_b & u_b^2 \end{bmatrix} \tag{14a}$$

where: ρ_{ab} is the correlation coefficient between parameters a i b.

Because the matrix $\boldsymbol{U}_Y^{-1}$ is symmetrical, and $\boldsymbol{U}_Y = \boldsymbol{U}_Y^T$, $\boldsymbol{U}_Y^{-1}\boldsymbol{U}_Y = \mathbf{I}$ – unit matrix, then: $\boldsymbol{cU}_Y\boldsymbol{c}^T = \boldsymbol{X}^T\boldsymbol{U}_Y^{-1}\boldsymbol{U}_Y(\boldsymbol{X}^T\boldsymbol{U}_Y^{-1})^T = S_{xx}$, $\boldsymbol{dU}_Y\boldsymbol{d}^T = \boldsymbol{1}^T\boldsymbol{U}_Y^{-1}\boldsymbol{U}_Y(\boldsymbol{1}^T\boldsymbol{U}_Y^{-1})^T = S$, $\boldsymbol{cU}_Y\boldsymbol{d}^T = \boldsymbol{dU}_Y = \boldsymbol{X}^T\boldsymbol{U}_Y^{-1}\boldsymbol{U}_Y(\boldsymbol{1}^T\boldsymbol{U}_Y^{-1})^T = S_x$ and elements of matrix $\boldsymbol{U_{ab}}$ simplify to:

$$u_a^2 = \boldsymbol{A}\ \boldsymbol{U}_Y\ \boldsymbol{A}^T = \frac{S}{\Delta}; \quad u_b^2 = \boldsymbol{B}\ \boldsymbol{U}_Y\ \boldsymbol{B}^T = \frac{S_{xx}}{\Delta}; \quad \rho_{ab}u_a u_b = \boldsymbol{A}\ \boldsymbol{U}_Y\boldsymbol{B}^T = -\frac{S_x}{\Delta} \tag{14b-d}$$

The standard uncertainty of the regression line $y = ax + b$ for one-dimensional variable Y can be find from the formula:

$$u_{ab}^2(y(x)) = \left[\frac{\partial y}{\partial a}, \frac{\partial y}{\partial b}\right]\boldsymbol{U_{ab}}\left[\frac{\partial y}{\partial a}, \frac{\partial y}{\partial b}\right]^T = [x, 1]\begin{bmatrix} u_a^2 & \rho_{ab}u_a u_b \\ \rho_{ab}u_a u_b & u_b^2 \end{bmatrix}\begin{bmatrix} x \\ 1 \end{bmatrix} \tag{15}$$

and after both multiplications:

$$u_{ab}^2(y(x)) = u_a^2x^2 + 2x\rho_{ab}u_a u_b + u_a^2 == \frac{1}{S} + \frac{S}{\Delta}\left(x - \frac{S_x}{S}\right)^2 \tag{15a}$$

This equation determines the uncertainty type A of the variable Y as a function of two correlated quantities ax and b. According to the GUM convention [1], the combined uncertainty $u(y)$ is the geometric sum of the uncertainties of types A and B:

$$u(y) = \sqrt{u_{ab}^2(y) + u_B^2(y)} \tag{16}$$

Formula (16) applies to cases where the coordinates y_i of the all points are measured with the same instrument and under the same conditions. In this case, it is not necessary to consider separately uncertainties of the type B for y_i of each measured point and $u^2_{regA}(y) = u^2_{ab}(y)$.. Function of type B uncertainty, i.e. $u_B(y)$ in the y range of straight line shall be determined from the maximum permissible error $E_{\max\,\%}$ of the meter. It is the standard deviation of uniformly distributed errors with a width of $\pm E$, i.e.:

$$u_B(y) = u_B(y_0) + (y - y_0)E_{\max}/\sqrt{3} \tag{16a}$$

In the most cases in the measurement practice $u_B(y_0) << u_B(y{-}y_0)$ and it is negligible.

Parameters a and b of the regression line depend on the coordinates of the y_i of n measured points. The effective number of degrees of freedom is $n{-}2$. Extended uncertainty U_P of the regression line, e.g. about level of confidence $P = 0.95$, is determined by the combined uncertainty $u(y)$ multiplied by coverage factor $k_{0,95}$. For the small number n of points, the Student's t-distribution is used, and following formula is obtained:

$$U_{0,95} = k_{0,95}u(y) = t_{0,95,n-2}\sqrt{u^2_{ab}(y) + u^2_{By}} \tag{17}$$

In rare cases, the y_i coordinates and some or all of the measurement points are measured with different measures and/or under different influenced conditions. Then the sequence of numerical operations in determining the expanded uncertainty is

$$\begin{aligned} u_i &= \sqrt{u^2_{Ai} + u^2_{Bi}} i = (1, \ldots, n); \\ u_{ab} &= u_{ab}(u_i(ax + b)); \\ U_y(ax + b) &= t_{0.95,n-2}u_{ab}(ax + b) \end{aligned} \tag{17a-c}$$

The equation of the regression line considering its uncertainty band therefore is:

$$y = ax + b \pm U_{0.95} \tag{18}$$

and finally:

$$y = ax + b \pm k_{0.95}\sqrt{u^2_{regA}(y) + u^2_B(ax + b)} \tag{18a}$$

The regression straight-line parameters formulas for correlated values of $\boldsymbol{Y}$ variable and their special cases are summarized below in in the left column of Table 1.

In the absence of autocorrelation, the covariance matrix $\boldsymbol{U_Y}$ and its inverse $\boldsymbol{U_Y^{-1}}$ reduce to the diagonal form and dependencies for simple regression parameters are simplified. They are compiled in the right column of the Table 1. They are a unique example of patterns for the size of correlated. Uncertainty parameters and bands for non-correlated data Y with uncertainty u_B the authors examine in detail in the previous work on the uncertainties of the straight-line determined by the regression method [19].

For variables with correlated values and an unspecified covariance matrix, a singularity may appear $\Delta = 0$, that does not occur when the matrix is positively or negatively definite. Correctly defined correlation coefficients guarantee a positive determination of the matrix, i.e. all positive eigen values of its own, like all major minor determinants. In the regression for uncorrelated values of measurement points, solutions for values of a and b in regression equation are unambiguous, as in formulas (9a,b) always the denominator $\Delta > 0$. In the literature on linear regression, neither information about the existence of this peculiarity nor tips – how to avoid it was found. There is a lack of analysis of this issue, together with a proposal for further works.

Table 1. Parameters of regression line for autocorrelated and noncorrelated variable $\boldsymbol{Y}$

Parameter			
Parameter	measured: $Y = [y_1, \dots y_i, \dots y_n]^T$ with uncertainty u_i^T; known $\boldsymbol{X} = [x_1, \dots x_i, \dots x_n]$, $u_i(x_i) = 0$		
Variable Y	**Correlated**		**Uncorrelated**
Covariance matrix Y and its inverse	$\boldsymbol{U_Y} = \begin{bmatrix} u_1^2 & \cdots & \rho_{y1n}u_1u_n \\ \vdots & \ddots & \vdots \\ \rho_{y1n}u_1u_n & \cdots & u_n^2 \end{bmatrix}$	$\boldsymbol{U_Y^{-1}}$	$\boldsymbol{U_Y} = \begin{bmatrix} u_1^2 & \cdots & 0 \\ \vdots & \ddots & \vdots \\ 0 & \cdots & u_n^2 \end{bmatrix}$, $\boldsymbol{U_Y^{-1}} = \begin{bmatrix} u_1^{-2} & \cdots & 0 \\ \vdots & \ddots & \vdots \\ 0 & \cdots & u_n^{-2} \end{bmatrix}$ $\boldsymbol{U_Y}$, $\boldsymbol{U_Y^{-1}}$ only diagonal elements
Criterion	SK	$[a\boldsymbol{X} + \boldsymbol{b} - \boldsymbol{Y}]^T \boldsymbol{U_Y^{-1}} [a\boldsymbol{X} + \boldsymbol{b} - \boldsymbol{Y}] \to \min$	$SK = \sum_{i=1}^{n} (y_i - ax_i - b)^2 u_i^{-2} \to \min$
Regression line parameters	Slope $a = \frac{SS_{xy} - S_xS_y}{S\,S_{xx} - (S_x)^2} \equiv \frac{\Delta_a}{\Delta}$, abscissa $b = \frac{S_yS_{xx} - S_xS_{xy}}{S\,S_{xx} - (S_x)^2} = \frac{\Delta_b}{\Delta}$ where: $\Delta = S\,S_{xx} - (S_x)^2$, $\Delta_a = SS_{xy} - S_xS_y$, $\Delta_b = S_yS_{xx} - S_xS_{xy}$		
Auxiliary parameters	$\Delta = \frac{1}{2}\sum_{i=1}^{n}\sum_{j=1}^{n}(v_iz_j - v_jz_i)^2 \lambda_i^{-1}\lambda_j^{-1}$ [a]		$\Delta = \frac{1}{2}\sum_{i=1}^{n}\sum_{j=1}^{n}(x_i - x_j)^2 u_i^{-2}u_j^{-2}$
	S	$[1, \dots, 1]\boldsymbol{U_Y^{-1}}\boldsymbol{1} = \sum_{i=1}^{n}\sum_{j=1}^{n}[u_y^{-1}]_{ij}$	$\sum_{i=1}^{n} u_i^{-2}$
	S_x	$\boldsymbol{X^T\, U_Y^{-1} 1} = \boldsymbol{1^T U_Y^{-1} X}$	$\sum_{i=1}^{n} x_iu_i^{-2}$
	S_{xx}	$\boldsymbol{X^T\, U_Y^{-1} X}$	$\sum_{i=1}^{n} x_i^2u_i^{-2}$
	S_y	$\boldsymbol{Y^T U_Y^{-1} 1} = \boldsymbol{1^T\, U_Y^{-1} Y}$	$\sum_{i=1}^{n} y_iu_i^{-2}$
	S_{xy}	$\boldsymbol{X^T\, U_Y^{-1} Y} = \boldsymbol{Y^T\, U_Y^{-1} X}$	$\sum_{i=1}^{n} x_iy_iu_i^{-2}$
	S_{yy}	$\boldsymbol{Y^T\, U_Y^{-1} Y}$	$\sum_{i=1}^{n} y_i^2u_i^{-2}$
Sensitivity matrix	$\boldsymbol{AB} = [\boldsymbol{A}, \boldsymbol{B}]$ $\boldsymbol{A} = \frac{\partial a}{\partial Y} = \frac{Sc - S_xd}{\Delta}$, $\boldsymbol{B} = \frac{\partial b}{\partial Y} = \frac{S_{xx}d - S_xc}{\Delta}$		$\boldsymbol{AB^T} = [a_i, b_i]$ $a_i = \frac{\partial a}{\partial y_i} = \frac{1}{u_i^2}\frac{Sx_i - S_x}{\Delta}$, $b_i = \frac{\partial b}{\partial y_i} = \frac{1}{u_i^2}\frac{S_{xx} - S_xx_i}{\Delta}$
Covariance matrix and its elements	$\boldsymbol{U_{ab}} = \begin{bmatrix} u_a^2 & \rho_{ab}u_au_b \\ \rho_{ab}u_au_b & u_b^2 \end{bmatrix} = \boldsymbol{AB\, U_Y\, AB^T}$ $u_a^2 = \frac{S}{\Delta}$, $u_b^2 = \frac{S_{xx}}{\Delta}$, $\rho_{ab}u_au_b = -\frac{S_x}{\Delta}$		
Uncertainties type A of regression straight line	Standard	$u_{ab}^2 = u_a^2x^2 + 2x\rho_{ab}u_au_b + u_b^2$, $x_{min} = \frac{-\rho_{ab}u_b}{u_a}$, $u_{abmin}^2 = (1 - \rho_{ab}^2)u_b^2$ $u_{ab}^2 = \frac{1}{S} + \frac{S}{\Delta}\left(x - \frac{S_x}{S}\right)^2$, $x_{min} = \frac{S_x}{S}$, $u_{abmin}^2 = \frac{1}{S}$	
	Extended	$U_{ab} = t_{95,n-2}\; u_{ab}$	
Equall values of all uncertainties $\boldsymbol{u_i = u}$	Matrix $\boldsymbol{U_Y}$ is replaced by correlator $\boldsymbol{R}$ $\boldsymbol{U_Y} = u^2\boldsymbol{R_Y}$ and $\boldsymbol{U_Y^{-1}} = \boldsymbol{R_Y^{-1}}/u^2$		$u_a^2 = \frac{u^2}{\sum_{i=1}^{n}(x_i - \bar{x})^2}$, $u_b^2 = \frac{u^2}{n} + \frac{u^2\bar{x}^2}{\sum_{i=1}^{n}(x_i - \bar{x})^2}$, $\rho_{ab}u_au_b = -\frac{u^2\bar{x}}{\sum_{i=1}^{n}(x_i - \bar{x})^2}$, where $\bar{x} = \frac{1}{n}\sum_{i=1}^{n} x_i$, $u_{ab}^2 = u^2\left(\frac{1}{n} + \frac{(x - \bar{x})^2}{\sum_{i=1}^{n}(x_i - \bar{x})^2}\right)$
Variance estimator	$\widehat{u^2} = \frac{-a\boldsymbol{X^T\, R_Y^{-1} Y} - b\boldsymbol{Y^T\, R_Y^{-1} 1} + \boldsymbol{Y^T\, R_Y^{-1} Y}}{n-2}$		$\widehat{u^2} = \frac{-a\sum_{i=1}^{n} x_iy_i - b\sum_{i=1}^{n} y_i + \sum_{i=1}^{n} y_i^2}{n-2}$

*[a]This pattern is obtained after the transformation of vectors **1** and $\boldsymbol{X}$ respectively to vectors $\boldsymbol{v} = [v_1, \dots, v_n]^T$ and $\boldsymbol{Z} = [z_1, \dots, z_n]^T$, i.e. $\boldsymbol{1} = \boldsymbol{Hv}$ and $X = \boldsymbol{HZ}$. It is carried out using a linear operator $\boldsymbol{H}$, i.e. the orthogonal matrix diagonalizing the covariance matrix $\boldsymbol{U_Y^{-1}}$. Matrix $\boldsymbol{H}$, as symmetrical has actual own values λ_i^{-1}, and λ_i are own values of the covariance matrix $\boldsymbol{U_Y}(i = 1, \dots, n)$.

3 Kinds of Correlation of the Measured Ordinates of Points

3.1 Points with Uncorrelated Ordinates and Equal Uncertainties

For the same value of absolute uncertainties $u_i = u$ of y_i of the measured points, the covariance matrix is:

$$\boldsymbol{U_Y} = u^2\boldsymbol{R_Y} \tag{19}$$

where: the correlator $\boldsymbol{R_Y}$ is a square symmetrical matrix size n x n, whose elements are correlation coefficients ρ_{yij} between values y_i and y_j for $i, j = 1, \ldots, n$.

In this case, for a given correlation coefficients, the equation for a regression line will always be the same, since the values of its a i b from formulas (9a,b) do not depend on the uncertainty u, but only on the correlator matrix. The regression line for points with given coordinates and the same relative uncertainties y_i of all points has other parameters than for the same absolute uncertainties.

For unknown but equal all absolute uncertainties u, the estimator of variance $\widehat{u^2}$ defined as a minimum of squares of distances divided by n-2 degrees of freedom:

$$\widehat{u^2} = u^2 \frac{-aS_{xy} - bS_y + S_{yy}}{n-2} = \frac{-a\boldsymbol{X}^T\boldsymbol{R}_Y^{-1}\boldsymbol{Y} - b\boldsymbol{Y}^T\boldsymbol{R}_Y^{-1}\mathbf{1} + \boldsymbol{Y}^T\boldsymbol{R}_Y^{-1}\boldsymbol{Y}}{n-2} \tag{20}$$

3.2 Correlation in a Series of Measurements of the Ordinate of a Single Point

In a series of m repeated measurements $y_{i1}, \ldots y_{ik}, \ldots y_{im}$ ($k = 1, \ldots m$) of the ordinate y_i of a single point with abscissa x_i, the autocorrelation may occur. With the same values of uncertainties $\sigma_{ik} = \sigma = \text{const}$ for each of the obtained ordinate values y_{ik} and correlated their results with the correlation coefficients, from ρ_{i1} for the adjacent points to $\rho(i, m-1)$ for the farthest points, the local covariance matrix of the size m x m is obtained, given in the Eq. (21).

$$\boldsymbol{U}_{yi} = \sigma^2 \begin{vmatrix} 1 & \rho_{i1}\ \rho_{i2} \cdots\cdots & \rho_{im-1} \\ \rho_{i1} & 1 \quad \rho_{i1}\ \rho_{i1} \cdots\cdots & \\ \rho_{i1} & \rho_{i1}\ 1 \quad \cdots\cdots\cdots & \\ \cdots & \cdots\cdots\cdots\cdots & \rho_{i2} \\ \cdots & \cdots\cdots\cdots\ 1 & \rho_{i1} \\ \rho_{im-1} & \cdots\cdots\ \rho_{i2}\ \rho_{i1} & 1 \end{vmatrix} \tag{21}$$

In the covariance matrix $\boldsymbol{U_{yi}}$, the greatest positive correlation occurs for terms close to the mean value of y_i. The real type A uncertainty u_{yi} of this point is greater than from the classical formula for m number of repeated measurements of y_i.

In assessing the uncertainty u_{Ai} of the mean value of y_{im} results of the point with abscissa x_i, instead of the number of measurements m, the effective number $n_{ieff} < m$ must be used described by the following formula [5, 6, 13, 14]:

$$n_{ieff} = \frac{m}{1 + 2\sum_{k=1}^{m-1} \frac{m-k}{m} \rho_{ik}} \tag{22}$$

where: m – number of autocorrelated measurements, ρ_{ik} – value of autocorrelation function for y_{ik} measurement.

If the adjacent results in the series of repeated measurements the tested point Q_i are correlated with the constant coefficient $\rho_{ik} = \rho$, then the effective number of measurements of this series is:

$$n_{ieff} = \frac{m}{1 + \rho(m-1)} \tag{22a}$$

Table 2 shows an example of the effective uncertainties u_{yi} for m-fold measurements of the ordinate y_i of the tested point with abscissa x_i and three values of the correlation coefficient ρ between successive measurements. In each of the three series, $m = 20$ measurements with the same uncertainty σ_i were made. Table 2 shows that with the increase of the correlation coefficient ρ, the standard uncertainty of the series increases, because the effective number of measurements decreases from 20 to 1.9 and 1 for the correlation coefficients of 0, 0.5 and 1, respectively. The effective uncertainty for uncorrelated observations will be about 5 times less than when they are fully correlated.

Table 2. Effective uncertainties of measurement series of y_i with three correlation coefficients

Correlation coefficient	ρ	$\rho = 0$	$\rho = 0,5$	$\rho = 1$
Effective number of measurements	n_{ieff}	$n_{ieff} = m = 20$	$n_{ieff} = \frac{2m}{m+1} = 1,9$	$n_{ieff} = 1$
Effective uncertainties of a series of measurements y_i of point (x_i, y_i)	$\frac{\sigma_i}{\sqrt{n_{ieff}}}$	0,22 σ_i	0,72 σ_i	σ_i

3.3 Correlation Between Ordinates of the Various Measured Points

To determine the covariance matrix describing the measurements of the correlated ordinates y_i, y_j of points (x_i, y_i), $(x_j y_i)$ with the same uncertainties $u_i = u_j = u =$ const, the autocorrelation function can also be used. In the matrix (21), the uncertainty σ of single measurements in the series is replaced by the uncertainty of the measured points. Autocorrelation is a symmetrical function rapidly decreasing from the maximum value to zero, and its initial arms can be approximated with a straight line [5, 14, 15]. Usually it is enough to consider only the neighboring values of the measured ordinate y_{i-1}, y_i, y_{i+1} of points with known abscissa x_{i-1}, x_i, x_{i+-1}, i.e. $\rho_1 = \rho$ and $\rho_k = 0$ for $i = 2, \ldots, n-1$. For n points equally remote and the same uncertainties u and correlation coefficients ρ of the covariance matrix $\boldsymbol{U}_Y$ from (23) of size n x n has all correlation coefficients $\rho_k = \rho$ for $i = 1, \ldots, n$-1. When the influence of autocorrelation is negligible, the $\boldsymbol{U}_Y$ matrix reduces to the main diagonal:

$$\boldsymbol{U}_Y = u^2 \begin{vmatrix} 1 & \rho_1 \ldots \rho_k \ldots \rho_{n-1} \\ \rho_1 & 1 \quad \ldots\ldots \rho_k \ldots \\ \ldots & \ldots\ldots\ldots\ldots \rho_k \\ \rho_k & \ldots\ldots\ldots\ldots\ldots \\ \ldots & \rho_k \ldots\ldots 1 \quad \rho_1 \\ \rho_{n-1} & \ldots \rho_k \cdots \rho_1 \quad 1 \end{vmatrix} \tag{23}$$

4 Numerical Examples

Simulated numerical examples will be shown and quantifiable the theoretical relationships given.

4.1 Regressions Line and Uncertainty Bands for Points of the Same Value of Absolute or Relative Uncertainty

Table 3 shows coordinates of 10 measured points with the same value uncertainties.

Table 3. Coordinate of measured points and equal absolute and relative uncertainties of y_i

Symbol	Coordinates of points (x_i, y_i) and uncertainties of y_i									
x_i	1	2	3	4	5	6	7	8	9	10
y_i	2	1	4	3	6	5	8	7	10	11
$u_i(y_i)$	0.5									
$\delta_i(y_i)$	5%									
$u_B(y_i)$	$0.01(1+y)\%$									
Regr.lines	for $u_i = u$: $y = 1{,}48\,x - 0{,}067$					for $\delta_{ui} = \delta$: $y = 0.978x - 0{,}463$				

The y_i values are the averages of multiple measurements of the ordinates y_i of each point with the abscissa x_i. In order to identify the effects of single factors, simplified cases will be analyzed. It was assumed that the absolute uncertainties u_i or the relative uncertainty δ_{ui} of the measurements of the ordinate values y for all points are the same, and their abscissas x_i are constant and known exactly. In order to compare with the results for uncorrelated ordinates, the same data were adopted as for experiment I from the Table 1a in the first our paper [18]. Considerations are illustrated by Fig. 2a, b.

Figure 2a shows the line and its uncertainty band (blue lines) determined by the linear regression method for the data from Table 3 in the absence of correlation. It also shows the uncertainty of the ordinates of the measured points (red lines). The figure shows that the 95% expanded uncertainty region changes along the regression line such that its width decreases from the initial value to a minimum near the center of the regression line, and then widens back to the original initial size.

At the top in Fig. 2b there are the graphs of Gaussian probability density distributions (pdf- red curves) for the ordinates of the measured points on both sides of the regression line. The pdf of the y_i ordinates are for 95% coverage with standard deviation $u = 0.5$. Below, the probability density curves for y around the regression line inside its uncertainty band (blue 3D solid) are shown. In the middle of the range of the regression line, pdf is the narrowest and therefore is the highest. The pdf maximum (narrowest pdf) occurs for the minimum uncertainty bandwidth in the middle part of the straight line. When moving away from this maximum in opposite directions, the maximum value of the pdf function decreases, and the distribution widens as the uncertainty band boundaries move away from each other.

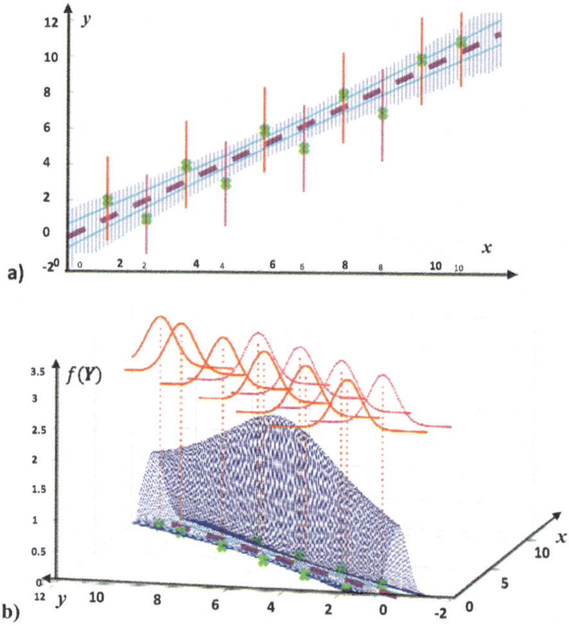

Fig. 2. **a)** Graphs of regression line with band of its extended uncertainty type A. **b)** Gaussian probability density distributions (pdf-s) with uncertainty $u = 0{,}5$ equal for all measured points and pdf-s $f(y)$ of y uncertainties along the regression line in 3D view

Figure 3a shows two regression lines with their type A extended uncertainty bands type A $\pm\ U_{0.95}$. They are calculated for the ordinates (x_i, y_i) of 10 points from Table 3 measured 20 times each with equal absolute uncertainties $u_i = u$ (blue line $y = 1{,}48\ x - 0{,}067$ runs above) or with equal relative uncertainties $\delta_{ui} = \delta_u$ (red line $y = 0.978x - 0{,}463$ running below) and for three different correlation coefficients for y_i.

Figure 3b shows sets their $U_{0.95}$. With the increase of the correlation coefficient between the results of repeated measurement observations, extended uncertainty U_A increases [5, 6, 15]. In Sect. 3.2 it was shown that the effective number of observations of a sample with 20 measurements for the correlation coefficients 0, 0.5 and 1 decreased from 20 to 1.9 and to 1. The narrowest uncertainty band is obtained for non-correlated observations, i.e. for $\rho = 0$. If such measurements are made for the ordinates y_i of all 20 measurements with the same uncertainty, then for the correlation coefficient $\rho = 1$, the uncertainty band of the regression line is over four times wider than in the absence of correlation. Negative coefficients ρ do not occur in these measurements.

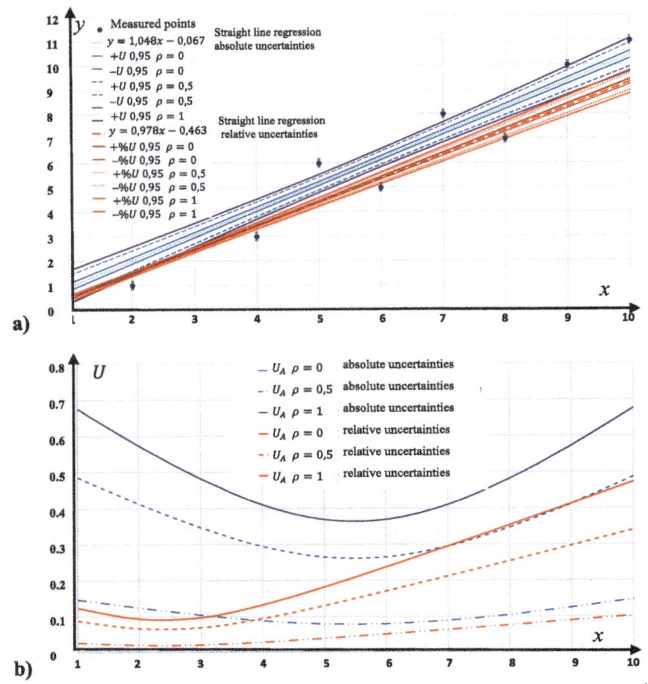

Fig. 3. **a)** Regression lines with expanded uncertainty bands type A, **b)** half height of uncertainty bands $\underline{U}_A$ for three values of the correlation coefficient ρ for 10 points series from Table 3 measured 20 times each

4.2 Comparisons of Uncertainty Bands for Uncorrelated and Correlated *Y* Values

Figure 4a, b compares the type A uncertainty bands for two regression lines with uncorrelated and correlated *y*-coordinates. The line with correlated data is described by the covariance matrix (21) with $u_i = 0, 5$, $\rho = 0{,}5$ for neighboring points $\rho_1 = \rho$ and $\rho_k = 0$ for the remaining points (where $k = 2,\ldots, 9$).

The graphs in Figs. 4a, b show that a slight correlation between the ordinates of the measured points slightly changes the width of the uncertainty band, and the consideration of type B uncertainty described in Table 3 broadens the band of expanded uncertainty at the end of the regression straight-line range.

4.3 Effect of the Correlation Coefficient Value

Figure 5 compares two regression lines and uncertainty bands for the covariance matrix $\boldsymbol{U}_Y$ (21) with the parameters $u_i = 0, 5$, $\rho_1 = 0, 6$; $\rho_2 = 0{,}5$; $\rho_3 = 0{,}4$; $\rho_4 = 0{,}2$; $\rho_5 = 0{,}1$ and $\rho_k = 0$ for others $k = 6,\ldots, 9$. Equation of the regression line for correlated points is $y = 0.981x + 0.417$. There is a significant increase (over 30%) in the width of uncertainty band, when there is a autocorrelation of points described by the covariance matrix. On the other hand, type B uncertainty has a greater impact on the width of band

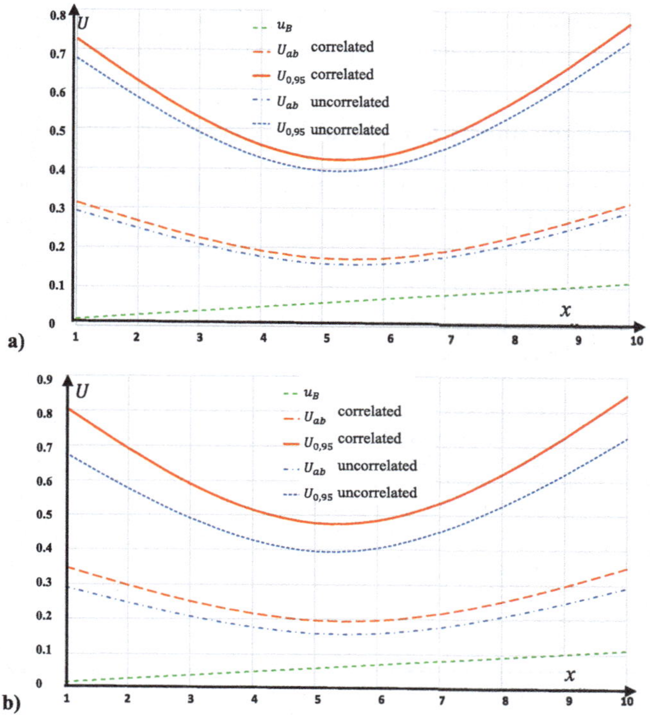

Fig. 4. a, b. Uncertainty functions of regression lines of uncorrelated and correlated *y*-coordinates of nearest points and covariance matrix $\boldsymbol{U_Y}$ (21): a) $y = 1{,}044x\text{-}0{,}028$ and correlation $\rho = 0{,}1$; b) $y = 1{,}023x + 0{,}129$ for correlation $\rho = 0{,}3$

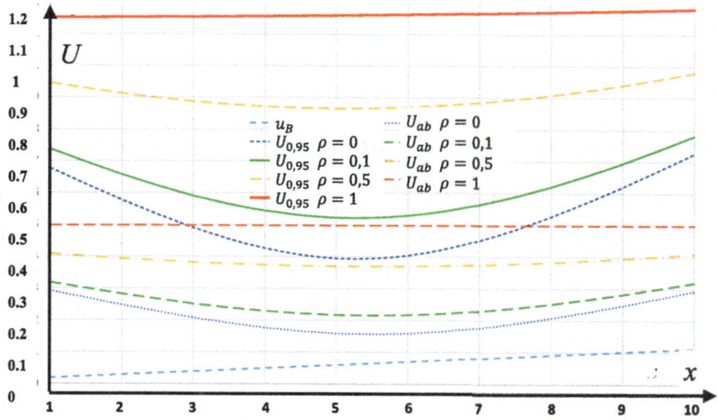

Fig. 5. Uncertainty bands for regression line $y = 1,048x - 0,067$ for correlated and non-correlated measurements point for three covariance matrixes $\boldsymbol{U_Y}$ from (21) for $u = 0.5$, $\rho_k = \rho$ for $i = 1, .., 9$ $\rho = 0.1$, $\rho = 0.5$ and $\rho = 1$ with uncertainty of type B

determined without correlation than on the band with correlation, because with the data considered here of the experiment I, an increase in uncertainty by 1% is negligible.

4.4 Comparing of the Impact of Full and Non-correlated Ordinates y_i

The regression lines and their absolute uncertainty bands for full correlation ($\rho = 0.99 \approx 1$) of the measured ordinates y_i of the points with the covariance matrix $\boldsymbol{U}_Y = u^2|1|$ and for the lack of correlation ($\rho = 0$) are given in Fig. 6.

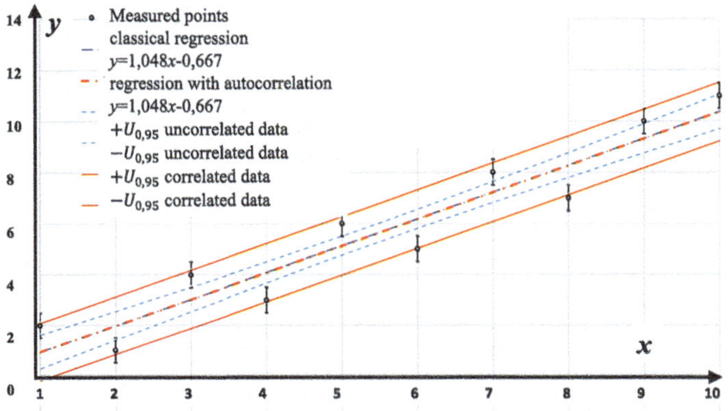

Fig. 6. Regression lines and uncertainty bands for measuring points with uncertainties marked at uncorrelated and correlated points with $\rho \approx 1$ for covariance matrix $\boldsymbol{U}_Y$

It is shown that with the full correlation, i.e. for $\rho = 1$, hyperbolic boundaries of the uncertainty band become straight lines parallel to the regression line. The band constriction in the middle of the regression line range disappear. Parameters of this regression line do not change with the increase of equal coefficients in the covariance matrix.

4.5 Effect of Correlation Between Ordinates y_i of Measured Points

The graphs in Fig. 7 concern the courses of uncertainty U_{ab} (type A) and uncertainty u_B for the regression line obtained for the same values of the correlation coefficients between the y_{ik} coordinates of each of the repeatedly measured points of $x_i = \text{const}$, i.e. according to their covariance matrix (21). With the presence of positive correlation, an increase in the width of the uncertainty band is observed.

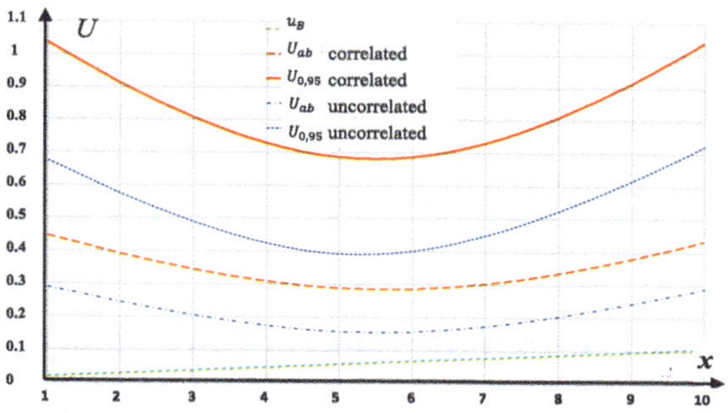

Fig. 7. Uncertainties for uncorrelated and correlated y_i of points with covariance matrix $\boldsymbol{U_Y}$ (21)

4.6 Regression Lines for Constant Absolute or Relative Uncertainties of Measured y_i

In Figs. 8a, b the regression lines with uncertainty bands for the correlated and uncorrelated ordinates y_i, measured with the relative uncertainty $\delta = 1\%$, were compared.

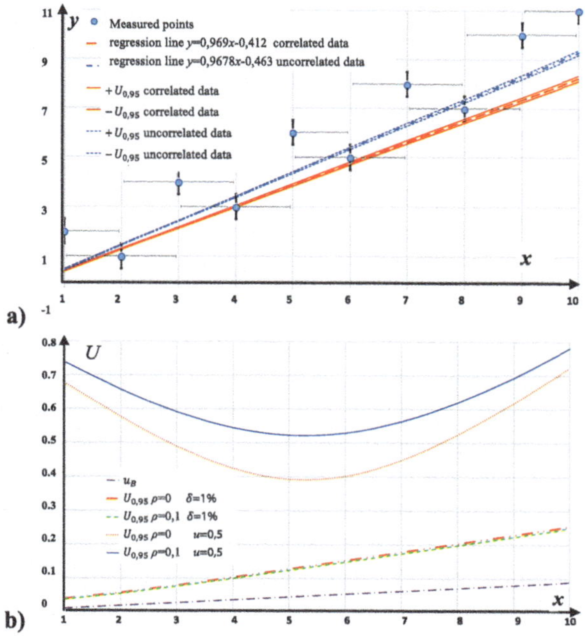

Fig. 8. a, b. Cases of equal relative uncertainty $\delta = 1\%$ for uncorrelated and correlated y_i of measured points for the covariance matrix $\boldsymbol{U_Y}$ of (21) with $\delta = 1\%$, $\rho_k = 0, 1$for $i = 1, \ldots, 9$ regression lines, b) uncertainty $U_{0.95}$ together with uncertainty of type B

For small the same values of correlation coefficients, such a straight line (Fig. 8a) deviates from their direction for the same ordinates without correlation. Its slope was reduced from 0.978 to 0.869. The boundaries of the uncertainty band slightly widen for absolute uncertainties and the boundaries of the band for relative uncertainties remain practically unchanged. The linearly increasing type B uncertainty u_B causes a disturbance of the symmetry of limits of extended uncertainty $U_{0.95}$ band in relation to the center of the regression line and a shift it towards higher values of the independent variable x. As a result, this uncertainty bandwidth increases with x, for both correlated and uncorrelated measuring points.

5 Correlation of *Y* within the Band of Uncertainty

Apart from the correlation between y_i of the measured points' ordinates, y_i of points belonging to the regression line $y = ax + b$ are also correlated with each other. This is because the covariance matrix of the two its points $y_i = ax_i + b$ (for $i = 1, 2$) is determined by the variance propagation equation:

$$\begin{bmatrix} u_{y_1}^2 & \rho_{y_1y_2}u_{y_1}u_{y_2} \\ \rho_{y_1y_2}u_{y_1}u_{y_2} & u_{y_2}^2 \end{bmatrix} = \begin{bmatrix} \frac{\partial y_1}{\partial a} & \frac{\partial y_1}{\partial b} \\ \frac{\partial y_2}{\partial a} & \frac{\partial y_2}{\partial b} \end{bmatrix} \begin{bmatrix} u_a^2 & \rho_{ab}u_au_b \\ \rho_{ab}u_au_b & u_b^2 \end{bmatrix} \begin{bmatrix} \frac{\partial y_1}{\partial a} & \frac{\partial y_2}{\partial a} \\ \frac{\partial y_1}{\partial b} & \frac{\partial y_2}{\partial b} \end{bmatrix}$$
$$= \begin{bmatrix} u_a^2x_1^2 + 2x_1\rho_{ab}u_au_b + u_b^2 & u_a^2x_1x_2 + (x_1 + x_2)\rho_{ab}u_au_b + u_b^2 \\ u_a^2x_1x_2 + (x_1 + x_2)\rho_{ab}u_au_b + u_b^2 & u_a^2x_2^2 + 2x_2\rho_{ab}u_au_b + u_b^2 \end{bmatrix} \quad (23)$$

where: $\frac{\partial y_1}{\partial a} = x_1$, $\frac{\partial y_1}{\partial b} = 1$, $\frac{\partial y_2}{\partial a} = x_2$, $\frac{\partial y_2}{\partial b} = 1$.

For y_i at points $i = 1, 2$ squares of uncertainty u_{y_i} and correlation coefficient of y_1 and y_2 are as follows:

$$u_{y_i}^2 = u_a^2x_i^2 + 2x_i\rho_{ab}u_au_b + u_b^2 \quad (i = 1, 2) \quad \text{(23a,b)}$$

$$\rho_{y_1y_2} = \frac{u_a^2x_1x_2 + (x_1 + x_2)\rho_{ab}u_au_b + u_b^2}{u_{y_1}u_{y_2}} \quad \text{(23c)}$$

If the measurements relate to the determination of a regression line describing the characteristics of an instrument, process or device and it will be used in the presence of unknown part of systematic errors that cannot be eliminated under variable influencing conditions, then the type B uncertainty must also be estimated. Usually it is assumed that it results from the maximum permissible error $E_{\max}$ of digital meters as a linear function $u_B(y) = u_{B0} + \delta(y - y_0)$ [1], or as a function of x as $u_B(x) = u_{B0} + \delta_B(b - y_0) + \delta_B ax$. The combined uncertainty u_c of the regression line is larger than the uncertainties u_{ab} of type A. The combined uncertainty $u_c(x)$ of the regression straight line is described by the formula obtained after ordering the words, is

$$u_C^2(x) = u_{ab}^2 + u_B^2 = (u_a^2 + a^2\delta_B^2)x^2 + 2\big(\rho_{ab}u_au_b + a\delta_B(u_{B0} + \delta_B(b - y_0))\big)x + u_b^2 + (u_{B0} + \delta_B(b - y_0))^2 \quad (25)$$

Then from (25), the effective values of combined uncertainties u_{aC}, u_{bC} and the correlation coefficient ρ_{abC} with the influence of type B uncertainties, will be respectively:

$$u_{aC}^2 = u_a^2 + a^2\delta_B^2, \qquad u_{bC}^2 = u_b^2 + (u_{B0} + \delta_B(b - y_0))^2,$$
$$\rho_{abC} = \frac{\rho_{ab}u_au_b + a\delta_B(u_{B0} + \delta_B(b - y_0))}{u_{aC}u_{bC}}. \quad \text{(26a,b,c)}$$

Squares of combined uncertainty in points x_i for $i = 1, 2$ are

$$u_{yCi}^2(x) = (u_a^2 + \delta^2 a^2)x_i^2 + 2x_i\Big(\rho_{ab}u_a u_b + a\delta^2(1+b)\Big) + u_b^2 + \delta^2(1+b)^2 \tag{27}$$

Figure 9 shows the bands of relative and absolute uncertainties described by (27) for absolute uncertainties of $u = 0.5$ and for relative 1% of correlated variables $\rho_k = 0, 3$ for $i = 1, \ldots, 9$ with an increasing linear uncertainty $\rho_k = 0, 3$ for $\delta = 1\%$, $u_{B0} = \delta_B$, $y_0 = 0$. It is shown also that in the band narrowing area, the uncertainty for the correlated values y_i increases almost twice as compared to the uncorrelated y_i with the same absolute uncertainties and the influence of type B uncertainty makes slightly asymmetric the uncertainty bands.

The resulting relative uncertainties decrease from several dozen percent at the beginning of the regression line to less than 10% at the end of its range. For relative uncertainties, the bands of correlated and uncorrelated variables almost coincide. They are slightly greater for the values of uncorrelated variables than for correlated variables. The initial relative values increase linearly to the level below 0.3, while the relative values do not exceed the level of 2–3%. The influence of the linearly increasing type B uncertainty is practically imperceptible.

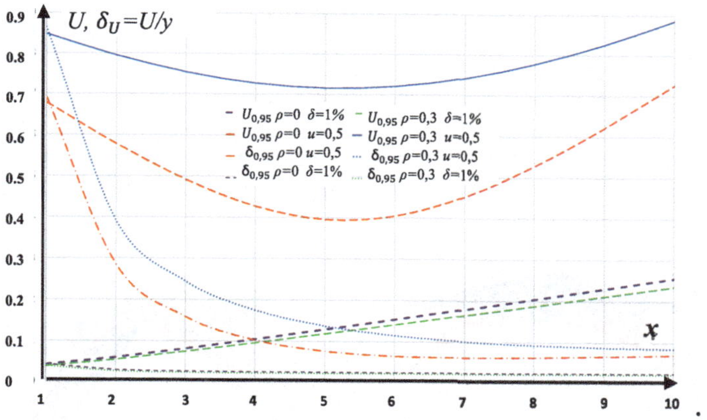

Fig. 9. Relative and absolute uncertainties for type A and type A and B correlated variables $\rho_k = 3$ for $i = 1, .., 9$ and noncorrelated variables with considering type B as linear uncertainty $\delta = 1\%$ of (27) for $u = 0{,}5$ and $\delta = 1\%$

The correlation coefficient between the variables y_1 and y_2 from (23c) of the regression line considering the influence of uncertainty of type B (27) has the form:

$$\rho_{y_1 y_2} = \frac{\begin{array}{c}(u_a^2 + a^2\delta_B^2)x_1x_2 + (x_1 + x_2)\rho_{ab}u_a u_b + \\ +\delta_B(u_{B0} + \delta_B(b - y_0)) + u_b^2 + (u_{B0} + \delta_B(b - y_0))^2\end{array}}{u_{yC_1}u_{yC_2}} \tag{28}$$

Due to the linear dependency of the size of the y from x these variables are fully correlated, which means that the same correlation coefficient (28) may variables x_1 i x_2.

Figures 10 and 11 represent the relationship between correlation coefficient $\rho_{y_1 y_2}$ from (28) between any ordinates y of the regression line with uncertainty due to (27). Figure 10 shows that the correlation coefficient as a function of the variable x_i depends on the mutual position of the points, regardless of whether the measured ordinates y are correlated or not and whether the linearly increasing uncertainty of type B was considered. The largest value of the coefficient $\rho_{y_1 y_2} \to 1$ is for closest points and decreases as they move away. Figure 11 shows the dependence of the coefficient $\rho_{y_1 y_2}$ as a function of two variables x_1 and x_2 for percentage values 1%, 3%, 5% of the relative uncertainty type B. This coefficient has the value $\rho_{y_1 y_2} = 1$ for $x_1 = x_2$ on the a straight-line and it symmetrically decreases with respect to this points on the regression line move away. The surfaces corresponding to the B-type uncertainties fall over each other with an increase or decrease in the value of $\delta_B = 1\%$, 3% and 5%, respectively, with no correlation and with full correlation between measured points.

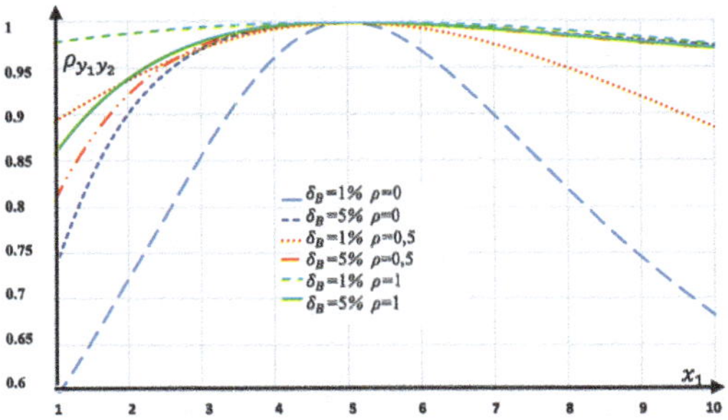

Fig. 10. Correlation coefficient $\rho_{y_1 y_2}$ of two y values of the regression line as a function of x_1 (without influence of the uncertainty type B, i.e. $\delta_B = 0$) for different values of x_2 if $u = 0{,}5$

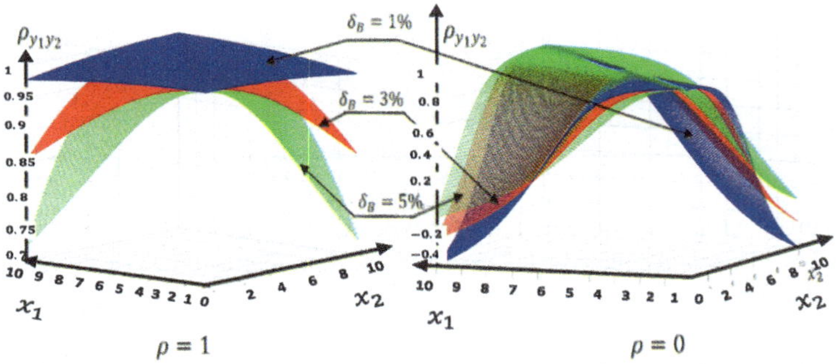

Fig. 11. Three dimensional (3D) charts of the correlation coefficient $\rho_{y_1 y_2}$ as a function of x_1 and x_2 for correlated and noncorrelated variable y of measured points and different relative uncertainties of type B

6 Straight Line for Two Repeatedly Measured Measurement Points

The simplest case is to determine the line and its uncertainty band for the number of measuring points $n = 2$ [16]. This line runs through the mean values of the results of multiple measurements of both points, and the dependence of the uncertainty as a function of the location of the selected value of the Y variable is a weighted convolution of Y_1 and Y_2 variables, taking into account their correlation. Uncertainties in points x_1 and x_2 are denoted by $u_i = u(y_i)$ for $i = 1, 2$ respectively, and the correlation coefficient between their variables Y - as ρ. After adopting the notations $\varepsilon = u_1/u_2$ and $\gamma = \frac{1}{1-\rho^2}$, the covariance matrix and its inverse are as follows

$$U_Y = \begin{bmatrix} u_1^2 & \rho u_1 u_2 \\ \rho u_1 u_2 & u_2^2 \end{bmatrix} \text{ and } U_Y^{-1} = \frac{1}{1-\rho^2}\begin{bmatrix} 1/u_1^2 & -\rho/u_1 u_2 \\ -\rho/u_1 u_2 & 1/u_2^2 \end{bmatrix} \tag{29}$$

The parameters (7a-f) determined for only two measured points are:

$$\begin{aligned} S &= 1^T U_Y^{-1} \cdot 1 = \gamma\left(\tfrac{1}{u_1^2} + \tfrac{1}{u_2^2} - \tfrac{2\rho}{u_1 u_2}\right) = \gamma u_1^{-2}\left(1 + \varepsilon^2 - 2\rho\varepsilon\right), \\ S_x &= X^T \cdot U_Y^{-1} \cdot 1 = 1^T U_Y^{-1} \cdot X = \gamma u_1^{-2}\left(x_1 + x_2\varepsilon^2 - \rho\varepsilon(x_1 + x_2)\right), \\ S_{xx} &= X^T \cdot U_Y^{-1} \cdot X = \gamma u_1^{-2}\left(x_1^2 + x_2^2\varepsilon^2 - 2\rho\varepsilon x_1 x_2\right), \\ S_y &= Y^T U_Y^{-1} \cdot 1 = 1^T U_Y^{-1} \cdot Y = \gamma u_1^{-2}\left(y_1 + y_2\varepsilon^2 - \rho\varepsilon(y_1 + y_2)\right) \\ S_{xy} &= X^T U_Y^{-1} \cdot Y = Y^T U_Y^{-1} X = \gamma u_1^{-2}\left(x_1 y_1 + x_2 y_2 \varepsilon^2 - \rho\varepsilon(x_2 y_1 + x_1 y_2)\right) \end{aligned} \tag{30a-e}$$

$$\Delta = S S_{xx} - (S_x)^2 = \gamma^2 u_1^{-4} \varepsilon^2\left(1 - \rho^2\right)(x_2 - x_1)^2 \tag{30f}$$

From (9a,b) the slope a and the intercept b of the straight-line equation for two points are obtained as follows

$$\begin{aligned} a &= \frac{S \cdot S_{xy} - S_x S_y}{\Delta} = \frac{\gamma^2 u_1^{-4}\varepsilon^2(1-\rho^2)(y_2-y_1)(x_2-x_1)}{\gamma^2 u_1^{-4}\varepsilon^2(1-\rho^2)(x_2-x_1)^2} = \frac{y_2-y_1}{x_2-x_1}, \\ b &= \frac{S_y \cdot S_{xx} - S_x S_{xy}}{\Delta} = \frac{\gamma^2 u_1^{-4}\varepsilon^2(1-\rho^2)(x_2 y_1 - x_1 y_2)}{\gamma^2 u_1^{-4}\varepsilon^2(1-\rho^2)(x_2-x_1)^2} = y_2 - a x_2 \end{aligned} \tag{31a,b}$$

This is the same regression line as for two points with coordinates without correlation. From (12 b, c, d) the following variances and the non-diagonal element of the covariance matrix U_{ab} are determined:

$$\begin{aligned} u_a^2 &= \frac{S}{\Delta} = \frac{u_1^2 + u_2^2 - 2\rho u_1 u_2}{(x_2-x_1)^2}, \quad u_b^2 = \frac{S_{xx}}{\Delta} = \frac{x_2^2 u_1^2 + x_1^2 u_2^2 - 2\rho u_1 u_2 x_1 x_2}{(x_2-x_1)^2} \\ \rho_{ab} u_a u_b &= -\frac{S_x}{\Delta} = \frac{\rho u_1 u_2 (x_1 + x_2) - x_2 u_1^2 - x_1 u_2^2}{(x_2-x_1)^2} \end{aligned} \tag{32a,b,c}$$

The uncertainty band of a regression line is determined from the uncertainty of correlated parameters a and b:

$$u_{ab}^2(x) = u_a^2 x^2 + 2x\rho_{ab} u_a u_b + u_b^2 \tag{33}$$

After substituting (32a,b) to (33) and grouping words for u_1^2, $2\rho u_1 u_2$ and u_2^2 we get:

$$u_{ab}^2(x) = u_1^2\left(\frac{x_2 - x}{x_2 - x_1}\right)^2 + 2\rho u_1 u_2 \frac{(x_2 - x)(x - x_1)}{(x_2 - x_1)^2} + u_2^2\left(\frac{x - x_1}{x_2 - x_1}\right)^2 \tag{34}$$

When a variable X shall be presented as a weighted average of two adjacent points with coordinates x_1 ix_2 that as $x = (1-k)x_1 + kx_2$ with variable k from the range $0 \le k \le 1$, which determines the position of the x, then the uncertainty u_{ab} will take a simpler form:

$$u_{ab}(k) = \sqrt{(1-k)^2u_1^2 + 2\rho(1-k)ku_1u_2 + k^2u_2^2} \quad (35)$$

The same form of dependency as the square root of the weighted sum of the squares of uncertainty for the ends of the interpolated range and the double product of uncertainty with weights and correlation coefficient, the authors were given for determination of uncertainty at the point x [15–17]. At work [16] demonstrated that the uncertainty band of a regression line for multi-point measurements is equivalent to two points with appropriately selected uncertainties and a correlation coefficient.

7 Summary and Conclusions

This is the second from a series of several authors works about the developed evaluation of the accuracy of linear regression methods from the point of view of their application in metrology and measurement technology. Cases for uncorrelated data of the measured Y-dependent variable are discussed in the paper [18] in this book. This second paper analyses generalization for correlated measurement data Y. Equations of their straight-line and uncertainty bands are determined here by a weighted linear regression method. An assessment of measurement uncertainty has been applied to the recommendations of the Guide GUM [1]. Type B uncertainty is so far ignored in the statistical literature on regression methods, and relative uncertainties characterizing digital measurements were also considered.

The equation of the regression line depends on values of the measured ordinates y_i and on the elements of their covariance matrix. The correlation of the dependent variable Y points is described by non-diagonal elements of the $\boldsymbol{U}_Y$ covariance matrix.

The influence of positive correlation between the ordinates of the measurement points was considered. Such correlation changes both: the parameters of the regression line equation and a widens of its uncertainty band. With an increase of the correlation coefficient of adjacent measured points, the slope of the regression line also increases. The broadest uncertainty band of this line occurs when the value of the variable Y is fully correlated, i.e. when $\rho = 1$. Then the regression line for correlated ordinates is identical to that for uncorrelated data.

The x and y variables of the regression line points are fully correlated as linearly dependent. With a proportional change in the uncertainty of all measured points, the line parameters are unchanged and depend only on the covariance matrix elements.

The width of the uncertainty band depends linearly on the absolute uncertainty, equal for all points, determined by non-diagonal elements of the correlator matrix. Equation of the regression line for ordinates of measured points uncorrelated and correlated with the same value of correlation coefficients, is the same.

For the same relative uncertainty of all points, also the same equation of regression line is obtained for different values of this uncertainty. But it is different than for the same absolute uncertainty. This also applies to the boundaries of the uncertainty bands.

The uncertainty bands of type A of the regression line determined from the measurements, similarly as for the uncorrelated variables, run hyperbolic and their cutoffs are symmetrical with respect to this line. For the same relative uncertainties, the limits of the uncertainty band of the regression line are split hyperbolas with tangents diagonal to the line. In the middle part of the band, there is a constriction with a width *S* depending on the parameters of the covariance matrix. With full correlation, i.e. for $\rho = 1$, the hyperbolic boundaries of the uncertainty band become straight lines parallel to the regression line. With the increase in the correlation coefficient between the ordinates of the tested points, the slope of the regression line decreases significantly.

Points within the uncertainty band are correlated more strongly if closer they are, and their correlation coefficient tends to 1. This applies to the entire range of *y*.

The type B uncertainty is estimated heuristically from the maximum permissible error *E* (of the acronym MPB). For the uniform distribution with a range 2*E*, uncertainty u_B is a standard deviation $E/\sqrt{3}$. Usually the error *E* depends linearly on $(y{-}y_0)$ inside the measured value range $y_{max}{-}y_0$ (e.g. for digital instruments). The u_B uncertainty is summed up geometrically with the type A uncertainty of the regression line obtained from measurements (root of the sum of squares). As a result, the resultant combined and expanded uncertainty bands widen with increasing *y* and *x*.

The accuracy of the regression line parameters improves with repeated measurements of the *y*-ordinate value of each point with abscissa x_i. The results of the observations in such a series are close to each other and it is necessary to consider the impact of autocorrelation. For a given point, the local covariance matrix or the formula for the effective number of measurements is used – works of Dorozhovets and Warsza [6, 13, 14], and Zięba [5].

It has been found that for variables with correlated values and an undefined covariance matrix, there may appear a singularity $\Delta = 0$ which does not occur when the matrix is positively specified. In the literature on linear regression, neither information about the existence of this singularity nor any hints on how to proceed was found. We intend to develop a detailed analysis of this issue. The next paper in this book is about uncertainties of regression methods in the measurement of both random variables *Y* and *X*.

References

1. GUM JCGM100: Guide to the expression of uncertainty in measurement. + S1: JCGM101:2008; Suppl.1. Propagation of distributions using a Monte Carlo method; + S2: JCGM102:2011; Suppl. 2 - Extension to any number of output quantities. BIPM Paris (2008)
2. Draper, R.D., Smith, H.: Applied Regression Analysis, 3rd edn. Wiley, New York (1973)
3. Piotrowski, J.: Theory of Physical and Technical Measurement. PWN-Elsevier, Amsterdam (1992)
4. Sanford, W.: Applied Linear Regression, 3rd edn. J, Willey and Sons (2005)
5. Zięba, A.: Analysis of Experimental Data in Science and Technology PWN 2013 (book in Polish. Extended English version by Cambridge Scholars Publishing, Newcastle GB in print)
6. Dorozhovets, M., Warsza, Z.L.: Uncertainty type A evaluation of autocorrelated measurement observations. Biuletyn of Military Tech. Acad. Warsaw **LVII**(2), 143–152 (2008)
7. Kessel R., Kacker R.: Correlation in uncertainty of measurement - a discussion of state-of-the-art techniques. In: Proceedings of XIX IMEKO World Congress Lisbon Portugal, p. 2352 (2009)

8. Elster, C., Toman, B.: Bayesian uncertainty analysis for a regression model versus application of GUM Supplement 1 to the least-squares estimate. Metrologia **48**, 233–240 (2011)
9. Bich, W., Cox, M., Michotte, C.: Towards a new GUM-an update. Metrologia **53**, 149 (2016)
10. White, D.R.: In pursuit of a fit-for-purpose uncertainty. Metrologia **53**(01), 107 (2016)
11. Amiri-Simkooei, A.R., et al.: Estimation of straight-line parameters with fully correlated coordinates. Measurement **48**, 378–386 (2014)
12. Chunovkina, A., Stepanov, A.: Estimation of linear regression confidence bands in case of correlated noise. In: Proceedings of the 12th International Conference Measurement 2019 in Smolenice Castle, Slovak Academy of Sciences, pp. 58–61 (2019). ISBN 978-80-972629-2-1
13. Warsza, Z.L.: Evaluation of the type a uncertainty in measurements with autocorrelated observations. In: Journal of Physics. Conference Series 459 Joint IMEKO TC1+TC7+TC13 Symposium: Measurement Across Physical and Behavioral Sciences, Genova, Italy (2013)
14. Warsza, Z.L.: Methods of extending the analysis of measurement uncertainty. Monograph. PIAP 2016 (in Polish with abstracts of chapters in English)
15. Warsza, Z.L., Puchalski, J.: Estimation of uncertainties in indirect parameter measurements of correlated quantities. In: Proceedings of 12th International Conference "Measurement 2019" Slovak Academy of Sciences, Slovakia, pp. 51–57 (2019)
16. Warsza, Z.L., Puchalski, J., Idzkowski, A.: Application of the vector method for estimating characteristic function based on measurements uncertainty at two control points. In: Proceedings of AMSA V 2019 18–20 Sept. Novosibirsk NSTU, pp. 60–73 (2019). ISSN 2313-870X (amsa.conf.nstu.ru)
17. Warsza, Z.L., Puchalski, J.: Estimation of uncertainties in indirect multivariable measurements Part 1. Case of correlated quantities. In: Szewczyk, R., et al. (eds.) Proceedings of Automation 2020, Towards Industry of the Future. AISIC Series, vol. 1140, pp. 309–325. Springer (2020)
18. Warsza, Z.L., Puchalski, J.: Dependence of the uncertainty bands of the regression line on uncertainties of types A and B of measured variable Y. Paper 32 in this book

Dose Measurement of Flocculants in Water Treatment of Stone Processing Plants

Igor Korobiichuk[1(✉)], Volodymyr Shamray[2], Valentyn Korobiichuk[2], Andrii Kryvoruchko[2], and Sergii Iskov[2]

[1] Łukasiewicz Research Network – Industrial Research Institute for Automation and Measurements PIAP, Warsaw, Poland
igor.korobiichuk@piap.lukasiewicz.gov.pl
[2] Zhytomyr Polytechnic State University, Zhytomyr, Ukraine

Abstract. In the publication, the concentration of chemical elements in samples of dirty and purified water from settling tanks of a stone processing enterprise was measured using the MP-AES device. The data obtained were compared with the standards of Ukraine and the EU. 180 samples of dirty water of 60 ml each were also taken to add flocculants. The sedimentation rate of stone sludge was measured as a function of the flocculant concentration in the solution with water and stone sludge. At each stage of the study, at flocculant concentrations of 0.25 g/l, 0.5 g/L, 1 g/L, 2 g/L, 3 g/l, we measured the time intervals during which the cut point between two phases passes the mark of 70 ml, 60 ml and 50 ml. The time for the cut point between the two phases to pass the marks on the flask was also recorded.

Keywords: MPEAS analysis · Water samples · Flocculants · Granulometric composition of sludge particles · Rock sludge settling rate · Dewatering equipment

1 Introduction

Natural facing stone is a reliable and durable building material. The decorative stone mining and processing industry is growing all over the world, and gross production of about one hundred million tons is expected to be achieved by 2020, of which approximately 50% is waste in various aggregate forms [1]. There are more than 1,600 stone processing enterprises in Ukraine. The main raw materials are strong igneous rocks, such as granites, gabbro, labradorites. During the processing of natural stone, natural stone waste is generated, which has various geometric sizes [2]. Large pieces of natural stone are processed into paving stones and crushed stone. Fine waste less than 1 mm in size is stored in open sludge accumulators. A huge amount of fine waste has accumulated and the issue of their disposal has turned into serious economic and environmental problems. In the course of processing natural stone, water is used to cool the tool and flush the sludge into the settling tank. Water circulates in a closed cycle: dirty water after cooling the tool settles in the sludge tank, where the solid phase settles, clarified water is re-fed to the processing machines.

R. Szewczyk et al. (Eds.): AUTOMATION 2021, AISC 1390, pp. 387–394, 2021.
https://doi.org/10.1007/978-3-030-74893-7_34

Usually, sludge tanks are cleaned several times a month; in this case, the sludge is removed from the sump in a wet state, stored on the sites where it settles. In this case, contaminated water enters the soil, which can lead to environmental pollution.

As practice shows, when cleaning sludge tanks from sludge, excess water is discharged by enterprises into rivers and lakes, which negatively affects the environment. Improper management of stone mining industry waste is the main reason for the increase in the total content of suspended solids in groundwater [3, 4]. Therefore, the analysis of water quality requirements in Ukraine and the EU is the first step towards improving water treatment technologies in Ukraine.

Modern directions of intensification of wastewater treatment processes are the improvement of existing technologies and the development of new effective treatment methods, the introduction of resource-saving low-waste technologies, modernization of existing methods and designs of water treatment devices, allowing to save material and natural resources, as well as dispose of valuable components of sludge waste.

Many countries, including Ukraine, have already developed, and in some cases implemented, local systems for treating chemically contaminated wastewater in order to create a closed system of water and sludge management [5]. However, there are still a number of unresolved issues related to the separation of polydisperse suspensions and further reduction of sludge moisture.

Environmental indicators of granite production in the European Union are analyzed using indicators that are given in the European Recommendation [6] on the life cycle of natural materials. Unfortunately, there are no such requirements in Ukraine.

Italian studies [7] distinguish two different categories of sludge: from carbonate rock and hard rock. Both categories are equally dispersed. Carbonate sludge does not contain impurities and consists of chemical elements and minerals that are contained in the stone (marble, limestone and travertine). Conversely, hard rock sludge can be characterized by a high content of heavy metals, which are components of tools used in the processing of natural stone. Since Ukrainian enterprises process mainly hard rocks, there is a need to study the metal content in the sludge at each enterprise, because there is a difference in the tool that is used for processing natural stone.

An important role is assigned both to the creation of a new type of dewatering equipment, which significantly exceeds the existing one in terms of technical and economic indicators [8], and to the scientifically based choice of methods for intensifying the processes of separation of polydisperse suspensions due to particle aggregation and the selection of the best parameters for the implementation of the flocculation process with minimal consumption of expensive chemical reagents. Solving these scientific problems will increase the efficiency of sludge dewatering and at the same time improve the technical and economic indicators of the process of separating moisture from sludge and disposing the solid phase waste. Thus, research, aimed at increasing the degree of separation of polydisperse suspensions of sludge waste by intensifying the process of solid particle aggregation, is an urgent scientific and practical task.

2 Research Materials and Methods

Water samples were taken at the inlet and outlet of the sludge tank at Granit LLC (Dovzhyk village, Zhytomyr region, Ukraine (Fig. 1)). The analyzed water samples were acidified (150 μL of HNO_3 per 10 ml of sample) to suppress the adsorption of metal ions on the vessel surface. 180 samples of dirty water of 60 ml each were also taken to add flocculants.

Fig. 1. Plan of the company from which the samples were taken

At the first stage, the chemical composition of water samples at the inlet and outlet of the sludge tank, as well as water samples that were purified using fluculants, was established. For this purpose, the MP-AES device was used (model: 4200, Agilent Inc., Santa Clara, California, USA) with an axial plasma view (Fig. 2) to determine the chemical composition of water. The principle of operation of the device is described in publications [9, 10].

At the second stage, the rate of sludge sedimentation depending on the concentration of flocculant in a solution with water and stone sludge was studied.

3 Results and Discussion

It is generally accepted when determining the state of the aquatic environment to conduct physical and chemical studies, as well as compare the results obtained with the sanitary standards of maximum permissible concentrations with the standards of Ukraine [11] and the EU [12].

Using MPAES analysis, the content of chemical elements in the water was determined before purification, after purification with the addition of flocculants and without addition. The results of the study are shown in Table 1.

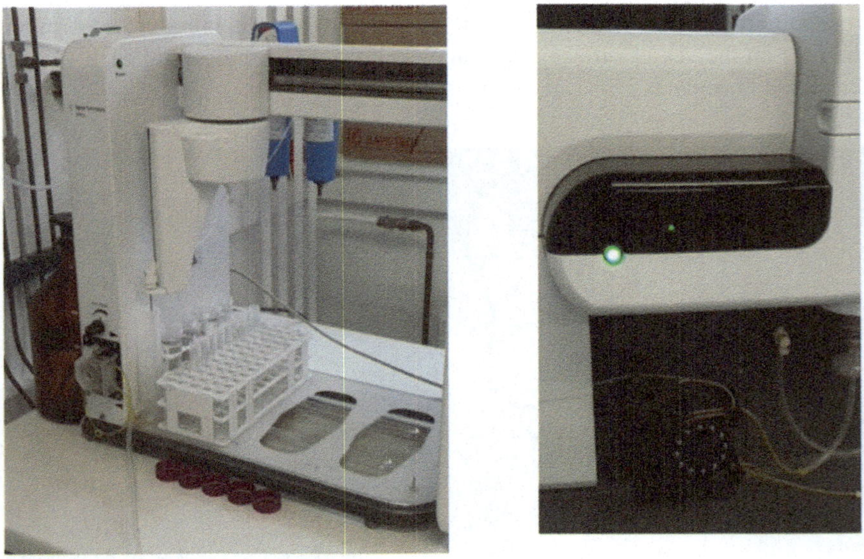

Fig. 2. Agilent 4200 MP-AES spectrometer with axial plasma view

Table 1. Concentrations of chemical elements present in water

Sample type	Al	Ca	Cd	Co	Cr	Cu	Fe	K	Mg	Mn	Na	Ni	Pb	Zn
	mg/l	mg/l	mg/l	mg/l	mg/l	mg/l	mg/l	mg/l	mg/l	mg/l	mg/l	mg/l	mg/l	mg/l
Water sample before flocculation	0.10	40.44	<0.013	<0.03	<0.001	0.03	<0.039	35.31	11.39	0.03	44.07	<0.035	<0.05	<0.018
Water sample after flocculation	0.11	36.89	<0.013	<0.03	<0.001	0.04	<0.039	33.12	10.25	0.02	43.01	<0.035	<0.05	<0.018
Water sample after settling	0.07	40.09	<0.013	<0.03	<0.001	0.03	<0.039	33.48	10.34	0.05	43.94	< 0.035	<0.05	<0.018

As can be seen from the Table 1, the content of elements such as Cd, Co, Cr, Cu, Fe, Mg, Mn, Ni, Pb, Zn is small, since these particles must be sedimented. Also, when the flocculant interacts with calcium oxide in water, we get the following precipitate (aluminum hydroxide and calcium sulfate), which explains the decrease in calcium content in water:

$$Al_2(SO_4)_3 + 3CaO + 3H_2O = 2Al(OH)_3\downarrow + 3CaSO_4\downarrow$$

The use of flocculants increases the efficiency of water purification compared to natural lighting of water by settling.

PH, electrical conductivity, and redox potential were also measured in water, which are shown in Table 2.

The concentration of chemical elements present in the water before and after purification meets the standards of Ukraine [11] and EU [12]. In general, the change in the above-mentioned water quality indicators after flocculation and settling is insignificant. Therefore, the process of water purification should be considered from the point of view of the sedimentation rate of fine particles.

For sludge collection, flocculants are most often used in such cyclone sludge tanks.

Table 2. Qualitative characteristics of water samples before and after its purification processes

Sample type	pH	Electrical conductivity, μS/cm	Redox potential U, mV	Temperature t, °C
Water sample before flocculation (cloudy water with stone sludge)	9.27	503	207	20
Water sample after flocculation	8.42	623	320	
Water sample after settling	8.38	621	307	

Flocculants are substances that can cause flocculation (the formation of fragile loose aggregates of indeterminate shape from small particles of the dispersed phase, which can be found in a suspended state in a liquid or gas medium).

The kinetics of floccule settling in the free (unlimited) precipitation mode was measured in a laboratory measuring cylinder with a capacity of 100 ml. Before the direct experiment, the type of flocculant and its concentration were selected. For this sludge, a flocculant based on aluminum sulfate ($Al_2(SO_4)_3$) was used, which is a white powder with grains of 1–2 mm in size (Fig. 3).

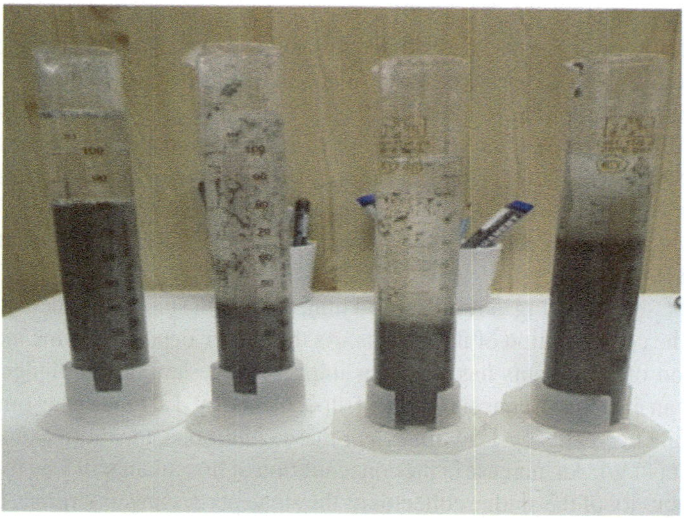

Fig. 3. Measuring cylinders for measuring kinetics of floccule settling

According to the technical characteristics of the treatment system, the flocculant must be added to contaminated water in the form of a liquid. This solution is added to the pipeline through which dirty water is pumped into the cyclone. Its amount can be adjusted by the metering pump.

A granulometric composition of sludge particles was performed, since this affects the rate of sludge settling, the results are shown in Fig. 4.

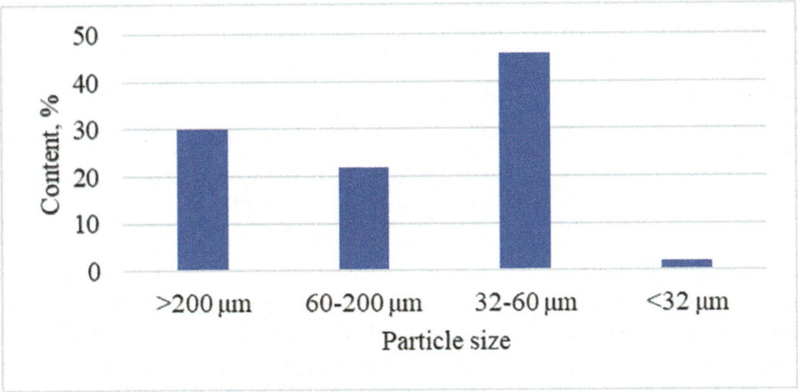

Fig. 4. Granulometric composition of sludge particles

Most of all, the sludge contains particles of 32–60 microns in size, which settle very slowly. Particles with a size greater than 0.2 mm, the concentration in the sludge of which is 30%, have a high settling rate and are the core for sticking smaller particles when using flocculants.

The sedimentation rate of stone sludge was measured as a function of the flocculant concentration in the solution with water and stone sludge. At each stage of the study, at flocculant concentrations of 0.25 g/l, 0.5 g/L, 1 g/L, 2 g/L, 3 g/l, we measured the time intervals during which the cut point between two phases passes the mark of 70 ml, 60 ml and 50 ml. The time for the cut point between the two phases to pass the marks on the flask was also recorded. In accordance with the results of the study, a graph of the dependence of the sludge sedimentation rate on the flocculant concentration was constructed (Fig. 5).

As can be seen from Fig. 5 for the first 4 min, the rate of sludge sedimentation doesn't depend on the concentration of flocculants. As the sludge deposition time increases, the concentration of flocculants has a greater impact on the amount of sludge deposited. Thus, with an increased concentration of flocculants by 12 times, the time spent on sludge settling decreases by 1.57 times. For these conditions, the rational concentration of flocculant is 1 g/l. An increase in the concentration of flocculant will lead to an increase in the productivity of the sludge collector of the stone processing enterprise, but this will require additional financial costs.

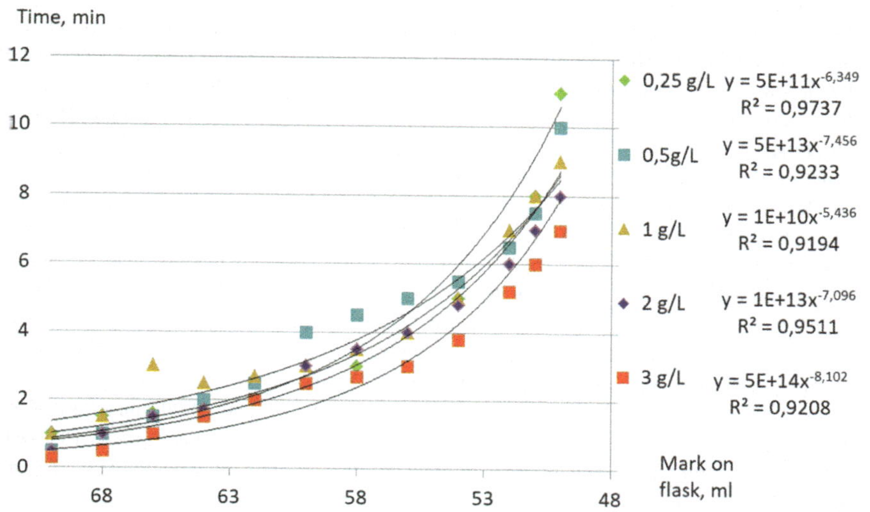

Fig. 5. The diagram of dependance of the rate of sludge sedimentation on the concentration of flocculant in a solution with water and stone sludge

4 Conclusions

The concentration of chemical elements present in the water before and after purification meets the standards of Ukraine and EU. In general, the change in the above-mentioned water quality indicators after flocculation and settling is insignificant.

Most of all, the sludge contains particles of 32–60 microns in size, which settle very slowly.

For the first 4 min, the rate of sludge sedimentation does not depend on the concentration of flocculants. As the sludge deposition time increases, the concentration of flocculants has a greater impact on the amount of sludge deposited. Thus, with an increased concentration of flocculants by 12 times, the time spent on sludge settling decreases by 1.57 times.

References

1. Careddu, N.: Dimension stones in the circular economy world. Resour. Policy **60**, 243–245 (2019). https://doi.org/10.1016/j.resourpol.2019.01.012
2. Korobiichuk, V.V., Sidorov, O.M., Sobolevskyi, R.V., Shlapak, V.O., Kryvorushko, A.O.: European integration: treatment of stone processing enterprises waste in Ukraine. Visnik Zhitomirskogo derzhavnogo tehnologichnogo universitetu. Zhytomyr: ZSTU, No. 1(79), 182–190 (2017)
3. l-Jabari, M., Sawalha, H.: Treating stone cutting waste by flocculation–sedimentation. In: Proceedings of the Sustainable Environmental Sanitation and Water Services Conference, 28th WEDC Conference, Calcutta, India (2002)
4. Korobiichuk, I., Davydova, I., Korobiichuk, V., Shlapak, V., Herasymchuk, O.: The influence of geological and anthropogenic factors on the change of the water quality parameters in the Kamyanka river within the City of Zhytomyr. In: Advances in Intelligent Systems and Computing, pp. 476–486 (2020)

5. Environmental Management—life Cycle Assessment—requirements and Guidelines. International Standard 14040:2006 International Organisation for Standardisation, Geneva (2006)
6. European Commission Building the Single Market for Green Products – Facilitating Better Information on the Environmental Performance of Products and Organisations COM (2013) 196 Final
7. Careddu, N., Dino, G.A.: Reuse of residual sludge from stone processing: differences and similarities between sludge coming from carbonate and silicate stones – Italian experiences. Environ. Earth Sci. **75**(14), 1–9 (2016)
8. Nasserdine, K., Mimi, Z., Bevan, B., Elian, B.: Environmental management of the stone cutting industry. J. Environ. Manag. **90**, 466–470 (2009)
9. Balaram, V., Dharmendra, V., Roy, P., Taylor, C., Kamala, C.T., Satyanarayanan, M., Kar, P., Subramanyam, K.S.V., Kumar Raju, A., Krishnaiah, A.: Analysis of geochemical samples by microwave plasma-AES. At. Spectrosc. **35**, 65–78 (2014)
10. Kamala, C.T., Balaram, V., Dharmendra, V., Satyanarayanan, M., Subramanyam, K.S.V., Krishnaiah, A.: Application of microwave plasma atomic emission spectrometry (MP-AES) for environmental monitoring of industrially contaminated sites in Hyderabad city. Environ. Monit. Assess. **186**, 7097–7113 (2014)
11. State sanitary norms and rules "Hygienic requirements for drinking water intended for human consumption" DSanPiN 2.2.4-171-10, Kyiv (2010)
12. Directive 2008/105/EC of the European Parliament and of the Council of 16 December 2008 on environmental quality standards in the field of water policy, amending Directive 2000/60/EC of the European Parliament and of the Council

Resonance Situation in the Gyro Suspension When Changing the Acoustic Beam Angle

Igor Korobiichuk[1](✉), Viktorij Mel'nick[2], and Serhii Fesenko[2]

[1] Institute of Automatic Control and Robotics, Warsaw University of Technology, Boboli 8, 02-525 Warsaw, Poland
igor.korobiichuk@pw.edu.pl
[2] National Technical University of Ukraine "Igor Sikorsky Kyiv Polytechnic Institute", Kyiv, Ukraine
vmm71@i.ua

Abstract. An analysis of the features of the resonant type for both the computational models – the approximate model (which has the appearance of an elementary plate) and the refined model (which has the appearance of a shell of considerable length), revealed some discrepancies in the results. The results obtained have been confirmed in experimental modeling of shell behavior. The reasons for the discrepancy of the results are thoroughly considered, and the values of the most dangerous angles of incidence of sound rays in relation to the surface of the shell are calculated.

Keywords: Floating gyroscope · Caustic zone · Angle of coincidence · Resonance · Buffer zone

1 Introduction

Along with the unconditional benefits of hypersonic movement, the rapid increase in aircraft speed to 20 Machs leads to the development and spread of physical phenomena – such as the powerful vibration of the aircraft body, problems in motion control, high temperatures (2000 °C) and other less significant factors [1]. In addition to these factors, the penetrating acoustic radiation of high intensity and wide frequency spectrum becomes crucial [2].

The study of the nature of the occurrence of the caustic zone in the polyaggregate suspension of the float gyroscope, the reasons for the formation of such zones in the suspension, and their development in time and space and the effect on the occurrence of additional errors floating gyroscope on the resonance wave coincidence level [3, 4].

Let us briefly describe the cause of the appearance of pressure pulsations on the surface of the flight product from the standpoint of the general theory of sound field generation by an unstable aerodynamic flow. The theory of pressure ripples originates from two classic works by M.J. Lighthill's is about the sound generated in an aerodynamic flow. They state that under field conditions the sound is acoustically equivalent to the sound generated in an ideal environment by a system of distributed sources, quadrupoles.

R. Szewczyk et al. (Eds.): AUTOMATION 2021, AISC 1390, pp. 395–401, 2021.
https://doi.org/10.1007/978-3-030-74893-7_35

Under the influence of acoustic pressures, the waves move at a local speed of sound relative to the environment. However, as a result of the movement of the flight products, the velocity of the surface pressure waves will already be determined by the sum of the local sound speed and the flight speed of the aircraft. And since the latter changes greatly, and the sound waves approach the flight product at different angles, it becomes apparent that the velocity of the pressure waves moving relative to the hull of the flight product can take many values.

The nature and degree of acoustic impact, as well as the choice of mechanical calculation model of the perturbed motion, are directly dependent on the ratio of product dimensions and half the sound wavelength. If this value is within one, then the acoustic load is equal to the evenly distributed, and the design model – to a rigid body on elastic bonds. If the size is several times more than half the length, then the external influence is considered as a wave, and the whole structure – as a system with distributed parameters.

In the domestic and foreign literature quite fully reflected the achievements of science in the fight against low and medium noise (not higher than 130 dB). As for the study of the properties of structures when exposed to high-intensity acoustic fields (160–180 dB), they have found a thorough reflection in only a few scientific works, mainly in relation to the elements of aircraft structures. Questions of acoustic stability of some types of on-board equipment and their components remain open.

Damage to structures from sound pressure has the appearance of irreversible deformation of surfaces. The peculiarity of acoustic loads is high frequency, so the process of accumulating fatigue damage is very fast. There are data indicating that the service life of a conventional structure at a pressure level of 160 dB is calculated in hours, and at a pressure of 170 dB – only seconds.

The most unpleasant factor in this issue is the possibility of manifesting local and global features of the operation of the equipment and the occurrence of additional errors, in particular at the resonant level. The dissertation focuses on the action of one of the specific influential external negative factors, namely, penetrating acoustic radiation on a floating gyroscope, the principal feature of which is the presence of a polyaggregate suspension. Under certain conditions penetrating acoustic radiation, as it turned out, becomes of paramount importance in terms of the negative impact on the onboard equipment. The preliminary quantitative and qualitative characteristics of this influence can be realized by means of computational models of elastic interaction of penetrating acoustic radiation with the elemental base of the aircraft, which should adequately reflect the conditions of operation of flight products, to this end, should outline the principles of design of the calculating model of the middle of the calculation the middle through the supports of the device, but also in view of the peculiarities of the spatial propagation of acoustic radiation.

2 Materials and Methods of Research

Let us dwell in more detail on the causes of the discrepancy between the results obtained from both calculation models.

We choose the refined calculation model as a circular cylindrical shell of unlimited length (Fig. 1). The study of short shells was carried out, for example, in [5].

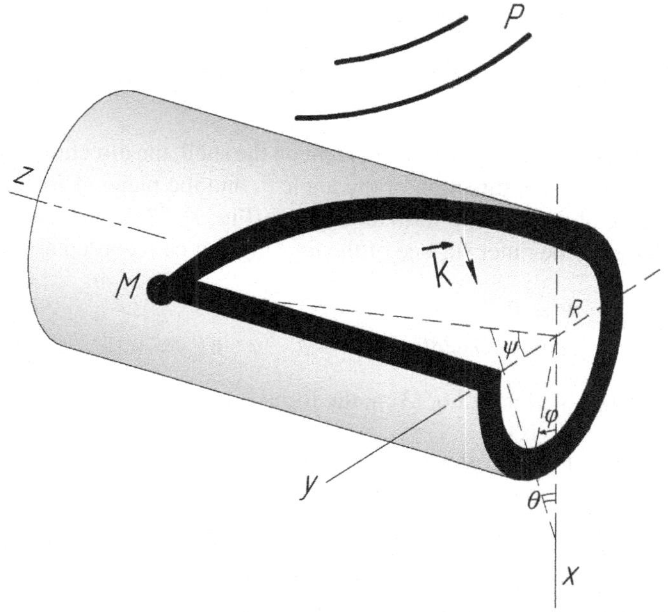

Fig. 1. The mechanism of action of an acoustic wave on the body of a float gyro. Refined calculation model

We build further considerations based on the assumption that the thickness of the shell 2δ is rather small compared to its radius R, that is, we will assume that $2\delta < \frac{R}{20}$ and less than one sixth of the bending wavelength, that is $2\delta < \frac{1}{6}\lambda_u$.

If only perturbations normally applied to the shell act on the shell, then the oscillatory motion of its surface will be reflected by the system of equations [6]:

$$\begin{aligned} &(2\delta E)^{-1}\nabla^4\gamma - R\frac{\partial^2 W}{\partial z^2} = 0; \\ &R\frac{\partial^2\gamma}{\partial z^2} + D\nabla^4 W + R^4 m_p \frac{\partial^2 W}{\partial t^2} = R^4 q_p, \end{aligned} \tag{1}$$

where W – radial elastic displacement of the shell surface; γ – function of stresses; $m_p = \rho\ 2\delta$ – specific surface mass; $q_p = P_1^{/} - P_3$ – sound pressure difference on both sides of the shell; ∇^4 – Laplace operator.

$$\nabla^4 = \frac{\partial^4}{\partial z^4} + \frac{\partial^4}{\partial \gamma^4} + 2\frac{\partial^4}{\partial z^2 \partial \gamma^2},$$

in dimensionless coordinates z and φ that determine the distance (in parts of radius R) in Fig. 1.

Using the auxiliary function F [7, 8]:

$$W = \nabla^4 F; \quad \gamma = RE2\delta\frac{\partial^2 F}{\partial z^2}, \tag{2}$$

we will bring system (1) to one equation –

$$D\nabla^8 F + R^2 2\delta E \frac{\partial^4 F}{\partial z^4} + R^4 m_p \nabla^4 \frac{\partial^2 F}{\partial t^2} = R^4 q_p. \tag{3}$$

Suppose that a plane sound wave is incident on the shell, the direction of propagation of which is with the radial direction of the angle θ, and the plane of incidence and the plane of the frame of the housing form an angle ψ (Fig. 1). Then, the sound pressure in the incident wave on the outer surface of the housing can be recorded as follows

$$\begin{aligned} P &= P_0 \exp i \\ &\times [\omega t - k_0 2\delta(z \sin\theta \sin\psi - \varphi \sin\theta \cos\psi)]. \end{aligned} \tag{4}$$

We look for the solution of Eq. (3) in the form

$$\begin{aligned} F &= F_0 \exp i \\ &\times [\omega t - k_0 2\delta(z \sin\theta \sin\psi - \varphi \sin\theta \cos\psi)]. \end{aligned}$$

If you do not take into account the scattering of sound on the surface shell part of the housing, then, taking into account the notation (2), the elastic displacements W of the surface in the plane of the frame can be stated as follows

$$\begin{aligned} W &= \left(P_1^/ - P_2\right) m_p^{-1} \omega^{-2} \\ &\times \left(\frac{D\omega^2}{m_p c_0^4} \sin^4\theta + \frac{E 2\delta}{m_p \omega^2 (2\delta)^2} \sin^4\psi - 1\right)^{-1}. \end{aligned} \tag{5}$$

Suppose that the body resistance of symmetric oscillations is much greater than the resistance to antisymmetric oscillations, i.e.:

$$|Z_a| \ll |Z_c|,$$

and the expression for sound permeability is converted to the form –

$$\tau_0 = \frac{1}{\left|1 + Z\frac{\cos\theta}{2\rho_0 c_0}\right|^2}, \tag{6}$$

where is the impedance $Z = \frac{P_1^/ - P_2}{i\omega W}$.

The ratio (6) for the plate can be practically applied to the shell [9, 10]. How It Comes from Expression (5)

$$\begin{aligned} Z &= i m_p \omega \left(1 - \frac{D\omega^2}{m_p c_0^4} \sin^4\theta - \frac{E \cdot 2\delta}{m_p \omega^2 R^2} \sin^4\psi\right) \\ &= i m_p \omega \left(1 - \left(\frac{\omega}{\omega_{gr}}\right)^2 \sin^4\theta - \left(\frac{\omega_{pp}}{\omega}\right)^2 \sin^4\psi\right), \end{aligned} \tag{7}$$

where $\omega_{gr} = 2\pi f_{gr}$; $\omega_{pp} = \frac{c_1}{R}$ – circular frequency of natural, purely radial, oscillations of the ring, arising from compression-stretching, with the wavelength λ_p becomes equal to the length of the circle of the ring ($c_1 = \sqrt{\frac{E}{\rho}}$ – speed of a circular wave) if the angle $\psi = \frac{\pi}{2}$ rad.

The magnitude of the permeability of the cylindrical shell will be:

$$\tau = \left[\left|1 + \frac{i}{\Delta_p}\left[1 - \left(\frac{\omega}{\omega_{gr}}\right)^2 \sin^4\theta - \left(\frac{\omega_{pp}}{\omega}\right)^2 \sin^4\psi\right]\right|^2\right]^{-1}, \tag{8}$$

where $\Delta_p = \frac{2\rho_0 c_0}{m_p \omega \cos\theta}$.

3 Results and Discussion

With an infinitely large radius of curvature of the shell, expression (8) goes into formula (6).

It also follows from expression (8) that the sound permeability of the housing can sharply decrease also at low frequencies, i.e. lower than, in contrast to the approximate plate – like design model.

When $\omega < \omega_{gr}$, then the first addition in the formula (8) becomes infinitesimal compared to one. Since then, subject to condition:

$$\omega = \omega_{pp} \sin^2\psi, \tag{9}$$

The shell part of the housing will become "acoustically transparent" (of course, in the absence of losses). This condition can also be considered as the equality of the trace of the length of the frame circle and the length of the circular wave on a plane parallel to the front of the incident plane wave (Fig. 2)

$$\frac{2\pi R}{\sin\psi} = \lambda_p \sin\psi. \tag{10}$$

Thus, in the cylindrical body, at low frequencies another wave coincidence, compared to the plate, occurs for circular (longitudinal) waves.

Interestingly, wave coincidence for bending waves can occur only at frequencies above f_{gr}, and for circular (longitudinal) waves, on the contrary, can occur only at frequencies not higher than ω_{pp}.

In more detail, there may be another match in the shell – a combination. This comes from formula (8)

$$\left(\frac{\omega}{\omega_{gr}}\right)^2 \sin^4\theta + \left(\frac{\omega_{pp}}{\omega}\right)^2 \sin^4\psi = 1. \tag{11}$$

Due to the diffuse sound field, the permeability coefficient of the instrument body is determined by averaging over Paris:

$$\tau_{\theta\psi} = \int_0^{\frac{\pi}{2}} \int_0^{\frac{\pi}{2}} \tau \sin 2\theta \; \partial\theta \; \partial\psi, \tag{12}$$

where τ the value is calculated by the formula (8). By integrating this expression, it is advisable to study the frequency ranges separately. For example:

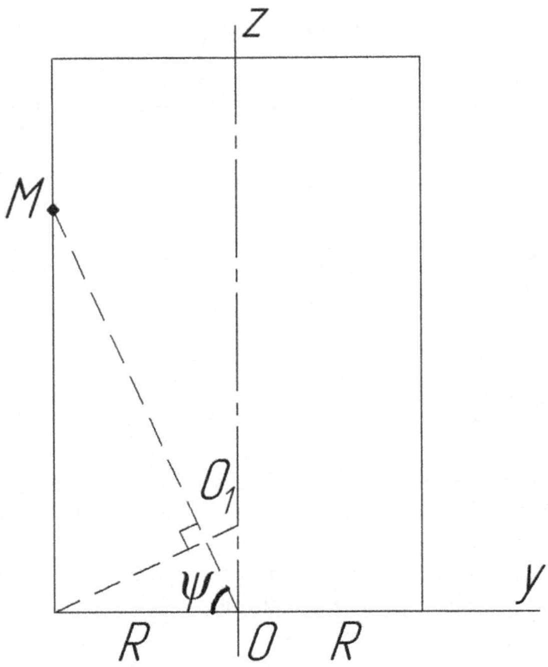

Fig. 2. Wave coincidence in the DUSU body for circular waves: $\omega < \omega_{gr}$

1) $\omega < \omega_{pp}$; $\omega < \omega_{gr}$;
2) $\omega_{gr} < \omega < \omega_{pp}$; $\omega_{pp} < \omega$; $\omega_{gr} < \omega$.

Through the use of an approximate (flat) calculation model, it can be found that the bending vibrations of the housing ($Z_a \ll Z_c$) are capable of generating wave coincidence only at frequencies ω, exceeding the cutoff frequency ω_{gr}. Moreover, each frequency corresponds to its angle of coincidence θ_c.

The refined computational model (in the form of a shell) revealed an additional resonance for the circular (longitudinal) wave ($Z_c \ll Z_a$) in the form of a coincidence of the trace of the length of the circle of the frame and the trace of the length of the longitudinal wave. Moreover, both of these resonances can occur at frequencies below the limit:

$$\begin{aligned} \omega &\ll \omega_{gr} = 33\,881\ \ rad\ s^{-1}\left(f_{gr} = 5395\,Hz\right); \\ \omega &\ll \omega_{pp} = 251\,998\ \ rad\ s^{-1}\left(f_{pp} = 40127\,Hz\right). \end{aligned} \tag{13}$$

where $\omega_{gr} = 2\pi f_{gr}$; $\omega_{pp} = \frac{c_1}{R} = \frac{1}{R}\sqrt{\frac{E}{\rho}}$; c_1 – velocity of circular (longitudinal) waves; ω_{pp} – radial self-oscillations of the ring of the frame ($Z_c \ll Z_a$).

For the conditions of the previous example, that is, taking the radius of the housing R equal 2,5 cm, and the velocity of the circular (longitudinal) wave in duralumin is equal $c_1 = 6300 - 5230$ ms^{-1}, the longitudinal wave resonance conditions in the housing can be set.

First of all, find out the cutoff frequency:

$$f_{gr} = \frac{c_1}{2\pi R} = \frac{6300}{2{\cdot}3{,}14{\cdot}0{,}025}$$
$$= 40127, 39(Hz) = 40, 127\,kHz$$

The value of the angle at which the phenomenon of wave coincidence occurs at low frequencies is calculated by the formula (10).

Thus, as the angle of incidence of wave ψ decreases, the frequency at which the coincidence effect is reduced also decreases:

$$2\pi R\frac{f}{c_1} = \sin^2\psi.$$

4 Conclusions

In order to find out the peculiarities of the development in the time of the resonance phenomenon, a program for calculating the resonance angle between longitudinal axis of the housing and the wave vector was created. The results of the change resonance values of the angle ψ, depending on the frequency of the circular wave f are given in analog form; the data indicate that the growth of the frequency of penetrating radiation produces an increase in the values of the resonant angle ψ.

References

1. Wang, L., Tian, F.B.: Recent progress of immersed boundary method and its applications in compressible fluid flow. Scientia Sinica: Physica, Mechanica et Astronomica **48**(9), 9 (2018). https://doi.org/10.1360/SSPMA2018-00191. Article no. 094703
2. Korobiichuk, I., Kachniarz, M., Karachun, V., Mel'nick, V., Fesenko, S.: Maintenance of masking and limited performance of fire-fighting equipment from means of aircraft location. Measurement: J. Int. Measurement Confederation **136**, 558–563 (2019). https://doi.org/10.1016/j.measurement.2018.12.102
3. Coulouvrat, F., Luquet, D., Marchiano, R.: Numerical model of sonic boom in 3D kinematic turbulence. In: AIP Conference Proceedings, vol. 1685 (2015). https://doi.org/10.1063/1.4934474. Article no. 090008
4. Luquet, D., Marchiano, R., Coulouvrat, F., El Din, I.S., Loseille, A.: Sonic boom assessment of a hypersonic transport vehicle with advanced numerical methods. In: 21st AIAA/CEAS Aeroacoustics Conference (2016). 13 p.
5. Karachun, V., Mel'nick, V., Fesenko, S.: Reduction of technological risks of flight operation by artificial formation of the buffer zone to penetrating acoustic radiation. Technol. Audit Prod. Reserves **4**(1(36)), 19–24 (2017)
6. Makarov, V.I., Fadeev, N.A.: On wave radiation by shells in the sound field. Acoust. Mag. **IV**(2), 261–263 (1960)
7. Shenderov, E.L.: Hydroacoustic wave problems. Sudostroyeniye, Leningrad (1972). 352 p.
8. Zaborov, V.I.: Walling sound insulation theory. Stroyizdat, Moskva (1962). 116 p.
9. Karachun, V.V., Mel'nick, V.M.: Acoustic radiation energy focus in a shell with liquid. In: Recent Advances in Systems, Control and Information Technology: Proceedings of the International Conference SCIT 2016, pp. 459–4632016https://doi.org/10.1007/978-3-319-48923-0_48
10. Kostyk, S., Shybetskyy, V., Povodzinsy, V., Fesenko, S.: Revealing special features of hydrodynamics in a rotor-disk film vaporizing plant. Eastern-European J. Enterp. Technol. **1**(6(97)), 28–33 (2019). https://doi.org/10.15587/1729-4061.2019.156649

Application of the Polynomial Maximization Method for Estimation Parameters in the Polynomial Regression with Non-Gaussian Residuals

Serhii Zabolotnii[1,2](✉), Oleksandr Tkachenko[1], and Zygmunt Lech Warsza[3]

[1] Cherkasy State Business College, Cherkasy, Ukraine
zabolotniua@gmail.com
[2] Kyiv National University of Technologies and Design, Kiev, Ukraine
[3] Łukasiewicz Research Network – Industrial Research Institute of Automation and Measurements PIAP, Al. Jerozolimskie 202, 02-486 Warszawa, Poland

Abstract. This paper considers the application of the polynomial maximization method to find estimates of the parameters of polynomial regression. It is shown that this method can be effective for the case when the distribution of the random component of the regression models differs significantly from the Gaussian distribution. This approach is adaptive and is based on the analysis of higher-order statistics of regression residuals. Analytical expressions that allow finding estimates and analyzing their uncertainty are obtained. Cases of asymmetry and symmetry of the distribution of regression errors are considered. It is shown that the variance of estimates of the polynomial maximization method can be significantly less than the variance of the estimates of the least squares method, which is a special case. The increase in accuracy depends on the values of the cumulant coefficients of higher orders of random errors of the regression model. The results of statistical modeling by the Monte Carlo method confirm the effectiveness of the proposed approach.

Keywords: Regression · Polynomial · High-order statistics · Non-Gaussian residuals

1 Introduction

Estimation of parameters of the polynomial regression is one of the earliest problems in statistical analysis. For more than two centuries, the polynomial regression model has become widespread in solving various applied problems of technical, geophysical, biomedical, economic and other aspects of human activity. In most cases, variations of the least square (LS) method are used to find the estimates of parameters.

The widespread adoption of the least square's method is due to two main factors. Firstly, the algorithmic simplicity of its implementation. Secondly, by the well-known fact that under a number of conditions, the estimates of the Ordinary Least Square (OLS)

R. Szewczyk et al. (Eds.): AUTOMATION 2021, AISC 1390, pp. 402–415, 2021.
https://doi.org/10.1007/978-3-030-74893-7_36

method are optimal in the sense of ensuring a minimum of their variance. One of these conditions is the normalization of the distribution law of the random component of the regression model. It is because the linear OLS estimator is equivalent to the Maximum Likelihood Estimator (MLE). In addition, the Gaussian probabilistic models a convenient idealization, as its application greatly simplifies the synthesis and analysis of statistical methods. However, such a model is not always adequate to real statistical data. In case of mismatch may be employed various approaches, in particular, the normalizing transform [1, 2] or modifiers robust least squares or maximum likelihood estimators [3].

When fulfilling the assumption that model errors are independent of predictors, an approach that is sometimes called adaptive can be applied [4]. Adaptability is understood in the sense that a series of successive refinement steps are applied. At the first stage, the parameters of the deterministic component of the model are estimated by a simple method that does not consider the specifics of the probabilistic distribution of statistical data. After removing the deterministic component, the type of the random component of the model is identified and estimates of its parameters are found. At the third stage, refined (adaptive) estimates of the parameters of the deterministic component of the model are found and are considered the probabilistic properties of the residual variable (errors). Various authors have shown that the application of this approach can significantly reduce the variance of parameter estimates [5–8].

The adaptive approach can be based not only on likelihood maximization, which requires determining the probability distribution law and finding estimates of its parameters. This is a separate and often computationally complex task. An alternative can be based on the use of a probabilistic description of regression models in the form of higher-order statistics, for example, moments or cumulants [9–13]. The use of this description in combination with the adaptive approach used in this work.

2 Purpose of the Study

Polynomial Maximization Method (PMM) is a relatively new method of statistical estimation, which is an integral part of the stochastic polynomial apparatus developed by Kunchenko [14]. The use of this mathematical apparatus provides new opportunities for constructing semi-parametric methods for processing non-Gaussian data. In addition to estimating the parameters, these can be tasks of testing statistical hypotheses [15], pattern recognition ("template matching") [16], posteriori [17] and on-line [18] disorders detections (change point problems) etc.

In a sense, PMM is ideologically similar to MLE. It also uses the principle of maximizing statistics from sample data in the neighborhood of the true value of the parameter being estimated. However, to form such a functional, the probability density distribution is not used, but a partial description based on higher-order statistics.

This study is a continuation of works [19, 20], which considered the use of PMM for solving the problem of adaptive estimation of parameters of linear regression with asymmetric and symmetric non-Gaussian residuals. Its main purpose is to generalize the results of these works for a more complex polynomial regression model, as well as to carry out a comparative analysis of the estimation efficiency by means of the statistical modeling by the Monte Carlo method.

It should also be noted that the notion of "stochastic polynomials" and the presence of the term "maximization of polynomials" in the name of the estimation method used in this work are not related to the polynomial type of the deterministic component of the regression model. These are completely different types of polynomials in terms of their functionality and practical application.

3 Mathematical Formulation of the Problem

Let there be a regression model of observations that describes the dependence of values of the target variable $\boldsymbol{Y} = \{y_1, y_2, ..., y_N\}$ from the predictor $\boldsymbol{X}\{x_1, x_2, ..., x_N\}$:

$$y_v = \sum\nolimits_{p=0}^{Q-1} a_p x_v^p + \xi_v, \; v = \overrightarrow{1, N} \tag{1}$$

where $\boldsymbol{\theta} = \{a_0, a_1, ...a_{Q-1}\}$ – a vector containing the parameters (coefficients of the power polynomial) of the deterministic component of the model; ξ_v – readings of the random component (error) of the model ($E\{\xi\} = 0$), which is a sequence of independent and equally distributed random variables.

The probabilistic properties of the error of the regression model differ significantly from the Gaussian law and can be described using a sequence of central moments μ_r or cumulants κ_r (cumulative coefficients γ_r). With this description, the second-order cumulant κ_2 coincides with variance μ_2 random component, and the cumulant coefficients of higher orders $\gamma_r = \kappa_r \big/ \kappa_2^{r/2}$ numerically describe the degree of difference from a Gaussian distribution.

The general task is to find estimates of the components of the vector parameter $\boldsymbol{\theta} = \{a_0, a_1, ...a_{Q-1}\}$ based on statistical analysis of multiple points (x_v, y_v), $v = \overrightarrow{1, N}$. In this case, the values of the parameters of the probabilistic component of the model are a priori unknown.

4 Results

To solve the posed problem using the we use the power transformations of sample values as basic functions, i.e. $\phi_i(y_v) = y_v^i$. In this case, the sequence of mathematical expectations is a set of initial moments of the corresponding order

$$\Psi_{iv} = E\left\{y_v^i\right\} = \alpha_{iv}, \; i = \overrightarrow{1, S}, \; v = \overrightarrow{1, N},$$

and functions $F_{(i,j)v} = \alpha_{(i+j)v} - \alpha_{iv}\alpha_{jv}$ called the centered correlants [14].

4.1 PPM-Estimates of Regression Parameters with Polynomial Degree S = 1

When using a stochastic polynomial of order $S = 1$ PMM Estimates of vector parameter elements $\boldsymbol{\theta}$ deterministic component of the regression model (1) can be found from the solution of the system Q equations of the form:

$$\sum\nolimits_{v=1}^{N} \left\{ k_{1,v}^{(p)} \left[y_v - \sum\nolimits_{p=0}^{Q-1} a_p x_v^p \right] \right\} = 0, \tag{2}$$

where $k_{1,v}^{(p)} = \frac{1}{\mu_2} \frac{\partial}{\partial a_p} \left[\sum_{p=0}^{Q-1} a_p x_v^p \right] = \frac{x_v^p}{\mu_2}, \; p = \overrightarrow{0, Q-1}$.

After certain transformations (2) can be represented in matrix form as a system of linear equations

$$\left(\boldsymbol{B}^T\boldsymbol{B}\right)\boldsymbol{\theta}^T = \frac{1}{N}\sum\nolimits_{v=1}^{N}\boldsymbol{B}^T\boldsymbol{Y}^T, \tag{3}$$

where $\boldsymbol{B}$ – Vandermonde matrix with elements $b_{v,p} = x_v^p$, $p = \overrightarrow{0,Q-1}$, $v = \overrightarrow{1,N}$.

Note that the resulting system (11) for finding PMM estimates $\hat{\boldsymbol{\theta}}_{(1)}$ parameters of polynomial regression (1) for $S = 1$ is equivalent to the linear system of OLS-estimates. This means that the accuracy of both methods is the same. Using (2) and (3), we can write down the variational matrix of estimates, which for asymptotic case (for $v \to \infty$) allows you to calculate their variance.

$$\boldsymbol{V}_{(1)} = \mu_2\left[\boldsymbol{B}\boldsymbol{B}^T\right]^{-1}. \tag{4}$$

As you know, such estimates are optimal (according to the criterion of minimum variance) in the situation when the errors of the regression model have a Gaussian distribution. In the case when the probabilistic nature of the errors differs from the Gaussian law, there are more accurate alternative estimation methods. Below we consider a new approach to nonlinear estimation of regression parameters (1), based on the use of higher-order power-law stochastic polynomials.

4.2 PPM-Estimates of Regression Parameters with Polynomial Degree $S = 2$

When using stochastic polynomials of order $S = 2$ PMM scores $\hat{\boldsymbol{\theta}}_{(2)}$ deterministic component of the regression model (9) can be found from the solution of the system Q equations of the form:

$$\sum\nolimits_{v=1}^{N}\left\{k_{1,v}^{(p)}\left[y_v - \sum\nolimits_{p=0}^{Q-1}a_p x_v^p\right] + k_{2,v}^{(p)}\left[(y_v)^2 - \left[\left(\sum\nolimits_{p=0}^{Q-1}a_p x_v^p\right)^2 + \mu_2\right]\right]\right\} = 0,\ p = \overrightarrow{0, Q-1} \tag{5}$$

where: the optimal coefficients $k_{i,v}^{(p)}$, $i = \overrightarrow{1,2}$ ensure the minimization of the variance of the estimates of the components of the desired parameter when using the degree of the polynomial $S = 2$.

These coefficients can be represented as functions depending on the central moments of the random component of the regression model:

$$k_{1,v}^{(p)} = \frac{\mu_4 - \mu_2^2 + 2\mu_3\left(\sum\nolimits_{p=0}^{Q-1}a_p x_v^p\right)}{\mu_2\left(\mu_4 - \mu_2^2\right) - \mu_3^2}x_v^p,\ k_{2,v}^{(p)} = -\frac{\mu_3^2}{\mu_2\left(\mu_4 - \mu_2^2\right) - \mu_3^2}x_v^p. \tag{6}$$

Substituting the coefficients (6) in (5), after certain transformations, the system of equations for finding estimates can be written in the form:

$$\sum\nolimits_{v=1}^{N}\left\{x_v^p\left[A_2\left(\sum\nolimits_{p=0}^{Q-1}a_p x_v^p\right)^2 + B_{2,v}\left(\sum\nolimits_{p=0}^{Q-1}a_p x_v^p\right) + C_{2,v}\right]\right\} = 0,\ p = \overrightarrow{0, Q-1} \tag{7}$$

where $A_2 = \mu_3$, $B_{2,v} = \mu_4 - \mu_2^2 - y_v\mu_3$, $C_{2,v} = y_v^2\mu_3 - y_v\left(\mu_4 - \mu_2^2\right) - \mu_2\mu_3$. (8)

Obviously, for the degree of the polynomial $S = 2$ PMM estimates can only be found numerically, for example, using the Newton-Raphson iterative procedure.

Using expressions of the optimal coefficients (6), it can be shown that the elements of the variation matrix of PMM estimates for $S = 2$ differ from the elements of the matrix of linear estimates (4) by a certain coefficient $\boldsymbol{V}_{(2)} = g_2\boldsymbol{V}_{(1)}$, which can be represented as

$$g_2 = 1 - \frac{\gamma_3^2}{2 + \gamma_4}. \tag{9}$$

The transition in expression (9) from the instantaneous to the cumulant description is due to the fact that the deviations of the values of the cumulant coefficients of higher orders γ_r from zero indicates the degree of deviation from a Gaussian distribution. In addition, higher order cumulant coefficients cannot take arbitrary values. In particular, for the cumulant coefficients of skewness and kurtosis, the inequality $\gamma_4 + 2 \geq \gamma_3^2$. Taking into account this inequality, based on the analysis of (9), we can conclude that the coefficient of variance reduction g_2 is a dimensionless quantity that belongs to the range $(0;\ 1]$. Thus, the relative decrease in the asymptotic variance of PMM estimates is the same for all components of the vector parameter. With an increase in the asymmetry of the distribution, the value of the relative decrease in variance can be quite significant and even asymptotically tend to zero as the absolute value of the asymmetry coefficient approaches the limits of the range of permissible values $|\gamma_3| \to \sqrt{2 + \gamma_4}$.

4.3 PPM-Estimates of Regression Parameters with Polynomial Degree S = 3 (Symmetric Case)

Analysis (9) shows that in the case of symmetry of the distribution (or at least for $\gamma_3 = 0$) the random component of the errors of the regression model of the PMM application at $S = 2$ is impractical. But in the case of symmetry (provided that the distribution is significantly different from the Gaussian law), the distribution can be applied by a stochastic log of a higher degree $S = 3$. In this case, PMM estimates $\hat{\boldsymbol{\theta}}_{(3)}$ are found from the solution of the system Q equations of the form

$$f_3^{(p)}(y_v, x_v) = \sum_{v=1}^{N}\left\{k_{1,v}^{(p)}\left[y_v - \sum_{p=0}^{Q-1} a_p x_v^p\right] + k_{2,v}^{(p)}\left[y_v^2 - \left(\left[\sum_{p=0}^{Q-1} a_p x_v^p\right]^2 + \mu_2\right)\right]\right.$$
$$\left. + k_{3,v}^{(p)}\left[y_v^3 - \left(\left[\sum_{p=0}^{Q-1} a_p x_v^p\right]^3 + 3\mu_2\sum_{p=0}^{Q-1} a_p x_v^p\right)\right]\right\} = 0, \quad p = \overrightarrow{0, Q-1} \tag{10}$$

where: the optimal coefficients $k_{i,v}^{(p)}$, $i = \overrightarrow{1,3}$ ensuring the minimization of the variance of the estimates of components of the required parameter, and can be represented in the form

$$k_{1,v}^{(p)} = \frac{3\left(\sum_{p=0}^{Q-1} a_p x_v^p\right)^2 (\mu_4 - 3\mu_2^2) + 3\mu_4\mu_2 - \mu_6}{\mu_2^2(\mu_4^2 - \mu_2\mu_6)} x_v^p,$$

$$k_{2,v}^{(p)} = \frac{-3(\mu_4 - 3\mu_2)}{\mu_2^{-2}(\mu_4^2 - \mu_2\mu_6)} \left(\sum_{p=0}^{Q-1} a_p x_v^p\right) x_v^p \qquad (11a,b,c)$$

$$k_{3,v}^{(p)} = \frac{\mu_4 - 3\mu_2}{\mu_2^{-2}(\mu_4^2 - \mu_2\mu_6)} x_v^p, \qquad p = \overrightarrow{0, Q-1},$$

Substituting the coefficients (11) into (10), after certain transformations, the system of equations for finding estimates can be written in the form:

$$\sum_{v=1}^{N} \left\{ x_v^p \left[A_3 \left(\sum_{p=0}^{Q-1} a_p x_v^p\right)^3 + B_{3,v} \left(\sum_{p=0}^{Q-1} a_p x_v^p\right)^2 + C_{3,v} \left(\sum_{p=0}^{Q-1} a_p x_v^p\right) + D_{3,v} \right] \right\} = 0, \; p = \overrightarrow{0, Q-1}, \quad (12)$$

where $A_3 = 1$, $B_{3,v} = -3y_v$, $C_{3,v} = 3y_v^2 - \frac{\mu_6 - 3\mu_4\mu_2}{\mu_4 - 3\mu_2^2}$, $D_{3,v} = y_v \frac{\mu_6 - 3\mu_4\mu_2}{\mu_4 - 3\mu_2^2} - y_v^3$.

Obviously, as in the previous case, for the degree of the polynomial $S = 3$ PMM estimates can only be found by numerically solving systems of Eqs. (12).

Using expressions for the optimal coefficients (11), it can be shown that the elements of the variation matrix of PMM estimates for $S = 3$ differ from the elements of the matrix of linear estimates (4) by the coefficient

$$g_3 = 1 - \frac{\gamma_4^2}{6 + 9\gamma_4 + \gamma_6}, \qquad (13)$$

which is a function of 4th and 6th order cumulant coefficients.

As already noted above, for the value of the cumulant coefficients of a higher order, there are certain ranges of acceptable values. In particular, for symmetrically distributed random variables, the constraints for cumulant coefficients of the 4th and 6th order are determined by the inequalities: $\gamma_4 > -2$ and $\gamma_6 + 9\gamma_4 + 6 > \gamma_4^2$. Taking into account these inequalities, we can conclude that the variance reduction coefficient g_3 is a dimensionless quantity and belongs to the range (0; 1]. Analysis (13) shows that the accuracy of PMM estimates for $S = 3$ can be significantly higher than the accuracy of linear estimates. In particular, when the value of the cumulant coefficients approaches the boundary of the region of permissible values $|\gamma_4| \to \sqrt{\gamma_6 + 9\gamma_4 + 6}$ the variance of PMM estimates tends asymptotically to zero.

It is necessary to note the universality of expressions (9) and (13), which describe the corresponding coefficients for reducing the variance of PMM estimates both for the estimated scalar parameter [14] and for the estimation of linear regression parameters [19, 20].

4.4 Algorithm for Adaptive PMM Estimates

The results of the theoretical analysis of the effectiveness of PMM estimates indicate the advisability of applying this approach only if the distribution of the random component of the regression model is significantly different from the Gaussian law. However, it is known that the inadequacy of the Gaussian model is not a critical factor from the point of view that OLS estimates still remain unbiased and consistent, although they cease to be optimal. Since the least square's method is inherently linear, the probable properties of the regression residuals practically do not differ from the properties of the original random component of the regression model [21].

Let us recall the fact that is important from a practical point of view that to obtain PMM estimates, information is used not on the distribution of regression errors, but their moment-cumulative description. At the same time, the solution of the problem of overcoming the a priori uncertainty regarding the probabilistic characteristics of the model is significantly simplified in comparison with the approach based on maximum likelihood. In addition, the obtained posteriori estimates of the cumulant coefficients of skewness and kurtosis can be used to test the hypothesis about the Gaussian law of distribution and the symmetry of regression errors [22].

Considering the above, we generalize the algorithm (elements of which are proposed in [19, 20]) for finding adaptive PMM estimates of the regression parameters:

Step 1 – finding linear OLS-estimates of regression parameters;
Step 2 – formation of regression residuals and finding posteriori estimates of moments/cumulants up to the 4th order;
Step 3 – testing the hypothesis about the Gaussian distribution of regression residuals (if confirmed, the algorithm ends);
Step 4 – testing the hypothesis about the symmetry of the distribution of regression residuals (if it is confirmed, go to step 6);
Step 5 – finding PMM estimates using a polynomial of degree $S = 2$ and completion of the algorithm;
Step 6 – finding estimates of the moments of the 6th order of regression residuals;
Step 7 – finding PMM estimates using a polynomial of degree $S = 3$ and the completion of the algorithm.

5 Statistical Modeling

To verify the theoretical results, the set of MATLAB/OCTAVIO compatible statistical modeling functions was modified, which was used in [17–20]. This toolkit, based on repeated tests by the Monte Carlo method, allows for a comparative analysis of the accuracy of various methods of statistical estimation, as well as to investigate the properties of PMM estimates of parameters when the statistical data model is non-Gaussian.

Since the value of the coefficients of the ratio of the variance of estimates is used as a comparative criterion of efficiency, and the theoretical values of the variance themselves are calculated for the asymptotic case, the experimental values $\hat{g}_S$ will obviously differ from theoretical ones in a certain way. The degree of discrepancy depends on the size of the samples used to find parameter estimates and the number of experiments performed. M according to the Monte Carlo procedure.

To carry out statistical modeling of the estimation of the parameters of regression dependences of the form (1), a quadratic polynomial with a vector of parameters was used as a deterministic component of the regression model $\boldsymbol{\theta} = \{1, 2, -4\}$ that were assumed to be unknown and subject to assessment.

When using PMM of degree $S = 2$ as an asymmetric random component (error) of the regression model, a sequence of independent and equally distributed random variables with exponential, lognormal, gamma, and Weibull distribution is used (see Table 1). When using PMM of degree $S = 3$, random variables with uniform, trapezoidal, triangular and Laplace distribution are used as a symmetrically distributed random error (see Table 2). In this case, the values of the parameters of the random component of the regression model, which are necessary to find the adaptive PMM estimates, were considered a priori unknown, and in accordance with the adaptive algorithm, their posterior estimates were used.

Table 1. Results of statistical modeling of Monte Carlo with the degree of the polynomial $S = 2$

Distribution		Theoretical values			Results of statistical modeling Monte Carlo								
		γ_3	γ_4	g_2	$\hat{g}_2$								
					$n = 20$			$n = 50$			$n = 200$		
					a_0	a_1	a_2	a_0	a_1	a_2	a_0	a_1	a_2
Gamma	$\alpha = 0.5$	2.83	12	0.42	0.44	0.38	0.37	0.43	0.35	0.35	0.43	0.36	0.36
Exponential (Gamma, $\alpha = 1$)		2	6	0.5	0.57	0.53	0.54	0.54	0.46	0.46	0.53	0.47	0.47
Gamma	$\alpha = 2$	1.41	3	0.6	0.64	0.64	0.63	0.63	0.59	0.6	0.63	0.59	0.59
	$\alpha = 4$	1	1.5	0.71	0.75	0.74	0.75	0.73	0.72	0.71	0.74	0.72	0.71
Lognormal $\sigma^2 = 0.1$, $\mu = 1$		1	1.86	0.74	0.77	0.76	0.76	0.75	0.74	0.74	0.74	0.75	0.75
Weibull $\alpha = 1, b = 2$		0.63	0.25	0.82	0.91	0.85	0.85	0.84	0.84	0.84	0.83	0.82	0.82

Table 2. Results of statistical modeling of Monte Carlo with the degree of the polynomial $S = 3$

Distribution		Theoretical values			Results of statistical modeling Monte Carlo								
		γ_4	γ_6	g_3	$\hat{g}_3$								
					$n = 20$			$n = 50$			$n = 200$		
					a_0	a_1	a_2	a_0	a_1	a_2	a_0	a_1	a_2
Uniform		1.2	6.9	0.3	0.55	0.53	0.54	0.41	0.4	0.4	0.35	0.35	0.35
Trapezoid	$\beta = 0.75$	−1.1	6.4	0.36	0.61	0.58	0.58	0.47	0.47	0.46	0.41	0.41	0.41
	$\beta = 0.5$	−1	5	0.55	0.75	0.73	0.74	0.66	0.66	0.66	0.59	0.58	0.58
	$\beta = 0.25$	−0.7	2.9	0.76	0.95	0.94	0.94	0.86	0.86	0.87	0.81	0.82	0.82
Triangular		−0.6	1.7	0.84	One	One	One	0.95	0.94	0.94	0.9	0.9	0.89
Laplace		3	30	0.86	1.4	1.66	1.72	0.86	0.86	0.85	0.83	0.82	0.82

The set of experimental values of the coefficients, the variance ratio was obtained at $M = 10^4$. To find the refined values of adaptive PMM estimates by solving systems of polynomial equations, we used the Newton-Raphson numerical procedure.

The results presented in Table 1 and Table 2 show that the experimental values of the ratio of PMM variances to the variance of OLS estimates differ to a certain extent from the theoretical ones. But as noted earlier, this is due to the fact that the theoretical values g_2 and g_3 calculated based on formulas (9) and (13) obtained for the asymptotic case.

In addition, the accuracy of obtaining adaptive PMM estimates is affected by the uncertainty of the posterior estimates of parameters (moments of the corresponding order) of regression residuals, which also decreases with increasing sample size. However, in general, the analysis of the presented data indicates the adequacy of analytical calculations, since the maximum relative discrepancy between the theoretical and experimental values of the coefficients is the variance ratio at $N = 200$ does not exceed 3%.

Another important result of statistical modeling is the confirmation of the theoretical property, which is that the relative increase in the accuracy of PMM estimates is the same for all components of the vector parameter (coefficients of the deterministic component of the polynomial regression). Moreover, it is practically independent of the order Q the deterministic component of the regression model (1). This property is also confirmed by the significant correlation between the results of this study and the results of statistical modeling of finding estimates of the parameters of linear one-factor regression, given in [19, 20].

6 Experiments with Real Data

As an example, for testing the proposed adaptive procedure for finding PMM-estimates parameters of polynomials regression, we used the Auto MPG data set from the UCI repository. This set of data represents the dependence of fuel consumption in the urban cycle on the different characteristics of cars [23].

6.1 Experiment 1 (Asymmetrical Case – PMM $S = 2$)

In the first experiment, we solve the problem of estimating the parameters of the dependence of fuel consumption (parameter *MPG* – Mile Per Gal) on the weight of the car (parameter *Weight*). Figure 1, (a) presents experimental data for $N = 392$ cars (considering the missing data) of different. A visual analysis of this figure suggests the existence of a quadratic relationship $MPG = a_0 + a_1 Weight + a_2 Weight^2$.

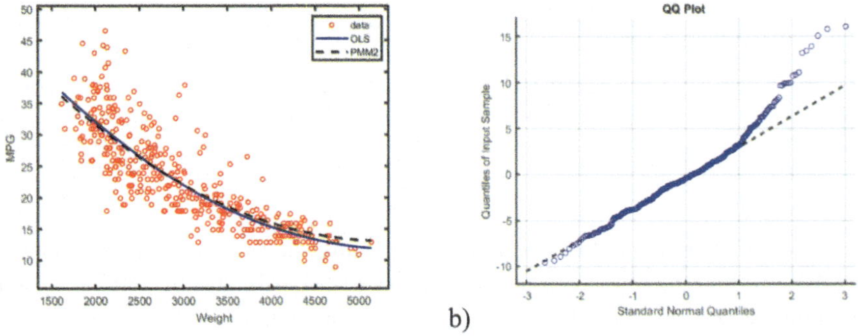

a) b)

Fig. 1. Construction of the model fuel consumption depending on the weight of cars: a) Experimental data and regression estimates, based on OLS and adaptive PMM2; b) QQ plot of OLS-regression residuals

OLS estimates (with 95% confidence bounds) of this model parameters have values

$$\hat{a}_0^{(1)} = 62.3\ (56.4, 68.1);$$
$$\hat{a}_1^{(1)} = -18.5 * 10^{-3}\left(-22.4{*}10^{-3}, -14.6{*}10^{-3}\right);$$
$$\hat{a}_2^{(1)} = 1.7 * 10^{-6}\left(1.1{*}10^{-6}, 2.3{*}10^{-6}\right).$$

The asymmetry of the error distribution of the model is visually visible in Fig. 1 (b), where presents a Q-Q plot of regression OLS-residuals. The hypothesis of the Gaussian of OLS-residuals is also refuted by the Yarki-Bera test ($JBSTAT = 93.2$ at the threshold value $CV = 5.8$ at a fixed significance level $alfa = 0.05$).

Estimated values $\hat{\gamma}_3 = 0.8$ and $\hat{\gamma}_4 = 1.8$ regression OLS-residuals (taking into account a sufficiently large volume of initial data N), allow us to estimate the magnitude of the decrease in the variance estimates (PMM for $S = 2$ compared to OLS) at the level $\hat{g}_2 = 0.83$. As result PMM2 estimates (with 95% confidence bounds) of the parameters of this model are the values of:

$$\hat{a}_0^{(2)} = 60.7\ (55.3, 66);$$
$$\hat{a}_1^{(2)} = -18 * 10^{-3}\left(-21.5{*}10^{-3}, -14.4{*}10^{-3}\right);$$
$$\hat{a}_2^{(2)} = 1.7 * 10^{-6}\left(1.1{*}10^{-6}, 2.2{*}10^{-6}\right).$$

6.2 Experiment 2 (Symmetrical Case – PMM $S = 3$)

In the second experiment, we solve the problem of estimating the parameters of the dependence of fuel consumption (parameter MPG) on the power of the car (parameter $Horsepower$). A visual analysis of Fig. 2, (a) suggests the existence of a quadratic relationship $MPG = a_0 + a_1 Horsepower + a_2 Horsepower^2$.

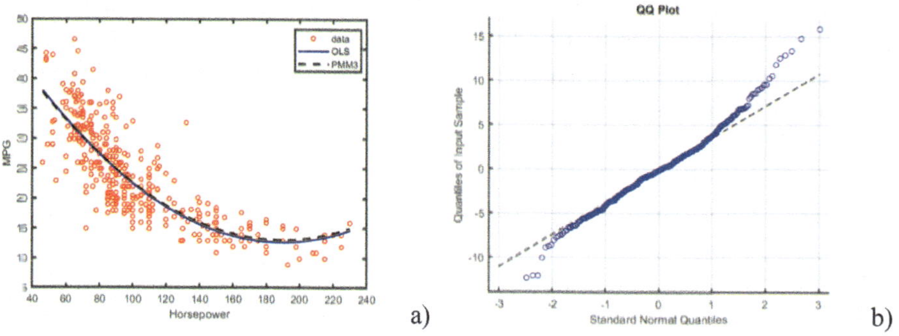

Fig. 2. Construction of the model fuel consumption depending on the horsepower of cars: a) Experimental data and regression estimates, based on OLS and adaptive PMM3; b) QQ plot of OLS-regression residuals.

OLS estimates (with 95% confidence bounds) of the parameters of this model are the values of:

$$\hat{a}_0^{(1)} = 56.9\ (53.36, 60.44);$$
$$\hat{a}_1^{(1)} = -46.6 * 10^{-2}\left(-52.7{*}10^{-2}, -40.5{*}10^{-2}\right);$$
$$\hat{a}_2^{(1)} = 12.3 * 10^{-4}\left(9.9{*}10^{-4}, 14.7{*}10^{-4}\right).$$

The hypothesis of the Gaussian of OLS-residuals is refuted by the Yarki-Bera test ($JBSTAT = 30.6$ at the threshold value $CV = 5.8$ at a fixed significance level $\alpha lfa = 0.05$). However, visual analysis of the Q-Q plot of regression OLS-residuals presented in Fig. 2 (b) indicates the symmetry of the distribution. This is numerically confirmed by the closeness to zero of the value of the skewness coefficient $\hat{\gamma}_3 \approx 0.2$. In this situation, for adaptive refinement of estimates, it is advisable to use PMM at degree $S = 3$.

Estimated values $\hat{\gamma}_4 = 1.3$ and $\hat{\gamma}_6 = -1.9$ regression OLS-residuals (taking into account a sufficiently large volume of initial data $N = 392$) according to (18), allow us to estimate the magnitude of the decrease in the variance estimates (PMM for $S = 3$ compared to OLS) at the level $\hat{g}_3 = 0.89$. As result PMM3-estimates (with 95% confidence bounds) of the parameters of this model are the values of:

$$\hat{a}_0^{(2)} = 58.2\ (54.9, 61.5);$$
$$\hat{a}_1^{(2)} = -48.9 * 10^{-2}\left(-54.7{*}10^{-2}, -43.1{*}10^{-2}\right);$$
$$\hat{a}_2^{(1)} = 13.2 * 10^{-4}\left(10.9{*}10^{-4}, 15.3{*}10^{-4}\right).$$

6.3 Bootstrap Simulation

To verify the correctness of parameter estimation regression models, a statistical experiment based on the bootstrapping method. With the help of built-in MATLAB bootstrap resampling tools of the original sample was multiplied by 10^4 bootstrap samples with the return. For each of them, we found OLS and PMM estimates for quadratic regression parameters. The empirical distribution of these estimates is presented in Fig. 3 and Fig. 4 in the form of boxplot-graphs.

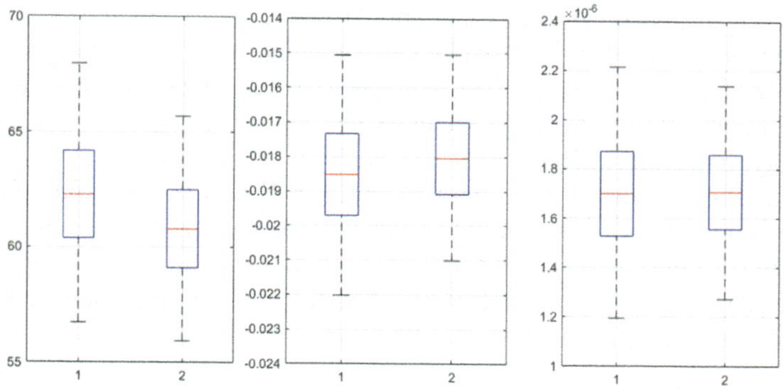

Fig. 3. Empirical distribution of bootstrap OLS (1) and PMM2 (2) estimates of regression

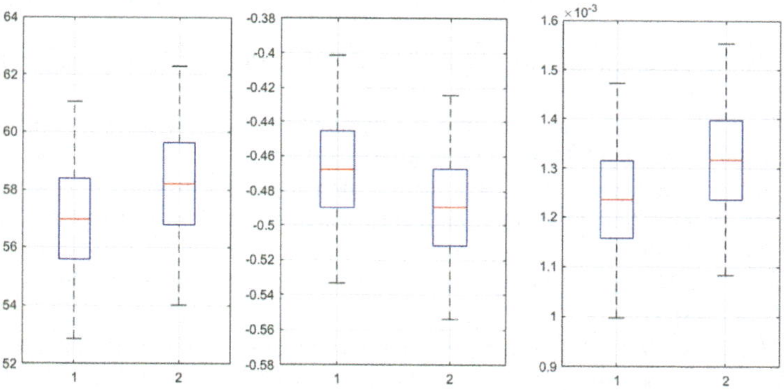

Fig. 4. Empirical distribution of bootstrap OLS (1) and PMM3 (2) estimates of regression

The presented results confirm the difference between the corresponding OLS and PMM parameter estimates, which consists in the shift of the centers of their confidence intervals and the change in the width.

$$MPG = a_0 + a_1 Weight + a_2 Weight^2$$

$$MPG = a_0 + a_1 Horsepower + a_2 Horsepower^2$$

7 Summary and Conclusions

The totality of the results obtained confirms the possibility of effective application of the Polynomial Maximization Method for solving problems of regression analysis in situations where the regression residuals has a distribution that differs significantly from Gaussian idealization.

The presented results generalize the approaches to the estimation of linear in the parameter regression models, a particular case of which is polynomial regression. Note that the analytical expressions obtained in this work for finding estimates of the parameters of polynomial one-way regression can also be applied for the case of a multivariate linear or polynomial regression model.

The next stage of the study should be the distribution of the proposed adaptive approach based on the use of stochastic polynomials and a partial description by higher-order statistics for the case of nonlinear regression models. In addition, it is interesting to compare the efficiency of the adaptive estimates of the Polynomial Maximization Method and the Maximum Likelihood Estimation.

References

1. Johnson, N.L.: Systems of frequency curves generated by methods of translation. Biometrika **36**, 149–176 (1949)
2. Box, G.E.P., Cox, D.R.: An analysis of transformations. J. Roy. Stat. Soc. Ser. B **26**, 211–246 (1964)
3. Huber, P.J., Ronchetti, E.M.: Robust Statistics. Wiley, Hoboken (2009). https://doi.org/10.1002/9780470434697
4. Stone, C.J.: Adaptive maximum likelihood estimators of a location parameter. Ann. Stat. **3**(2), 267–284 (1975). https://doi.org/10.1214/aos/1176343056
5. Schechtman, E., Schechtman, G.: Estimating the parameters in regression with uniformly distributed errors. J. Stat. Comput. Simul. **26**(3–4), 269–281 (1986). https://doi.org/10.1080/00949658608810965
6. Zeckhauser, R., Thompson, M.: Linear regression with non-normal error terms. Rev. Econ. Stat. **52**, 280–286 (1970)
7. Islam, M.Q., Tiku, M.L., Yildirim, F.: Nonnormal regression. I. Skew distributions. Commun. Stat. Theory Methods **30**(6), 993–1020 (2001). https://doi.org/10.1081/STA-100104347
8. Tiku, M.L., Islam, M.Q., Selçuk, A.S.: Nonnormal regression. II. Symmetric distributions. Commun. Stat. Theory Methods **30**(6), 1021–1045 (2001). https://doi.org/10.1081/STA-100104348
9. Van Montfort, K., Mooijaart, A., de Leeuw, J.: Regression with errors in variables: estimators based on third order moments. Stat. Neerl. **41**(4), 223–237 (1987)
10. Dagenais, M.G., Dagenais, D.L.: Higher moment estimators for linear regression models with errors in the variables. J. Econ. **76**(1–2), 193–221 (1997). https://doi.org/10.1016/0304-4076(95)01789-5
11. Cragg, J.G.: Using higher moments to estimate the simple errors-in-variables model. RAND J. Econ. **28**, S71 (1997). https://doi.org/10.2307/3087456
12. Gillard, J.: Method of moments estimation in linear regression with errors in both variables. Commun. Stat. Theory Methods **43**(15), 3208–3222 (2014). https://doi.org/10.1080/03610926.2012.698785
13. Giacalone, M.: A combined method based on kurtosis indexes for estimating p in non-linear L p-norm regression. Sustain. Futures **2**, 100008 (2020). https://doi.org/10.1016/j.sftr.2020.100008
14. Kunchenko, Y.: Polynomial Parameter Estimations of Close to Gaussian Random Variables. Shaker Verlag, Aachen (2002)
15. Zabolotnii, S.W., Martynenko, S.S., Salypa, S.V.: Method of verification of hypothesis about mean value on a basis of expansion in a space with generating element. Radioelectron. Commun. Syst. **61**, 222–229 (2018). https://doi.org/10.3103/S0735272718050060

16. Chertov, O., Slipets, T.: Epileptic seizures diagnose using Kunchenko's polynomials template matching. In: Fontes, M., Günther, M., Marheineke, N. (eds.) Progress in Industrial Mathematics at ECMI 2012, pp. 245–248. Springer, Cham (2014). https://doi.org/10.1007/978-3-319-05365-3_33.
17 Zabolotnii, S.V., Warsza, Z.L.: Semi-parametric polynomial method for retrospective estimation of the change-point of parameters of non-Gaussian sequences. In: Advanced Mathematical and Computational Tools in Metrology and Testing X, pp. 400–408 (2015).https://doi.org/10.1142/9789814678629_0048
18. Zabolotnii, S.W., Warsza, Z.L.: Semi-parametric polynomial modification of CUSUM algorithms for change-point detection of non-Gaussian sequences. In: Electronic Proceedings of XXI IMEKO World Congress "Measurement in Research and Industry", Prague Czech Republic, September 2015, vol. 30, pp. 2088–2091 (2015)
19. Zabolotnii, S., Warsza, Z., Tkachenko, O.: Polynomial estimation of linear regression parameters for the asymmetric pdf of errors. Advances in Intelligent Systems and Computing, vol. 743, pp. 758–772. Springer (2018). https://doi.org/10.1007/978-3-319-77179-3_75
20. Zabolotnii, S.W., Warsza, Z.L., Tkachenko, O.: Estimation of linear regression parameters of symmetric non-Gaussian errors by polynomial maximization method. Advances in Intelligent Systems and Computing, vol. 920, pp. 636–649. Springer (2020). https://doi.org/10.1007/978-3-030-13273-6_59
21. Cook, R.D., Weisberg, S.: Residuals and Influence in Regression. Monographs on Statistics and Applied Probability (1982). https://doi.org/10.2307/1269506
22. Jarque, C.M., Bera, A.K.: A tests of observations and regression residuals. Int. Stat. Rev. **55**, 163–172 (1987)
23. Quinlan, J.R.: Combining instance-based and model-based learning. In: Proceedings of the Tenth International Conference on Machine Learning, pp. 236–243 (1993)

LTspice Implementation of Gyrator-Capacitor Magnetic Circuit Model Considering Losses and Magnetic Saturation for Transient Simulations of Switching Mode Power Supplies Utilizing Inductive Elements with Cores Made of Amorphous Alloys

Roman Szewczyk[1(✉)], Oleg Petruk[1], Michał Nowicki[1], Anna Ostaszewska-Liżewska[1], Aleksandra Kolano-Burian[2], Piotr Gazda[1], Adam Bieńkowski[3], Paweł Nowak[1], and Tomasz Charubin[1]

[1] ŁUKASIEWICZ Research Network – Industrial Research Institute for Automation and Measurements PIAP, Al. Jerozolimskie 202, 02-486 Warsaw, Poland
rszewczyk@onet.pl
[2] ŁUKASIEWICZ Research Network – Institute of Non-Ferrous Metals, Sowińskiego 5, 44-100 Gliwice, Poland
[3] Faculty of Mechatronics, Warsaw University of Technology, św. A. Boboli 8, 02-525 Warsaw, Poland

Abstract. Modelling the transient simulation of magnetic circuits is the crucial process for development of modern switching mode power converters. Such models are especially important for newly developed magnetic materials, such as amorphous alloys. Paper presents the gyrator-capacitor model of inductive element with core made of high permeability, Co-based amorphous alloy. Parameters of the model were identified during Nedler-Mead method based optimization process. Presented results of modelling confirm that model can be used for SPICE simulations; however, the accuracy of losses modeling is limited in the wide range of frequencies.

Keywords: Gyrator-capacitor model · SPICE modelling · Amorphous alloys

1 Introduction

Electronic components with inductive cores, such as coils and transformers, play important role in switching mode power converters [1]. Development of switching mode power converters is the crucial part for development of electromobility systems [2], such as Polish electric car IZERA [3] or mobile robots. As a result, recently we observe increasing interest in optimization of power conversion systems and development of magnetic materials for cores of inductive components.

R. Szewczyk et al. (Eds.): AUTOMATION 2021, AISC 1390, pp. 416–424, 2021.
https://doi.org/10.1007/978-3-030-74893-7_37

Efficient development of power conversion systems requires efficient optimization during the development process. However, in spite of significant efforts, cores of inductive components seem to be the most difficult part from the point of view of accurate modelling. Such cores are highly nonlinear due to magnetic saturation. Moreover, magnetic hysteresis occurs in inductive cores, what makes modelling even more sophisticated.

The lack of accurate models is especially visible considering newly developed soft magnetic materials, such as amorphous alloys [4]. In spite of theoretical efforts to for development of models describing magnetic hysteresis [5], the model of inductive component with core made of such material, suitable for transient simulations required in practical application, seems to be not previously presented.

Paper is trying to fill this gap. For efficient modelling of inductor with core made of Co-based amorphous alloy, gyrator-capacitor [6] model was adapted. Proposed model was implemented in LTspice environment. It should be highlighted, that LTspice is considered as the most efficient, freely available modelling tool for switching mode power conversion modelling [7]. As a result, proposed model creates new possibilities in efficient and accurate modelling of inductive cores for switching mode power converters development.

2 Gyrator Concept

The concept of gyrator was first introduced in 1948 by Bernard D. H. Tellegen [8]. Gyrator is linear and loss-less element, which can invert the current–voltage characteristic of electrical network. From the practical point of view, gyrators can be developed utilizing two operational amplifiers [9]. As a result, the use of gyrator enables substitution of real inductive component (such as inductor) by gyrator and the electric network including capacitor and resistance. Such circuit was proposed by David C. Hamil in 1993 [6], creating new possibilities of transient simulation of circuits including inductive components.

Figure 1 presents the symbol of gyrator proposed in 1948 by Bernard D. H. Tellegen (Fig. 1a) as well as practical implementation of gyrator in LTspice environment (Fig. 1b). This implementation utilizes two voltage controlled current sources as well as two resistors with resistance 100 GΩ. These resistors don't influence on gyrator's characteristics, however, they increase the numerical stability of the gyrator's model.

The operation of gyrator is determined by the following equations [10]:

$$i_1 = N \cdot v_2 \tag{1}$$

$$i_2 = -N \cdot v_1 \tag{2}$$

where i_1, v_1 are respectively the current and voltage on the left side of gyrator, and i_2, v_2 are respectively the current and voltage on the right side of gyrator.

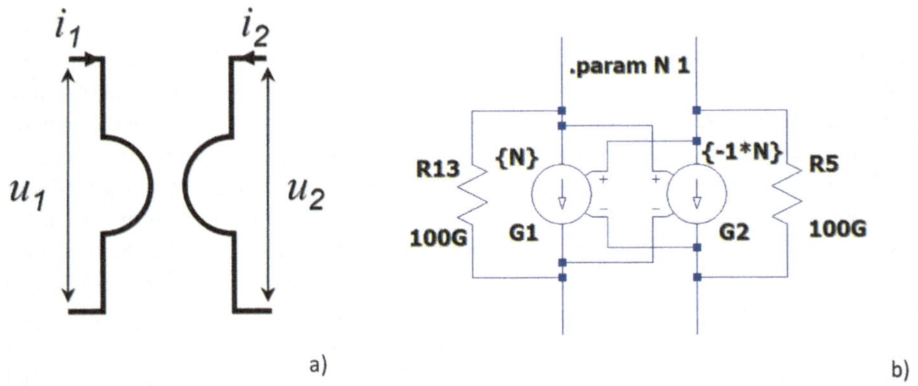

Fig. 1. Gyrator: a) symbol, b) LTspice implementation consisting two voltage dependent current sources

3 Implementation of Gyrator-Capacitor Model of Inductive Component

The gyrator-capacitor model of real inductor proposed by D. C. Hamill [6] and developed by M. Eaton [11] in 1994 consists capacitor C, nonlinear voltage source V_B and resistor R_L. Capacitance of capacitor C is given by the following equation [6]:

$$C = \frac{\mu_o \mu_r S_e}{l_e} \tag{3}$$

where μ_0 is magnetic constant, μ_r is permeability of magnetic core, S_e is effective cross-section of magnetic core and l_e is magnetic path length. Voltage on nonlinear voltage source V_B reproduces saturation in nonlinear characteristic of inductive core. Voltage V_B is dependent on potential difference on capacitor C and is given by following equation [11]:

$$V_B(V_c) = b \cdot (V_c)^n \tag{4}$$

where b is parameter determining the location of saturation in core's magnetization curve and n is the odd integer value. Magnetic core's losses are represented by resistor R_L. It should be highlighted that value of loses increases with frequency what is in line with physical processes observed in the magnetic core of inductive element.

The LTspice implementation of described gyrator-capacitor model is presented in Fig. 2. Parameters of the model are stored in separate variables, what clarifies the implementation and simplifies the variable update process.

Figure 3 presents the model of simple hysteresis graph enabling verification of the proposed model and identification of its parameters. Proposed hysteresis graph consists of arbitrary controlled sine wave current source, implementation of gyrator-capacitor model of coil as well as instrumental amplifier with amplification $k = 1$ and integrator with Laplace transmittance $1/s$. As a result, for given sine wave current amplitude a with frequency f, the magnetic $B(H)$ hysteresis loop can be plotted as dependence of variable

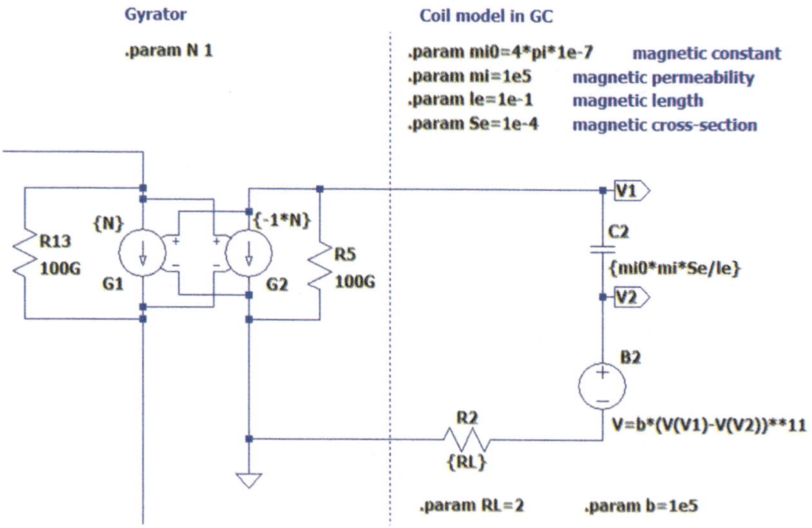

Fig. 2. LTspice implementation of the gyrator-capacitor model of inductive coil considering material permeability (described by parameter "mi") as well as both saturation (described by parameter "b") flux density and loses (described by parameter "RL"). Two 100 GΩ resistors added for numerical stability of the model

B on variable H. Both variables are given in volts, whereas they are scaled in Tesla (T) and amperes per meter (A/m).

Proposed model of inductive component may be easily integrated with OCTAVE [12], the open-source MATLAB alternative. During such integration, OCTAVE script may modify model's parameters presented text values in.asc file describing electronic circuit. Next, LTspice can work as external solver working in text mode. Results of transient simulation may be acquired by OCTAVE using LTspice2OCTAVE.m function proposed by Marcin Mleczko [13].

4 Accuracy of the Model

Parameters of presented gyrator-capacitor model for ring-shaped core made of $Co_{58}Fe_4B_{13}Si_{13.5}Mo_{1.5}$ amorphous alloys in as-quenched state [14] were identified during the Nedler-Mead method-based optimization process. The optimization target function was defined as:

$$T_{100} = \sum\nolimits_{i=1}^{n} \left(B_{meas}(i)|_{100Hz} - B_{sim}(i)|_{100Hz}\right)^2 \tag{5}$$

where $B_{meas}(i)|_{100Hz}$ are the results of measurements carried out at frequency $f = 100$ Hz, whereas $B_{sim}(i)|_{100Hz}$ were the results of simulation. The results of modelling are presented in Fig. 4.

Parameters of the model obtained in optimization process are presented in Table 1. Results of simulation are in good agreement with the results of measurements. This agreement is quantitatively confirmed by coefficient of determination $R^2 = 0.9955$.

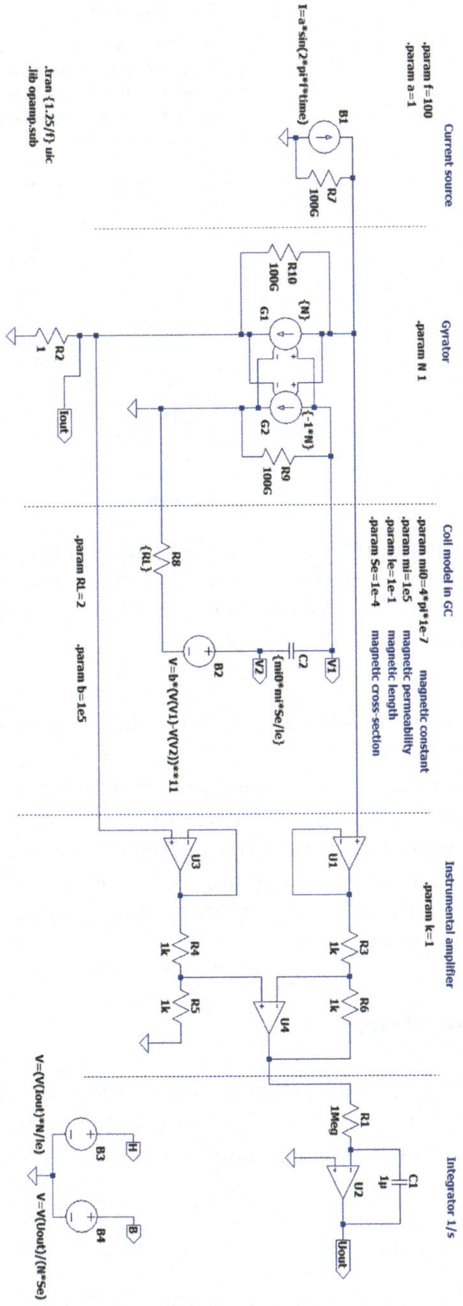

Fig. 3. LTspice implementation of the gyrator-capacitor model of inductive coil connected to a model of simple hysteresis graph consisting of current source, instrumental amplifier (amplification $k = 1$) and integrator (with transmittance $L(s) = 1/s$)

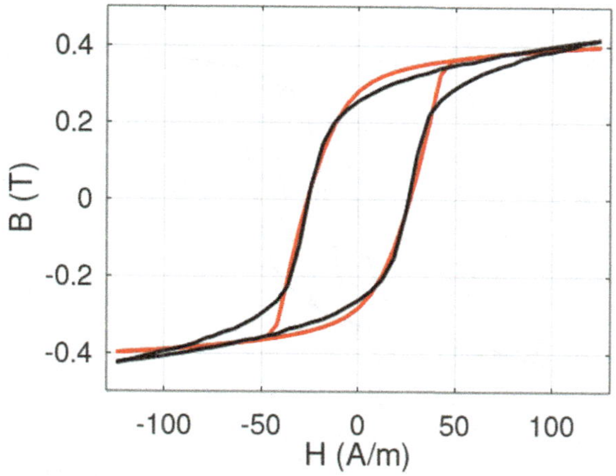

Fig. 4. The results of modelling of $B(H)$ characteristics of ring-shaped core made of $Co_{68}Fe_4B_{13}Si_{13.5}Mo_{1.5}$ amorphous alloys in as-quenched state, at frequency $f = 100$ Hz (red line) in comparison to the results of measurements at the same frequency (black line)

On the other hand, the quality of modelling decreases for modelling in significantly different frequency. To minimize this unfavorable effect, the optimization was carried out for modified target function:

$$T_{100,1000} = \sum\nolimits_{i=1}^{n} (B_{meas}(i)|_{100Hz} - B_{sim}(i)|_{100Hz})^2 + \ldots \\ + \sum\nolimits_{i=1}^{n} (B_{meas}(i)|_{1000Hz} - B_{sim}(i)|_{1000Hz})^2 \tag{6}$$

where $B_{meas}(i)|_{1000Hz}$ and $B_{sim}(i)|_{1000Hz}$ are the results of measurements and simulation carried out at frequency $f = 1000$ Hz. The results of such enhanced modelling are presented in Fig. 5.

Also, for a simultaneous simulation for two frequencies, results of modelling are in moderately good agreement with the results of measurements confirmed by coefficient of determination $R^2 = 0.\ 9456$ and $R^2 = 0.9562$ for frequencies $f = 100$ Hz and $f = 1000$ Hz respectively. The limited quality of modelling over the different frequencies clearly indicates that the magnetic core losses model described by single resistor is not sufficient. Core loss model should take into account eddy current losses, anomalous loses and losses due to magnetic hysteresis. These losses, different in their nature, should be described by different loss elements in gyrator-capacitor circuit.

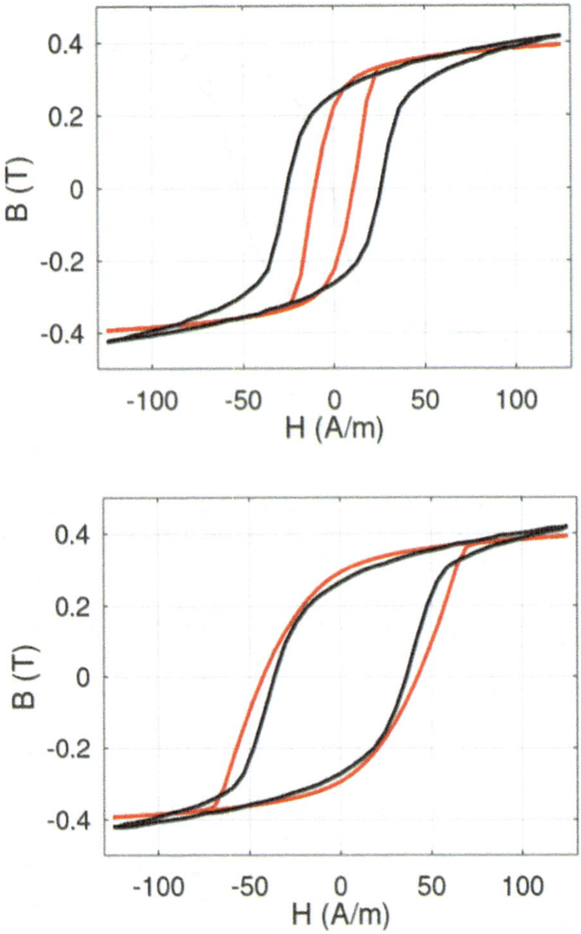

a)

b)

Fig. 5. The results of modelling of *B*(*H*) characteristics of ring-shaped core made of $Co_{68}Fe_4B_{13}Si_{13.5}Mo_{1.5}$ amorphous alloys in as-quenched state (red line) in comparison to the results of measurements (black line) at frequencies: a) $f = 100$ Hz, b) $f = 1000$ Hz

Table 1. Parameters of the gyrator-capacitor model determined in Nedler-Mead method-based optimization process

Parameter	Description	for $f = 100$ Hz	for $f = 100$ Hz and $f = 1000$ Hz
μ	Represents magnetic permeability μ of the core	17 743	12 933
b	Describe saturation of the magnetic core	3 522	124.5
n		11	11
$R_L(\Omega)$	Describes losses in the core	58.53	14.58
$R^2 f = 100$ Hz	Coefficient of determination	0.9955	0.9456
$R^2 f = 1000$ Hz	Coefficient of determination	n.a	0.9562

5 Conclusions

Results of modelling presented in the paper clearly indicate that gyrator-capacitor model has a great potential for modelling the transient states in switching mode power supplies. Proposed solution reproduces moderately well the magnetic characteristics of inductive elements with core made of $Co_{68}Fe_4B_{13}Si_{13.5}Mo_{1.5}$ amorphous alloys in as-quenched state. The quality of modelling is confirmed by coefficient of determination exceeding 0.94 in all cases. Such quality is sufficient for engineering modelling of switching mode power supplies with inductive elements with cores made of modern soft magnetic materials.

On the other hand presented results clearly indicate the need of further development of presented gyrator-capacitor model. This development should be focused on improvement of nonlinearity description in the model (considering recent works on development of models of anhysteretic magnetization curves [15]) as well as on taking into account the physical background of losses occurring in cores of inductive elements, such as eddy current losses, anomalous loses and losses due to magnetic hysteresis.

This work was partially realized within research project No. 2/Ł-IMN/CŁ/2020 entitled “Specialized high-performance energy conversion systems for electromobility and robotics applications” (SPECKON) co-financed by the targeted grant of the Łukasiewicz Research Network.

References

1. Tse, C.: Complex Behavior of Switching Power Converters. CRC Press, Boca Raton (2004)
2. Hayes, J., Goodarzi, G.: Electric Powertrain : Energy Systems, Power Electronics and Drives for Hybrid, Electric and Fuel Cell Vehicles. John Wiley & Sons, Incorporated, Newark (2017)
3. https://izera.com/
4. Hasegawa, R.: Applications of amorphous magnetic alloys in electronic devices. J. Non-Cryst. Solids **287**, 405–412 (2001). https://doi.org/10.1016/S0022-3093(01)00633-0
5. Liorzou, F., Phelps, B., Atherton, D.: Macroscopic models of magnetization. IEEE Trans. Magn. **36**(2), 418–428 (2000). https://doi.org/10.1109/20.825802
6. Hamill, D.: Gyrator-capacitor modeling: a better way of understanding magnetic components. In: Proceedings of 1994 IEEE Applied Power Electronics Conference and Exposition - ASPEC94 (1994). https://doi.org/10.1109/apec.1994.316381
7. Alonso, G., Spencer, J.: LTspice: Worst-Case Circuit Analysis with Minimal Simulations Runs (2020). https://www.analog.com/en/technical-articles/ltspice-worst-case-circuit-analysis-with-minimal-simulations-runs.html
8. Tellegen, B.D.H.: The Gyrator, a New Electric Network Element. Philips Res. Rep. **3**, 81–101 (1948)
9. Antoniou, A.: Gyrators using operational amplifiers. Electron. Lett. **3**(8), 350 (1967). https://doi.org/10.1049/el:19670270
10. Lambert, M., Mahseredjian, J., Martinez-Duro, M., Sirois, F.: Magnetic circuits within electric circuits: critical review of existing methods and new mutator implementations. IEEE Trans. Power Delivery **30**(6), 2427–2434 (2015). https://doi.org/10.1109/TPWRD.2015.2391231
11. Eaton, M.: Modeling magnetic devices using the gyrator re-cap core model. In: Proceedings of NORTHCON 1994 (1994).https://doi.org/10.1109/northc.1994.638970
12. https://www.gnu.org/software/octave/index

13. https://octave.1599824.n4.nabble.com/attachment/4693440/0/LTspice2Octave.m
14. Szewczyk, R., Bieńkowski, A., Kolano, R.: The influence of thermal treatment on the properties of Co-based magnetoelastic sensing elements. Physica Status Solidi (c) **1**(12), 3681–3685 (2004). https://doi.org/10.1002/pssc.200405532
15. Szewczyk, R.: Validation of the anhysteretic magnetization model for soft magnetic materials with perpendicular anisotropy. Materials **7**(7), 5109–5116 (2014). https://doi.org/10.3390/ma7075109

Estimation of the Parameters of Anhysteretic Curve of Isotropic and Anisotropic Magnetic Materials on the Base of Initial Anhysteretic Permeability Measurements

Roman Szewczyk(✉)

ŁUKASIEWICZ Research Network – Industrial Research Institute for Automation and Measurements PIAP, Al. Jerozolimskie 202, 02-486 Warsaw, Poland
rszewczyk@onet.pl

Abstract. Determination of parameters of anhysteretic magnetization curve plays the key role in modelling the characteristics of soft magnetic materials. This curve can be efficiently used to describe the characteristics of inductive components in SPICE or FEM modelling. Moreover, anhysteretic magnetization curve is the base of Jiles-Atherton model of magnetic hysteresis loop. Paper presents the method of identification of parameters of anhysteretic curves of soft magnetic materials without or with different types of anisotropy. Due to the fact, that calculation of anhysteretic curve of anisotropic materials requires numerical integration, optimization-based method is proposed. Moreover, open-source MATLAB/Octave scripts are provided to verify and validate proposed method for variety of anisotropic materials.

Keywords: Anisotropic magnetic materials · Anhysteretic magnetization curve · SPICE modelling · FEM modelling

1 Introduction

Accurate description of the nonlinear behavior of soft magnetic materials in SPICE (Simulation Program with Integrated Circuits Emphasis) or FEM (Finite Elements Modelling) is still one of the most problematic tasks in engineering modelling [1, 2]. Recently developed solutions often focus on neglecting the hysteresis and describe the magnetization characteristic on the base of anhysteretic magnetization curve [3]. Anhysteretic magnetization curve can be measured experimentally, using constant magnetic field biased demagnetization [4], utilizing robust and reliable method presented by M. Nowicki [5].

On the other hand, efficient modelling requires not only experimental shape of anhysteretic curve, but also its mathematical model. Such model was presented by D. C. Jiles and D. L. Atherton in 1986 [6] together with useful method of determination of model's parameters [7]. However, presented previously method is suitable only for isotropic materials, such as soft ferrites for switching mode power supplies [8]. In the case of electrical steels [9] or modern anisotropic amorphous [10] or nanocrystalline [11] soft

R. Szewczyk et al. (Eds.): AUTOMATION 2021, AISC 1390, pp. 425–430, 2021.
https://doi.org/10.1007/978-3-030-74893-7_38

magnetic materials, the adequate method of identification of anhysteretic curve model's parameters were not presented previously.

This paper is filing above gap. Proposed method enables identification of anhysteretic curve model's parameters on the base of two hysteresis loops measured in different directions. In addition, on the base of proposed method, the average anisotropy energy density K_{an} can be estimated for given material. Quantitative information about this parameter is valuable also from technological point of view, enabling optimization of material's processing technology.

2 The Model of Anhysteretic Magnetization Curve for Isotropic and Anisotropic Materials

The idea of the anhysteretic magnetization model was first presented by D.C.Jiles and D. L. Atherton [6]. The idea of this model is based on modelling the anhysteretic magnetization curve of paramagnetic material. In such a case the anhysteretic magnetization curve M_{par} in the material is described on the base of Boltzman distribution [12]:

$$M_{par} = M_s \frac{\int_0^{\pi} e^{\frac{-E_m(\theta)}{k_B T}} sin\theta \cdot cos\theta \cdot d\theta}{\int_0^{\pi} e^{\frac{-E_m(\theta)}{k_B T}} sin\theta \cdot d\theta} \tag{1}$$

$$E_m(\theta) = -\mu_0 \cdot m_{at} \cdot H \cdot cos\theta \tag{2}$$

where M_s is the saturation magnetization of paramagnetic material, T is the temperature of material, θ is the angle between the atomic magnetic moment m_{at} and direction of magnetizing field H and k_B is Boltzmann constant.

In the case of isotropic, ferromagnetic material, in Eq. (2), the atomic magnetic moment m_{at} is substituted by the average magnetization of domain m_d, given as [6]:

$$m_d = \frac{M_s}{N} \tag{3}$$

whereas magnetizing field is substituted by effective magnetizing field H_e [6]:

$$H_e = H + \alpha \cdot M \tag{4}$$

considering interdomain coupling α.

Due to the fact, that antiderivative in Eqs. (1)–(2) occurs, anhysteretic magnetization curve $M_{iso}(H)$ of isotropic ferromagnetic material can be presented in the form of Langevin's equation [6]:

$$M_{iso} = M_s \left[\coth\left(\frac{H_e}{a}\right) - \frac{a}{H_e} \right] \tag{5}$$

where a is given as:

$$a = \frac{N \cdot k_B \cdot T}{\mu_0 \cdot M_s} \tag{6}$$

In the case of anisotropic materials Eq. (1) can be presented as [13, 14]:

$$M_{aniso}(H) = M_s \left[\frac{\int_0^{\pi} e^{\frac{E(1)+E(2)}{2}} \sin\theta \cos\theta \cdot d\theta}{\int_0^{\pi} e^{\frac{E(1)+E(2)}{2}} \sin\theta \cdot d\theta} \right] \tag{7}$$

In the case of an axial anisotropy (typical for e.g. magnetic field annealed amorphous alloys) [15]:

$$E(1) = \frac{H_e}{a}\cos\theta - \frac{K_{an\_a}}{\mu_0 M_s a} sin^2(\psi - \theta) \tag{8}$$

$$E(2) = \frac{H_e}{a}\cos\theta - \frac{K_{an\_a}}{\mu_0 M_s a} sin^2(\psi - \theta) \tag{9}$$

Whereas for a cubic anisotropy (typical for electrical steels) [16]:

$$E(1) = \frac{H_e}{a}\cos\theta - \frac{K_{an\_a}}{\mu_0 M_s a}\left(cos^2(\psi - \theta) \cdot sin^2(\psi - \theta) + \frac{sin^4(\psi - \theta)}{4} \right) \tag{10}$$

$$E(2) = \frac{H_e}{a}\cos\theta - \frac{K_{an\_c}}{\mu_0 M_s a}\left(cos^2(\psi + \theta) \cdot sin^2(\psi + \theta) + \frac{sin^4(\psi + \theta)}{4} \right) \tag{11}$$

where K_{an_a} represents average energy density of uniaxial anisotropy, and K_{an_c} represents anisotropy constant K_1 of cubic anisotropy [16]. In both cases, the angle ψ is between the easy axis of anisotropic material and direction of the external magnetizing field H.

It should be highlighted, that Eq. (6) has no antiderivative. For this reason the anhysteretic anisotropic magnetization curve should be calculated on the base of numerical integration. Due to the occurrence of trigonometric functions, Gauss-Kronrod integration method is recommended for this purpose [17].

3 Identification of Parameters of Anhysteretic Curves for Both Isotropic and Anisotropic Materials

Considering the identification of parameters of anhysteretic curves, it should be highlighted that for both isotropic and anisotropic materials, for $H = 0$, a singularity occurs. However, it can be easily calculated considering the fact, that:

$$M_{aniso}(0) = \lim_{H \to 0} (M_{aniso}(H)) \tag{12}$$

as well as that:

$$\lim_{H \to 0} (H_e(H)) = H \tag{13}$$

Considering these assumptions, it was calculated previously, that for isotropic materials [7], the initial permeability of anhysteretic curve μ_{ah_i} is given as:

$$\mu_{ah\_i} = \lim_{H \to 0} \left(\frac{M_{aniso}(H)}{H} \right) = \frac{M_s}{3a} \tag{14}$$

The initial permeability of anhysteretic curve μ_{ah_i} can be estimated by differential permeability μ_r for $B = 0$ (as it is presented in Fig. 1). In addition both saturation flux density B_s and saturation magnetization M_s can be easily estimated from the shape of experimental magnetic hysteresis loop. As a result, the parameter a can be easily calculated from Eq. 11. Finally, domain coupling α can be estimated from shape of measured anhysteretic curve or from its estimation from the hysteresis loop [18].

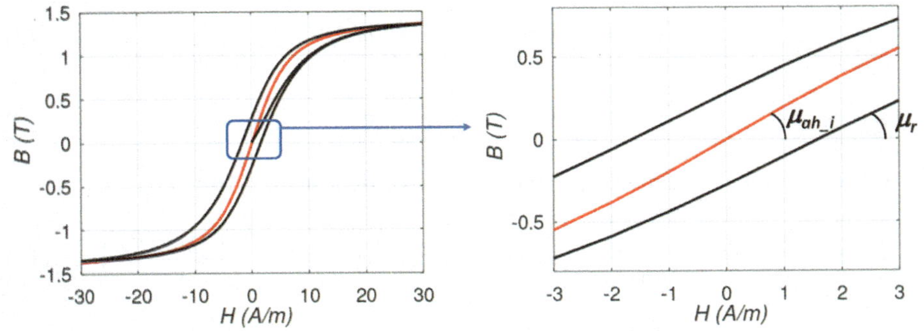

Fig. 1. Initial anhysteretic permeability μ_{ah_i} and differential permeability μ_r for $B = 0$ marked on the shape of magnetic hysteresis loop (black line) and anhysteretic magnetization curve (red line)

In case of anisotropic magnetic materials, the problem of magnetic anhysteretic curve is much more sophisticated. Due to the fact, that Eq. (7) has no antiderivative, values of K_{an_a} or K_{an_c} should be identified in optimization process. Let's consider the case for anisotropic material, were initial permeability of anhysteretic curve μ_{ah_ia} was estimated from two different hysteresis curves measured in the parallel and perpendicular direction from the point of view of magnetization easy axis ($\psi = 0$ and $\psi = \pi/2$). In such a case the set of equations can be stated:

$$\mu_{ah\_ia}\big|_{\psi=0} = \lim_{\mathrm{H}\to 0}\left(\frac{M_{aniso}(H)|_{\psi=0}}{H}\right) \tag{15}$$

$$\mu_{ah\_ia}\big|_{\pi/2} = \lim_{\mathrm{H}\to 0}\left(\frac{M_{aniso}(H)|_{\psi=\pi/2}}{H}\right) \tag{16}$$

In Eqs. (15) and (16), the limit for H $\to$ 0 as well as integral in M_{aniso} function should be calculated numerically. Equations should be solved using optimization algorithm, such as Nedler-Mead method [19].

The example of scripts for numerical determination of average anisotropy energy constants K_{an_a} and K_{an_c} as well as model's parameter a is provided in the repository www.github.com/romanszewczyk/JAmodel in subdirectory "*Automation2021b*".

For verification of efficiency of provided scripts, the average anisotropy energy density and parameter a were calculated for both axial and cubic anisotropic material. Following assumptions were taken: $\mu_{ah\_ia}\big|_{\psi=0} = 10000$, $\mu_{ah\_ia}\big|_{\psi=\pi/2} = 8000$, $\mathrm{B_s} = 1$ T. Results of calculations are presented in Table 1.

Table 1. Parameters of anhysteretic magnetization curve of anisotropic magnetic material calculated for the following assumptions: $\mu_{ah\_ia}\big|_{\psi=0} = 10000$, $\mu_{ah\_ia}\big|_{\psi=\pi/2}8000$, Bs = 1 T

Anisotropy	Average anisotropy energy density K_{an} (J/m^3)	Parameter a (A/m)
Axial	12.42	29.60
Cubic	43.56	28.30

4 Conclusions

Method presented in the paper enables estimation of parameters of model for anhysteretic magnetization curve of anisotropic materials with both axial and cubic anisotropy. Due to the fact, that there is no antiderivative in the function stating anhysteretic curve for anisotropic magnetic materials both integration and calculation of the limit were carried out numerically.

On the base of presented equations and its numerical implementation, average anisotropy energy density and parameter a can be calculated on the base of two hysteresis loops measured for two different directions. As a result, presented solution enables efficient estimation of parameters required for both numerical modelling of magnetization curve in SPICE and FEM applications, as well as enables better, quantitative understanding of magnetization processes in soft magnetic materials with axial or cubic anisotropy.

References

1. Mészáros, I., Vértesy, G.: Modelling of normal magnetization curves of soft magnetic alloys. Mater. Sci. Forum **659**, 429–434 (2010). https://doi.org/10.4028/www.scientific.net/msf.659.429
2. Górecki, K., Rogalska, M., Zarębski, J., Detka, K.: Modelling characteristics of ferromagnetic cores with the influence of temperature. J. Phys.: Conf. Ser. **494**, 012016 (2014). https://doi.org/10.1088/1742-6596/494/1/012016
3. Szewczyk, R.: Technical B-H saturation magnetization curve models for SPICE, FEM and MoM simulations. J. Autom. Mob. Robot. Intell. Syst. **10**(2), 3–8 (2016). https://doi.org/10.14313/jamris_2-2016/10
4. Nowicki, M.: Anhysteretic magnetization measurement methods for soft magnetic materials. Materials **11**, 2021 (2018). https://doi.org/10.3390/ma11102021
5. Szewczyk, R.: The method of moments in Jiles-Atherton model based magnetostatic modelling of thin layers. Arch. Electr. Eng. **67**, 27–35 (2018). https://doi.org/10.24425/118989
6. Jiles, D.C., Atherton, D.L.: Theory of ferromagnetic hysteresis. J. Magn. Magn. Mater. **61**, 48–60 (1986). https://doi.org/10.1016/0304-8853(86)90066-1
7. Jiles, D.C., Thoelke, J.B., Devine, M.K.: Numerical determination of hysteresis parameters for the modeling of magnetic properties using the theory of ferromagnetic hysteresis. IEEE Trans. Magn. **28**(1), 27–35 (1992). https://doi.org/10.1109/20.119813
8. Szewczyk, R., Bieńkowski, A.: Magnetoelastic Villari effect in high-permeability Mn–Zn ferrites and modeling of this effect. J. Magn. Magn. Mater. **254–255**, 284–286 (2003). https://doi.org/10.1016/s0304-8853(02)00784-9

9. Liu, J., Zhang, H., Shan, N., Sha, Y., Zhang, F., Zuo, L.: Improvement of texture and magnetic properties in 4.5 wt.% Si grain-oriented electrical steels. Mater. Res. **22**(suppl 2) (2019). https://doi.org/10.1590/1980-5373-mr-2018-0756
10. Najgebauer, M.: Scaling-based analysis and modelling of power losses in amorphous and nanocrystalline alloys. Acta Physica Polonica A **131**(5), 1225–1228 (2017). https://doi.org/10.12693/aphyspola.131.1225
11. Hawelek, L., Warski, T., Wlodarczyk, P., Polak, M., Zackiewicz, P., Radon, A., Wojcik, A., Kolano-Burian, A.: Effect of co substitution on crystallization and magnetic behavior of Fe85.45−xCoxCu0.55B14 metallic glass. Materials **13**(4), 919 (2020). https://doi.org/10.3390/ma13040919
12. Kittel, C.: Introduction to Solid State Physics. Wiley, Hoboken (2004)
13. Ramesh, A., Jiles, D.C., Bi, Y.: Generalization of hysteresis modeling to anisotropic materials. J. Appl. Phys. **81**, 5585 (1997). https://doi.org/10.1063/1.364843
14. Szewczyk, R.: Validation of the anhysteretic magnetization model for soft magnetic materials with perpendicular anisotropy. Materials **7**, 5109–5116 (2014). https://doi.org/10.3390/ma7075109
15. Ramesh, A., Jiles, D.C., Roderik, J.: A model of anisotropic anhysteretic magnetization. IEEE Trans. Magn. **32**, 4234–4236 (1999). https://doi.org/10.1109/20.539344
16. Baghel, A.P.S., Kulkarni, S.V.: Hysteresis modeling of the grain-oriented laminations with inclusion of crystalline and textured structure in a modified Jiles-Atherton model. J. Appl. Phys. **113**(4), 043908 (2013). https://doi.org/10.1063/1.4788806
17. Laurie, P.D.: Calculation of Gauss-Kronrod quadrature rules. Math. Comput. **66**(219), 1133–1146 (1997). https://doi.org/10.1090/s0025-5718-97-00861-2
18. Chwastek, K., Szczyglowski, J.: Identification of a hysteresis model parameters with genetic algorithms. Math. Comput. Simul. **71**(3), 206–211 (2006). https://doi.org/10.1016/j.matcom.2006.01.002
19. Nelder, J.A., Mead, R.: A simplex method for function minimization. Comput. J. **7**(4), 308–313 (1965). https://doi.org/10.1093/comjnl/7.4.308

Author Index

R. Szewczyk et al. (Eds.): AUTOMATION 2021, AISC 1390, pp. 431–432, 2021.
https://doi.org/10.1007/978-3-030-74893-7

The manufacturer's authorised representative in the EU is Springer Nature Customer Service Centre GmbH, Europaplatz 3, 69115 Heidelberg, Germany. If you have any concerns regarding our products, please contact ProductSafety@springernature.com

Printed and bound by CPI Group (UK) Ltd, Croydon, CR0 4YY
30/09/2026
02225292-0009